Information Technology for Travel and Tourism

SECOND EDITION

Gary Inkpen

LONGMAN

Information on the products and services presented in this book has been supplied by the companies concerned, and the author has made every effort to ensure that this material is accurate and is presented correctly. Most of the companies concerned have reviewed the text and have had an opportunity to modify the material written about them. It must also be stressed that the book does not represent any kind of survey of all available products on the market. The items that have been included are simply examples of systems available to travel agents in the UK at the time of writing.

It is for these reasons that the author and publisher cannot be held responsible for any errors or misrepresentations concerning the descriptions of products and services mentioned in this book. Products described here may have changed since they were originally documented; for up-to-date information and pricing details the reader is advised to contact any company in which they are interested directly.

Addison Wesley Longman Limited
Edinburgh Gate, Harlow
Essex CM20 2JE, England
and Associated Companies throughout the world

© Gary Inkpen 1994, 1998

The right of Gary Inkpen to be identified as author of this work has been asserted by him in accordance with the Copyright, Designs and Patents Act 1988

All rights reserved; no part of this publication may be reproduced, stored in any retrieval system, or transmitted in any form or by any means, electronic, mechanical, photocopying, recording, or otherwise without either the prior written permission of the Publishers or a licence permitting restricted copying in the United Kingdom issued by the Copyright Licensing Agency Ltd, 90 Tottenham Court Road, London W1P 9HE.

First published by Pitman Publishing 1994
Second edition published by Addison Wesley Longman 1998

British Library Cataloguing in Publication Data
A catalogue entry for this title is available from the British Library

ISBN 0-582-31002-4

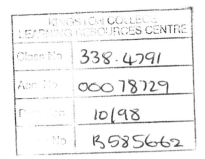

Set by 35 in Sabon 10/12 and News Gothic
Produced by Addison Wesley Longman Singapore (Pte) Ltd
Printed in Singapore

Contents

Preface	vii
Acknowledgements	x

1 The travel and tourism environment — 1
Introduction — 1
The industry players — 2
   Governments — 2
   Suppliers — 2
   GDSs — 2
   Retail outlets — 4
   Consumers — 4
Types of travel and tourism — 4
   Leisure travel — 4
   Business travel — 5
   Payment and funding — 5
   Bureau-de-change — 6
Legal and regulatory — 6
   The European Community — 6
   GDS and CRS regulatory issues — 7
Trade bodies — 9
   ABTA's ABTECH — 10
   IATA — 11
   TTI — 12
   UNICORN — 16
   TTI's RESCON — 17
   HEDNA — 19
   GBTA — 22
   GEBTA — 23
   NAITA — 23
   ENTER and IFITT — 25
Structural environmental issues — 28

2 Tourism — 30
Introduction — 30
The United Kingdom — 31
   The BTA — 31
   TICs — 39
   The Canterbury TIC — 40
Ireland — 44
   The original Gulliver system — 44
   The new Gulliver — 48
Other tourism support systems — 52
   Australia's ETAS visa system — 52

3 Suppliers — 56
Introduction — 56
Airlines — 56
   The airline business — 56
   The airline CRS — 57
   Fares and fare distribution — 65
   Ticketing — 68
Hotels — 77
Tour operators — 83
   Cosmos — 84
Rail companies — 86
   Rail travel in the UK — 86
Information Suppliers — 97
   Reed Travel Group — 97
Product distribution — 103

4 Distribution systems — 106
Introduction — 106
   What is a GDS? — 107
Airline GDSs — 107
   Amadeus — 109
   Galileo — 115
   Sabre's Travel Information Network — 131
   Worldspan — 151
   The INFINI GDS — 162
HDSs — 162
   Pegasus Systems — 162
   Utell — 166

Marketing on the GDSs	173
Static information	173
Dynamic information	173
Hotel Systems Support Services Limited (HSSS)	176

5 The Internet — 177

Introduction	177
Marketing on the Internet	178
Web site presence	179
Booking engine	180
Targeted marketing	181
Internet marketing related issues	183
Disintermediation	185
Travel agents	185
Tour operators	192
Distribution systems	193
Tourism	195
The new intermediaries	196
Expedia	196
Travelocity	206
Worldspan	211
Travelweb	216
Suppliers' Web sites	222
British Midland	223
Marriott	229
Utell's Hotelbook	234
Interfacing supplier systems to the Internet	237
Business travel on the Internet	239
American Express' AXI	240
ResAssist	251
Travelnet	253
Tourism on the Internet	254
The British Tourist Authority	255
Gulliver	264
Travel information on the Internet	267
World travel guide on-line	267
Weissmann travel reports	269

6 Networks — 270

Introduction	270
Video-conferencing	270
Meetings management	271
Techniques	272
The standards	273
The equipment	273
Video-conferencing communications	274
The benefits	274
Conclusion	275
The future	276
Electronic mail	276
Electronic data interchange (EDI)	277
Teletext	279
Videotex	283
Why Videotex?	284
Suppliers' Videotex Systems	286
The problem with Viewdata	298
Communications networks	299
Concert	299
AT&T	301
Imminus	303
Conclusions	313

7 Travel agents — 315

Types of travel agents	315
The business travel centre (BTC)	316
The high street travel agency	316
Inplant	316
Bureau	317
Bureau-de-change	317
The independent high street travel agent	317
Automation of agencies	317
Front-office systems	319
Back-office systems	324
The bank settlement plan (BSP)	326
Financial systems	337
Administration services for hotel commission	339
Functions of the agency management system	342
Accounting	342
Management information	345
Marketing	347
Agency management systems	348
Sabre's Travel Information Network (STIN)	348
Icanos	358
Voyager	360
ICC's travel systems	365
TARSC	371
Other agency management systems	375
Conclusions	375
Epilogue	376
Appendix	377
Index	381

Preface

Although this book is called a new edition, in fact it more closely resembles a complete re-write; little of the first edition of *Information Technology for Travel and Tourism* has survived. The reason for this has been the rapid developments in the three key subject areas: travel, tourism and technology. For example, the Internet was not even mentioned in the first edition, yet here it has a chapter all of its own. Also, the term 'disintermediation' was hardly talked about back in the early 1990s, whereas in this edition it is a thread which runs throughout the entire text. As a result of some excellent feedback, the new edition has also been expanded in scope. It therefore now includes material on tourism, and has a greater emphasis on hospitality related technology and the strategic use of IT within travel and tourism.

No doubt the reader of this second edition will detect some recent travel-related technologies which are not discussed. I'm afraid that this is unavoidable, given the nature of the subject; new technologies and industry developments are constantly being introduced. However, I believe that much of the information contained in the book will remain relevant and pertinent to your chosen study areas – at least until the next edition arrives!

The book is written primarily for those studying an undergraduate course in travel, tourism or some other leisure-related subject. It could also be of interest to post-graduates and students of other courses who are interested in how IT can be applied within a specific industry. In addition to its academic readership, it would also allow practitioners in the industry to gain an overall understanding of the technologies used to support travel and tourism. In particular it would be an effective way for someone working in one sector of the industry to learn about how IT is used in another sector, outside of their own. Ideally, the book should be read cover-to-cover if a good understanding of the subject is to be gained. However, it could equally well be used as a source of reference material for specific course topics, and I have included a comprehensive index to support this.

I have to admit that the book is written from a UK perspective. Having said this, many of the systems and technologies described in the book are becoming increasingly global in nature. For example, at least one of the four main global distribution systems (GDSs) included in the book (Amadeus, Galileo, Sabre and Worldspan) should be applicable in most countries of the world. Wherever possible I have tried to include systems which have either been developed in other countries or which have global applicability (e.g. the Australian ETAS system, Ireland's Gulliver and the travel agents' Voyager system which is also distributed in the Asia–Pacific region). Other deployments of technologies such as the UK's BSP exist in a similar form in other countries. So, for these reasons, I hope the book will be useful to students in other countries and international educational establishments.

The book is organized around my view of the industry's structure, which I have shown pictorially in Figure 1.1 (page 3). This sets the scene for the text by positioning within the supply chain the regulatory bodies, travel suppliers, intermediaries, delivery mechanisms and different types of consumer. The book therefore progresses from a general description of the industry, through tourism and, via travel supplier systems, into the Internet and other means of distribution. It finishes up with an analysis of travel agency systems. A chapter summary is as follows:

- **Chapter 1** – I think one of the least understood areas, from the viewpoint of both the technological and business sectors, is the structure of the industry. In the first chapter I have therefore tried to set the scene for the remainder of the book by creating a structure within which the new technology fits. This chapter also includes some information on the leading standards' bodies and other travel organizations which are so essential to the successful development, direction and application of new technologies in travel and tourism.

- **Chapter 2** – This is a completely new chapter which specifically addresses how technology has been successfully applied by tourist offices and destination service organizations. This is a rapidly developing area and I have used the experiences of the British Tourist Authority and the Irish Tourist Board to illustrate how the innovative use of IT can make destination service organizations more effective, support the promotion of in-bound tourism and help travellers plan their trips better.

- **Chapter 3** – In my view it is essential to understand how the suppliers of travel products and services have used technology for their own in-house automation purposes, before evaluating the more challenging topic of distribution. So, this chapter focuses on the various technologies which different suppliers in each of the major areas have used. Where relevant, I have provided some limited historical background.

- **Chapter 4** – This chapter analyses the world's major travel distribution systems, which in the main are the GDSs. However, many other types of distribution system are also discussed, including HDSs (hotel distribution systems) and CRSs (central reservation systems). I begin my exploration of the marketing aspects of distribution technologies at the end of this chapter, with an analysis of marketing on the GDSs.

- **Chapter 5** – I hope you will agree that the Internet deserves a chapter of its own. It has had a profound impact on the travel and tourism markets already and is set to further revolutionize travel and tourism distribution in the future. So, I start this chapter with a general discussion of Internet marketing, sectoral disintermediation and other Internet-related issues. I then explore some leading new intermediaries such as Expedia and Travelweb before analysing several innovative supplier Web sites including British Midland and Marriott. Finally, I investigate how the Internet is being used to support destination service organizations in the context of tourism by looking at the VisitBritain site and the revamped Gulliver.

- **Chapter 6** – This chapter examines the various communications technologies, other than the Internet, which are used to distribute travel and tourism information to end users. It covers video-conferencing, EDI, e-mail, teletext, videotex and value-added network services such as those run by Imminus and AT&T.

- **Chapter 7** – This chapter actually bears a passing resemblance to material in the first edition! Basic terminology is defined, including such terms as back-office systems, the front office, accounting, consumer marketing for travel and MIS. However, it has been completely updated to include a new POS technology (such as software robots and point-of-sale assistants), the latest changes to the UK's BSP, and examples of four travel agency system products.

It is always difficult to achieve a good balance between quantitative definitions of what certain technologies are and how they work, and the more qualitative discussions concerning issues and strategies. In this context, I have tried to focus on how the new technologies are being deployed rather than describe the intricacies of how they work. When considering new technologies such as the Internet I have included discussion of the major issues and the factors which may influence their future directions. I think this should help lecturers and other thought-leaders to set the scene for further in-depth debate on subjects related to technology in travel and tourism.

Finally, I would like to thank my family, friends, publisher and business colleagues (see long list of acknowledgements!) for their help and support. Without them I would not have been able to write this book. I hope you enjoy it and that it helps to stimulate new and innovative ways of using IT within the world of travel and tourism.

Gary Inkpen
Hove, March 1998

Acknowledgements

It would not have been possible for me to write this book without the participation of some leading industry figures. These individuals gave me their time, their views and some very interesting material. I would therefore like to publicly acknowledge and thank the following people for their help:

Tony Allen	Sea Container Services (and TTI)	Gail Gillogaley	Teletext
Barbara Austin	BTA	Terri Godwin	Galileo UK
Graham Barnes	ABTECH (and The LINK Initiative)	Neil Goram	Cosmos
		Martin Gregg	Microsoft
Neil Beck	Worldspan	Peter Grover	Amadeus
Alan Boyce	HSSS	Phillip Hart	TravelWeb
Guy Briggs	ICC Travel Systems	Trevor Heley	Novus
Victor Brophy	Bord Failte Eireann (and Gulliver)	Peter Horder	HBC Consulting
		Peter Joel	Murray and Company
Sally Brown	American Express	Karsten Karcher	Imminus
Dimitrious Buhalis	University of Westminster (and IFITT)	John Levene	American Express
		Ron Muir	NAITA
Des Butler	British Midland	Howard Needham	BTA
Philip Carlisle	GBTA	Daryl Nurthen	ABTA
Michael Cleeve	MCA	Peter O'Conner	Institut de Management Hotelier International
Mike Cogan	Equinus	Peter O'Shea	BT
Kerry Costello	TARSC	Julian Palmer	AT&T
Stuart Coulson	Worldspan (now with Gradient Solutions)	Ashvin Pathak	ICC Travel Systems
		Alison Pickering	Reed Travel Group
Peter Cowan	BT	Terry Rattray	Utell
Hiliary Cox	Canterbury TIC	Natalie Rawlinson	Icanos
Alex Dalgleish	American Express	Linda Richards	HSSS
Chris Dawes	Imminus	Mike Ruck	Imminus
Gideon Dean	TravelWeb	Lynn Sugars	BTA
Peter Dennis	Marriott (now with Time)	Sarah Taylor	Teletext
Alyson Dombey	Partners in Marketing	Bob Teerink	Sabre
Bob Dunbar	BR Business Systems	Mark Walker	Sabre
Ray Eglington	Country Wide Porter Novelli Voyager	Hannes Werthner	University of Vienna (and IFITT)
Dennis Eyre			
Brett Gilbert	Addison Wesley Longman	Alan West	IATA (and BSP UK)

1
The travel and tourism environment

This chapter discusses the structure of the travel and tourism industry and the regulatory framework in which it operates. Although the focus is very much on the UK, I have also included some examples of overseas organizations. In many countries, the trade and industry structures and bodies are similar to the UK.

Introduction

People travel for many reasons. Some need to travel for business purposes, some for leisure or recreational purposes. But whatever the reason for travelling, people are doing more of it now than they ever used to and what is more, the trend is upwards yet further. Future growth is due to socioeconomic reasons including more leisure time, increased levels of disposable income, the globalization of business and the natural attraction of travelling around the world to see new sites or visit old friends in far away places.

If you want tangible evidence of our desire to travel then just ask any person in one of the following situations: a big lottery winner, a student who has just finished a degree course, someone who is about to retire from work or a businessperson who needs to export a product in order to expand the company. They will all invariably include travel in their plans. So, it's a big and fast growing industry and one in which a large amount of effort goes into making those travel plans; and plans are generated from the communication of information. Travel and tourism is therefore an information business. A business in which access to up-to-date information on an extremely wide range of topics is required instantly.

Now, if you put this highly visible, rapidly growing and complex activity into the context of rapid change in distribution channels caused largely by technology, then you have an interesting story. A story that in many ways has been told by technological innovation in the way information is accessed and used by travel suppliers. From the use of computers by airlines to automate the booking of seat sales in the 1950s to the travel agency use of personal computers (PCs) linked to those airline systems in the 1990s, the change has been dramatic: and today, we are at the beginning of an even more revolutionary era that could see the boundaries of the booking process pushed right out to the travellers themselves via the latest Information technology (IT) paradigm, the Internet.

The Internet is both an opportunity and a threat. It is an opportunity for people to use technology to make travel planning and execution easier and more informed. However, it is a threat to intermediaries such as travel agents and tour operators. It's a big subject, the Internet, and I'll discuss it in more detail later in the forthcoming chapters. For the moment though, it is sufficient to say that the Internet is one of the most powerful movers and shapers of the travel and tourism industry both at the present time and, no doubt, in the future.

However, before we dive into some of the issues that surround IT in travel and tourism, I would like to spend a few pages and a little of your time discussing the environment in which the business operates. It is essential that you understand who the key players are, what the current distribution channels are and how the whole lot is regulated by government and trade bodies. Although it may sound boring, the legal and regulatory aspects of the industry are key to how IT is used to support it. Only by gaining a clear understanding of the industry's structure and the environment in which it exists can you hope to gain a clear perception of the pivotal role that IT has to play within the field of travel and tourism.

The industry players

I think a vital first step in gaining an insight into IT for travel and tourism is to understand the overall structure of the industry: the supply-side travel companies, the types of demand-side consumers and the distribution channels that support it all. Figure 1.1 shows the major structural elements of the travel and tourism industry, as I see it from an IT perspective. The architecture of this diagram is important and I need to define the major elements in greater depth, chiefly because I have structured a large portion of the book around it! So, let's examine each of the players in a little more detail.

GOVERNMENTS

Governments have two main areas of influence within the field of travel and tourism. First, they set the regulatory structure of the industry and, second, they promote inbound tourism to their areas as a means of growing the local economy; and this applies at all levels of government: at the trading block level (e.g. the European Community), at the national level and at the local government level.

- **Regulatory structure** Much of the regulatory structure within which travel and tourism operates, is set by governments at the trading block and national levels. A primary role of most governments is to protect the citizens of its country by setting the ground rules for the travel and tourism businesses. Although such regulations are usually focused on controlling the safety of travel and protecting the commercial interests of travellers, they often have a significant impact on how technology is used within the industry. Most of us have heard of the data protection act; well this is just one example. I'll discuss this act and other regulations that have an impact on IT in more detail in the latter part of this chapter.
- **Tourist authorities** Governments at virtually all levels have an interest in promoting inbound tourism. This is because tourism is a powerful factor in developing and growing an economy. It is a means of bringing foreign currency into an area, supporting and growing infrastructure, strengthening national suppliers and boosting local trade. There are many ways in which governments can encourage the growth of tourism and besides advertising and promotion, one of the most cost effective means is the use of modern technology. Applications of IT in tourism are covered in more detail in Chapter 2.

SUPPLIERS

Travel suppliers provide the actual services required by individuals and groups as they move around the globe. Examples of these suppliers include transportation companies, sightseeing services and accommodation properties; but there are many other variations. Each travel supplier company will invariably operate its own technology: firstly, to automate its in-house operations and, secondly, to use this as a platform for distributing its product to retailers and consumers. It is the precise way in which these systems are made available to end users, whether they be individual travel consumers or travel agents, that is at the very heart of IT in travel and tourism. Consequently Chapter 3, which examines this topic in detail, covers a great deal of ground.

GDSs

Global distribution systems (GDSs) distribute reservation and information services to sales outlets around the world. Incidentally, in my terminology, a computerized reservation system (CRS) is used solely by an airline or a hotel chain and is quite different from a GDS, which distributes more than one CRS to users who are usually travel agents. However, more of that in Chapter 3, which deals

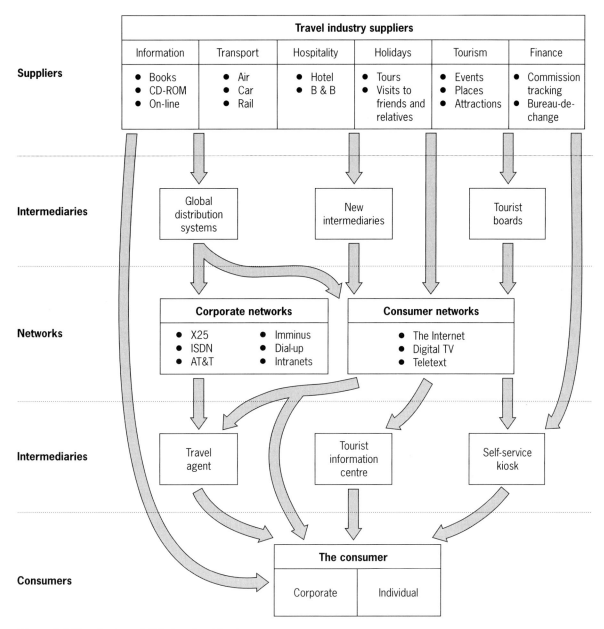

Figure 1.1 The structure of IT in travel and tourism

with suppliers. GDSs are companies that are invariably owned by groups of travel suppliers, each of which shares a common business interest. The principal GDSs are owned by airlines, although there are also a few that focus entirely on distributing hotel booking facilities and these are known as hotel distribution systems (HDSs). GDSs and HDSs provide their users with links to many types of travel supplier and travel-related services. At present the GDSs distribute their reservations and information systems to travel agents, using a variety of computer and telecommunications technologies. However, as we shall see later, this infrastructure is shifting rapidly and both the means of distribution and the ultimate end users are changing very rapidly.

RETAIL OUTLETS

The retail outlet, in terms of the travel industry, is where the rubber meets the road. It's where the consumer physically comes face-to-face with a seller or purveyor of travel products and services. Although some suppliers operate their own direct customer retail locations in city high streets, the majority of face-to-face selling outlets fall into two broad categories: travel agents and tourist offices.

- **Travel agents** Travel agents are the intermediaries between the travel suppliers and the travel consumers. They position themselves as experts on a wide range of travel service companies and areas of the world. They give advice to travel consumers on, for example, where are the best places to go on holiday, what are the best travel suppliers for an individual's specific needs and how airlines and hotels can best meet the needs of business travellers. Travel agents derive their income largely from commissions paid by the travel suppliers. So, travel agents use IT for two main purposes: (a) to access travel supplier systems for their customers, and (b) to automate in-house operations, thus minimizing costs in a low margin business.
- **Tourist offices** There are two main types of tourist organizations: public and private. The public organizations usually concentrate on promoting a country or region to the general public and other interested parties. Their principal objective is to encourage inbound visitors to their area. Private organizations, by contrast, are often the owners of attractions such as theme parks and their focus is on persuading people to visit them. This whole area is covered in more detail in Chapter 2. Both types of tourism organization are nowadays heavily reliant on IT. For example, they rely on large data bases of factual information and communication networks to disseminate the information to their customers that are invariably actual or potential tourists; and increasingly, self-service kiosks are being used to interface these information sources with people.

CONSUMERS

These are the individual travellers like you and me. These travellers may be travelling for leisure purposes, i.e. going on holiday, they may be travelling on business for their employer or they may simply be travelling to visit a location for some other purpose, e.g. to visit friends and/or relatives. Travel consumers may either: (a) purchase their travel arrangements from travel agents, or (b) obtain them direct from travel suppliers. Likewise, they may obtain their advice from tourist organizations or from travel agents; and increasingly these days, they may obtain both travel advice and booking services via the Internet. There are many related issues here and these are discussed in more detail in Chapter 5.

Types of travel and tourism

There are two main types of travel and tourism that are worth exploring and defining, if only so that we can all use some common terminology throughout the remainder of the book. These distinctions in the types of travel are important because each is often supported by a different type of technology. The two main types are leisure travel and business travel.

LEISURE TRAVEL

Leisure travel is the term used to describe the type of travel undertaken by an individual either on holiday or at least, travelling for pleasure. A category known as visits to friends and relatives (VFR), is playing an ever increasing role in leisure travel. The main leisure product is, however, the packaged holiday that is assembled and marketed by specialist tour operators. However, the product range is vast and covers anything a personal traveller might demand of a travel agent. Leisure travel is remarkably different from business travel in many ways and so is the technology used to support it. Generally speaking leisure travel is less automated than business travel although, as we shall see during the course of this book, the situation is changing rapidly. The main characteristics of leisure travel are as follows:

- Lower transaction volumes.
- Highly seasonal.
- Personal customer contact.
- Higher margins.
- Wide product range.
- More complex bookings.
- Variety of bookings.
- Positive cash flow.
- Late booking trend.
- Brochure racking is key.
- Multiple payments from customers.
- Timetable controlled booking process.
- Long booking period prior to departure.
- High-street location is key.
- Unsophisticated travellers as customers.
- Long booking time.

BUSINESS TRAVEL

Business travel is the provision of travel products and services to companies. The individual employees of these companies travel as part of their jobs; jobs like sales representatives, mining and exploration engineers, buyers and senior management. The main products sold by business travel agents to companies are airline tickets, hotel arrangements and car hire. There are some important characteristics of business travel that have a bearing on the technologies used to support it, for example:

- High volume of air sales.
- Low margin.
- Remote customer contact.
- Highly automated.
- High pressure.
- Travel policy.
- Management information.
- Repetitive itineraries.
- Negative cash flow.
- Step function growth.
- Price sensitive.
- Sophisticated travellers as customers.

You may also have heard of a market known as 'travel and entertainment'. This is really another term for business travel but includes company expenditure; including, for instance, fuel on company cars, business entertainment expenses and meals incurred while away on business. The T&E market, as it is known, is alone worth approximately £18 billion in the UK.

PAYMENT AND FUNDING

As with any business, the way in which the customer pays is very significant; and nowhere is this more so than in travel. Leisure travel is more straightforward in that the payment methods follow similar lines as used throughout the retail sector. However, business travel uses one or two special instruments of its own.

The classic method is for the agent to send the customer an invoice and a monthly statement that is eventually paid. However, what does 'eventually' mean? It usually means that the travel agent will not be paid for at least one month. Well, with a reasonable size company with staff who travel extensively, the resulting negative cash flow is therefore a major problem for the agent. If the outstanding balance is not monitored carefully, then the funding costs can become prohibitive for the agent.

Also, from the customer's viewpoint, i.e. the company that employs the business traveller, it is often necessary to provide a cash advance for each trip. This is needed to cover local miscellaneous expenses such as taxis, tips and snacks.

The T&E market has therefore been identified as a prime target for the major credit and charge card companies. These card issuers see an opportunity to sell their cards to companies as a solution to the agents billing problems and as a means of channelling a lot more spending onto their cards.

They therefore developed card products especially for this area. For example:

- **Corporate card**   A plastic card that is carried by individuals who travel on business or who at least incur business expenses. The plastic card may be either a credit card or a charge card. A credit card gives the holder the option of spreading the payments for billed amounts over a period of several months (with the application of an interest charge). By contrast, a charge card stipulates that the entire month's billings must be paid in full, with nothing carried forward to the next month (otherwise a penalty fee is imposed).

    The card fees for a corporate card are settled by the company who also often guarantees liability for expenditure, to the card issuing company. There are several ways in which companies can be billed for expenditure on the card including, for example: (i) central billing to an accounting department, or (ii) individual billing to cardholders themselves.

- **Lodge card**   A credit or charge card for which only a single card is issued and is held in the company's safe. It is used for charging all centrally controlled expenses. For example, air tickets and rail tickets. At the end of each month all transactions for all employees are included on a statement that is sent to the central administration department within the company.

    This statement often contains a great deal of management information that is needed for control and reconciliation purposes. The lodge card statement is therefore a powerful source of management information that is collected by a form of partnership between the card issuing company, supplier systems and an associated travel agent. Some of these lodge card products are now quite sophisticated and even offer the option of management information on magnetic media, i.e. either disk, tape or via data transmission.

BUREAU-DE-CHANGE

The currency services business, which I have entitled bureau-de-change, is not undertaken by the vast majority of travel agents. In fact, outside of the major multiples probably less than 1 per cent operate their own bureau-de-change. The reason I have included this aspect of the business is that: (a) it is an integral part of the services that a traveller expects, and (b) I feel sure that the current lack of support in this area is set to change thus allowing far more travel agents to provide such a service in the future.

The bureau-de-change business revolves around the buying and selling of foreign currency and travellers' cheques. One of the key determinants of profit in this business is the rate used to exchange one currency for another. The aim is to buy at a high rate and sell at a low rate thus making a profit on what is called the spread of the exchange rate. Another good source of profit is the commissions charged on each transaction. The characteristics of this business are as follows:

- Highly seasonal.
- High volume.
- Low margin.
- Sensitive to exchange rates.
- Lengthy end of day balancing tasks.
- Risk of fraud.
- Slightly impersonal service.

Legal and regulatory

Many parts of the travel and tourism industry are regulated by government departments. Governments in this context refer to trading blocks, national governments and local governments. Each sets and enforces its own rules and regulations that have a direct bearing on travel and tourism.

THE EUROPEAN COMMUNITY

The European Commission, an integral part of the EC has a number of regulatory bodies, the lead one of which is called the European Civil Aviation Conference (ECAC). ECAC is in the process of implementing an 'open skies' policy for its member countries. This is equivalent to the deregulation that was introduced in the USA in 1978. The deregulation or liberalization of the airline industry is causing a great deal of change within Europe and is opening up new aspects of competition among airlines operating in the area.

One of the areas that is being opened up or liberalized is the airline fare setting process. It used to be the case that airline fares did not change very much. So in the fairly recent past, when an airline wanted to change a fare or introduce a new one, it would apply to its civil aviation authority (CAA), for approval. The CAA would liaise with its counterpart in the other country involved on the route for which the fare applied and after about five days or more, respond to the requesting airline who would then be free to introduce the new fare. All that has *changed* with the introduction of the first phases of the open skies policy. A new fare can now be filed with the appropriate regulatory body and seats can be sold using that new fare within a period of 24 hours.

What makes this more difficult to handle is the complexity of the fare structures and the widespread use of CRSs by travel agents. A fare is not just a price. It comprises two main parts: (i) a standard fare definition which is known as IFARES; and (ii) the various rules, i.e. up to 30 paragraphs of complex text, that govern the application of the fare as expressed in a standard format known as 'Resolution 100'. In Europe, over 75 per cent of airline reservations originate from CRS terminals (in the USA this figure is around the 98 per cent level). So, with a large number of fare changes coupled with the need by airlines to distribute their fares to each of the worlds major CRSs, one can appreciate the enormous volumes of information and the demands for extremely fast response times to changes.

This means that the marketing activities of airlines are now far more competitive than they used to be with the result that there has been a significant downward pressure on prices. This is good news for the consumer, but puts a lot more pressure on the airlines. This is in fact an area where IT has an increasingly important part to play. I will be discussing some of the IT-based products that support this activity in Chapter 3.

Finally there is the CRS code of conduct, which the EC aims to issue in the near future (see the section later in this chapter for a more detailed discussion of the ECAC code of conduct). This code of conduct will set out some rules of the game in terms of how CRSs and GDSs can operate across national boundaries. One of the guiding principals is reciprocity. This means that if, for example, the semi-autonomous business research environment (Sabre) provides access to the reservation system of say Air France in the USA, then Amadeus should provide access to the reservation system of American Airlines in France. This example illustrates reciprocity using two airlines that are almost national carriers. In such cases reciprocity is not usually an issue. But it does become an issue if a smaller airline is involved, say a domestic carrier in France that wants its flights to be bookable via a dominant USA GDS. The code of conduct is really aimed to protect these small airlines against the power of the CRSs and GDSs that are controlled by the global airlines.

GDS AND CRS REGULATORY ISSUES

A fundamental question that you may be asking yourselves is: 'Why do the airlines pool their reservation systems whereas other travel principals like the tour operators for instance, do not?' Why is it that when booking a flight the agent can access just one system that shows the availability of virtually all other airlines operating flights on the route in question? When booking a package each operator's system must be accessed separately in turn in order to compare products and prices. What makes airlines different in this context? Well, the answer is that the airlines are regulated and the other travel principals are not (also see the section on regulatory bodies in this chapter).

As a direct result of this regulation, CRSs are required to show unbiased displays of all airline's flights. In other words an airline's CRS cannot be used to discriminate against other airlines. To help you understand this better, let me first describe what a biased display consists of. A biased display is one that is displayed at the time that the travel agent is making the decision about which airline will be used for a customer's itinerary. The display that helps the agent do this is the availability display. An availability display is really how the airline CRS answers the question: 'What flights are available between City 1 and City 2 on a specified date?' It is highly likely that there will in fact be several flights operating between the two cities on the date specified and therefore the CRS will need to display a list of these flights. If the

CRS is biased then the display will show the host airline flights first and all those of its competitors next or even on a separate screen so that if the agent wants to view these then an additional entry must be made. Humans, being what they are, i.e. lazy, the flights further down the list will not be looked at so frequently and as for making a separate inquiry entry, forget it. So the travel agent's choice is biased in favour of the host airline.

Some of you may think that is fair enough. However, the European Commission would not agree. It would say that a large airline can use its very size to squash its competitors who are not able to compete on equal terms. Smaller airlines cannot afford to develop and run their own reservation systems on a par with the mega carriers. While they may run their own computer systems and even their own reservation systems, they cannot expect to achieve the same level of penetration of the travel agency network as a large national airline. They are forced to either load their inventory of flights into the large airline CRSs or to establish on-line links to these CRSs. If the large airline CRSs then bias the availability displays against them then they stand little chance of fair competition, which is not good for the consumer. So the European Commission stipulates that airline displays must be neutral in the way in which they display availability lists. To achieve this, the large airline CRSs show their availability in a sequence that puts the flights in an order on the screen that is based firstly on those flights with the nearest departure time to that required by the customer. Outside of this parameter, the sequence of flights displayed is random. It is also important to emphasize that the unbiased display regulations apply only to the first availability screen shown. Other follow-on screens that form part of the same booking may be as biased as the airline feels is appropriate.

The issues of bias and unfair practices have been debated by the European Commission for some time now. Guidelines were originally published in December 1992 and the Commission is still awaiting ministerial approval as at the time of writing in mid-1993. The new codes of conduct were due to have been implemented on 1 January 1993 but a decision is awaited from Brussels. This lack of progress has, however, not stopped American Airlines from raising the issue of domestic unfair practices in Spain and France where it is contended that Amadeus discriminates unfairly against Sabre. However, it appears that there is agreement on many areas of discriminatory practice and it is only a matter of time before a workable code is in force.

Incidentally, one aspect of this code that I haven't mentioned so far is the rules governing the purchase of PCs from CRSs. It used to be the case that a CRS could demand that a travel agent purchase the reservations PC from them. Not so any more. The new EC rules are expected to outlaw this as a kind of restrictive practice. Travel agents will therefore be free to purchase their PC equipment from a supplier of their choice and use it along with software provided by the CRS for reservation purposes. It will, however, be incumbent on the travel agent to ensure that the specification of the PC meets the minimum guidelines set by the CRS for successful access to their system.

In the USA a similar situation exists and the US Department of Transport regulates CRSs to ensure they are unbiased. In fact the USA is ahead of the EC in this respect because the US Department of Transportation concluded its long-awaited guidelines on a code of practice aimed at eliminating anti-competitive activities, at the end of 1992. So if airlines are unable to use their availability displays to win business from their competitors, then how can they use this massive investment in technology to their maximum advantage? The answer is that they try to make their systems the very best in terms of functionality. So in the USA, although an agent can book a flight in American Airlines' Sabre CRS just as easily as in United Airlines' Apollo system with neither system showing any bias one way or the other, each airline will argue that the features of its system make it superior to the other. Therefore, the final question: 'If the CRS displays are unbiased then why should an airline care how many agents are connected to its system?' It seems the answer to this is that there is evidence to prove that if an agent is using the CRS of a particular airline then the agent will sell more seats on that airline. This is probably due to the fact that there are many routes where unbiased displays are irrelevant and where the agent will more naturally choose the flight on the display than any other.

The (ECAC) code of conduct for CRSs and GDSs

One of the most significant accomplishments of the European Civil Aviation Conference (ECAC) has been the adoption by its members of a code of conduct for CRSs in Europe. Incidentally, the term CRS, besides referring to an airlines computer system also embraces what we now call GDSs. In this section, I have used the term CRS because that is the terminology in which the regulation was originally couched. Once ratified by the European Commission's council of transport ministers, this code becomes the official policy of the European Community. ECAC recognizes that CRSs are becoming extremely sophisticated with the advance of new technology. Increased sophistication can, however, also bring with it problems: problems of distortion of competition between airlines; problems of fair treatment for travel agents; and, most importantly, problems for the consumers of air transport, who after all are what the aviation system is all about. They are entitled to be given the best quality information from which to make their travel choices. The dangers of excessive powers in the hands of owners of mega-CRSs, such as those operational in the USA and Europe, must be guarded against. Also, the negative aspects of bias and unfairness need to be discouraged actively. Conscious of these various problems, ECAC decided in April 1987 to embark on the preparation of a code of conduct that would govern the operation of CRSs in Europe.

The UK (CAA)

The Civil Aviation Authority (CAA) is a UK Government department that regulates airline activities to, from and within the UK. It negotiates with other foreign governments in order to agree how routes between the two countries are to be shared by their airlines. This is more complicated than it perhaps at first sounds. Take for instance flights between the USA and the UK. First of all there are several UK airlines that are represented by the UK Government, which itself no longer owns an airline as it once did with British Airways. The UK Government's objective is to negotiate the maximum number of flights that UK airlines can operate to USA gateway cities. A gateway city is a major airport that feeds a large geographic region of the USA.

So, the CAA might, for example, wish to have say four flights each day to New York, which is a principal USA gateway. But the US Department of Transportation (the USA equivalent of the CAA) might want its airlines that fly to the UK to have six flights from London Heathrow to New York Kennedy airport. First of all there is the need to agree on the number of flights, then there is the need to agree on the departure times of the flights granted to the UK and USA airlines and then finally when this has been agreed the CAA has to decide which UK airline is to be allocated which flight and from which airport. With the CAA negotiating with each country that has one or more airlines that fly into the UK, you can probably imagine the size and complexity of the task.

All I have talked about so far is flights between the major gateway cities of each country. There is, however, another dimension that addresses the flights within a country. At present non-UK domestic airlines are not permitted by the CAA to operate flights between cities in the UK. This seems fair enough, after all UK airlines are not allowed to operate between cities in the USA, for example. This whole subject is known as 'cabotage'. But if airlines are truly to compete with each other then why shouldn't BA, for example, operate flights between Paris and Nice and why not in turn allow Air France to operate UK domestic flights between London and Glasgow? This environment is known as true cabotage and is something that the European Commission is planning to implement.

Trade bodies

In addition to the regulatory activities undertaken by the UK Government and the European Commission, there are several important trade bodies or associations that play crucial roles in the orderly development of technology within the travel business. While it is not possible to cover all of these in depth, I have selected the following organizations for further exploration because they appear to be relevant to the use of information technology in travel and tourism in particular:

- **ABTECH** The Association of British Travel Agents (ABTA), is the principal trade body for UK agencies in general. ABTECH is a technology sub-group within ABTA that provides help and guidance on technology matters to UK travel agents.
- **IATA** The International Airline Transportation Association provides a single point of contact to the trade on behalf of all the world's airlines. For travel agents with a high volume of air ticket sales, a licence from IATA is essential if tickets are to be produced in-house.
- **TTI** The Travel Technology Initiative is a group with a diverse membership of travel suppliers, system companies and agents whose principle aim is the standardization of computer booking and ticketing systems.
- **HEDNA** The Hotels Electronic Distribution Network Association is a world-wide group whose members are mainly hotels. The aim is to encourage the development of electronic bookings for members via several distribution networks.
- **GBTA** The Guild of British Travel Agents has a membership of agents with a high concentration of business travel. The GBTA has its own technology experts and works with leading system suppliers to foster products that benefit its members.
- **GEBTA** The Guild of European Business Travel Agents is, as the name implies, a European-wide version of the GBTA.
- **NAITA** The National Association of Independent Travel Agents has members from among the smaller high street agencies in the UK.
- **IFITT** The International Federation of Information Technology and Tourism arranges the annual ENTER conference that addresses key issues in the field of travel and tourism.

In the following paragraphs I will give you a brief overview of these organizations. For further information you may wish to approach them directly and I have therefore included their contact details in the Appendix. Some of the other organizations that I have not been able to cover in depth include the Institute of Travel and Tourism, which although primarily concerned with training programmes, launched a quality initiative in 1997 aimed at travel system suppliers. Then there is the World Travel and Tourism Council that does much to stimulate travel and tourism at a global level and may in the future initiate IT-oriented initiatives.

ABTA's ABTECH

ABTECH is a UK-based independent technology advisory service for ABTA members and others. ABTA is the UK's trade body for travel agents and tour operators. It was formed in 1950 with just 100 members and in 1955 became a company limited by guarantee. Today it has a membership of 3,000 with over 7,000 travel branches. It has both a commercial and a regulatory role:

- **Commercial** ABTA maintains a high public profile with regular appearances in the press, on radio and on television. Over 80 per cent of the UK public now recognize the ABTA name and are aware of the financial protection offered by its members. A public information service handles thousands of enquiries each week and a consumer affairs service operates in liaison with member organizations. ABTA's Travel Training Company is a wholly owned subsidiary aimed at raising standards throughout the industry through education and training.
- **Regulatory** ABTA maintains a code of conduct governing the relationships: (a) between tour operators and travel agents, and (b) between ABTA members' and their clients. Apart from inspecting all members' accounts, it also has rules on financial protection of members' clients specifying individual bonding requirements for travel agents and tour operators. ABTA has two wholly owned insurance subsidiaries through which cover may be obtained to support the scheme of financial protection.

ABTECH is supported by Autofile, British Telecom, Galileo UK, Owners Abroad, Sabre's Travel Information Network, Voyager Systems and NatWest Streamline. The products and services offered by most of these companies are presented in the remaining chapters of this book. ABTECH is operated by a travel industry consultancy called The Link Initiative. ABTECH's work programme is determined by a project board comprising representatives of the supporting companies and chaired by Dr Graham Barnes who reports on progress to ABTA. The primary aims and objectives are:

- To improve continually the unbiased technology advice available to ABTA member tour operators, agents and other organizations agreed by the project board.
- To be accessible to relevant suppliers as a source of feedback on planned products, thereby exerting a positive influence on developments.
- To understand better the ways in which technology is used and can be used in future, in the travel industry.
- To act as a focus for relevant co-operative projects that move the industry forward.

One of the ways in which ABTECH carries on a dialogue with agents is via a help line. The ABTA Helpline enables ABTA members and others to source unbiased information and advice on travel industry technology free of charge. The Helpline is a nation-wide telephone service that allows callers to telephone ABTECH at a local call rate no matter where they are located in the UK. A large and growing database of travel system suppliers is maintained by ABTECH and details can be provided to callers on request. The ABTECH Helpline is on (0345) 581339.

IATA

IATA is really best described as the trade association of the world's airlines. It has its headquarters in Montreal, Canada, and is active in many areas of the airline industry from ticketing through to air traffic control. The principal areas in which IATA is active are:

- **Ticket standardization** This is concerned with standardizing the usage and format of airline tickets around the world. The widespread use of a common and standard ticket helps promote bank settlement plan (BSP) type schemes and increases the operating efficiencies of airlines in general. This benefits the passengers directly and indirectly in availability and acceptance of standard tickets world-wide.
- **Tariff rules** Not to be confused with pricing or fare setting, this activity is aimed at defining the rules that govern how each tariff, i.e. air fare, is to be applied and sold to travellers. Rather like the conditions of use really.
- **Agency programmes** These are IATA's accreditation and appointment activities in the travel agency and cargo sectors. It is important to note here that IATA does not itself offer customers protection. So that when, for example, a travel agent is IATA licensed and displays the familiar IATA sticker on the door or window of the agency, it does not imply any form of consumer protection. It simply designates the travel agent as meeting certain standards set by the IATA airlines. Agents meeting these standards can, subject to specific airline authorization, sell tickets on behalf of airlines.
- **Bank settlement plan (BSP)** This is concerned with standardized airline settlement procedures for ticket sales from travel agents in countries or areas that have decided to implement a BSP (see Chapter 7 for a full description of BSP).
- **Inter-line billing** This is how the airlines do business with each other. When a passenger buys a ticket with several sectors on it, it is quite possible that each will be flown by a different airline. The inter-line billing process supports the distribution of the ticket sale revenue collected by the carrier that issued the ticket, to the other airlines involved in the traveller's itinerary.
- **Taxation** It is perhaps not surprising that this is a core activity of IATA because running a global airline business is no trivial matter from a tax viewpoint. There are therefore many issues dealt with in this area of IATA activity. On behalf of a large number of airlines, IATA works closely with the various government agencies around the world.
- **Technical** There is a special department that deals with air traffic control (ATC) matters on behalf of airlines. Again, in this area, there is a great deal of liaison with government departments and regulatory bodies.
- **Marketing** A special department within IATA exists to co-ordinate market research activities, such as in-flight surveys, that are designed to obtain marketing information from travellers. The ultimate objective is to obtain information

needed by airlines that will help them to improve service levels and deliver more accurately what the customer actually wants.
- **Data processing and systems** This is an IATA run group that is available to member airlines for special IT related projects on a chargeable service bureau basis.

IATA is run by a director general with a number of senior directors reporting to him. There is, however, also an executive committee, the membership of which is drawn from very senior members of airline management. For example, Sir Colin Marshal and Bob Crandal are representatives. The policies of the association are agreed at the annual general meeting (AGM). All members have a vote and can therefore influence the various conferences where policy detail is developed. Unanimity voting applies at conferences. The following are examples of some conference activities:

- **Passenger agency services** This conference focuses on distribution matters such as travel agent accreditation and the effective implementation of BSP around the world.
- **Passenger services** Activities concerned with reservations, ticketing and interlining.

Other sub-committees and *ad hoc* working groups support the objectives of the conferences. Without IATA accreditation, a travel agent is not allowed to issue BSP standard airline tickets. In order to gain IATA accreditation an agent has to demonstrate that it meets certain criteria aimed at establishing its ability to carry out the functions of an IATA recognized agent. One such requirement is to satisfy IATA that it is capable of looking after stocks of blank airline tickets that are, after all, an important form of value. If blank ticket stock falls into the wrong hands as a result of theft or fraud, for example, it can be instrumental in defrauding the airlines of substantial funds. The travel agent must therefore have a secure office and a safe in which to store blank ticket stock. IATA also needs to be assured that the travel agent is fully capable of writing tickets and accounting for them. One way to satisfy IATA on this score, is for the travel agent to send staff on the appropriate fares and ticketing courses. This is an excellent way for staff to attain specific levels of competence.

Figure 1.2 The TTI logo

TTI

The TTI (Fig. 1.2) is a not-for-profit European industry user group that is one of the most important standards bodies in the travel industry. TTI promotes the development and use of open systems within travel, tourism and leisure. It has been successful in many areas, particularly the development of UNICORN and RESCON (more of which later), and also in the area of ferry standardization using EDI. RESCON, the UNICORN standard (as used in the AT&T Istel 'Ferry' service) and the development of the ATB2 standard for non-scheduled airline operations are probably the best and most prolific examples of successful standardization in the UK travel industry. It is therefore worth examining TTI in a little more detail.

The members and structure of TTI

The Travel Technology Initiative was created in 1987 by seven founder member companies, several of whom had previously originated a tour operators working committee known as the Holiday Systems Group e.g. Redwing, Owners Abroad, Fastrak. By 1996 this had grown to almost 100 members, 15 per cent of which were European based companies. The 12 shareholders of TTI are: Airtours plc, Brittany Ferries, Cosmos plc, Eurostar (UK) Ltd, First Choice plc, Hoverspeed, Imminus, Irish Ferries, P&O European Ferries, Stena Line, System Aid Technology Ltd and Travellog Systems. Each of the founding members contributed investment funds for the development of the TTI.

In 1995 TTI actively grew its membership. It formalized a structure and set about pursuing activities that would deliver real value to members. Membership was opened up to other categories, besides shareholder, for example: executive, corporate, associate and trade body. Each membership category pays a fee commensurate with the benefits enjoyed. The following is a summary of the various membership categories:

- **Shareholders** These companies have a stake in TTI and consequently play a full part in the development of the group's plans. Company representatives attend all regular board and executive committee meetings and are entitled to vote.
- **Executive** This category of membership allows representatives to attend executive committee meetings, participate in working parties and general meetings. Executive members receive technical specifications free of charge.
- **Corporate** Corporate members may participate in general meetings and receive important specifications at a substantial discount. They also receive release notes and technical briefs.
- **Associate** There are over 55 technology and travel companies who have joined TTI on an associate membership basis. The advantage for these companies is that they are kept fully up-to-date with all TTI's activities and plans for a very modest investment.
- **Trade association** Representatives of other travel trade associations may become members of TTI for a low fee and yet still enjoy many of the benefits, such as one call per month to the help line and access to many of the important publications.

In order to ensure that these levels of membership mesh effectively, TTI is run within a well organized structure. The board meets quarterly and consists of shareholder members; it provides overall strategic direction and control, agrees the organization's budget and sets fee levels. A management committee with members from the TTI executive, meets monthly and manages the budget. The technical standards working party meets prior to the management committee and addresses all technical development projects. Within this overarching structure, several working committees undertake research and development activities both within the membership and throughout the industry in general. They then make recommendations to the board.

One of the key success factors of TTI is that participation on working committees is open to all executive members. Contributions to working committee efforts are given without cost or charge-out from the members involved. It is in this way that TTI is able to leverage the strengths of its members and co-ordinate the activities of working committees as quickly and effectively as possible.

The formation of TTI

The founder members of TTI felt strongly that as travel industry technology progresses and advances in the coming years, it is necessary to have some central forum where users and suppliers could meet and discuss developments on neutral ground. These were early days in the formation of the systems we take for granted today. But much of what is now available would not have been possible in such a short time without the establishment of standards and guidelines at an early stage.

While all the founder members were in competition with each other in their own marketplaces, as holiday companies or system suppliers they knew only too well that co-operation was needed if technological developments in all areas of the travel industry were not to suffer. The industry has so often seen the damage that competing standards have done to systems developments. On the domestic front, how much more advanced would video be today if so much time, money and energy had not gone into rival recording methods, e.g. Betamax and video home systems (VHSs)? In such circumstances one system will eventually dominate the market, but the price is an enormous waste of resources.

All the founder TTI members shared a vision of the benefits that technology could bring to every sector of the travel industry. Equally, many feared the nightmare of disparate development, competing standards and the eventual writing-off of dead-end developments. TTI was formed to ensure that everyone involved in the industry could put all their efforts into the achievement of that vision, safe in the knowledge that a totally impartial body was there to monitor progress and keep all its members fully informed of individual developments.

The objectives of TTI

In order to carry out its work effectively the TTI set out its key objectives at an early stage in its development. All the work done by TTI since this time has been based on and has adhered to these objectives. They are:

- To use technology to maximize business opportunities and improve margins for all.
- To develop standards for technology in travel.
- To ensure systems are developed to a common standard and thus to reduce development costs.
- To allow all users to invest in technology, secure in the knowledge that it will be supported.
- To achieve consensus through discussion and co-operation.

TTI believes these objectives to be as valid today as when they were first set out and experience has proved that they have been an important guideline to all TTI's efforts in bringing rival players together in a spirit of mutual co-operation for the overall furtherance of technological developments within the travel industry. Having said this, it is natural that questions get asked from time-to-time about how TTI members see the role of their organization within the industry. It might therefore be worth mentioning what TTI is not. The objectives of TTI are *not* to:

- Seek self-interest for members or to dominate the industry.
- Impose standards – neither hardware nor software.
- Force compliance.
- Refuse to allow TTI's products to be distributed to non-members.

In this way TTI hopes to make it clear that it is there to help, to encourage development and to act as a forum for all suppliers and users of systems (Table 1.1) with the overall objective of achieving a more professional and profitable industry for all concerned.

The current activities of TTI

TTI is active in many areas within the travel industry. It organizes two general meetings each year for its members and specially invited guests. It holds executive meetings and board meetings that govern the overall direction of the group. TTI also produces two regular publications: (i) *News and Views* – a 12–16 page publication issued twice each year (currently January and August), which airs topical views on industry matters; and (ii) *Working Together* – a monthly news-sheet circulated to all members. Also, the TTI Web site, which is hosted by BT, may be found at **http://www.tti.org**

But it is the activities of the working parties that spearhead the progress that TTI delivers to its members and the industry. These have so far concentrated on the following main areas:

1. The identification of secure document printing as a requirement for the industry.
2. The development of a standard ticket – initially this will be a manual ticket but the focus of attention is now shifting towards the ATB2 area and other forms of electronic ticketing.
3. The continuing development of UNICORN electronic data interchange (EDI) messages.
4. The playing of a leading role in the United Nations' EDI for reservations and confirmation Expert Group No. 8 (UN EDIFACT EEG8) working committees (see UNICORN below for more details).
5. The establishment of EDI as the base for future technological developments.

To ensure that all this work proceeds apace, TTI now has two main working committees covering: (a) general management, and (b) standards. The management committee deals with the general running of the TTI 'business' as well as contracts, licences, public relations and other commercial activities. The standards committee is involved in the development of future standards. This committee is currently looking at invoices, confirmations, EDI messages for IT reservations and interfaces with CRSs and GDSs. Special interest groups are also supported by working committee members as appropriate, e.g. liaison with the Retail Tour Operators Group (RETOG) – see below.

In addition to these committees, all of whom are of course manned by people with full-time jobs of their own, TTI has also appointed a well respected independent consultant, Equinus, as a source of technical expertise and to co-ordinate and help drive forward all the developments including the provision of administrative support. While all the work of TTI is carried out by the working committees, the overall direction is set by a business plan that encompasses all TTI activities and runs for a 12 month period. The overall business objectives of TTI for the near future are:

Table 1.1 TTI membership benefits

Benefits	Membership categories				
	Shareholder	Executive member	Corporate member	Associate	Trade association
Attends board meetings	✓				
Attends executive committee	✓	✓			
Votes at executive committee	✓				
Eligible to chair TTI meetings	✓				
Receives minutes of executive meetings	✓	✓			
Participates in working parties	✓	✓			
Receives minutes of working parties	✓	✓			
Participates in general meetings	✓	✓	✓	✓	✓
Technical specifications Free					
A 50 % discount	✓	✓	✓		
A 20 % discount				✓	✓
Release notes and technical briefs	✓	✓	✓		
TTI publications Free					
A 50 % discount	✓	✓	✓		
A 20 % discount				✓	✓
Free access to TTI helpline	✓	✓	✓		
Telephone support (maximum calls per month)	Unlimited	Unlimited	4	1	1
Free copy of *News and Views*	✓	✓	✓	✓	✓
Free copy of monthly circular	✓	✓	✓	✓	✓
TTI seminars Free places, 2					
A 50 % discount	✓	✓	✓		
A 20 % discount				✓	✓
TTI special meetings – guaranteed place	✓	✓	✓	✓	✓
Preferential rates at TTI promoted events	✓	✓	✓	✓	✓
Link from TTI web page	✓	✓	✓	✓	✓

- **Ticket issue**   TTI intends to continue developing the specification for ATB2 ticketing.
- **Confirmations and invoices**   TTI will be looking at the definition of certain EDI messages and the design of a standard invoice layout. It will also give consideration as to whether documents should show a TTI logo or whether retailers could or should be able to issue such documents on their own customized stationery that may not adhere to the standard layout.
- **EDI messages for tour operating**   In this highly complex area, TTI will be involved in liaison with the EEG8 group of the EDIFACT UN working committees, to assist with and drive the development of these messages. Consideration will be given to the need to develop TTI-specific messages if EEG8 is not progressing quickly enough for TTI members' needs.
- **EDI modules for PCs**   TTI has defined the functionality required for a PC to handle all

anticipated TTI standards. This is known as RESCON (see below for details). This in turn means that the PC should be able to:
- Hold ticket, invoice and confirmation data prior to printing.
- Convert messages from screens into TTI–EDI format to send to hosts and reconvert responses for screen display.
- Provide by-product data capture for back office systems.
- TTI will also produce a list of approved PC plus software that meet all the above requirements. However, it must be stressed that TTI will not itself provide hardware or software.
- **CRS interface** Airline CRSs and GDSs are now key players in travel technology. TTI will pursue dialogues with all the CRSs and GDSs in order to encourage the development of standard interfaces between the present TTI hosts and CRS networks through EDI. If possible, this will include standard methods of ticket and documentation printing and the use of automated ticket and boarding passes (ATBs) and generic reservation screens.
- **Other EDI** The task in this area is to consider the development of standards for messages in other system areas including reservations in those fields not already covered, plus check-in messages.
- **Standards** TTI has already set up and will continue to maintain a central library of all hardware and software available to handle TTI functionality, identifying in each case whether that product has the TTI 'seal of quality'.
- **Millennium** TTI have launched a millennium initiative. The aim of this is to increase awareness of the date change issues involved and to provide a help pack with check list for companies. Some key issues for supplier systems are: will they release the room on the correct date over the millennium period, will they continue to issue tickets on the correct date, and will systems provide a compliance certificate similar to that demanded by BT?

Information on all these activities will be made available, not only to members, but also to non-members on request. These and other subjects that may arise are being examined by the working committees with founder and participating members attending meetings to develop the projects. In addition, associate members, while they do not attend executive meetings, are encouraged to help and make their views known either directly to the relevant committee or to the TTI retained consultant.

TTI encourages new members to join and to both: (a) spread the use of standards, and (b) assist in the development of new standards. In particular, more smaller retail members are currently sought. All TTI standards are placed in the public domain for all to use and review as they see fit. In order to join TTI an organization or individual need simply write to the membership secretary.

UNICORN

UNICORN is a set of standards for EDI messages used within the travel and tourism sectors of industry. It started out as a TTI initiative that aimed to use EDI technology to simplify and standardize the way in which bookings were handled across different supplier host systems. This was a particularly relevant issue for Germany's START system when it faced a costly development exercise resulting from the interconnection of its multi-access switch to several cross-channel ferry companies. Since that time, UNICORN has grown in scope and complexity to cover a wide range of inter-system communications, including the secure printing of ATB2 tickets (see Chapter 3).

It is now the leading global standard for system-to-system communication via EDI *message sets* within the travel industry. In this context a *message set* is simply a standard format that governs a two-way conversation between two computers. There are now over 130 different such message pairs that are standardized under the UNICORN banner. Messages that, for example, pass between a travel agent and a tour operator, such as: (i) show me the latest availability for a certain tour, (ii) book me the tour currently displayed, (iii) sell me a travel insurance policy, and (iv) print me a ticket. There are two other standards that are particularly relevant to the development of UNICORN, one is the UN's EDIFACT and the other is IATA's Passenger and Airport Data Interchange Standards (PADIS). Each is described more fully as follows:

- **UN's EDIFACT** UNICORN is similar to the UN's standard known as EDIFACT. This is the world-wide standard agreed by the UN known more fully as: *EDI* *f*or *a*dministration, *c*ommerce and *t*ransportation. The main differences between these two seemingly overlapping standards is that EDIFACT focuses on batch messages, i.e. messages with long intervals of time between the originating message and the response; whereas UNICORN is used primarily in cases where an interactive dialogue between two computer systems is involved. Where computers communicate interactively, the interchange of message pairs is said to be in *fast batch* mode or, put another way, using *interactive dialogue*.

    Having said all that, EDIFACT also incorporates an interactive dialogue standard known as I-EDI or interactive EDI. However, this is a generic set of standards that applies to virtually all industry and commerce sectors. As such, it is not specific to the travel industry, whereas UNICORN is designed solely for use within travel and tourism.
- **IATA's PADIS** This is an IATA board that has developed a set of standards, mainly for computer systems that need to communicate information concerning air freight, weight and balance calculations and some passenger information, with other computers. IATA has worked with the UN to submit some PADIS messages for approval as UN/EDIFACT. In fact, just lately, there have been many PADIS messages introduced into EDIFACT.

The European Commission is keen to see the convergence of world-wide standards. It has therefore provided part-funding for the migration of UNICORN standards to make them fully compliant with I-EDI, within the UN/EDIFACT sphere. Other standards bodies have also been involved with TTI including the European Board for EDI Standardization (EBES), which runs an important sub-group, EEG8, that focuses on travel, tourism and leisure (formerly known as the MD8 group).

TTI has consequently been pretty busy over the past few years, just with the development and promotion of UNICORN. This activity has, however, paid dividends in terms of standardization and has governed the core activities of TTI since its inception.

UNICORN has therefore been developed using a phased approach, over several years. It is for this reason that there are different versions of UNICORN, each of which has been released at different times over the past few years. It is nevertheless the case that each version is totally *backwards compatible* with earlier versions. This means that if, for example, you have implemented Version 4.1 then you will not have to change your systems if you then migrate to Version 4.2 at some point in time. The various releases of UNICORN are summarized as follows:

- **Version 4.1** This was (in the first quarter of 1997) the current version. RESCON is an integral part of Version 4.1 (see below).
- **Version 4.2** This version was released in February 1997 and includes support for rail message pairs.
- **Version 5.0** This version will be totally compliant with UN/EDIFACT for I-EDI and was scheduled for release during the first half of 1997.

When a company purchases UNICORN (which in the first quarter of 1997 cost £500), they receive a developers pack. This contains all required documentation specifying the standards and may in the future contain a relational data base of message formats and software libraries.

TTI's RESCON

RESCON, which stands for *res*ervation *con*firmation, is an outstanding example of what can be achieved by standardization. The problem that TTI set out to tackle was one that travel agents were all too familiar with. This was the tiresome task of keying a customer's booking data twice; once into the videotex reservation system and then again, into the agency's own back office system. The root cause of this duplicate keying effort was the different formats used by tour operator systems to record a customer's booking. Let's take a look at this in more detail.

Most tour operators have developed front-end communication interfaces for their booking systems based on videotex technology (sometimes called viewdata – see Chapter 6 for more details). However, travel agents are increasingly using PCs that can emulate a videotex terminal and also provide the added benefits of general office automation and GDS access. The trouble is, that once a travel agency user makes a booking in a tour operator's videotex system, the very same information must be manually re-keyed into their own back office or agency management system. It is this system that then prints the itinerary for the customer as well as other relevant information on the booking. Booking information includes the following fields of data:

- Lead passenger name.
- Address.
- Flight times.
- Itinerary dates and times.
- Other passenger names.
- Booking reference.
- Accommodation details.
- Financial information.

Now, many agency management systems provide an automated interface to some of these tour operator systems. This is done by special applications software that maps each field from the tour operator's videotex screen display into a standard booking record, as defined by the agency management system supplier. However, it's not that simple because each tour operator's system is different and this means that the booking information as displayed on the final videotex confirmation screen is different; in other words it is non-standard.

Agency management system suppliers tried to get around this problem by developing several maps, one for each major tour operator system. That was all very well until the tour operator changed their booking screen, which does happen from time to time. However, such changes cause havoc to system suppliers that provide booking confirmation maps. Each map had to be changed to reflect the new format. Invariably, changes to these maps meant that the travel agent had to install a new version of software. Some suppliers tried to make the mapping process more responsive by providing the agency with the means to modify the maps themselves. But this is a complex technical process and one that is fraught with difficulties to trap the unwary. Even the Global Trade Initiative (GTI) failed to address the problem (see Chapter 6). Something had to be done.

The answer was provided by TTI. They worked closely with some major tour operators, travel agency users and system suppliers over a period of nine months. Some of the EDI work was salvaged from the abortive GTI initiative. A TTI working group was formed and met on many occasions. Members eventually jointly agreed on a common standard for a booking confirmation. A common booking record format based around the EDI standard was then developed. This enables all booking fields to be identified and labelled thus enabling different computer systems to recognize and process the content of the message. Even TTI non-members, Thomson, Galileo UK and Lunn Poly, participated in the development of the standards.

The resultant structured messages are not videotex specific and are in fact generic to any type of computer system: and to support this standard booking record format, some special new software known as a *message handler* was written. To use RESCON the *message handler* software must be used by both the tour operator's and the travel agency's PC systems.

Once installed, videotex bookings are processed as normal. The only difference is that travel agents can choose to have the booking details downloaded to their own system when confirming a booking. This takes only a couple of seconds, leaving the travel agent's system to do what it wants with the information – typically storing the booking data within its customer record. From here it can be retrieved at any time and fed through to the payment system of the tour operator. It can also be used to print ATOL receipts (now a legal requirement for package holidays) and seat-only bookings; all of this can be done without any re-keying by the travel agency. The benefits may be summarized as follows:

- Elimination of duplicate keying by travel agents; once into the videotex booking system and again into their agency management system.

- Improved accuracy because data is recorded by the tour operator's system and downloaded electronically without human intervention, into the travel agent's system.
- Enhanced customer service due to the automatic printing of accurate reports such as the ATOL customer receipt.
- Reduction in the tour operator's efforts associated with answering accounting queries and reconciling the travel agent's payments.
- Elimination of system maintenance tasks associated with maintaining tour operator booking screen maps for conversion by agency management systems.
- Improved security because RESCON uses error detection and correction transmission techniques that can identify and correct errors due to such things as *line noise*.
- Establishment of the basic building blocks for future secure printing of items such as travel tickets, invoices and receipts.

Live usage of RESCON commenced in April 1995. It is now used by the tour operator Cosmos with 140 Co-op Travelcare branches and by Eurostar UK and Galileo. The technical infrastructure of RESCON was developed in nine months for a shared investment of £20,000. This was undertaken by TTI and 12 leading travel companies: Airtours, A. T. Mays, Best, Co-op Travelcare, Cosmos, First Choice, Going Places, Sunworld, Lunn-Poly, Stena, Thomas Cook and Thomson. RESCON involves no licence fees or maintenance charges and has been developed as an *open system* for use throughout the travel industry.

The Retail Tour Operators Group (RETOG) has now been re-formed to work with TTI on the future development and enhancement of RESCON. Many tour operators have already built RESCON functionality into their systems. Others recognize that RESCON could be the platform for future growth and development that they have been seeking for many years. In particular, tour operators are actively considering the following areas:

- **Update messages** Even when a passenger books a holiday, there are often times when the tour operator needs to communicate some changes to the package. RESCON could be enhanced so that the tour operator's system could transmit a message to the travel agent's system, which could then amend the booking details held locally. Eventually, the travel agent could then notify the customer.
- **Quote message** There are many cases where a customer will peruse the tour operator's systems with the help of the travel agent. In some cases a quotation is required before a final confirmation is made by the customer. RESCON could be enhanced to transmit a quote automatically to the travel agent's system for local printing and transmittal to the customer.
- **Printing messages** There are many items that the tour operator would like to see automatically produced by the travel agent's own system. Some examples include confirmation advices, tickets and baggage tags. RESCON could be enhanced to print these securely on behalf of the tour operator.

TTI has achieved a high degree of success in the development of IT standards within the worldwide travel industry. Besides the widespread use of UNICORN and RESCON, there are many other examples of TTI's success. Take, for example, the Greek Government, which has recently stipulated that UNICORN should be used nationally in order to control the overbooking of ferries; and also the French tourism industry where calls for UNICORN to be supported by supplier systems are widespread.

HEDNA

HEDNA is a not-for-profit organization that was formed in 1991 by hoteliers to promote and accelerate the use of GDSs and other electronic means for the booking of hotel reservations. It is now recognized as the major conduit for training and the dissemination of information relating to all electronic distribution issues for the hotel industry. The mission of HEDNA is to increase hotel industry revenues and profitability from electronic distribution channels by:

- Optimizing the use of current technology.
- Influencing the development of current and emerging electronic distribution channels.
- Education.
- Providing an opportunity for open exchange among members.

Programmes

The following is a brief overview of the major activities of HEDNA that are undertaken on a regular basis. These programs may change from time-to-time so as to recognize the needs of members:

- **HEDNA meetings and conferences** Provides networking opportunities to gain insight into how peers, suppliers and customers view the industry, how business is conducted and how it can be improved.
- **HEDNA GDS educational materials** The curriculum of HEDNA University is available to cater to all audiences in three formats: self-guided workbooks, self-paced computer training and presentation style. All formats have been combined into a kit and are available on diskettes for company personalization.
- **HEDNA GDS pocket reference guide** Formats at your fingertips. This handy guide lists general hotel availability, description and sell formats for all of the GDSs. It is ideal for sales managers who deal directly with the agency community.
- **HEDNANet** HEDNA's Internet Web site that facilitates member and non-member access to all HEDNA resources including educational materials, meeting schedules, membership information, new minutes of meetings, etc. (also see **http://www.hedna.org** – a Web site for use by HEDNA members).
- **Technology showcase** Provides hands on product demonstrations of current and emerging technologies for hotel distribution.

Structure

HEDNA's officers consist of a president, vice president, secretary and treasurer. These officers, along with five elected members and three invited members, compose the board of directors. Elected board members are hotel company executives and each serve two-year terms. Invited members serve for one year. The executive director, who is responsible for the HEDNA staff, is an ex-officio member of the board.

Associations such as HEDNA need to find an efficient way of dealing with regular day-to-day administration. One solution is to out-source these functions to a specialist service company. Hakanson & Company, an association management company, has consequently been engaged to provide HEDNA with services that include: office-of-record, financial management, membership development and records management, publication and materials dissemination and other association management services as needed.

HEDNA Europe, Middle East and Africa (HEDNA EMEA) was formed in 1994 and is responsible for the organization's activities within this region, using the main HEDNA group as an effective transatlantic communications channel to North America. This allows developments in this area to be disseminated and shared throughout all HEDNA Chapters. This sub-group is known as the HEDNA EMEA Chapter and is officially recognized by the HEDNA main board. It does not, however, maintain a separate membership from HEDNA and adheres to the core mission statement but with special emphasis on education, conferences and seminars.

Support and enthusiasm for HEDNA has grown substantially in the late 1990s and evidence of this is the formation of European regional sub-groups. The first such HEDNA sub-committee was formed in Italy during 1996. In order to support the growth of HEDNA sub-committees in this way, a standard operating procedure has been produced that defines how to set-up, manage and run a regional HEDNA Chapter.

Committees

The strength of the HEDNA organization is the opportunity to participate on various industry committees, which work to further HEDNA's mission. Members are welcomed and invited to participate in any of the HEDNA committees, which include:

- **GDS** There is a separate committee for each GDS. Each committee serves as a focus group where representatives of the GDSs meet with HEDNA members to standardize, simplify and accelerate the use of their systems in the booking of hotel reservations.
- **Standards** The focus of this committee is on hotel industry standardization as it applies to GDSs and other electronic distribution channels.

The principal areas where HEDNA supports the development of standards includes, but is not limited to, EDI message formats, keywords and the identification of data fields that need to be shown on the computer screens of systems used by hotel properties and chains.

- **Education** This committee addresses the issues that relate to the planning and execution of HEDNA University sessions. Also included are any other training and educational needs identified by HEDNA, such as the preparation and execution of training material.
- **Leisure** A separate committee has been formed to consider the issues that relate to the distribution of leisure-oriented products and package tours through GDSs and other electronic channels.
- **Request for proposal (RFP)** This committee addresses standards that govern the transfer of electronic data for the RFP process that is used throughout the accommodation industry.
- **Meetings** Separate committees plan and execute HEDNA meetings. The tasks undertaken include, but are not limited to: site selection, contract negotiation and on-site co-ordination.
- **Technology** This committee is responsible for HEDNA's Internet site, called HEDNANet. One of the tasks related to this is the effective co-ordination with each of the other HEDNA committees to ensure that the information displayed on the HEDNA Web site is: (a) current, and (b) available to all HEDNA members. The committee also monitors emerging technologies and keeps HEDNA members abreast of the latest developments in hotel distribution.
- **Marketing** The marketing committee helps develop and implement the overall HEDNA marketing plan. These activities include: membership recruitment, communications, public relations and surveys.

Membership

The membership of HEDNA has grown significantly since the organization's foundation. Membership is open to any entity or individual whose actions, goals and purposes are consistent with the mission of the organization. There are three types of members:

- **Principal member** A principal member may be either: (a) a lodging or hospitality company, or (b) any company that manages the sale of rooms through electronic distribution systems for other lodging and/or hospitality companies. This category of membership has full voting rights and is reserved primarily for hotels. A fee of US$500 per year is payable by all principal members, of which there are currently 139.
- **Allied member** This category of HEDNA member comprises companies or individuals who provide, or desire to provide, products and services to principal members. This is a non-voting associate membership costing US$1,000 per annum that is designed mainly for GDSs, CRSs and independent consultants. There are currently 31 allied members.
- **Associate member** These are companies or individuals who are not eligible for principal or allied membership, e.g. travel agencies, tour operators, educational institutions. This is a non-voting associate membership costing US$100 per year. There are currently 36 associate members.

HEDNA brings together participants from every aspect of the electronic hotel booking process. In the autumn of 1996 HEDNA's membership totalled over 200, including hotel companies, hotel representation companies, major travel organizations, consultants, travel technology service vendors (including all major GDSs), all hotel switching companies and international travel media. Evidence of HEDNA's success is the consistent support from its member companies. This support is repeatedly demonstrated through their willingness to provide staff time for HEDNA related activities.

The benefits of HEDNA membership are many and varied. There is, for example, the opportunity to attend world-wide conferences, technology briefings and meetings of peers, as well as access to the HEDNA educational kit and annual market survey reports. Plus the ability to experience direct involvement with GDSs and influence their future direction along with the chance to benefit from the tools and presentations that help optimize the use of current technology. Membership also provides access to new technologies, such as the Internet's HEDNANet site.

Code of conduct

An organization such as HEDNA must be careful to avoid any accusation of being a cartel or acting in a monopolistic way. It is primarily for these reasons that a definitive code of conduct is so important to HEDNA. So, in accordance with regulations stipulated by the USA antitrust laws and Europe's Treaty of Rome, the following code of conduct is strictly enforced at all HEDNA gatherings:

1. HEDNA does not permit the discussion or exchange of information relating to:
   – Room rates.
   – Surcharges.
   – Conditions.
   – Terms or prices.
   – Allocating or sharing of customers.
   – Refusing to deal with a particular supplier or class of suppliers.

   Neither serious or flippant remarks are therefore permitted on any of these topics at HEDNA meetings.
2. HEDNA does not issue recommendations on any of the above subjects or distribute to any of its members any publication concerning such matters. Information exchange or discussions that directly or indirectly attempt to fix purchase or selling prices is not permitted to take place.
3. Standards or certification requirements for membership must give equal conditions to all similar parties.
4. All HEDNA-related meetings shall be conducted in accordance with previously prepared and distributed agendas.
5. All HEDNA-related meeting agendas will contain standard topics including:
   – Educating the HEDNA membership.
   – Improving GDS functionality and making technical progress.
   – Improving distribution of the hotel product.
   – Standards of display and codes used in the GDSs and other electronic hotel booking media.
6. The objective is free exchange of ideas consistent with the HEDNA mission statement, to the benefit of the travel industry.

Achievements

HEDNA has made some real progress in achieving its aims. In addition to establishing a vital forum for the exchange of ideas relating to the electronic hotel booking process, HEDNA members are proud of many important accomplishments. These include:

- **Improved GDS bookings** Identifying and addressing impediments to electronic booking through extensive discussions between hotel, travel agency and GDS representatives.
- **The Hotelier's GDS educational manual** Developed in co-operation with Cornell University School of Hotel Administration, this publication is a comprehensive manual for the electronic reservation environment. In addition to serving as the primary text for HEDNA University, it is used at Cornell as a teaching resource in its hotel management programmes.
- **HEDNA University** This is a one-day educational event that is run by HEDNA. In slightly over one year, there have been more than 400 graduates of the HEDNA University programme. 'University days' have been held in Europe, Canada and the USA and are taught by volunteer members of the organization.
- **HEDNA GDS educational kit** A three-part self-guided training programme. This is available in either a computer-based version or as self-guided work books with a teacher's guide for group applications. The kit provides:
   – An introduction to GDSs.
   – A description of hotel staff's roles in the electronic booking process.
   – A comprehensive summary of GDS marketing and sales opportunities.
- **HEDNA electronic reservation statistical survey** This is a survey that is conducted annually to identify trends in electronic hotel bookings. Results are available to HEDNA members.

GBTA

The Guild of Business Travel Agents (GBTA) was formed in 1967 by six London-based travel agents who felt their needs were not being fully represented by their national association, which was more interested in the mass market and leisure business. Today, the Guild has 43 members, which include high street names like the country's giant multiples: American Express, Thomas Cook, Hogg Robinson, Carlson Wagonlit and the UK's leading independent business travel specialists.

Between them they generated an annual turnover in 1995 of £6.0 billion, i.e. more than £16 million each day, employed 15,000 people in over 1,500 outlets; and were collectively responsible for handling over 75 per cent of all business house air traffic generated by agents in the UK.

The Guild was officially incorporated in 1967 and became a limited company in 1987. It effectively represents the industry's good housekeeping seal of approval. Membership is strictly by invitation only, with each new member undergoing a rigorous vetting procedure before being allowed to join the ranks. All hold IATA licences and are members of ABTA.

Central to the Guild's philosophy is customer care. It constantly and vigorously strives to raise standards and the quality of travel for its clients in a myriad of ways. Prime among these is an ongoing dialogue with influential organizations like the European Commission, BAA plc, the CAA, airlines and hotel groups, in addition to governmental lobbying at the highest level. The Guild has its parliamentary consultant who advises and helps address key issues like new airport terminals and runways, improved road and rail links to prime hubs and a wide array of other matters of concern to the business traveller.

Apart from keeping up the pressure of lobbying in key areas, the Guild has embarked on a campaign to create even greater professionalism and quality in the industry. It initiated the first ever qualifications specifically geared to the business travel industry. The six month City and Guilds course leading to a Certificate of Business Travel (CBT) was introduced in 1991 and the first candidates to gain a CBT, received their awards in the early part of 1992. There is also a supervisory and a management level certificate.

The Guild believes that as the market continues to become more competitive, particularly with the advent of the single European Market, the necessity for high standards and superlative quality will be paramount. Technology is a vital ingredient in the Guild's campaign of excellence and is essential to the development of international travel and the infrastructure required to support the business traveller. The Guild employs a top consultant to ensure that members are at the forefront of technological progress and that new GDS and airline electronic check-in systems meet members' and their clients' needs well into the next century.

The Guild is a pioneer in professionalism and its client companies are assured of the highest standards of service in every sphere of commercial travel. At a time when international business has never been more fiercely competitive, reliance on professionally qualified and equipped agencies like those within the GBTA is quintessential.

GEBTA

The GBTA's interest is, however, not confined just to the UK alone. The GBTA helped to guide the formation of the Guild of European Business Travel Agents (GEBTA), which has an increasingly powerful role to play in Europe as a whole. GEBTA now has ten national guilds representing Austria, Belgium, Denmark, France, Ireland, Italy, The Netherlands, Portugal, Spain and the UK. GEBTA speaks with a unified voice on issues like value-added-tax (VAT) on European air fares, air traffic control, airline competition, etc. It has full-time representation in Brussels for liaison with the EC.

NAITA

In travel, just like any other retail business, it is the large businesses that can obtain the most favourable deals with their suppliers. The reason for this is the bulk purchasing power that the large buyer brings to the negotiating table. This is one reason why the large travel multiples have been so successful in winning new business based on very favourable pricing of their products and services. A large multiple can, for example, go to a hotel chain with the knowledge that it represents a substantial portion of the travellers who stay at the hotels in that chain. They can therefore negotiate a reduced room rate for their customers, based on a certain level of booking volume. Once this special rate has been agreed it is up to the travel agency multiple to ensure that its branches actually deliver the number of bookings necessary to obtain the favourable rates as negotiated. This is what is termed buying power.

Well, that's fine for the large multiple agencies who used their buying power to enjoy a substantial competitive advantage over their smaller independent travel agency competitors for many years. However, the smaller independent agents soon realized what was going on and considered different ways in which they could organize themselves into a group that could represent a major buying force. This was one of the founding principles of the National Association of Independent Travel Agents (NAITA).

NAITA was formed in 1978 to represent independent travel agents carrying on business from branches or offices located in the UK, including Northern Ireland, the Channel Islands and the Isle of Man. It currently consists of member companies with a total branch network of over 700 offices employing 6,000 staff and is the largest retail travel agency group in the country representing almost 10 per cent of the UK travel agency population. Its membership has a combined turnover of some £1.8 billion of which £938 million is derived from leisure travel and £862 million from business travel.

Membership of NAITA is open to professional travel agents who meet the following criteria: (a) they must be members of ABTA; (b) they must have a minimum turnover of £500,000 if IATA licensed or £750,000 if non-IATA; and (c) they must have less than 25 branches at the time of joining NAITA. NAITA members currently cover the full range of travel agent activity with a substantial number of member companies choosing to specialize in either the leisure travel or business travel sectors.

NAITA is governed by a board consisting of one executive director and five non-executive directors. The executive director is a full-time salaried employee of the Association, while the non-executive directors are elected by the existing membership from within the membership. The board is chaired by a president who once again is elected by the existing membership from within the membership. The objectives of NAITA are to:

1. Promote and develop the trade, commerce and general interests of members of the Association.
2. Promote relations between members and others in the travel business and to provide the means for negotiations, discussions and liaison with other bodies concerned with the development of or in any way connected with the travel business throughout the world.
3. Inform and make public by way of television, radio, advertising or otherwise the activities of members and of the Association, and generally do all such things as may be necessary to raise and protect the prestige, reputation and status of members and of the Association.
4. Develop, investigate, improve and establish technological developments, marketing methods, discounting structures, 'money back' guarantee schemes, special insurance policies, bonding requirements and all or any other matters that may improve or affect the business of the members.
5. Promote and advance the cultural and educational aspects of travel, and the education and instruction of persons concerned or intending to be concerned with travel and travel agencies.

The strategy of NAITA is to develop the business of the small independent travel agency. In 1992 NAITA introduced the ADVANTAGE consumer brand to promote the qualities of its membership, together with the range, quality and competitiveness of its key operators (tour operators, airlines, car hire, cruise, ferry, hotels, etc.) to the travelling public. The continued development of the NAITA marketing strategy that enables quality operators and travel agents to promote jointly the highest levels of service, choice and competitiveness to the consumer is regarded as a primary objective.

In recognition of the importance of information technology in the modern travel industry, NAITA makes various IT facilities available to its member travel agents. These facilities include bulk purchase terms on PC hardware and both industry standard and tailor-made software. Bulk purchase deals such as these enable an organization to use the potential buying power of its combined members to negotiate favourable rates and prices on IT products with selected suppliers. After all, this is just what the multiples are able to do on behalf of their own branch networks. NAITA also has arranged a teletext programme that enables individual independent travel agents to participate in this growing form of travel advertising at realistic cost levels. All these teletext pages contain an 0990 telephone number and callers are automatically and evenly distributed using the latest communications technology to participating members' offices.

Additionally, NAITA operates 'closed user groups (CUGs)' on both Galileo and Worldlink enabling its member travel agents to access restricted NAITA commercial data, relating to the leisure and business travel markets. I will explain CUGs in more detail later on, in Chapter 6. For the moment though, a brief description is as follows. A CUG is a portion of a large data base available to many users, which is closed off to all except a restricted set of registered users. So, in the case of Galileo, for example, there is a portion of information storage on the core system that only NAITA travel agents can access. Within this, there are certain pages that can only be created and maintained by NAITA head office staff. The pages of information created in this way are only available to travel agents who are registered NAITA members. Access is controlled by the Galileo system, which checks the user identification entered during the sign on process against a table of authorized NAITA travel agents. I'll be explaining all these terms later in the book.

Through such CUGs, NAITA members can communicate with one another, and with the NAITA head office. Using the information stored within the CUG, NAITA travel agents can provide their clients with 24 hour global emergency assistance facilities and access to preferred hotel, car hire and air seat rates. Such facilities enable NAITA independent travel agents to operate as a national chain of agencies in the UK, with global connections to their international associates for client servicing.

ENTER AND THE IFITT

The International Federation of Information Technology and Tourism (IFITT) organization has its origins in the highly successful series of ENTER conferences that were first initiated to help guide the development of tourism in Austria. In general, IFITT aims to promote international discussion on the subject of information technologies in the field of tourism. As such, one of its primary objectives is the organization of the annual ENTER conference that is held each year in an international setting. Information and communication systems now form a global network that have a profound influence on both the tourism and leisure industries. Reservation systems, distributed multi-media systems, mobile working places and electronics markets are all manifestations of this phenomenon. Advances in the use and deployment of tools, technologies and methodologies that have facilitated the efficient networking of information systems in the tourism industry and their economic and organizational impacts are all prime subjects for discussion within the Federation.

The initial development of the ENTER conference first started in 1993 during some preliminary research and development work being undertaken for the Tirol Werbung – the Tyrolean Tourist Board. The objective of this work was to produce a new tourist information system (TIS), aimed at promoting inbound tourism to Austria. As part of the project's design efforts, it was recognized that a forum was needed for discussing electronic media and their influence on tourism, especially in the following areas:

- The relationship between the internal and external management of tourism related information. Organizations need to rationalize how information is stored internally for use by its employees. However, this information, or at least parts of it, may also need to be communicated or made available to external parties. So the way in which sub-sets of internally stored information is distributed throughout the public domain is a key issue for players in today's travel and tourism industry.
- The characteristics of travel and tourism products as they flow through the industry's value chain from suppliers, through intermediaries to consumers. Each stage offers several opportunities for activities such as re-branding, price mark-up, inventory access and product marketing.
- The practical ways in which travel and tourism information may be stored electronically and distributed via telecommunications networks

throughout the value chains. With the rapid evolution of IT and scientific advances, completely new ways of managing and distributing information are now available.
- The inter-relationship of travel and tourism disciplines, which are now being increasingly integrated via IT. The separate study of tourism, management science and computer science are, for example, rapidly becoming inextricably linked.
- The lack of an international platform for exploring travel and tourism issues in the context of IT. At the time, in 1993, there was no international conference where these kind of issues could be aired, where people in the industry could learn from each other and where information on new developments could be shared for the benefit of the world-wide travel and tourism industry in general. IT experts knew of the potential of the Internet but were less aware of the practical applications of the technology.

A group of academics from the universities of Vienna, Switzerland and Germany worked with the Tyrolean Tourist Board to arrange a conference that would address the areas identified above. What was needed was a conference that would not necessarily focus on just topical products, but that would instead concentrate on discussing longer term trends in the industry and look ahead to what might happen in the future. The name ENTER was given to the fledgling conference – a name derived simply from the 'enter' key on all PC keyboards. Many organizations were contacted and invited to participate in the preliminary ENTER conference programme.

The structure of the ENTER conference was formed in the very early stages of its life. It was decided that each year the conference would have a single major topic or theme around which all presentations, papers and discussions would be organized. Speakers would be invited from leading travel and tourism companies and academic institutions. The conference would be segregated into several activities, each pursued on parallel tracks:

1. A scientific programme with a call for papers, a rigorous reviewing process and the publication of conference proceedings.
2. An applied programme that was to be a commercially oriented series of presentations focusing on IT in travel and tourism; a significant feature of which would be 'the ENTER debate'.
3. An exhibition of IT oriented products and services aimed at travel and tourism.
4. An annual Web award given to the best Internet Web site designed for a travel or tourism application in any part of the world.

The first three conferences were organized in close conjunction with the Tirol Werbung (the Tyrolean Tourist Board) and Congress Innsbruck. All these activities were also strongly supported by several other Tyrolean tourism institutions. The first ENTER conference was held in 1994 on the subject of the new electronic markets in travel and tourism. This theme was again repeated in 1995, and then in 1996 the topic of business process re-engineering and related standards was addressed. This conference was attended by over 400 people from 21 countries, and had more than 500 people visit the exhibition stands. The applied programme featured 15 presentations, eight workshops and two panels. The scientific programme offered 31 papers covering a wide range of topics.

The 1996 conference was highly successful and made it apparent to the participants that IT has led to a paradigm shift in the tourism industry. It identified a variety of trends such as the way in which individuals are beginning to build their own customized tours, the development of remote destinations, the emergence of global competition and the problems experienced by concentrating power in a few large tourism operators. These factors suggest that IT can be utilized to re-engineer the processes in order to ensure a fair allocation of the profit margins generated among tourism sectors. The conference also concluded that a culture of constant innovation had emerged with structural improvements in the travel and tourism industries arising from information management on a global scale. It was at this conference that the first Web award was given. This fostered an in-depth discussion of the Internet and World Wide Web in relation to the potential offered to travel and tourism product design, development, promotion, distribution and delivery, throughout the world.

For 1997, the organizers recognized the need for the conference to expand into the international arena and therefore decided to hold the conference outside Austria for the first time. In order to accomplish this and to move the organization of ENTER forward, a more formal and broadly based structure was needed. It was at this time, in the Spring of 1996, that IFITT was formally founded. Participation in IFITT was open to decision makers, researchers and scientists in the tourism industry and academia. In fact anyone agreeing with the following mission statement:

> The IFITT will, in the best professional tradition, seek to develop and maintain the integrity and competence of individuals and organizations engaged in information and communication technologies and tourism. The IFITT will promote the free interchange of information about these fields among specialists and the public. The mechanism which will allow this to happen will be the ENTER conference which will be an annual event open to members and non-members.

The IFITT was therefore created with the prime objective of organizing and running the annual ENTER conference. However, several other objectives related to the deployment of IT within the travel and tourism field were also to be included within the IFITT's remit. By this overt mechanism, the IFITT strives to support the concerted development and growth of IT in the field of global travel and tourism. It is a prime objective of the IFITT to become a lively forum for discussion and exchange of ideas. It aims to establish an international body of knowledge and to contribute to the process of theory building in this growing field. As an inter-disciplinary federation it is placed between information systems and tourism research activities, such that both fields may influence each other. The IFITT will promote the free interchange of information about these fields among specialists and the public. Finally, the IFITT can act as a well-informed consultant to governments and other industry decision makers.

As one of its first tasks, the IFITT decided to work with an international publisher to create a journal on IT in travel and tourism. This journal is still being developed and is therefore not yet publicly available, i.e. as at mid-1997, although when it is published IFITT members will be able to obtain copies at a discount. A market newsletter is also published regularly and this provides an overview of technological developments in the field. Linked to this is a news service run by IFITT which is based on electronic publishing and is available to members. All ENTER conference papers are published on the ENTER Web site at **www.tis.co.at/enter** and may be downloaded by anybody with access to the Internet. The IFITT also supports special interest groups, e.g. it provides guidance on the formation of academic courses with curricula that include IT in tourism, and arranges speakers for the ENTER conference programme.

An IFITT management committee decides the group's future strategies and addresses political issues. This committee also chooses the main forthcoming conference topics and the subjects that are addressed by both the scientific and applied programmes. The IFITT membership fees for 1997 were £50 for an individual member and £560 for an organization. Students are eligible to join the IFITT at a reduced fee. Table 1.2 shows the IFITT management structure as at 1997.

In its first year, 1997, the IFITT had over 80 members, many of whom attended the inaugural committee meeting that was held after the ENTER conference in Edinburgh. The Edinburgh conference spanned a three day period from 22 to 24 January, 1997. Subsequent IFITT general assembly meetings will be held immediately following each ENTER conference. It is at these annual sessions that management committee members are elected and the next conference site is selected. Attendees at these sessions are encouraged to contribute their ideas for the development of the next conference.

Edinburgh was the location for ENTER's first conference outside Austria and was a resounding success. Billed as the Fourth International Congress on Information and Communications Technology in Tourism, its theme was 'travel and tourism – the challenge of the digital economy'. The conference offered a stimulating programme of keynote presentations by speakers of international standing, a carefully structured programme of presentations on topics directly relevant to the tourism industry, workshops for detailed discussion of specific topics, a programme of scentific/research

Table 1.2 The IFITT management committee – 1997

Management committee	IFITT role	Contact location
Hannes Werthner	President	University of Vienna, Austria
Josef Margreiter	Vice President	Tyrolean Tourist Board, Austria
Roger Carter	Treasurer	Edinburgh & Lothians Tourist Board, Scotland
Walter Schertler	Secretary	University of Trier, Germany
Kari Aanonsen		Norwegian Computing Centre, Norway
Dimitrios Buhalis		University of Westminster, England
John Rafferty		Irish Tourist Board, Ireland
Eva Hafele	Managing Secretary	IFITT, Austria – e-mail: **ifitt@tis.co.at**

papers, a special lunch for the prestigious ENTER Tourism Web Awards and an exhibition featuring tourism applications using new technologies. It was attended by 400 fee-paying delegates and many more non-fee-paying participants. The Edinburgh conference supported 30 on-site exhibitors, each of which demonstrated and promoted their products and services to delegates throughout the three day conference. The official language of the conference was English and all proceedings were published by Springer-Verlag Wien New York immediately following the end of the conference (ISBN: 3-211-82963-6).

Structural environmental issues

It is almost given that the dual subjects of (a) information technology, and (b) travel and tourism, are subject to rapid environmental change. The IT area itself is changing at an ever increasing pace as new hardware and software technologies evolve. In parallel with this, and in my view partially as a result of this, travel and tourism are changing rapidly on their own accounts. So, when you put these two together, along with several major infrastructural changes that are about to descend upon us, you have a recipe for potential disaster. The infrastructural changes that I am specifically talking about here are the turn of the century, which itself has spawned several other related software time bombs, and European monetary union. Let's consider each in more detail:

- **The turn of the century** Also known as the Y2K issue (although this is not strictly correct because one 'K' is 1,024), this is something that will impact every area of every business and in fact anything that uses a miniature computer processor. I won't go into the details of what the Y2K issue is because I am sure that most people know all about it by now. Suffice to say that it is caused by legacy systems, which were developed in the 1950s, 1960s and even 1970s, only allowing two numeric characters for the representation of the year within a date field. The problem is: How will software applications process the contents of this year field, following the 1 January 2000? There is a very real danger that insufficient action is currently being taken to address the problems that may occur at the turn of the century as computer systems of all types process inadequate date fields.

  The problem is obviously far reaching, but let's consider the travel and tourism related issues. While many large companies have dedicated individuals or even whole teams to tackle the problem, small travel agents may not even be aware that a problem exists within their own offices. Many of these are still running old PCs based on the 486, 386 and even 286 Intel processor chips. Most of these will not process applications correctly that depend upon the date for their actions. In other words, most applications won't work properly after 1 January 2000. Even the latest generation of Intel's Pentium and comparable chips from other hardware suppliers use something called a basic input/output system (BIOS), which may not be Y2K compliant.

One of the best ways for the travel industry to tackle these problems is to co-ordinate their actions with one of the professional trade bodies that support IT-related subjects. In the UK, two good examples are ABTA's ABTECH organization and the TTI, both of which are presented earlier in this chapter. These organizations can provide guidance on various software packages that can detect those elements of a computer system that are not Y2K compliant. Although these software packages are not 100 per cent accurate in their judgements on compliance, they do provide at least some guidance on whether a user's PC has a Y2K problem.

But the problem is far wider than just point-of-sale systems. It extends to any device that uses a computer chip to control its actions. This includes virtually all forms of transport, such as aircraft, trains and coaches, and also extends to devices used throughout the travel and tourism industries to control such devices as hotel lifts: and because the size and scale of software changes usually turn out to be at least double that originally estimated, time is fast running out for these changes to be made.

For example, the Y2K date problem is compounded by several other related problems. The date of 9 September 1999 will occur before the millennium and this may spawn problems of its own. This date translates numerically into 9/9/99, which was often used by early programmers to indicate an upper limit for a date field. Who knows what will happen when this date is reached? Then there is the added complexity introduced by the fact that the year 2000 is also a leap year. Under normal circumstances this might not pose a major problem for computer software but with the added Y2K issues, it could well serve to compound the problems faced by the IT community tasked with resolving this issue.

- **European monetary union**   Although this may be put off for a few years, it will no doubt happen one day. But what if it happens during the year 2000? The scale of software changes throughout the industry would be enormous and would occur all at one time: these changes are in any event non-trivial. At the simplest level, they equate to introducing another completely new currency. At the highest level they represent an opportunity for companies in the travel and tourism sector to exploit the benefits of eliminating exchange rate fluctuations from their marketing and pricing activities. To some, however, this will be seen as a drawback because many companies derive a revenue stream from currency conversions that are an integral part of their business. I discuss the implications of the Euro in more depth in Chapter 8 – Financial Services.

So, the travel and tourism industries are both complex and subject to rapid change; and as I hope you will observe from this chapter, the industry is highly dependent upon IT for its success and profitable growth. Now, having reviewed the environment, it is time to move on to explore how IT is used in the field of tourism to foster travel by holiday-makers and business people, from both national and regional perspectives.

2
Tourism

Introduction

This chapter is dedicated to exploring the ways in which IT is used to support tourism. In my view, the two main factors that characterize tourism are: (a) information, and (b) marketing. Both factors rely heavily upon IT for their successful deployment. New technology can store and process information extremely efficiently. This means that information management is easier and less costly than it used to be several years ago. Marketing relies on information in the first instance to identify potential consumers, establish the need for new products and project future income streams. But marketing relies on IT in far more basic ways than this. Tourism in the modern era is highly dependent upon IT as a new media channel through which products and services can be distributed to individual consumers and companies. Take for instance the tourism life cycle:

1. **First stimulus**   The very first stimulus to travel is pretty hard to identify accurately. From an individual's perspective, it could be as straightforward as an advertisement or television programme or it could be something as remote as a conversation with a friend who described a travel experience that appealed to the prospective tourist. However, it is increasingly possible that the first stimulus for travel may be attributable to the Internet; and this is a pure marketing tool for companies and countries wishing to promote their product or destination to potential travellers around the world.
2. **Information**   Once a person has been stimulated to embark upon a trip, either for personal or business reasons, the first thing the person will be seeking is information that describes all aspects of the planned trip. Information on alternative travel arrangements and the destination itself is needed. In the early stages of the tourism life cycle, this information is almost purely concentrated on facts and figures that describe the destination for the visitor. Besides booklets, pamphlets and telephone conversations, this information can be conveyed in a number of new ways. For example, it can be recorded on compact disk lead-only memory (CD-ROM) disks and viewed on a PC, viewed on an Internet site or accessed via a self-service kiosk. The possibilities are extremely varied and set to increase over the next few years as technology becomes ever more sophisticated.
3. **Booking**   Booking systems used to be the domain of the travel agent. However, with the growth of the Internet some booking systems are being distributed directly to consumers and large companies. Also, in the context of tourism, the most sought after products are bed and breakfast accommodation and tickets for local events. Both are now increasingly available from either tourist offices using computer systems supported by large data bases of supplier information or Internet sites that provide this information direct to the consumer.
4. **Travelling**   Once the tourist embarks on the trip, IT is there once again to make the tourist's life easier. Airlines allow travellers to use self-service machines for check-in, air carriers offer passengers computer games and information displays, hotels offer their guests automatic check-in with in-room Internet access

and desk-top PCs, tourist offices provide self-service information machines, many of which also support accommodation booking services and payment collection.

I hope that from this introduction, you can see that IT is widely used in almost every part of the travel and tourism environment. Not only is it used so widely but many companies and organizations actually rely upon it for their profitability and success. I'll be covering the Internet in Chapter 5 so for the remainder of this chapter I focus on how IT is used to support tourism organizations both at the governmental level and at the local inbound/outbound level. I have taken the UK and Ireland as prime examples of how IT is being used to support tourism-related activities. Let's take the UK first of all and start with the British Tourist Authority (BTA).

The United Kingdom

The UK has a well established infrastructure to support and develop inbound tourism. There are three main components of this infrastructure: (i) the BTA, which promotes Great Britain as a destination to overseas countries and foreign visitors; (ii) the English Tourist Board (ETB), which promotes England as a holiday destination for domestic holiday-makers; and (iii) tourist information centres TICs, which are scattered throughout the destination locations of the UK and offer information and local booking services to tourists. I'll be describing each of these in turn, but let's start with the BTA.

THE BTA

The UK Government, like many others around the world, recognizes the value of tourism to its domestic economy. This is clearly illustrated by some basic UK statistics, for example: 23.6 million people visited the UK from overseas in 1995 and these visitors generated £12 billion of expenditure. By the year 2000, the BTA estimates that overseas visitors' spend in the UK will amount to £18 billion a year, 40 per cent more than in 1996. So, in order to promote Britain to overseas countries as a place to visit, the UK Government supports a dedicated body, i.e. the BTA.

The BTA's mission may therefore be simply stated as: 'to promote inbound tourism to the UK'. However, the supporting business objectives are more targeted than that. Some important sub-objectives are, for instance: (i) to increase the amount of money that overseas visitors spend while in the UK, and (ii) to encourage as even a balance of expenditure as possible, both geographically across the region and seasonally throughout the year.

In order to achieve its objectives the BTA deploys a considerable proportion of its resources on the collection and dissemination of information to consumers and to a variety of companies and organizations in the business of bringing visitors to Great Britain. Examples include inbound travel companies, parts of the UK travel trade, travel suppliers, business travel conference organizers, overseas tour operators and incentive travel companies. Consequently the BTA, like other tourism organizations, is a highly information intensive activity; one for which the opportunities offered by IT are particularly important.

Any discussion of the UK Government's tourism organizations must include both the BTA and its sister group, the ETB. These two organizations work hand in glove within the field of UK tourism. The ETB focuses primarily on encouraging domestic consumers to take their holidays at home in the UK, whereas the BTA's attention is very much on encouraging overseas consumers to visit the UK. Within the context of IT, both organizations share a common interest that is fundamental to their core promotional activities. This is the collection and dissemination of tourist information on the UK as a destination: because it is the ETB that plays a major role in collecting and vetting such information, this is a good place to start our review of IT within the UK's tourism promotion activities.

The ETB

The primary objective of the ETB, is to promote England as a holiday location for domestic consumers. The ETB accomplishes this not through marketing activities of its own but via strategic partnerships and the development of jointly funded promotional programmes. A secondary, but equally important objective, is to raise standards of product/service quality throughout the industry in the UK. Also important to ETB activities is the feedback of information to suppliers on consumer likes, dislikes and views. Feedback such as this is vital to the development of new markets, products and services by travel suppliers.

The ETB achieves these objectives through ten regional tourist boards (RTBs) located around the country. Each of these RTBs is an independent autonomous company in its own right and is not controlled directly by the ETB organization. RTBs are usually formed as limited companies and derive their membership from local authorities and business people. It is always the case that one of the RTB board members is an ETB director. Generally speaking, the primary role of an RTB is to co-ordinate the local development of tourism in its area, but in the context of doing this it also supports ETB marketing plans and therefore is eligible to receive some ETB funding for these activities. Other RTB sources of funding are derived from membership of RTB and from commercial activities. One of the key roles played by the RTBs is the collection of information on local tourism products and services for use in promotional programmes and a variety of publications. I'll explain the IT-related aspects of this in more detail in a moment. But first, let's just consider the main items of UK destination information that are collected and stored in an unbiased way by the RTBs. These are as follows:

- **Accommodation** This is categorized into *serviced* and *non-serviced* accommodation. *Serviced* accommodation includes hotels and bed and breakfast locations, whereas non-serviced accommodation covers self-catering and rented cottages.
- **Attractions** There are over 5,000 attractions in the UK that range from theme parks, such as Alton Towers, to castles and stately homes. Both private and public attractions are included here.
- **Events** This includes specially organized activities or shows that either occur just once or are ongoing. Examples include The Chelsea Flower Show, millennium shows and village fetes.
- **Activities** In particular, English language schools are promoted as a source of incoming tourism. These may be either full-time or part-time courses at colleges or schools in Britain. Other examples are activity based holidays.

So, in summary, the ETB tries to encourage people to spend their holidays in their home country and the BTA encourages overseas visitors to come to the UK for their holidays. Thus, as you can see, both organizations are largely complimentary; and this is one key reason that they work so closely together. Another reason is that they both share a common interest in maintaining accurate and up-to-date information that describes the UK as a tourist destination. The collection and storage of information is therefore fundamental to the activities of both the BTA and the ETB; as well as, for that matter, any tourist organization. So, before we go any further, it is important that you understand how this information is managed.

Information

What the BTA and the ETB both have in common is a need for information that describes their home market, which is of course the UK. For historical reasons, it is the ETB that has spearheaded the collection and recording of information about England. The ETB accomplishes this in close collaboration with the geographically dispersed offices of the RTBs: this is where IT plays an important role in the promotion of tourism in England. In fact, IT is set to become the single most important critical success factor for both the BTA and the ETB in the medium to long term future. So, it is important to understand what information we are talking about here and how IT can be used to distribute this information to consumers and inbound travel companies in the UK and around the world, both now and in the future.

Let's start with the information itself. The ETB collects and stores a variety of textual data on some 40,000 hotels and other types of accommodation, 5,000 tourist attractions and 8,000 events occurring in the UK at any one time. It does this in close collaboration with the RTB, Fig. 2.1, as follows:

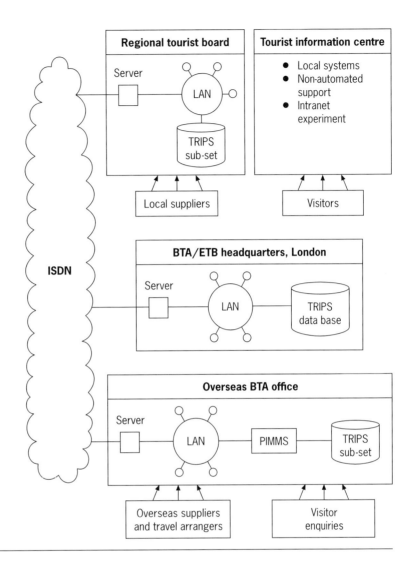

Figure 2.1 The UK tourism information network

- The information needs are determined and standard questionnaires for each type of data are designed and pro formas of these are distributed annually by the ETB to each of the ten RTBs. The RTBs pre-print the forms locally. They are designed to capture all items of pertinent information concerning an accommodation supplier, an attraction or event that is planned to be held somewhere in England. The objective is to collect this kind of information as close to the source as possible.
- The questionnaires are distributed to local suppliers for completion and returned to the local RTB office. All information is vetted by the local RTB officers who check the forms for correctness and ensure that data are presented in the appropriate 'house style'.
- Completed questionnaires are manually keyed by local RTB staff into their client/server local area network (LAN) based computer system, which contains a single region copy of the data base. Overnight, any amendments are uploaded via communication lines to the central tourism resource information processing system (TRIPS).
- TRIPS is the main component of the ETB and BTA's London head office system. This uses a client/server architecture with standard PCs as clients and multiple servers, one which controls a Novell based LAN and the other which

TOURISM 33

is the TRIPS data base server based on a DEC VAX hardware platform.
- The data collected from the RTBs are therefore input to a central application that performs validation checks before updating the central TRIPS data base held on the DEC VAX. This is an application written in a hierarchical data base language.
- The result of the RTB data collection exercise is therefore updated regional and central TRIPS data bases. Access to TRIPS is also provided to remote locations like the BTA office in New York, using wide-area network (WAN) technology and terminal emulation software (a Windows front-end is planned for 1998).
- The TRIPS data base also incorporates a flexible report generation facility. This can be used to generate customized reports on demand, for local printing. Overseas locations can receive extracts of the TRIPS data base for local processing on their own PCs.
- In addition to the regular annual updates (quarterly for events), described above, the information is also kept current by the RTBs on an ongoing basis. Each time an important change is detected, the software communicates the update overnight.
- Information on English language schools and activity holidays is stored on separate data bases. These small data bases, consisting of just a few thousand entries, are managed by the Microsoft Access software package.

The information used to create the TRIPS data base is used for many purposes. For example, sub-sets of the information are placed on the British Airways BABS system (see Chapter 3 for a description of BABS). This data is currently keyed manually from TRIPS into the DRS section of BABS (although more sophisticated update software is being planned). Other information systems are also under consideration by the BTA and ETB. These offer consumers an on-line 'searchable' data base facility (see Chapter 5 – The Internet).

Despite the fact that TRIPS has been developed over many years, it remains a cornerstone of Great Britain's tourism activities for two main reasons: (a) it is the only repository of up-to-date country-wide information on tourism in Great Britain, and (b) it has adapted to an increasingly sophisticated end-user population by the continual development of new and easy-to-use interfaces: over the years, this information has been recorded and disseminated in a remarkably consistent way. This data base has therefore been the underlying reason for much of the success of the UK's inbound tourism promotion activities since the early 1980s. Much of this success has been achieved because:

- Promotional publications are quick and easy to produce from a comprehensive and reliable source of information on the UK, e.g. leaflets, guide books, pamphlets.
- Sub-sets of the data base are held in most overseas BTA offices, thus allowing immediate answering of queries on the UK by, for example, telephone service centres.
- Information can be provided quickly and efficiently to overseas tour operators wishing to construct package holidays to the UK.

However, despite the clear evidence of this success it is nevertheless sometimes said within the ETB that it is still easier for a UK consumer to book a holiday in Spain than in the UK itself; and from the BTA's viewpoint, it is still difficult, for example, for a Dutch cyclist club to book bed and breakfast accommodation and other services in small- to medium-sized establishments (SMEs) over the course of a two week tour of the UK. There is therefore a pressing need for local UK products and services to be booked easily by the consumer and inbound suppliers. Although this is an issue that has vexed the BTA for many years, the Internet holds the promise of an ideal solution (see Chapter 5 for a review of tourism and the Internet). However, having said this, there are no plans for a commercial booking engine to be supported by any Great Britain tourism organization at present.

A related issue is: How can the BTA increase the level of support given to overseas companies to help them build package tours that include visits to the UK? The challenge for the British tourism organizations is therefore to use IT to communicate with consumers and provide interfaces with supplier booking systems. The following paragraphs explain just a few of the innovative IT-based initiatives currently under consideration by UK tourism organizations.

The teletourism experiment

A key challenge for the BTA and other tourist organizations is how to capitalize on the gradual integration of what has historically been two separate information storage technologies: (i) data bases of textual information, and (ii) libraries of graphical bit mapped images (including photos). In the new multi-media environment, however, some significant benefits can be derived from the shared organization of linked text and image data. This is required, for example, to answer disparate queries from consumers on a variety of subjects in a flexible way – sometimes to show a picture of any castles linked to say Arundel; at other times to show pictures of all Norman castles and where they are located in England. When these multi-media technologies are linked with the new high speed telecommunication technologies, some really interesting possibilities begin to appear.

One such possibility was the teletourism initiative that was funded by a European Community grant (Reference DG14 – the Communications Directorate of the EU). The overall objective was to seek ways in which IT could be used to enhance the interaction between central tourism organizations and their remote offices. Specifically, the grant was aimed at encouraging the growth of Europe's Integrated Services Digital Network (Euro-ISDN), which is now a telecommunications standard supported by the EC. However, at the time the experiment was commissioned, ISDN existed in several slightly different 'flavours' across Europe. It was hoped that experiments like teletourism would identify and therefore help resolve these wrinkles in the ISDN standards.

The experiment involved two information suppliers: the BTA and the tourist office for Tuscany in Italy. The idea was twofold: (i) to provide an information and booking service for tourists wishing to stay in London apartments, and (ii) to promote Tuscany villas and holiday homes to European tourists visiting Italy. Five remote user sites in Paris, Madrid, Brussels, Dublin and the British Travel Centre in Regent Street, London, could all access the system via ISDN communications technology. Visitors to these five European locations could request a tourist information officer to help them use the teletourism PC to browse a data base and possibly book accommodation (the system was not designed to be intuitive to allow consumers to use the system sufficiently themselves). On-line help desk support is always available to all users of the service during normal office hours.

In order to learn as much as possible within the constraints of a limited investment in time and money, only a small number of suppliers and locations were included in the experiment. This was accomplished by sharing information recorded on an on-line multi-media data base, using video-conferencing techniques and by providing a rapid response booking capability. Taking each component in turn:

- **Multi-media data base** Tourism officers in the five remote tourist offices used the ISDN link to access a searchable multi-media data base stored on the BTA's server in London. The end user (in this case the tourism officer), uses a map based graphical user interface (GUI) to specify the kind of property in which the consumer is interested. The search criteria may include several parameters such as the number of rooms, price, facilities, location, period of construction and many others. A drill-down facility enables a great deal of relevant information to be retrieved from the data base, which can be searched until a property of interest is found.

  Properties of interest to the consumer can then be listed and pictures of the apartments can be viewed. These are stored in a library of images on the BTA's London server. The images show, for example, a wide area map of Greater London; a local area street map; pictures of the exterior of a selected apartment; views of the inside, including specific rooms and views from certain rooms. Naturally, each image is accompanied by the relevant text that describes it in full. All images and text retrieved from the server, are transmitted to the remote overseas office and displayed on the local screen.

- **Video-conferencing** For more in-depth information, ISDN is used to support video-conferencing (see Chapter 6 for more information on video-conferencing). This allows the prospective tourist to talk and see a person, i.e. a tourism information officer, at the remote end of the line in order to discuss requirements in more detail.

The video-conferencing link also supports some key interactive tools: (i) the shared use of resources such as a virtual 'white board', viewable by both the caller and called parties – this can be used, for example, to draw maps or diagrams; and (ii) the use of interactive form-filling and spreadsheet techniques to help develop ideas using work-group technologies.

- **Booking support** Finally, when the consumer (in this case either a traveller or a tour operator) wishes to make a booking, the system automatically records the booking information. This is then passed to an information officer at the London Tourist Board, where an actual booking is made. Bookings are currently made manually either: (a) via the telephone, or (b) using on-line reservation terminals in the London Tourist Board offices. Confirmations are faxed back to the end user in the remote tourist office.

The teletourism experiment was judged to be a success by its users. It was well received and consumers generally liked using it. However, in order for some real benefits to be derived, the system needs to be used more heavily. The only problem here is that the BTA's high volume of bookers for London apartments tend to originate in the USA and Middle East – both of which are of course outside the EC area and hence outside the scope of the experiment. But in terms of the original objectives, the experiment has: (a) identified a few small ISDN variations that in the end proved immaterial, and (b) demonstrated a working pan-European tourism application that made good use of ISDN technology.

The next steps in the development of this experiment are to extend the system to enable more of the TRIPS data base to be accessed, particularly other parts of the accommodation section. Also, it is recognized that for consumers and suppliers to derive real benefit, more supplier booking systems need to be automatically connected. Although this could be done via a communications switch of some sort, the resultant costs are expected to be high.

The BTA is therefore looking at other options, such as a 'booking by fax' service, which could prove to be more cost effective. This works by having the server automatically 'fill-in' an electronic form, based on the information entered by the end user as part of the dialogue. The electronic form is sent to the appropriate booking system and a reservation is made. Once confirmation is received, the computer automatically formats a fax message and transmits this back to the end user.

Although a 'booking by fax' service could well prove an excellent approach for larger suppliers that have their own computer systems, it is not always quite so suitable for smaller accommodation providers (sometimes known as SMEs). Most SMEs do not have computer systems that can automatically respond to electronic booking messages and many do not even have fax machines. However, it is the smaller guest houses and bed and breakfast locations that are often of prime interest to an overseas consumer. This is where the BTA faces a problem – it is not a commercial organization and is therefore not able to enter into the business of providing either a booking capability or a reservations service for these SMEs. However, it is within the BTA's remit to help and encourage such tourism related developments. One of the key IT-related issues that the BTA needs to address over the short to medium term future is therefore: How to support electronic bookings for SMEs?

There are many ways in which the teletourism experiment could be developed and extended. However, the real question is: What technology can best deliver comprehensive tourist information on the UK to consumers and inbound travel companies in all parts of the world and also support a wide range of supplier booking systems? The Internet is one obvious alternative that is worthy of some careful consideration. The BTA's use of this new channel is explored in more detail in Chapter 5.

Overseas BTA offices

From its headquarters in Hammersmith, London, the BTA is responsible for more than 40 overseas territories. Each of these is responsible for promoting Great Britain, which for the purposes of tourism encompasses England, Scotland, Wales and Northern Ireland (by special contract), to both consumers and the travel trade in other countries around the world. The BTA's overseas offices vary enormously in size. There are two main types of office: a dedicated BTA owned and run office, and a representative office:

- **BTA owned offices** The largest of these is in the USA where the New York office employs 60 staff. This dedicated tourism operation houses a nation-wide toll-free telephone service centre with 15 staff and a high speed telecommunications line connecting it to the BTA's London head office. The transatlantic line provides New York with direct access to the TRIPS data base of tourism information on Great Britain. This technological infrastructure enables the BTA's New York office to support an average of 0.5 million calls per year. Other large BTA offices include Frankfurt and Paris, both of which employ approximately 20 staff.
- **BTA's representative offices** In smaller overseas markets, BTA uses representative offices that are often shared with other UK organizations. These offices are therefore not usually wholly owned or staffed directly by the BTA. In many cases these representative offices are independent companies contracted to the BTA for the supply of tourism services in their local market.

The focus of all BTA's overseas offices' efforts is very much on the provision of information to consumers and the travel trade: this information currently takes the form of paper-based publications that describe the UK to overseas visitors and suppliers in such a way that it does not favour any particular company or commercial activity. Many overseas BTA offices stock around 300 brochure titles, each of which describes a specific region and/or aspect of the UK in a particular language. The cornerstone of this two pronged promotional activity is the BTA's public information mailing management system (PIMMS).

PIMMS

This is a purpose-built system originally developed by the BTA in 1989 using Borland's Paradox programming language, for running under Microsoft DOS. It has since been re-developed within a Microsoft Windows 95 operating system environment using the Access data base product with Visual Basic as the programming language. This new version is being Beta tested, i.e. as at mid 1997, and will be rolled out to all BTA offices. WinPIMMS has been developed for Windows 95.

Over the past six or seven years, PIMMS has become a powerful marketing and promotional support tool for the BTA. It may be customized and configured for different countries. PIMMS is used by all overseas BTA offices to: (a) fulfil requests for information on Great Britain as a destination, and (b) capture marketing information to support future promotional activities. PIMMS is used as follows:

- Operators receive incoming telephone calls or walk-in visitors to the BTA office in another country. Most of these enquiries may be satisfied by sending the caller a brochure. These operators use a PC that has the PIMMS software loaded and running. The main menu comprises the following options:
    - *New request* Enquiries are divided into two main classes: (i) those received from consumers considering visiting the UK, and (ii) those from travel trade companies constructing inbound travel products to the UK. PIMMS provides customized support for each type of enquiry, as follows:
        - *Consumers* Consumer details are keyed into PIMMS by the operator. The operator enters a check mark against the publications that seem most relevant to the consumer's enquiry. The system also records marketing information such as the consumer's interests, where he/she saw the BTA advertisement and some demographic information. Finally, the consumer's name, address and other contact details are captured.
        - *Travel trade* PIMMS enables the BTA offices to build up local data bases containing full contact details of all major travel trade suppliers in their countries. The operator is provided with a variety of selection parameters that can be used to access the data base and select the

appropriate company. Once selected, the process is similar to that described above for consumers. The operator enters the relevant details that describe the company's enquiry, while at the same time capturing many items of useful marketing information. The company data base also allows the operator to track the number of brochures requested to-date.

For both types of enquiry, the operator may access a range of information on UK locations, products, suppliers and services. These are all stored locally by PIMMS using information derived from the TRIPS data base.

- *Request query* This function is used to investigate a consumer's claims that information previously requested has not yet been received. The original request, as logged by PIMMS, can be viewed.
- *Labels* Mailing labels for affixing to envelopes may be selected, sorted and printed onto special stationery.
- *Business* This is a marketing support function and allows the PIMMS data base of contacts to be segmented and analysed in detail.
- *Archive* Old contact information may be removed selectively from the PIMMS data base and transferred onto floppy diskette. This frees up space on the PCs hard disk yet allows the information to be kept off-line for possible future use.
- *Maintenance* This allows data base items to be viewed, modified, deleted or added. Examples of data that may be maintained in this way are travel company profiles and the brochures that are available for distribution.
- *Events* This is the scaled-down version of the TRIPS data base and describes UK-based special events that are displayed in date or location sequence.
- *Accommodation* Similar to the above, this is a sub-set of the TRIPS accommodation data base and shows a wide range of hotels and SMEs.
- *Exit* Allows the user to close PIMMS down and exit from the system.

• At some pre-determined time during the day, the operator may request the PIMMS system to print mailing labels, a personalized letter and a picking list of paper-based promotional material, e.g. brochures. This is used to mail the appropriate brochures and other material to the enquirer.

As mentioned above, an extract of the TRIPS data base is stored within the local PIMMS system. This sub-set of TRIPS is crucial to the system's effectiveness. Periodically, a special extract of the TRIPS data base is taken and either: (a) transmitted from London to the overseas BTA offices by e-mail, or (b) stored onto floppy diskette and mailed to the overseas BTA offices. In either case the information received from head office is entered into the local PIMMS system. This process is undertaken quarterly for those items that describe UK events and annually for the 30,000 or so items that describe accommodation suppliers. Other categories of information are being considered for PIMMS including: UK attractions, activity holidays and English language schools.

Physical information distribution is a key issue for the BTA. The reasons for this are: (a) brochures cost a great deal to design and produce; (b) the logistics required to support brochure fulfilment are considerable, e.g. obtaining stocks from a central supplier, storing supplies of brochures locally, organizing and fulfilling promotional mailings; and (c) the cost of mailing brochures to consumers and trade contacts is a significant expense. For these reasons, the BTA continues to investigate other approaches to promotional efforts. These include centralized fulfilment and use of the Internet (see Chapter 5 for a more detailed evaluation of these opportunities).

PIMMS can also be accessed by other means including, for example, telephone voice response systems and consumer activated kiosks. Automated voice response systems support unattended brochure requests. The way this works is as follows. First of all, the PIMMS system is linked up to the telephone via a purpose-built voice response computer loaded with special software. Incoming calls are automatically answered by the voice response computer that controls the dialogue with the caller. Callers use their telephone keypad to select options from a spoken list. The end result of this dialogue is a brochure request that is logged by the PIMMS computer, just as though the call had been taken by a human operator. Similar but more sophisticated techniques are being considered that involve housing the PIMMS system and its human interface within a multi-media kiosk. This would then be used by walk-in clients of the BTA office or possibly passers-by in a public area, depending upon where the kiosk is situated.

However, in all cases, the BTA uses available technology to the greatest extent possible in each country. For example, overseas BTA offices use their in-house ISDN service to communicate with London and other sister offices. Also, other PC software packages are used including Autoroute for road journey planning and British Rail's timetable data base, which is stored on CD-ROM. Additionally, fax-back technology is currently used by the BTA in Tokyo, Los Angeles and New Zealand.

So, as you will have gathered from this brief overview of the technologies used by the BTA, it is an IT intensive activity. Like so many other tourist offices around the world, information is at the heart of everything the BTA does; and much of the information used by the BTA, which is in fact gathered and recorded by its counterpart the ETB, is also used by TICs.

TIC

The national UK Government is not the only public body to recognize the importance of tourism to the economy. Local authorities have also recognized how important it is for a local community to attract visitors to their area. The vehicle to support inbound tourism to an area of the UK is the tourist information centre (TIC). There are over 540 TICs in England. In most cases, they are independent bodies but they are usually either run by local authorities or are at least heavily influenced by them. There is an enormous variety of TICs, ranging from a small mobile van to a high street shop front. The ETB sets certain standards for TICs that, if met, allow them to participate within the TIC network and use its standard branding. So, although the ETB can influence the TICs, the ETB has no direct control over their activities.

Given this local sphere of control, it is not therefore surprising that TICs are very responsive to local needs. However, on the other side of the coin two basic problems lurk: queries from tourists who are travelling from place to place in the UK are difficult for TICs to answer satisfactorily; and (ii) there is precious little standardization of information storage methods used by TICs, from a national perspective. Let's consider each problem in turn:

- **Local/national information** The tourists who visit TICs usually require information about local events or accommodation. But there are also many who are visiting the TIC location as part of a grand tour of the UK (or at least a regional tour). These people require not just information on the locality but also on items such as: how to plan their itinerary, where to go next, how to book their next accommodation stop and what attractions to visit outside of the TIC area. Unfortunately, TICs do not have access to the BTA's TRIPS data base and have to rely on paper-based publications for such information.
- **Technolology** The technology used within TICs has been extremely focused on meeting local needs. For example, within each TIC there are often one or more local information and booking systems that cover the surrounding area's products and services. The aims of these systems are to increase the level of bookings and to encourage tourists to use the region's suppliers. Consequently the number of disparate systems used within the TICs around the country have grown like topsy. Most have been independently developed without any overall standards or a common technical architecture.

The ETB's challenge is therefore to capitalize on the various local systems used by TICs and thereby provide access to the TRIPS data base of UK products and suppliers. With so many different TIC systems out there, this poses a real problem. However, in 1989, the ETB identified a key opportunity to do something about this challenge by encouraging TICs to adopt a common standard for information recording and retrieval. This is how the ETNA initiative was launched.

ETNA

The objective of ETNA was to establish a common standard for the recording of information used by TICs. Although the ETB has no direct control over the TICs, it does have considerable influence on their activities. This influence arises primarily from the TRIPS data base, which is owned by the ETB–BTA and would be extremely useful to TICs. If only some way could be found to distribute the TRIPS data base to all TICs on a periodic basis, then this would enable them to provide an enhanced level of service to tourists.

The general approach used by the ETB to pursue the ETNA initiative was: (a) to commission the development of a standard information retrieval program that would run on most of the TICs existing PC equipment, and (b) to distribute the TRIPS data base among the TICs on a periodic basis. This is how it was done:

- A special purpose program was written in a fourth generation computer programming language called Paradox (marketed by a software company called Borland). This was developed as a prototype application and supported functions such as data base load, data base update, information retrieval and printed reports.
- The ETB identified three or four computer re-sellers with whom to work in collaboration. This entailed: (a) the ETB providing the re-seller with the Paradox application software together with a sub-set of the TRIPS data base on floppy diskette, and (b) the re-seller marketing this packaged service to each TIC.
- A common standard was agreed for recording a sub-set of the TRIPS data base on computer files. This sub-set contained all of the information that was relevant to the TICs and was structured using Paradox data base tools. So, some standardization was achieved and this enabled the system re-sellers to import TRIPS data and update their data base systems. This in turn enabled them to maintain the data and sell the packaged system to their customers.

The ETNA initiative was all about standards. So, in the final analysis, although the system worked, and is still providing a useful service to over 150 TICs, the standards setting objectives were not totally achieved. The main reasons for this were: (i) there was little consistency among the TICs in terms of technology – some TICs were even using unsupported systems, and (ii) the re-sellers were too small and either could not devote the necessary resources to promote and develop ETNA further in-line with technological progress. However, the experience gained as part of ETNA was to prove invaluable to the ETB when it came to consider the opportunities offered by the new technology paradigm – the Internet (see Chapter 5 for more information on this topic).

THE CANTERBURY TIC

Having set the scene for TICs, I thought it might be helpful to examine an actual TIC in more detail. The Canterbury TIC is a good example because first of all, it is a pretty typical TIC and, secondly, it has participated in the ETB Internet experiment (see Chapter 5 for more details). It is a very busy TIC, which handled 308,216 visitors over the period April 1996 to March 1997, and is located in the city centre, close to the Canterbury Tales visitor attraction. The main visitors' service area is located on the ground floor and other departments, including the accommodation booking service unit, are on the two floors above.

The Canterbury TIC is a private company that is a wholly owned subsidiary of the Canterbury Chamber of Commerce. The historic building that houses the TIC offices are provided by the city council which also subsidizes the operation with a grant covering approximately 25 per cent of outgoings. However the TIC is still responsible for all other expenses that arise from its operations. The Canterbury TIC is an information provider that has two main customer populations: (i) outbound travellers who are residents and local businesses interested in UK destination areas serviced by other TICs; and (ii) inbound visitors from outside the local area, including other TICs and overseas tourists, who are interested primarily in the Canterbury area. The terminology I have used for the remainder of this section therefore refers to inbound tourists to Canterbury and outbound tourists travelling to other parts of the UK. Let's take each type of tourist in turn and examine the services that the Canterbury TIC provide.

Residential and local business information services

The primary revenue earner, and the service that is growing fastest of all, is the accommodation booking service that is located on the first floor of the Canterbury TIC. This provides information and booking services directly for local accommodation and via other TICs for destination visitors. Taking each in more detail:

- **Inbound visitors** These usually comprise potential visitors to the city who wish to stay locally. Most of these requests are received over the telephone from other TICs and from walk-in visitors. However, the Canterbury TIC also receives direct contacts from customers who either telephone or send their inquiries by fax, post or e-mail.
- **Outbound visitors** These are visitors who may, for example, be touring the country and stopping off at other UK locations as dictated by their own flexible itineraries. These people usually visit the Canterbury TIC in person to arrange their accommodation. The TIC network offers a very popular service called book a bed ahead (BABA), to meet this requirement.

Naturally, the Canterbury TIC has specialist knowledge of accommodation services within its own locality. The office maintains its own detailed data base of hotels, bed and breakfast houses and many other hospitality suppliers within the Canterbury area. Much of this information is stored on a PC using a simple word processing application. This is used mainly to support the inbound accommodation business. The Canterbury TIC can, however, also book accommodation for outbound tourists who wish to stay anywhere in England, Scotland, Wales or Northern Ireland, via its links with other TICs around the country. Links that are prime candidates for the effective deployment of computers and telecommunications. But before we can consider how new IT can be used by TICs to support their inbound and outbound accommodation businesses, it is important to first of all understand the current booking process in more detail. This works as follows:

- Customers may contact the Canterbury TIC by either: (a) walking into the office if they are in the locality; (b) if they are in another part of the country, requesting their local TIC to telephone Canterbury; or (c) telephoning themselves directly. They give details of what accommodation they are seeking to the Canterbury TIC representative. The accommodation supplier is sought in the following ways:
  - *Inbound visitors* The Canterbury TIC representative will consult the local file of hotels and bed and breakfast establishments. A dialogue with the visitor enables a choice to be made and a telephone call to the proprietor usually secures the required accommodation.
  - *Outbound visitors* For immediate departures to outbound areas, the TIC representative will contact the servicing TIC in the destination area by telephone. Some TICs will not take advance bookings and will only deal with customers who wish to book for the current or next day's stay. The Canterbury TIC representative will advise the remote TIC of the customer's requirements.
- At this point, the customer may be asked to either wait in the office or call back later while the accommodation required is sought, booked and confirmed.
- For outbound visitors, it can be two hours or more before the remote TIC calls back to advise details of the booking that have been made for the customer. Sometimes the information is faxed back to the Canterbury TIC.
- The Canterbury TIC receives confirmation of the accommodation and provides the customer either with a locally produced booklet supplied by the hotel or a copy of the information received from the remote TIC.

This process is both cumbersome and labour intensive and is not conducive to the efficient handling of high numbers of enquiries. Over the period April 1996 to March 1997, for example, the Canterbury TIC received 25,864 contacts from visitors by mail, telephone and other means. Attempts have been made by other TICs to overcome these problems as part of the ETNA initiative (see above section). ETNA attempted to encourage third party computer companies to develop information systems based on the TRIPS data base format, for use by TICs around the country. This was, however, only partially successful and not all TICs could afford the investment in this new technology. Another unfortunate problem was that the initiative resulted in at least two incompatible systems: the CTV system and the Integra system. Despite the fact that both systems can use a version of the TRIPS data base for their information feeds, they are otherwise incapable of being linked or shared between TICs. However, the Canterbury TIC uses no such system at present.

The long term aim is to replace these third-party systems and the manual processes used within non-automated TICs, with a new system based on Intranet or Internet technologies. At present, the ETB is experimenting with an Intranet service that has the potential to automate many of the information servicing functions performed by TICs (see Chapter 5 for a description of the ETB's Intranet experiment). A final version of the experimental Intranet system could eventually be used to replace much of the above telephone-based processes between TICs. A customer's requirements, whether for inbound or outbound services, would first be understood by the TIC representative. The representative would then select the appropriate part of the country from the ETB's Intranet data base using simple menus and search engines. The details of accommodation alternatives could either be described to the customer or a print-out of the relevant pages could be given to the customer. Along with this, a map page showing how to get to the area desired and how to find the accommodation shown, could be produced. However, in terms of booking the accommodation, no alternative has yet been identified to using the telephone. This is primarily because, although an on-line booking system would be quite possible from a technical angle, the commercial and practical issues have not yet been resolved fully by UK tourism organizations.

At present, the ETB's experiment has provided only a single PC that is available for use by the Canterbury TIC staff on the ground-floor servicing area. Because this currently operates rather slowly and is shared by all staff, it has not been possible to process the high volumes of enquiries that were originally expected of the experiment. However, exposure to the functions provided by the experimental Intranet system has enabled the staff within the Canterbury TIC to appreciate the opportunities that are possible with this type of technology. Given an improvement in the system's performance and availability, some substantial benefits could be realized in terms of reduced time and lower telephone expenses. The only possible downside in this new IT-based environment would be some loss of human expertise and knowledge of local accommodation situations. Such knowledge can really only be learned easily by talking to an expert.

Besides accommodation services, visitors and residents need to know what national and major events are taking place around the country in destination areas. At present, the Canterbury TIC stores details of destination events on its PC word processor, which also incorporates a diary facility. Storing information in this way has a number of benefits: (a) it is relatively simple to maintain the information; (b) selected data can be printed and given to visitors; and (c) due to its simplicity, the system can be used by most TIC staff to service their inbound customers with very little training. However, ready access to local information on events in other areas around the country relies almost totally on reference documents. Other TICs and local authorities, for example, publish information booklets on events and places of interest within their local areas, but these are only distributed locally. The only exception to this is London events, which are more widely circulated. The ten regional tourist boards publish major events in their brochures, which are distributed across the country. Then there is a book called *BTA Events* that covers the whole country and is issued to all TICs. Finally, there are approximately 30 booklets that cover events and places of interest, e.g. historic houses and gardens, that are part of the TRIPS reference kit. All these publications are all stocked within a central library by the Canterbury TIC. This is the main reference source that enables TIC staff to answer customer's enquiries such as: What's on in specific geographical areas around the country?

Again, the ETB's Intranet experiment supports an events section that has the potential to replace the use of the reference library. The problem is that usage of the Intranet for events is subject to the same kinds of problems that I mentioned above, i.e. slow system response and limited access to the PC; and because it is perhaps even more essential for a system to deliver immediate responses to enquiries for information on events, the Intranet is seldom used in this area. Nevertheless, exposure to the Intranet experiment has enabled the Canterbury TIC staff to appreciate fully the opportunities offered by the Intranet. There is common agreement that provided: (a) the practical problems mentioned above can be solved, and (b) funding levels can be increased to meet the required IT expenditure, then the Intranet could be an effective answer to a TIC's event servicing problems. This could improve customer service times while simultaneously reducing a TIC's expenditure on items such as the printing and distributing of local event booklets.

Visitors

The second major information servicing function performed by the Canterbury TIC is provided to inbound visitors to the Canterbury area. Enquiries are received from all corners of the world by the Canterbury TIC. For example, in response to tourists' enquiries the TIC mailed over 10,000 Canterbury guides to people who asked about the area last year and in June received just under 1,000 letters and 4,000 phone calls. Over 200,000 Canterbury accommodation guides were distributed in 1997. At present only a handful of e-mail enquiries are received each day, although this is expected to increase with the launch of the VisitBritain site (see Chapter 5). Looking to the long-term future, the VisitBritain site could enable many of these enquiries to be satisfied by the customers themselves, always providing the site contained up-to-date and comprehensive information.

One of the most time-consuming tasks for the Canterbury TIC is booking local accommodation for walk-in customers, telephone callers and requests received from other origin TICs around the country. At present this is handled manually via the telephone. Looking to the future, a solution that has been identified by Canterbury TIC's staff is the 'touch vision' system currently being used in Ireland. This system provides an automated booking service with payment processing functions. It can be implemented either as a self-service automated teller machine (ATM) type machine or via a TIC operated PC. Public access systems (PAS) such as this have significant potential for TICs like the one in Canterbury. The initial thrusts for PAS like this are thought to be mainly: (a) to provide after-hours customer servicing functions, and (b) to support Internet bookings. However, the main obstacle is the investment required to purchase the hardware and system software.

Given a booking system such as this, TICs could use it to provide event booking and ticketing functions. The PAS would need to be capable of storing allocations for certain popular events, e.g. London shows. However, for local events and festivals, the Canterbury TIC usually holds its own ticket stock. There are substantial benefits to be gained from automating event ticketing. For example, a single book of event tickets can be worth £6,000 or more. The manual control of this kind of value poses risks and processing costs that would be largely eliminated if an automated system could be used, especially one that is accessible directly by customers using some form of self-service machine.

In summary, the Canterbury TIC, like its sister TICs around the country, is an information service that is highly labour intensive. There are significant opportunities for these TICs and the services that they offer visitors to be streamlined using new technology. The ETB's Intranet experiment has enabled some of these opportunities to be identified, but before it can be considered a practical tool for widespread use by TICs it needs to address the following basic problems:

- The response times need to be quicker by far. TIC representatives typically have just one minute to deal with a customer's enquiry. Any 'point-of-sale' system must be substantially quicker to use than a reference book.
- The PCs providing the Intranet or Internet service must be readily accessible to all TIC servicing staff on the front counter.
- The information service must be permanently on-line so that TIC representatives do not have to set up a communications link and log-on to the Intranet each time they have an enquiry. The experimental service currently times-out and this causes much annoyance to users.
- The unit costs of the service should be fixed and not time-based because this is a disincentive to more widespread and frequent use within

the TIC, which has to operate within a very tight budget. TIC representatives need to feel relaxed about browsing the information on the data base in order to answer a customer enquiry.
- The information service needs to support the building of a customer itinerary that can then be routed to other TICs for booking purposes.
- A full staff training plan, involving all TIC servicing representatives, needs to be given prior to the information service being implemented.

The BTA and ETB organizations are well aware of these and other issues that have arisen from the Intranet experiment. They have already addressed many of them during the development of the VisitBritain site. The remainder will only be addressed by greater investment in IT among TICs. However, given the stringent expenditure levels of many of these organizations, such expenditure will have to be offset by savings in other areas, increases in revenues and/or a demonstrably higher level of customer service. In my view the approach being taken by the UK tourism organizations in tackling these issues will deliver sound results in the future. The necessary expenditure offsets to the investment required are readily identifiable in terms of: (i) staff time savings; (ii) reduced brochure printing and distribution costs; (iii) increases in the volume of inbound tourism expenditure; (iv) reduced losses incurred as part of event ticketing; and (v) improved customer servicing, e.g. high quality printed maps and personalized itineraries.

Ireland

Tourism is vital to the Irish economy. In 1995, for example, 4.2 million people visited Ireland and as a direct result of this, their visits generated around I£1.5 billion in foreign earnings. Tourism in Ireland represents over 6.4 per cent of gross domestic product (GDP) and supports employment for one in every 13 workers. It is not therefore surprising that the Bord Failte Eireann, i.e. the Irish tourist board, and the Northern Ireland Tourist Board (NITB) have focused their attentions on increasing the numbers of inbound tourists and the resulting yields in the field of tourism. One prime example of a strategic initiative that makes excellent use of IT is the Gulliver project.

Gulliver is the name given to an initiative aimed at promoting tourism to the island of Ireland. Its goal is to establish an infrastructure that can provide the main channel of distribution for information and reservations on all major aspects of tourism in Ireland. More specifically, Gulliver's objectives were:

- To make it easier for tourists to choose Ireland as a destination.
- To improve visitor servicing while in Ireland.

Gulliver was developed as a joint venture between Bord Failte Eireann and the NITB. It was funded from a variety of sources including Bord Failte Eireann, the NITB, the International Fund for Ireland and an EU development grant. Following a technical feasibility study, Gulliver was developed over the period 1990 to 1991, piloted in early 1992 and went live later that same year. It has grown to become a leading example of how IT can be used to support the development of tourism for a country. However, this has not been achieved without some considerable growing pains. We can learn a lot from how Gulliver was developed, so let's start with an overview of how the original base system was constructed.

THE ORIGINAL GULLIVER SYSTEM

Gulliver is Ireland's national tourism information data base, which is used by the travel industry and by visitors to learn all there is to know about Ireland and its services. Virtually all sectors of the tourism industry play important roles in Gulliver and enjoy the benefits of this state-of-the-art system. Participants are drawn from each major sector, i.e. accommodation suppliers, tour operators, tourism offices and tourists themselves.

The structure of Gulliver mirrors that of virtually any major travel-related distribution system. It has a large data base at its core that is fed with information supplied by service providers and their systems. The products thus focused at the centre are then distributed to end users via various technologies and channels. Let's take each element of the Gulliver system in turn, as it was originally developed in 1991 (Fig. 2.2):

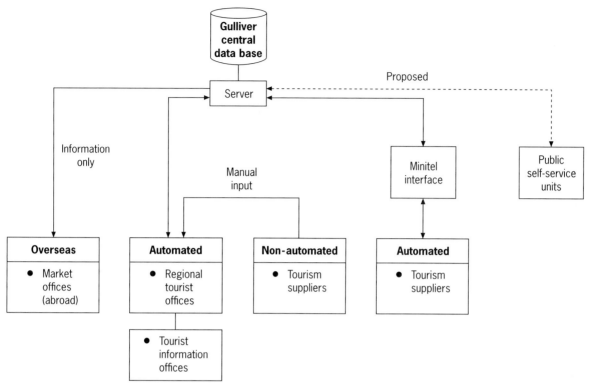

Figure 2.2 Gulliver's original design

- **Data base engine** At the core of the original Gulliver system was a powerful data base engine, driven by a Digital Equipment Corporation (DEC) VAX 6320 computer. This server platform was used to store all tourism related information about Northern Ireland and Eire. This information covered all 32 counties of Ireland and included: all properties approved by the tourist boards; calendars of upcoming events; general destination information, e.g. attractions, places of interest, restaurants, pubs, night-life and cultural events; specialist activities and services; useful contact names and addresses; sports and leisure facilities; a route planning facility, including maps and route details; and finally, tourist attractions. Detailed information on all of these subjects was stored on the Gulliver data base. The data base was distributed on-line to Gulliver's population of users in Ireland and around the world.

  Ireland's accommodation suppliers are small and widely dispersed with few multiple chains. Over 85 per cent of all accommodation units are SMEs, each with fewer than 30 rooms on average. The range of styles and facilities is consequently enormous, which means that the data base will always be large and highly dependent upon an efficient search engine.

- **Suppliers** Gulliver stores details on a wide range of suppliers, which fall into the following major sectors: (a) accommodation – all approved hospitality providers, including hotels, guesthouses, bed and breakfasts and self-catering; (b) premises – supplier information on prices, descriptions, directions, facilities and nearby attractions; and (c) transport services – including bus and rail schedules. To participate in Gulliver, an approved accommodation supplier must pay an annual fee that is inclusive of membership of the respective tourism organization as well as Gulliver itself. SMEs pay only an annual Gulliver membership fee that is related to the type and size of their premises. However, these fees are relatively low and amount to less than I£300 per year, even for the largest of properties.

Accommodation suppliers that are registered by the Bord Failte Eireann will automatically be included on the Gulliver data base. The information shown about each will include facilities, location and contact details. However, rooms are not available for on-line booking purposes unless the supplier is a Gulliver member. Each Gulliver member may register their property on either a request or an allocation basis:

- *Request* A supplier may simply provide a narrative description of the rooms on offer. However, these may not be booked automatically by Gulliver. To make a booking, tourist offices and other outlets must somehow contact the supplier directly to check availability and request a reservation. This is usually done over the telephone and can be long-winded, laborious and costly. It also makes the premises less attractive to intermediaries, for booking purposes.
- *Allocation* A supplier which is a Gulliver member may allocate a specific number and type of rooms from its inventory for Gulliver to onward sell directly to prospective purchasers. The supplier decides the number of rooms, the room type and the dates during which they are available. Gulliver also provides suppliers with a great deal of flexibility in terms of the date ranges and associated room rates for which allocations are displayed. For example, a higher number of rooms can be shown as available for a reduced price during periods when bookings are known to be low. The allocation method is preferred because it allows a purchaser to check availability on-line, make a reservation and obtain an immediate electronic confirmation, direct from the supplier (NB: advance bookings are confirmed by post).

All bookings originating from tourist offices are subject to a 10 per cent commission fee. When a booking is made in a tourist office, the visitor pays a minimum of 10 per cent of the total price, as a deposit. The tourist office retains its commission from this deposit and pays the balance where appropriate to the product supplier. Bookings made by tour operators are voucher based. The transaction value, which is not shown on Gulliver, is based on pre-contracted rates agreed between the two parties, i.e. on the one hand to the accommodation supplier, tourism product provider or in either case, their marketing representative and, on the other hand, to the individual tour operator.

No matter what type of participation is chosen by a supplier, the information is provided to Gulliver either electronically or manually. Electronic supplier information can be keyed into videotex terminals that feed Gulliver, or updates may be transmitted direct from certain property management systems (PMSs) owned by suppliers, i.e. the leading two PMSs in Ireland. Non-automated suppliers provide information to Gulliver offices over the telephone, fax or by letter, where they are keyed into the system.

- **Distribution channels** The tourism information recorded on Gulliver's central data base by Irish suppliers is distributed to end users by a variety of means using several different technologies. These end users are located both domestically and in international markets:
  - *Domestic* Gulliver's end users in Ireland include regional tourism organizations, TIC and, in the future, tourist activated kiosks located in airports, shopping centres, ferries and other transport terminals. The focus here is on providing inbound tourists with the information they need to travel around Ireland, assimilate the culture and use the services of domestic suppliers.
  - *International* Overseas, Gulliver's end users in the international area include: (a) the market offices of Irish tourist boards located in key foreign countries; (b) GDSs, such as Sabre, Galileo, Worldspan and Amadeus, that distribute Gulliver to travel agents selling bespoke holidays to independent travellers to Ireland; and (c) videotex systems, such as Minitel in France and Prestel in the UK. Overseas tour operators also use the information stored in Gulliver's central data base to help them build packaged holiday products to Ireland.

Originally, direct access to Gulliver was provided by dumb terminals connected to local controllers located in the end users' offices. These terminal controllers were themselves connected by leased data lines operating at 9600 b/s, to regional DEC Microvax minicomputers. Each regional mini was then connected via a high speed data line to the Gulliver central host system.

The original Gulliver system was introduced in 1992. It was rolled-out to suppliers and users in both the domestic and overseas markets. Gulliver's store of local, regional and national tourist information was available 24 hours each day, 365 days of the year. For the first time it provided prospective tourists to Ireland with advance planning and itinerary assembly facilities as well as on-line booking functions. Then, in 1994, following a period of two years' operational use, a review of the Gulliver system was undertaken. This enabled several technological problems to be identified:

- **Poor end-user performance during peak times** During the peak summer season the volume of transactions processed on the central Gulliver computer was extremely high. This placed a heavy burden on the host DEC VAX computer and provided end users with slow response times to their enquiries.
- **High telecommunications costs** The dedicated high speed data lines used by Gulliver were very expensive to install and run. These lines had a high fixed monthly rental that did not vary with regard to the volume of traffic carried. A related problem was the dependency upon these lines. If one failed, it affected a great number of users and prevented access to many parts of the Gulliver system.
- **Obstacles to international roll-out** A combination of poor performance times and high telecommunication costs meant that it was not attractive to extend Gulliver to overseas tourist offices, and this was precisely the key area where Irish tourism needed to concentrate.
- **Minitel problems** Suppliers did not take to the French Minitel system. One of the main reasons for this was that suppliers in Ireland did not embrace Minitel nor, for that matter, any other kind of technology. The result was that Minitel was only used for accessing Gulliver occasionally, when a supplier's data needed updating. There were two main consequences: (a) users quickly forgot how to use the system and a high ongoing training cost was incurred by Gulliver, and (b) the usage pattern also gave the impression that Gulliver was more costly than it really was. If, for example, suppliers had been using Minitel for other purposes, then the data line costs would have been spread over several applications and would not have been laid solely at the door of Gulliver.

  The end result was that Gulliver became unpopular with smaller Irish suppliers who were the very people the system was aimed at. After a short while SMEs reverted to making telephone calls and sending facsimile messages in order to communicate updates to the regional tourism centres. The consequences of this were that manual workloads increased and information could only be updated during normal office hours.
- **Out-dated technology** The original Gulliver system used a simple text-based approach to recording, storing and displaying information. However, rapid developments in new technologies, such as multi-media, CD-ROMs and the Internet, made Gulliver look increasingly out-of-date. What was ideally needed was a GUI that provided high quality images, graphics, sound and even video. But given Gulliver's legacy systems architecture, such a change in approach would put a further burden on the already over-worked central computer.

In summary, the review concluded that from a technological viewpoint, a centralized on-line system was not appropriate to achieve the objectives of the project. Subsequent investigations into Gulliver's data storage patterns and access profiles also revealed an interesting picture. The data stored about suppliers could be segregated clearly into static and dynamic elements. Static data represented around 90 per cent of the Gulliver data base and described a supplier's products and services. This type of data hardly ever changed but was accessed a great deal. Dynamic data represented far less in terms of volume because it described supplier's availability and rates. However, dynamic data was accessed only infrequently, i.e. when an actual booking needed to be made. This was one of the most significant findings of the review and its resolution became a cornerstone of the new Gulliver system.

THE NEW GULLIVER

The re-engineered Gulliver system was re-launched in 1996 and is still being enhanced and rolled-out around the world, i.e. as at mid-1997. Gulliver uses IT to link the supply and demand sides of Ireland's tourism industry. The supply side comprises service providers and the demand side is the marketplace. It is, however, the supply side that is particularly important to a tourism organization. Gulliver's supply side has the linkages to update allocations received from suppliers, keep prices up-to-date and receive both booking notifications and cancellations from users on the demand side. It is sufficiently flexible to support the wide range of automation options adopted by service providers in Ireland. For example, Gulliver supports service providers that use: (a) their own individual systems, e.g. Gullink, see below; (b) PC-based systems, e.g. Lilliput, again, see below; and (c) no automated system at all, e.g. dumb Minitel terminals and FaxLink. A diagram of the new Gulliver system is shown in Fig. 2.3.

The common factor that both the supply and demand sides share, is a data base of tourism information. This data base, which is the cornerstone of Gulliver, is controlled by a powerful server. One of the most important components of this server is its message router (sometimes also known as a message processing engine), which is the name given to the software used to control the server's dialogue with the outside world. It does this via several communications technologies that are supported by Gulliver, e.g. X25, ISDN, leased lines and PSTN or dial-up. The message router incorporates mailbox features and acts as a single source of information for the central system and also serves as a destination for all incoming messages. This allows external users with a variety of systems, e.g. Minitel, FaxLink, Lilliput and Gulliver, to communicate via message pairs with the Gulliver central system. The message router also provides a firewall that insulates the central system from on-line users thus protecting the integrity of the data base and providing the required degree of security needed by Gulliver. The technical architecture around which Gulliver is built may be summarized as follows:

- A single master data base that is maintained centrally in the Gulliver headquarters. This is updated via information received from a variety of sources by forms or e-mail. Quality control procedures safeguard the information that is applied to the master data base.
- Copies or snapshots of the data base, i.e. static copies, that are distributed to users periodically (usually daily). This allows users to store their own local version of the Gulliver data

Figure 2.3 The re-engineered Gulliver system

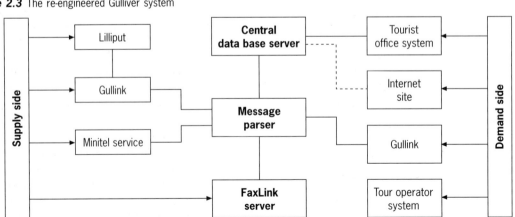

base and thereby enjoy fast response times to enquiries.
- Direct access to the data base is limited to specific 'as needed' cases. In other words, users only connect to the central system to update static data; submit queries to the dynamic data base, e.g. availability requests; and process certain transactions. They do not access the central data base for on-line information queries.
- Transactions are processed as electronic messages that fall under certain classifications: (a) price and allocation changes that originate in PMSs and from Minitel terminals, used by accommodation suppliers; (b) booking and cancellation notifications; and (c) availability and booking requests. A front-end mailbox is used to control the processing of these messages.
- Links to the data base server are effected using a variety of communications methods including leased lines, ISDN, X25 and PSTN, i.e. dial-up.

The original centralized approach was therefore replaced by a distributed client/server architecture. This uses Microsoft NT 4.0 for the networking functions and SQL Server data base software to provide a new interface to Gulliver's central information repository. This new approach minimizes telecommunications costs and reduces the load on the central server. The development tools used during the construction of the Gulliver system are shown in Table 2.1. The application functions are generally partitioned as follows:

Table 2.1 The tools used to develop Gulliver

Development area	Tools used
Operating system	Microsoft Windows NT
Data base	Microsoft SQL Server
	Microsoft Access
	ODBC Compliant
	Stored procedures
User interface	Microsoft Visual Basic
	Microsoft standards
Programming	Microsoft C++
Local data base update	SQL Server Replication
Remote management	Microsoft SMS

- **Client functions** In this context, 'clients' are PCs located in demand centres, e.g. tourist information offices, tour operators and suppliers. These demand centres use PCs to access locally stored static data. A tourism information office will, for example, browse the locally stored static data base of products that are of interest to a customer (or tourist). This data base is stored either on the user's PC itself or on the office's server, i.e. if there are several PCs in a single office. Several properties and services may be viewed and discussed with the tourist in this way before availability becomes an issue. When an availability enquiry needs to be made, the central dynamic data base held by the Gulliver central system is accessed.

Client PCs interact with the host Gulliver data base server using standardized message pairs. Messages may be of four main types: (i) enquiry messages that request the server to look up an availability of a particular property; (ii) response messages that are in effect answers to previously sent enquiries; (iii) changes to centrally held supplier data that is assembled on client PCs and sent to the server by service providers, and (iv) messages sent from the central system, which update the locally held data base of static supplier information. This makes the dialogue between the client and the server particularly efficient; and because the client site pays only a local call to access the server, costs are minimized. A family of special software products has been developed to support this messaging protocol. Two of these are branded Lilliput and Gullink, and their functions are summarized as follows:

- *Lilliput* The Lilliput software has been designed for installation in supplier's client sites where PMSs are used. It enables supplier systems in remote locations to dial into the central site, gain access to the message router and exchange information with the Gulliver data base server.
- *Gullink* This is a communications library featuring an application program interface (API), which is used to connect remote host computer systems to Gulliver's message router. It is available in both DOS and Windows

Table 2.2 Gulliver's main functions

Core functions	Supply side linkages	Demand side linkages
Maintenance of the Gulliver data base	Maintenance of prices	Query processing of the accommodation data base
Replication of data base changes to outlying user sites	Maintenance of allocations	Query availability
Maintenance of availability information for supplier products and services	Receipt of bookings	Processing of bookings
	Receipt of cancellations	Query tourist information
Processing of transactions, i.e. message pairs		Route planning

versions and runs on any PC that uses a 386 or higher specification processor.

When a booking is made, Gulliver sends a notification message to the accommodation supplier. This is done via the central system mailbox, which formulates and transmits an electronic message containing the booking details to the hotel PMS. This is effected in most cases by means of dial-up communications from the Gulliver centre to the hotel location.

- **Server functions** Only when a user requires availability information or a when a booking needs to be made, is the central system accessed. When this occurs, the local PC uses ISDN or several other telecommunication options to contact the central server. Once connected, the client PC exchanges enquiry messages with the server. This usually results in a response being received by the client PC within eight seconds – far better than with the 'old' system. If the desired supplier service is available and needs to be booked, Gulliver responds with a confirmation number and this can be quoted to the customer: in the future, Gulliver will be accessible from overseas locations via the Sprint telecommunications network. The major functions of the Gulliver central system are as shown in Table 2.2.

As described above, the server does not actually process clients' enquiries in the classical real-time on-line sense. Instead, it uses some special purpose software to control server operations, similar to a kind of EDI protocol that supports ISDN, X25, PSTN or leased line access. This software controls the set of standard electronic message pairs, each with a specific purpose, e.g. bookings, cancellations, changes, as described above. Messages are processed on the server and the corresponding response is returned to the originating client PC in a standard format. In time, it is planned that Gulliver's messaging system will comply with the UNICORN standards (see Chapter 1 – The TTI). When this happens, it will be far easier for overseas tour operators to exchange bookings and cancellation request messages automatically with Gulliver. An inherent part of this software is a high level of security that prevents unauthorized access to the central Gulliver system. Gulliver's availability checking routines work like this:

1. The booking criteria are received as an incoming message from a user's workstation. This booking message comprises a set of general search criteria specified by the user. The search criteria may include, for example, date that the service is required (or date range), the number of people in the party, the sharing preferences, the *en suite* preferences and whether or not an accommodation plan that includes dinner is required.
2. Gulliver selects a sub-set of products from its data base that meets the stated requirement criteria.
3. If the accommodation units are not specified then Gulliver matches the customer's preferences on the number of people in the party, the sharing option and the *en suite* option.

4. Gulliver then searches the selected subset for the availability of accommodation within the highest preference combinations.
5. If a date range has been specified then the search is based on a combination of the number of days required and the start day.
6. The search continues until at least 20 alternatives are available. Once this has occurred, the availability options are examined for features that match the price and 'dinner inclusive' flags as set by the original enquiry.
7. The search results are formatted into a response message that is transmitted to the end user, e.g. the tourist office system, and displayed on the end-user's workstation.

The server also replicates the central data base so that multiple copies may be distributed and stored on regional servers. All tourism data is first received and stored on the central Gulliver data base. Once validated, the static element of this data is then available for copying and distribution to remote users. The central data base comprises a multi-processor Pentium PC with random array of inexpensive disk (RAID) storage – a technology that provides mass storage with a high level of data security and integrity.

In terms of supplier access, although Gulliver still supports Minitel, it has migrated many domestic suppliers to a PC-based solution. This works in the same way as described for the client environment (see above). Suppliers can use their PCs to dial into Gulliver's central computer and exchange booking messages. The software used by suppliers may be run as a stand-alone communications application or it may be embedded within third party software packages. This facility allows the Gulliver messaging protocol to be linked directly into the major PMSs used by accommodation suppliers throughout Ireland.

For the lower end of the SME supplier population, Gulliver also supports a fax-back service branded 'FaxLink'. For low volume users such as this, standard pre-printed fax forms are used by suppliers to record changes in allocations and rates. These forms are similar to national lottery 'playslips' and simply require pre-printed boxes to be hand-marked with a black pen. Once completed by hand, the forms are faxed to a central Gulliver collection point. Here, the fax forms are electronically scanned and the images thus recorded are transmitted into Gulliver where they are automatically processed. This involves Gulliver forming a response to suppliers on subjects such as reservations needed, cancellations or amendments.

Gulliver provides a comprehensive customer support service to all its customers whether they be large hotels, small bed and breakfasts, tourist offices, self-catering providers, hostels or cruising companies. The Gulliver Helpline is available on a local call rate and is manned by trained staff from 9 a.m. to 6 p.m., Mondays to Fridays in the Republic of Ireland and 9 a.m. to 5 p.m. in Northern Ireland. However, during the summer peak season, the Helpline is manned from 9 a.m. to 9 p.m., seven days a week.

The new Gulliver system was successfully piloted by a group of seven tour operators (two in Ireland, four in the UK and one in France). Once the concept had been proven, the entire system was converted to the new architecture and rolled out first to Northern Ireland's tourist offices and then to Eire's in time for the 1997 holiday season. The Gulliver system now forms a key component of the current re-branding and promotional campaign for Ireland as a tourist destination. This itself has fostered a new set of tourism initiatives, many of which are based on the Gulliver infrastructure, such as:

- **Multi-lingual call centres** All promotional material on Ireland will feature a localized freephone number. Callers to this number will be automatically routed to an Irish based multi-lingual call handling centre that will operate seven days each week. Operators in this centre will use the Gulliver system as its core technology to handle customers' enquiries, check availability and make reservations. The centre will also back-up major advertising campaigns to establish brand images and promote Ireland in the years ahead.
- **Public self-service kiosks** Kiosks, similar to ATMs, will be able to use Gulliver's new technology. These machines will operate on the same distributed client/server architecture as the Gulliver core system. Each kiosk will store a

copy of the static data base in order to support local enquiries. It will also provide other functions, including a credit card reservations facility. It is, however, planned that these kiosks will be operated by independent commercial companies and that added value will be generated via advertising, multi-media and virtual reality facilities. Locations under consideration include airports, ferry ports, railway stations and even the inbound ferries themselves.

- **Expanded information** The information stored on the Gulliver data base will be broadened to include a comprehensive calendar of events; listings of restaurants; well-known pubs; nightlife; sports and leisure activities; suggested sightseeing itineraries; information on transportation schedules, including bus and rail timetables; a five day weather forecast service; and a route planning service with geo-referenced data to display exact locations. All such information will be checked for completeness and accuracy by a new quality control system linked to the regulatory functions of the tourist board.
- **Minitel in France** Although the use of Minitel by Ireland's suppliers was disappointing, the system is nevertheless widely used throughout one of Ireland's leading inbound markets, i.e. France. In fact Minitel is the most widely used electronic distribution channel in France. Bord Failte Eireann have therefore supported the plans put forward by an international reservations bureau, to make Gulliver available via France's Minitel population.

Finally, the most important issue facing the future of Gulliver is that of ownership. This issue was first identified by Bord Failte Eireann following a strategic review that concluded that Gulliver should focus its efforts more acutely on the promotion of Ireland as a destination, to other countries. This has resulted in a downsizing exercise and the outsourcing of many non-core activities. However, it has nevertheless been recognized that the Gulliver project urgently needs additional resources to continue its development.

In an attempt to meet both of these strategic imperatives, Gulliver is in the final stages of seeking a business partner that will provide both financial and technical resources to the project. The intention is that this partner will purchase a majority share-holding in Gulliver thus effectively privatizing the project. This will allow the Gulliver team to: (a) continue developing the system, and (b) establish its place as one of the leaders in the field of electronic tourist information distribution systems. The commercial focus of the re-formed group will also protect Gulliver's investment in IT and thus help drive the Gulliver project forward into the future.

Other tourism support systems

Technology is being used in some very innovative ways to support and encourage tourism. Besides the information-based systems described above, countries are beginning to use IT to increase the productivity of their inbound tourism endeavours. A prime example of this is Australia's recently introduced automated visa application system, i.e. its electronic travel authority system (ETAS).

AUSTRALIA'S ETAS VISA SYSTEM

An often overlooked aspect of IT in tourism is border management technology. This can be an important factor that supports the costs effective growth and development of tourism to any country. After all, tourism can be encouraged by many factors and being able to obtain a visa quickly, easily and efficiently in the origin or outbound country is an integral part of this. Likewise, the processing of travellers through customs and immigration at inbound airports is both an important tourist servicing factor and an opportunity to reduce operating costs for the authorities involved. A particularly good example of a country that has tackled this area very successfully, is Australia.

All travellers to Australia, other than Australian and New Zealand citizens, require a visa or approval to travel and enter Australia. So, if tourism is to be encouraged it is essential that visitors and potential visitors can easily obtain their visas with the minimum of fuss. Because most visitors to Australia arrange their trips via travel agents, it is these intermediaries that bear the brunt of any red tape encountered by their customers. The old paper-based standard approach, which has been used for many years now, entails the use of a paper form that must be completed by the potential visitor and mailed to Australia House. A central clerical operation must then open envelopes, check application forms, use on-line government systems to verify immigration records, issue a visa slip (for inclusion in the visitor's passport) and finally, mail the completed documentation back to the applicant. In many cases the visa application is a step that many people either forget or leave to the very last minute. Consequently many potential visitors to Australia often begin their journey with a trek to London for a spot of queuing at Australia House to apply in person. All in all, quite a lengthy and laborious procedure.

The ETAS system was initiated by Australia's Department of Immigration and Multicultural Affairs (DIMA). The DIMA's objective was to issue visas to all potential visitors to Australia as quickly and efficiently as possible with the goals of modernizing visitor flows at inbound gateways, supporting tourism and maintaining the integrity of Australia's borders. This was becoming especially challenging with the growth in visitors to Australia. For example, over the period between 1980 and 1990, Australia's short-stay visitors grew from 905,000 to more than 2.2 million per year. In 1996 over 4.1 million tourists arrived in Australia; a further 10 per cent more in 1997. More significantly, the Australian Tourist Commission and the Tourism Forecasting Council predicted a growth to 6.9 million visitors each year by the year 2001. This translates to 4.7 million visas, which is almost double the 2.6 million visas issued over the period 1994 to 1995. Clearly a new approach was necessary if the high number of visitors was to be supported without jeopardizing the integrity of the universal visa system and avoiding increases in cost.

The new approach embodied several principals. First, the visa issuance process needed to be seamless and virtually invisible to the prospective visitor. Second, it needed to maximize the number of visa issuing points around the world. Third, it needed to streamline the outbound airline check-in procedures and the inbound customs and immigration processing tasks. Finally, the new system needed to be efficient and reduce the overall cost of maintaining the visa system for Australia. To do this, DIMA worked closely with travel agencies and airlines to determine their processing requirements and especially with about 120 agents that have been issuing visas under special agreement with the Australian Government. DIMA also wanted to use the existing technological infrastructure for the design of the new system.

The Australian Government has simplified all this and has introduced an automated visa application system called the electronic travel authority system (ETAS). It is a revolutionary new way of obtaining the necessary approval to visit Australia using a combination of the GDSs, a Societe Internationale de Telecommunication Aeronautiques (SITA) computer system located in Atlanta and an Australian Government computer system in Sydney. Application forms are no longer required for an electronic travel authority (ETA) and, being electronic, there is no physical representation of an ETA. In other words, the visa is purely electronic and does not appear as a visa label or a stamp in a visitor's passport. This solution enables ETAS to be used by many of the 80,000 or so IATA registered travel agents around the world, and the system is quick; the actual ETA issuance process takes just ten seconds.

The ETA streamlines both the outbound and inbound parts of a traveller's journey. When a traveller has been issued with an ETA by his/her local travel agent, the first time it is used is when the traveller checks in at the departure airport. The airline carrying the passenger on the outward leg of a journey enters the ETA reference details into the departure control system. This link enables check-in staff to verify that the passenger has a valid visa; and because airlines can be fined for carrying a passenger who does not have a valid visa for entry to Australia, this is a significant benefit to the airline.

The ETA also streamlines the processing of inbound passengers when the flight lands in Australia. In fact, ETAS more than halves the processing time for visitors arriving in Australia. Immediately after their flights land at an Australian international airport, the visitors will present themselves at an Australian immigration counter. They offer their passports to the immigration clerk. The visitors' passport numbers are entered into a Government computer and the computer checks to ensure that an ETA exists on the data base. If it does and the details check with the passports submitted, then the visitors are waived through. If any discrepancies are detected, a somewhat more lengthy process then commences (which I will not go into here). So, the ETA and its associated visitor's passport number, is a critical item of information upon which the success of the whole ETAS system is based. ETA data are logged by the Australian immigration computer via an ingenious three-way link-up using global communications networks.

The ETAS system comprises three inter-linked functions, each of which is supported by computers that are located in different parts of the world: (i) a front-end ETA application processing system supported by a travel agent's chosen GDS or CRS, (ii) an international data collection and routing system based on a SITA computer located in Atlanta, and (iii) a data base look-up and electronic funds transfer system supported by the Australian Immigration Department's computer in Sydney. No new hardware is needed by either the travel agents or the airlines involved. Let's take each one in turn:

- **The GDS PC front-end** ETAS is accessed via any one of the major GDSs and several CRSs. The travel agents use their terminals to access the TIMATIC data base (see Chapter 4 for a description of how GDSs use TIMATIC and what this information source is). Once into TIMATIC, the travel agents enter a special code, depending upon the type of visa required by the traveller. This could, for example, be either a visitor's or tourist's visa, a long-stay business visa or a short-stay business visa. Then, depending upon the type of visa, a special electronic form is displayed on the GDS PC screen. This has several fields that must be completed by the travel agents. There are two important facts to bear in mind here:
  1. ETA electronic visas are only available for certain nationalities. However, the list of countries supported by ETAS is very comprehensive and caters for the vast majority of Australia's inbound visitors.
  2. The ETAS system must be supported by the airline in which the visitor's passenger name record (PNR) is held. This is because the airline will need to have the functionality to communicate with the external computer systems that deliver ETAS via the travel agents' GDSs.

Once all fields on the GDS electronic form have been completed, the travel agent ends the transaction. This causes the electronic form to be sent via the GDS network to the SITA host system in Atlanta, USA. There are two items of information that are critical to ETA applications: (a) the visitor's passport number and (b) the visitor's nationality. It is these two items of data that determine a passenger's authority to travel at the arrival and departure airports in Australia, as described above. These fields must therefore be entered again by the travel agent in order to verify their accuracy. So, the travel agent's customer does not even have to visit the agent's office personally to obtain their visa. The whole process can be undertaken over the telephone, with the travel agent using a GDS terminal to enter the information and obtain the response from ETAS.

- **The request capture system (RCS)** The principal role of the SITA host computer in Atlanta, USA, is to support the message handling aspects of ETAS. It receives all ETA transactions from originating point-of-sale GDS PC terminals used in travel agencies around the world. The SITA host performs some initial checking of the transaction format and then routes the message to the Australian DIMA computer in Sydney, Australia.
- **The request processing system (RPS)** Computers in Sydney and Canberra process and check all ETA applications from travel agencies and other visa application points around the world. The RPS receives each transaction and firstly checks the data fields using a variety of verification routines. When the transaction has been verified as OK, the computer then performs: (i) an on-line DIMA data base look-up to determine the status of the visitor for immigration purposes; and (ii) for certain types of visa application, a funds transfer that is effected via links to remote credit card computers for the payment of visa fees. A suitable response is formulated and transmitted via

SITA and the GDS where it appears on the travel agent's GDS PC.

Approved ETA requests are shown on a screen that includes the following fields: today's date and time in Australia, the expiry date of the ETA and a message denoting 'ETA approved'. When a visitor already holds a current visa or ETA, the system will respond by giving full details of what is held on the immigration data base. The travel agent has the option to use this as the current authority to travel or replace it with a new one. Besides issuing ETAs, other ETAS functions supported are: (a) help – the entry of a question mark in any field will result in a response that explains how that field is completed; (b) enquiry – if a visitor's passport details are entered, ETAS will respond with a full status report showing whether or not an authority to travel exist; (c) history – the principal use of this function is to reverse a credit card payment for a long-stay business ETA within 36 hours of it being initiated; and (d) check-in – this can be used to see if any specific passenger of any nationality has a valid and current authority to travel in Australia.

This new tourism supporting initiative has been a great success. ETAS was introduced in Singapore, North America and Japan where it proved popular with visitors and travel agents. Approximately 60 per cent of all travellers out of Japan are now issued with ETAs. The UK, Ireland and other Western European nations started using the system in late 1996 and early 1997. This area represents about 20 per cent of inbound tourists to Australia and the number is growing by 10 per cent each year. By the end of April 1997 approximately 63 per cent of all visitors to Australia had access to an ETA prior to taking their trip. This was as a result of ETAS covering some 20 nationalities and about a dozen airlines.

The electronic processing of visas is highly efficient for the traveller, the travel agent, the airline, airports and also for the overseas offices of Australia's DIMA. The traveller may obtain an ETA from wherever the passenger purchases an airline ticket without having to complete an application form. The travel agents becomes a one-stop shop providing their customers with an instant service involving virtually no bureaucracy and administration. Airlines are able to verify passenger visa details on check-in thus avoiding fines and can issue emergency ETAs themselves to last minute travellers. Airports benefit from faster clearance of passengers through customs and immigration with a consequent reduction in the demands placed on airport resources. Finally, Australia's DIMA offices are able to process visas using less staff, which reduces operating costs and ultimately helps lighten the financial burden on Australia's taxpayers.

Australia plans to extend and enhance its border management technology even further into the future. Australia is planning to make ETAS available via a dedicated PC linked directly to the SITA computer, thus by-passing the GDS and CRS reservation systems. This is aimed principally at large organizations that do not necessarily have access to a reservation system. Also, plastic card technology is being considered as a solution to streamlining processes at the border. The Australian business Access (ABA) card is designed to be an 'added value' product for visiting business people. The card embodies a machine readable strip that assists holders in being streamed into priority zones at arrival airports in Australia. This card may eventually be used to support additional automated customs and immigration functions. Likewise, the APEC business card is also being evaluated as a means of promoting business mobility within the Asia Pacific area. It is intended to be used as an entry authority by accredited business people from participating APEC countries. In 1997 the APEC card was introduced on a trial basis for Australian, South Korean and Philippine nationals. The APEC card will also simplify entry procedures by streaming holders into lanes at airports.

3
Suppliers

Introduction

This is the chapter that describes the technologies that travel industry suppliers use for their own in-house automation and, increasingly, as their platform for product distribution to the travel and tourism markets. Many of these systems have evolved over time from their origins in peripheral back-office areas like administrative support, accounting and inventory control. They have been continually refined and developed until they now represent what is arguably a supplier's primary marketing capability. Suppliers are increasingly using these systems as a platform to distribute front-office customer servicing, sales support and data base marketing activities. Supplier systems have therefore become a critical resource upon which many companies now depend. It is important that you gain a good understanding of the different kinds of systems and how they are used by suppliers, before we start exploring how they are distributed to intermediaries and consumers in the following chapters.

Airlines

This section describes how airlines use IT to support flight operations and control seat inventories. Inventories are important because they represent the core product of an airline that is sold by their sales offices and by intermediaries, such as travel agents, to consumers and businesses. Understanding how they do this is a fundamental step towards gaining a full knowledge of GDSs and other distribution technologies that are covered in a later chapter of this book. The core technology used by airlines is known generically as the CRS. Most airlines now use IT to control all aspects of their operations including flight schedules, load and balance calculations, revenue analysis and crew scheduling. These computer systems also supports seat reservations and ticket printing functions, which are the prime focus of this chapter.

THE AIRLINE BUSINESS

It has been said that the airlines bring the world closer together, unite people, broaden their experiences and expand trade; and so, the potential market for air travel is huge. Having said this, the airlines have only recently emerged from the industry's worst ever decline in volume and profitability, which resulted from: (a) the economic recession experienced by most of the major industrialized nations of the world, and (b) an intensification in competition between airlines. However, this is now well behind us and many airlines are now becoming more profitable.

It is the increased level of competition that has made life really difficult for the airlines. This competition has been brought about by the progressive deregulation of air travel in the USA, Europe and other parts of the world. Incidentally, the policy of deregulation is also sometimes referred to as 'open skies'. Deregulation removes many of the bureaucratic restrictions on fare setting and makes it easier for airlines to fly new routes. This has the dual effect of increasing competition and reducing fares: it is the downward pressure on fares that has been at the root cause of many airline failures. As a result of this, there has been

Table 3.1 The world's top airlines (1996)

Rank	International		Domestic		Total global	
	Airline	Pax*	Airline	Pax	Airline	Pax
1	British Airways	26.6	Delta Airlines	89.6	Delta Airlines	97.3
2	Lufthansa	20.1	United Airlines	70.1	United Airlines	81.9
3	American Airlines	16.7	American Airlines	62.6	American Airlines	79.3
4	Air France	15.6	US Airways	55.4	US Airways	56.6
5	KLM	12.8	Northwest Airlines	43.5	Northwest Airlines	52.7
6	Singapore Airlines	11.8	All Nippon Airways	36.8	All Nippon Airways	39.4
7	United Airlines	11.7	Continental	31.9	Continental	35.7
8	Japan Airlines	11.2	TWA	21.3	British Airways	33.1
9	SAS	11.0	Japan Airlines	18.3	Lufthansa	33.1
10	Cathay Pacific	10.9	Japan Air System	17.8	Japan Airlines	20.9
11	Alitalia	10.7	America West Airlines	17.6	Korean Airlines	23.6
12	Northwest Airlines	9.1	Korean Airlines	16.1	TWA	23.3
13	Thai Airways	8.1	Lufthansa	13.0	Alitalia	23.1
14	Swissair	7.9	China Southern Airlines	12.9	SAS	19.7
15	Delta Airlines	7.6	Alitalia	12.5	America West Airlines	18.1
16	Korean Airlines	7.5	Ansett Australia	11.8	Japan Air System	17.9
17	Iberia	7.1	Alaska Airlines	11.1	Air France	16.4
18	Air Canada	6.9	Qantas	9.8	Qantas	16.0
19	Qantas	6.1	Malaysian Airline System	9.0	Iberia	15.3
20	Malaysian Airline System	6.1	SAS	8.7	Malaysian Airline System	15.1

* Ranked in millions of passengers per year
(*Source*: IATA members' rankings 1996: Top 20)

a fair amount of consolidation within the airline business as carriers seek to exploit economies of scale and acquire new routes. It has also seen the emergence of global mega-carriers with the size necessary to compete for business around the world. Table 3.1 shows the relative sizes of the top airlines as at 1996.

All of these factors have emphasized the need for airlines to respond ever faster to market dynamics: one of the key ways in which airlines are able to do this is by the effective use of IT from which has sprung the CRS and the GDS.

THE AIRLINE CRS

The term CRS seems to have crept into use like so many Americanisms (perhaps that is one itself!).

It simply stands for computer reservation system and is the term used to describe the technology that controls an airline's seat inventory for sales, marketing and ticketing purposes. Now, you may also have come across the term GDS. Well, a GDS is simply a network that distributes one or more participating CRSs in different countries around the world. But more on GDSs in Chapter 4: for the moment, let's try to understand how a single airline CRS works.

The introduction to the airline business in the first part of this chapter should have demonstrated that selling airline seats is a complex business. This should have enabled you to begin to appreciate some of the complexities faced by airlines in controlling the sales of seats on their aircrafts. The only practical way of achieving this level of

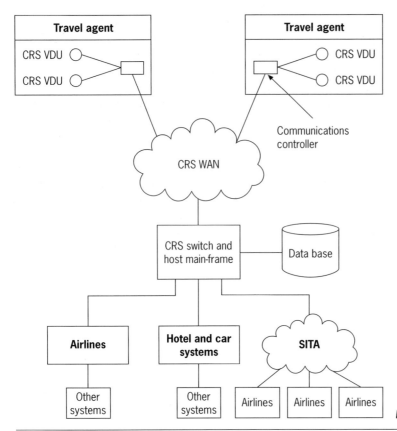

Figure 3.1 Diagram of a typical airline CRS

control is by the use of IT. This means using a large central computer to do all your processing, using a large central data base to store all the seats and their status, and providing access to the computer and data base via a wide area communications network. It is this combination of technologies that is known as a CRS.

Airline CRSs are an example of true legacy systems. These systems were the pioneers of computer applications in the 1950s and are now virtually indispensable to airlines because they enable their revenue streams to be maximized by efficient inventory control (an inventory in this context refers to an airline's stock of passenger seats that is available for sale). The way all this hangs together is shown in Fig. 3.1.

The agent terminals, which are usually PCs nowadays, are connected on-line, usually via leased telephone data lines, to the central host computer system or CRS. The host computer is almost always a main-frame with a massive data base attached. The main-frame host polls each travel agent terminal every second or so, to see if it has any messages to send. At the agency end this polling is controlled either by a dedicated communications controller or by a master PC. If the agent's controller has messages to send, it sends them to the main-frame for processing. This may involve looking up a passenger's PNR, assembling a fare quote or sending an enquiry to another airline. Eventually a response is formulated by the main-frame and when the poll allows, this message is transmitted back to the agent's terminal where it is displayed on the screen. The whole process is repeated and forms what is known as a dialogue. In certain cases there are special messages that are intercepted by the agent's controller, such as those that control the printers in the agency. Such messages do not appear on the screen but instead cause the printer to spring into action and produce an airline ticket or an itinerary.

Well, that covers the way in which airline CRSs 'talk' to travel agents, but there is also the communications between the airlines themselves. This communication comprises messages from one airline to another saying, for example, 'my flight number nnn is full' or 'I have sold a seat on flight number xxx', and so on. In other words there is a lot of message traffic between the airline's CRS computer systems. In fact, there is so much traffic that the airlines have clubbed together and set up two very large global communications networks to handle telecommunications on their behalf. In Europe, the SITA network handles inter airline message transmission whereas in the USA the AIRINC network is used. These networks were not originally designed to handle on-line traffic. Their method of operation was, and still is to a large extent, store and forward. In other words, a message is sent into the network where it is stored awaiting a spare communication channel down which to travel. So, the delivery time can vary between a few seconds and a few minutes (or even hours in certain cases!).

Now I used one or two terms in that preceding paragraph with which you may be unfamiliar. The terms that need a little more explanation are used liberally when any discussion of airline systems takes place. I think the buzz words and terminology are as bad as that used within the IT world. Anyway, the next section attempts to explain some of these terms in plain English.

- **The PNR** The PNR is perhaps one of the most critical pieces of information used to reference a passenger reservation in an airline CRS. The PNR may be used to reference a booking using an airline CRS terminal or via the telephone to a reservations operator of the airline concerned. An airline CRS holds one PNR for each passenger and for each trip. A busy business travel passenger, for example, could therefore have several PNRs in an airline CRS, one for each of the forthcoming trips being planned and organized by the travel agent.

  The code by which the PNR is referenced, is the primary key by which a booking is accessed. The code itself is simply a series of alpha and numeric characters. The PNR contains all the core data about the passenger and the associated booking record that is held on the CRS data base. In order to be able to retrieve a passenger's booking details you need to know the airline in which the reservation is held and the PNR reference. In a sophisticated CRS, a booking locator allows several PNRs relating to a passenger's trip to be referenced using just a single code. So in other words a booking locator is the master key to the PNRs stored in several airline CRSs, all of which relate to a single booking for a passenger or group of passengers.

- **Last seat availability** This is one of the most important features of an airline reservations system. Put simply, it is the ability of a CRS literally to sell the last seat available on a flight, immediately prior to the airline designating the flight as full. It may not seem such a big deal but consider the following situation in which you are a travel agent:
  - A business customer calls and requests a seat on a flight to a frequently visited city, say Paris, at the last moment. It really is an important meeting that he has to attend and he must catch the mid-day flight today.
  - You quickly access your high-tech CRS terminal and it shows the flight as full. You apologize but there is nothing more you can do.
  - However, the business traveller doesn't give up and he sneaks down to the travel agency around the corner just for one last try to get the mid-day flight to Paris.
  - This agency is using a different airline CRS and the traveller asks the same question about a flight to Paris today. The travel agent accesses the terminal and surprise, surprise: – 'Yes, a seat is available. Would you like me to book it for you sir?', asks the smiling travel agent.
  - Having booked his ticket, the traveller storms into his finance director's office in a rage and demands that he switch travel agent because the existing one doesn't know what is going on and can't get the flights the company needs.

A sorry tale but one that typifies how many a profitable business account has been lost to the competition. How could it possibly happen when both travel agents had the latest in CRS technology? The answer is to be found in the way an airline controls its seat reservations and in the precise way in which it is connected into the CRS systems of other airlines.

When a flight is nearly full, the airline operating the flight will close it out when there are only, say five seats, remaining to be sold. Closing the flight out means telling all other airlines that the flight is now full and is not available for further seat sales. This is accomplished by sending a message through the SITA network to all other airline CRSs with whom a sales agreement exists. This message is known as an availability status (AVS) message. An AVS message may either say 'no seats left' or it may say 'only four seats left' and then send out other AVS messages as the last few bookings arrive.

The reason AVS messages are sent is so that the primary airline can offer its own customers or its close partners, those last few precious seats. In my example above the first travel agent had an airline CRS that accessed the desired airline system indirectly via a separate link, possibly using SITA. The second travel agent was, however, using an airline CRS provided by the airline who had just the last few remaining seats to Paris on the day in question. It had closed out the flight to other airlines and only showed those last few seats to terminals that were connected directly to its computer and data base.

There are several different ways in which the airlines participate in each other's CRSs in the area of seat bookings. These are known as levels of participation and give rise to some commonly used terms that are important to understand fully if you are to grasp the essential differences between one CRS and another. So, let's look at the various levels at which CRSs participate in each other's systems.

Levels of CRS participation

This is all about how the CRSs interact with each other. The level at which a CRS participates in another CRSs system is governed by the type of connection between them and the relative functionality of the computer systems involved. The reason that this topic is so important to a travel agent is that: (a) it governs the type of display the

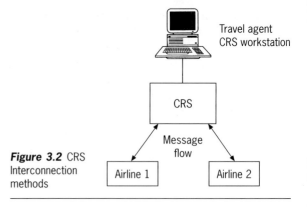

Figure 3.2 CRS Interconnection methods

travel agent sees on the CRS screen, (b) it determines the level of reliability that may be placed by the travel agent on the booking status displayed by the CRS, (c) it governs whether last seat availability is provided by the CRS or not, and (d) it determines the timeliness with which the reservation is made. From the airline's viewpoint the level of participation also determines the booking fees that are payable when a seat is sold through a CRS for one of the participating airlines.

The type of connection between the CRS and its participating airline systems (Fig. 3.2) is governed by several factors. The first of these is the message flow between the systems and the second is the relative speed of the telecommunications link between them. Let's look at these factors in a little more detail:

- **Message flow** This is the message flow between two interconnected CRSs. This message flow or dialogue may be one-way only or two-way. Most CRSs use a two-way message flow, whereby messages can flow from the participating airline to the CRS and from the CRS to the participating airline. These message flows can occur in one of two ways. They can be off-line or on-line.
- **Off-line** With an off-line connection between the participating airline and the CRS, message flows are separated by a substantial period of time. This may be anything from a few minutes to several hours. This delay is usually due to the passing of the messages over a store and forward communications network. The request is sent through the SITA or AIRINC networks jointly owned by the airlines.

- **On-line** An on-line connection between the participating airline and the CRS allows messages to be sent and received with only a few seconds or even fractions of a seconds delay. This is usually achieved by means of a dedicated high speed communications line connecting the two computers.

Although these factors are the principal ones that control the levels of participation, the relative sophistication of the CRS and the other airline systems involved also add dimensions of their own: and then there is the inevitable marketing involvement that is all about differentiating a CRS from its competitors. The result of all this is that there are a number of terms that describe the way bookings are made by CRSs. Some of these terms have become almost a standard throughout the industry while others have been modified slightly to form brand names offered by certain CRSs. The whole subject is therefore complex and difficult to understand without a lot of knowledge of the CRS technologies. It is for these reasons that there is a fair amount of confusion in this area. In order to try and overcome this, I have attempted to try and define the most commonly used terms that describe each level of CRS participation. Then when I come on to discuss specific CRSs and their big brothers, the GDSs, it should be easier for you to understand some of the proprietary brand labels given to their versions of participation levels.

- **Availability and phone** This is probably the most basic level of participation. It is based on a one-way off-line message flow. This enables the participating airline to send its flight schedules on a periodic basis to the host CRS. Travel agency users can view the airline's availability display as shown on the CRS screen but cannot book directly. To make a booking enquiry the agent must telephone the airline and talk to a reservation clerk.

    So, this level of participation is rather primitive and is labour intensive for the travel agent (and for the airline too). It is also inaccurate because the display shows the flights as though they are available whereas when the phone call is made, the flight may in fact be full.

- **On request** The message flow between the participating airline and the CRS is two-way, but is off-line. This allows the airline to send its flight schedules to the CRS on a periodic basis for display purposes. The travel agent can view the airline's availability display and can use the CRS to make a booking for selected seats. When the CRS receives a booking request such as this it will send a message to the airline saying: 'Can I sell this seat?' This message is sent via one of the airline networks, such as SITA, to the participating airline. Sometimes this stage of the booking process is known as 'sell and report'. At this stage the booking is held on the CRS, but is not confirmed. The booking is said to be 'on request' and the process itself is often referred to as 'hold and confirm'. A response is eventually received from the participating airline, which states whether the seat has been reserved or not. The CRS updates the PNR to show the status of the booking.

    This is better than the previous method but is still a long way short of being ideal. The travel agent must keep accessing the CRS to see if the booking has been confirmed or not. This is a labour intensive and non-productive task. During this stage the customer is in a state of limbo, not knowing whether or not a seat has been reserved on the flight.

- **Full availability** As with the previous method, the message flow between the participating airline and the CRS is two-way and off-line. This allows the airline to send its flight schedules to the CRS on a periodic basis for display purposes. The travel agent can view the airline's availability display as shown in the CRS system. The availability display shows the maximum number of seats that can be sold without reference to the participating airline. This maximum number of seats is usually four or seven. The travel agent can use the CRS to make a booking for selected seats, up to the maximum shown. At this point the travel agent can assume that the booking is confirmed and no further action is necessary. The CRS then formulates and sends a message to the participating airline saying: 'I have sold the following seats . . .' and this is used by the participating airline to update its seat inventory.

    When the flight begins to fill up, the participating airline will send a message to the

CRS reducing the maximum number of seats that are available for sale via the CRS, without direct reference to the participating airline. Eventually, the participating airline will send a 'no seats available' message to the CRS for the flight in question. In fact the participating airline will usually do this when there are only about four or five seats actually remaining. It holds on to these last few precious seats for its own special customers who book direct.

This form of access is sometimes also known as 'free sale', i.e. the seat is sold on the basis that no news is good news. In other words, the selling airline has permission to sell the operating airline's seats unless told not to do so. So, this form of participation does not provide the travel agent with true last seat availability; and there can be problems too. Sometimes because the message flow is off-line, the booking message from the CRS may get delayed or even lost! In these situations the participating airline may have sold all the seats on the flight before it receives the booking message from the host CRS. So, unfortunately, it is possible for a seat to be sold yet for there to be no room on the flight, again due to slow communications. In summary then, full availability is a method that is more open to potential problems and customer service issues, which often end up on the travel agent's plate.

- **Direct access** The message flow with direct access is two-way on-line. The participating airline sends its flight schedules to the CRS on a periodic basis, as before. Also, the availability display on the CRS will again show the maximum number of seats that may be booked without further reference to the participating airline. However, with direct access the travel agent can book as many seats as is required by the customer. When the travel agent makes a booking, the CRS enters into a computer-to-computer dialogue with the participating airline. The required number of seats are 'taken' straight out of the seat inventory of the participating airline system's computer. As the flight begins to fill up, the number of seats available will decline. However, even the last seat on the flight is available for sale via the CRS. This is a true last seat availability method and as such is superior to the other methods discussed so far. However, there are one or two problems that need to be guarded against. Some CRSs show the availability as at the time that the last message was received. So, the travel agent thinks that there are seats on the flight and proceeds with the reservation only to find that just prior to ending the transaction the CRS will respond to the effect that the flight is actually full. This can cause a bit of confusion between the travel agent and the customer. But all in all, direct access is a vastly superior method of participation than many others because the airline is in touch with the booking situation in real-time and can feel more confident that it has the true booking status of its flights.

- **Multi-access** This level of participation uses a two-way message flow and is on-line. Multi-access is, however, quite different from the other methods of participation discussed so far. It is the provision of reservation functions to multiple participating airline systems. In order to accomplish this multiple link-up, an on-line real-time communications link is required between the CRS and the participating airlines. On the face of it a multi-access airline system is good news for the travel agent because it provides direct access to a wide selection of airlines with last seat availability on each one. The availability displays show the actual number of seats remaining for sale, i.e. true last seat availability. Invariably, a CRS with multi-access will use a common language for all the inbound messages keyed by the travel agent. Outbound screen formats are, however, not converted and usually appear exactly as they were formatted by the host system (see below for a description of the term common language).

In order to use the participating airline's system the travel agent enters the two-letter code of the airline that needs to be accessed. The CRS then effectively connects the travel agent directly to the computer of the airline to be accessed. Once connected, the travel agent carries on a dialogue with the selected airline's computer system until a booking is made or the customer's enquiry is answered. Each airline available via multi-access is sometimes said to operate in a partition of the CRS. A partition is really a sub-division of the CRS's computer, which is dedicated to communication sessions with a particular airline system. Multi-access CRSs are therefore largely unbiased because the user is free to choose the CRS to which a connection is to be made.

But in the past there was often a price to pay for this multi-access; and the price was the need to keep a manual record of the booking locators or PNRs for your clients. You see, once a reservation was made in a particular airline's computer, the agent needed to know how to find that booking record again should the customer wish to modify the booking or should the agent wish to retrieve the record for ticketing. The agent will suffer a degree of embarrassment if the only response is: 'Well, I know your booking is in there somewhere!' Unfortunately, some of the earlier multi-access CRSs did not keep a log of which airlines the bookings were stored in.

Nowadays, however, most of the more sophisticated CRSs have a feature known as PNR indexing that aims to overcome this problem. The way PNR indexing works is by building a special program into the CRS computer itself. This program maintains a file of all PNRs created and indexes them by a variety of key fields. Put simply this means that the CRS builds a sort of index card that contains the passenger's name, the travel agent who made the booking and the airline in whose system the PNR is stored.

- **Answer back** This is based on a two-way message flow with on-line communications. The participating airline sends the CRS its flight schedules on a periodic basis. The travel agent can view availability and this will show a maximum number of seats bookable. However, the travel agent can book as many seats as are needed by the customer. As with direct access, the CRS then enters into a computer-to-computer dialogue with the participating airline and 'takes' the number of seats required directly out of the inventory (assuming the seats are available). The participating airline responds by sending a message back to the CRS; this is known as an 'answer back'. The CRS updates the PNR with the answer back code received from the participating airline.

  Besides providing last seat availability, this level of participation provides a comfort factor to both the travel agent and the customer. In the event of any dispute about whether or not the seat was actually booked and confirmed, the booking record, i.e. PNR, can be retrieved and the response code received from the participating airline can be used as evidence of confirmation.

- **Direct connect** The message flow is two-way and on-line. Once again, the participating airline sends its flight schedules to the CRS on a periodic basis. The travel agent can view availability and this will show a maximum number of seats bookable. However, the travel agent can book as many seats as are needed by the customer. As with direct access, the CRS then enters into a computer-to-computer dialogue with the participating airline and 'takes' the number of seats required directly out of their inventory (again, assuming the seats are available). However, in this case, the participating airline responds by sending its own PNR locator code back to the CRS. The CRS updates the PNR with the locator code received from the participating airline.

  Again, this provides true last seat availability. It also goes one step further than answer back, however, and is a true customer service advantage. Having the airline's PNR locator is a lot more reassuring for the customer and eliminates delays while travelling. Delays that may otherwise be experienced when passengers need to approach the participating airline directly to resolve a query or confirm a flight. In cases like this the participating airline can clearly see their own locator and use this to find the booking in their own system without delay to the passenger. Direct connect is also a boon in cases where there is a ticket on departure (TOD) to be collected at the airport by the customer. It simply requires the travel agent to send a copy of the booking to the airline desk for the PNR to be retrieved and ticketed without delay.

- **Direct connect availability** This is sometimes known as 'seamless availability'. It is based on an on-line two-way message flow between the CRS and the participating airline. With this level of participation, when a travel agent requests an availability display using the CRS terminal, the CRS accesses the participating airline's seat inventory directly. This allows the CRS to build a picture of the actual number of seats available for sale at that point in time. The CRS then proceeds to build an availability display for the travel agent that shows the actual number of seats available. The travel agent can then book any number of seats, up to the number shown as being available.

    When a booking is made the CRS then enters into a computer-to-computer dialogue with the participating airline and 'takes' the number of seats required directly out of the airline's inventory (assuming the seats are available). As with direct connect, the participating airline responds by sending its own PNR locator code back to the CRS. The CRS updates the PNR with the locator code received from the participating airline.

    This is the ultimate in CRS inter-operability. It allows the travel agent to enjoy the benefits of multi-access, e.g. actual availability displays, with the advantages of working within a single CRS environment, e.g. common language and central PNR referencing. The only problem is that the number of airlines with CRSs capable of supporting this level of functionality are somewhat small at the present time.

So, returning to my little case story above, the first travel agent did not have last seat availability access to the airline's seat inventory and was closed out before the last few seats were actually sold. Whereas the second travel agent around the corner was directly connected to that airline's system and therefore enjoyed the benefits of full last seat availability.

Common language and by-pass

Although the airline systems are very similar in many ways, they are nevertheless different enough to prevent a travel agent trained in one system from being able to simply use another airline's

Table 3.2 CRS native language examples

CRS	Availability request entry
British Airways	A30JUNLONNYC
American Airlines	130JUNLONNYC

CRS without a fair amount of re-training. One of the main reasons for this is the difference in entries that need to be made at the keyboard in order to get the CRS to perform the required function. Compare, for example, the entries for an availability display in British Airways' BABS and American Airlines' Sabre systems for flights between London and New York on 30 June (Table 3.2).

As you can see, the entries are similar but different in some basic respects. A common language is a set of frequently used commands, which have been standardized for a number of airline systems; like a sort of Esperanto of CRS languages. Common languages were developed mainly by the providers of multi-access systems. This made usage of the CRS simpler for users because they needed to remember fewer command codes in order to use a variety of airline CRSs. However, it would have been too complex a task to develop a common language entry for all commands in all airline systems and so the common language was restricted to only the essential and frequently used commands. This was quite acceptable because it was often useful to have access to the native commands of a particular airline's system. This kind of access to an airline's native commands was called 'by-pass'.

So, for example, in the world of multi-access, an agent first selects an airline CRS and requests connection to it. Then the common language is used to interrogate the airline's system to determine if its inventory can satisfy the customer's requirements. Finally, by-pass entries are used to access those functions that are provided by that airline only. Quite a wide range of capabilities; but a challenging training task, because the agent must learn the common language as well as the individual commands supported by all the other airlines. Another factor to consider here is that the host CRS system must perform a somewhat complex translation into the reservations language of the other airline CRSs being used. This can sometimes slow down the response times marginally and also restrict the set of functions available using the common language.

Pseudo city code

The pseudo city code is an important code that is assigned by a CRS to each user terminal when it is first set up (or initialized). This is the three character city code that is used to denote the location of the user. It is used by a CRS for many purposes including, for example, in constructing an availability display for a user. When a travel agent, in say, Sheffield requests an availability display for flights to New York using a CRS, the system first of all looks at the agent's pseudo city code. It then works out the country and uses this to select the departure airports for the itinerary. So, in this example the CRS might show Manchester, from which there are in fact direct flights to New York, or it might show London Heathrow or Gatwick. Other airports would be lower down the list. The sequence of the flights shown depends to a large extent upon the departure time specified. However, it would not show flights departing from, say, Paris. These departure airports are all worked out from the pseudo city code that is logged in the CRS core system for each user's terminal. It is often possible for the travel agent to use an atlas function in the CRS to find the nearest airport to the travel agent's city, if this was not known.

The pseudo city code also enables the CRS to determine default flight departure times. If for instance a departure time is not specified when making an availability enquiry, the CRS will have to make some assumptions on the departure times of flights that are to be shown on the display. For example, the default departure time for a UK traveller might be 1 p.m. The travel agent can, however, often alter these default departure times to suit the travel agent's own needs.

Finally, the pseudo city code allows different rates to be displayed for hotel and car companies that participate in a CRS. There are, for example, different sets of rental rates applied by many car rental companies, depending upon the country in which the customer is going to rent the vehicle. The rate for an economy car rented in the USA for instance would be one price whereas in Europe it would be another: if the pseudo city is used in conjunction with the terminal address, then different rates can be displayed for different travel agency users. It is thus possible for a multiple travel agency to have all of its negotiated corporate hotel rates shown on any of its branch CRS terminals whenever a hotel rate is accessed.

Shared travel agent access

Travel agents who are part of a multiple chain whether it be large or small, UK domestic or internationally spread, can have access to a common pool of client booking information. This must of course be agreed at the head office level and all branches must be in agreement on the principal of sharing data. The CRS can then alter its access security control system to allow travel agent A to access the bookings created by travel agent B. This can be a powerful servicing tool if used properly. It means that the customer can make a reservation in any travel agency in the group and be assured that the booking can be retrieved by another agent within the group. This is especially powerful if the traveller can be serviced while on a trip and in another country.

FARES AND FARE DISTRIBUTION

According to IATA, distribution costs form the greatest slice of airline expenditure. They amount to US $20 billion annually. Hence it is no surprise that airlines place great emphasis on getting information about their fares and schedules to the marketplace quickly and efficiently. While all airlines ultimately would wish to get their information right in front of the passenger and to take the booking directly from him or her, it is still true that the CRS and/or GDS account for most airline bookings today. Any services that efficiently supply data to these systems will be taken extremely seriously by airlines. Although this is principally an activity of prime interest to airlines and GDSs, it is nevertheless important for anyone in the field of travel and tourism to understand how this process is controlled if some of the characteristics of GDSs are to be understood fully. Take for instance the services offered to airlines by Reed Travel Group under the OAG brand name.

Two key services provided by Reed Travel Group help airlines control the distribution of their flight schedules and fares information to global CRSs and GDSs. These are known, respectively, as OAG Direct and OAG Genesis. In a deregulated environment, which has made fares extremely volatile as airlines vie with each other for passengers, a service such as Genesis is critical to an airline as it can help increase its market share by ensuring that the market is informed swiftly of new fare deals intended to beat the competition; and, it can help to reduce an airline's administration costs. Fares must be distributed accurately, widely and quickly to the world's population of GDS terminal users. Schedules, also, are changing more quickly as airlines focus increasingly on the market-competitiveness of flight timings and services, rather than on the operational aspects of schedules. It is key that an airline's schedules need to be updated in all of the world's GDSs as soon as is practically possible. Let's take each of these two services in a little more detail:

- **OAG Direct** Reed Travel Group is the leading distributor of airline schedules in the world, supplying all the major airline CRSs and all major GDSs with data. Reed operates one of the largest airline schedules data bases in the world, holding the schedules of more than 800 airlines and processing hundreds of thousands of schedule changes every month. A schedule only has to change by one minute to make a change happen on Reed's data base. Currently, more than 80 per cent of the data held is received from airlines by automated means, usually in the industry standard format (co-ordinated by IATA) called the (standard schedules information manual (SSIM). This standard format helps the industry to communicate schedule changes in a more co-ordinated way. Reed's twin roles in this area, i.e. in helping IATA set data standards and as a data intermediary between the airlines and the GDS, cannot be underestimated. It plays a major part in keeping the industry together.

  The current method of getting schedules data to GDSs has been used for many years and has been recognized as a reliable way of distributing schedules information to the marketplace. With increasing competition and schedules volatility, the length of time it takes for the information to get onto GDS screens (this can take up to two weeks), has provoked demands from airlines for the process to be accelerated dramatically. The cost to the industry of poor schedules synchronization between the various GDSs and airlines' own CRSs has been estimated at US $1 billion annually – an amount that airlines would be eager to slice off the US $20 billion they spend each year on distribution.

  Reed Travel Group's response to market demands for a swifter schedules distribution service is OAG Direct. This service enables schedules on the Reed database to be updated in an instant by airlines, and for those changes to be transmitted almost immediately to the GDSs. It uses a development of the SSIM standard format to communicate only the changes made to the data base. The key to this process is Reed's ability to process fully the necessary standard schedules messages (SSM) and *ad-hoc* schedule messages (ASM). However, airlines supplying data to Reed and not having the capability themselves to generate these standard messages can still use OAG Direct. Reed has developed the capability to transform schedules supplied in any format whatsoever into SSM and ASM for swift supply to the GDS.

  The benefits of the service are considerable. They arise principally because airlines can be sure that the schedules appearing on GDS screens are synchronized with their own reservations systems. This means they will not have to re-book passengers whose bookings were made using old schedules still displayed by the GDSs. Re-booking means inconvenience to the passenger and extra booking fee payments that airlines must make to GDSs. For the GDSs, up-to-date data are vital in their drive to compete in the cut-throat battle to gain market share. OAG Direct guarantees them up-to-date and accurate data.

- **OAG Genesis** The Genesis product (Fig. 3.3) provides a fares distribution service to the world's CRSs and GDSs, on behalf of airline

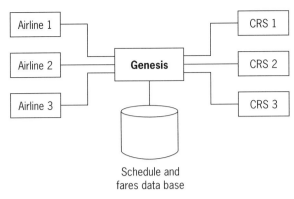

Figure 3.3 The Genesis system

customers. It is vital that GDSs and CRSs have access to the latest fares information, as they will fail to maintain market share if the fares offered by them prove to be incorrect. Reed Travel Group has become a leading player in the airline fares distribution business, providing the following GDSs with fares information:
– Amadeus.
– Galileo.
– Sabre.
– System One.
– Worldspan.

The data base that is used to record all this information is massive. It contains nearly one million individual fares and 5,000 fare rules; and it works pretty efficiently, because it handles hundreds of thousands of changes to the data base every month. The information is widely used because the combined population of travel agency users totals almost 500,000 GDS terminals world-wide, each of which obtains a proportion of its fares data via Genesis. These terminals generate more than 70 per cent of the world's airline seat sales.

Fares filing

Besides distributing fares to global CRSs and GDSs, the OAG Genesis service is also concerned with helping airlines to file the latest fares with the appropriate authorities at both ends of a route, using the automated tariff filing (ATF) facility. Each time an airline adds a new fare or changes an existing one on a route, it must report the details to each of the appropriate airline regulatory authorities in the countries affected. In the case of the UK, for example, if a flight starts or ends in the UK then its fare details must be reported to the CAA and several other similar government organizations in other countries.

Because the number of fares is enormous and the rate of change very high, this is a difficult and costly task for the airlines to handle themselves. In fact, with the EC's 'open skies' policy, it is going to be difficult increasingly for an airline to remain competitive without using some form of electronic fares filing service. Reed can help by providing an administrative infrastructure that can be used by airlines to remove a non-productive task and allow them to concentrate on sales and marketing activities. Genesis' automated tariff filing works like this:

1. The airline loads the fares information onto a PC using Genesis' airline tariff automated collection (ATAC) software. This is a flexible PC software package that allows fares information to be transmitted direct from a carrier to the Genesis data base in a recognized standard industry format. This information also can be transmitted direct from the airline system on a host-to-host link. Alternatively, information can be supplied manually.
2. If the fare is approved already, it will be loaded directly onto the data base for distribution.
3. If the fare is not yet approved, Genesis fares distribution, through its ATF facility, can process the filing application electronically with the relevant authorities.
4. Once stored within the system, fares are transmitted rapidly and in many cases daily to every one of the world's major CRSs and GDSs.
5. These CRSs and GDSs then provide access to the fares data via reservation terminals that are located in sales and servicing outlets around the world.

There are, in fact, two other companies – apart from Reed – that undertake this business: SITA FAIRSHARE and Airline Tariff Publishing Company (ATPCO). Because these are non-profit making companies owned by the airlines, Reed faces a tough challenge in competing for business. However, Reed's strong position, which arises from its highly developed data expertise, has provided it with a springboard from which to launch several other related products, e.g. see EasyRes in Chapter 6.

Management information

Reed can also provide valuable management information that is generated from its data bases. Given the size and comprehensiveness of Reed Travel Group's fares and schedules data bases, their suitability for providing insight into trends in the industry is obvious. Reed offers a specialist service called 'market analysis', which allows customers such as airlines, airports, aerospace companies and consultants to request specially tailored information relating specifically to their areas of interest. Market analysis is a really valuable service, for example, to airlines in today's business environment as it allows them to track competitors' activities and adjust their own strategies accordingly.

SITA's AIRFARE service

But Reed is not the only company to provide fares distribution and support systems to the airline market. SITA operate several services in this area as well. SITA is part of a group that operates the world's largest data communications network, servicing customers in 225 countries. The SITA AIRFARE product offers airlines and GDSs the capability to improve their service to customers greatly by providing access to systems providing automated quotations of passenger fares. This includes a function that supports the detailed itinerary pricing of all domestic and international journeys of up to 16 segments. The PRICECHECK product provides airline tariff departments with limited access to AIRFARE for the purpose of monitoring their own airline fare levels and rule interpretations in AIRFARE.

A related set of products branded 'fares data supply' also is provided by SITA. These products are derived from the AIRFARE system and are available to airlines, GDSs and tariff publishers for a variety of uses: (i) GDSs utilize SITA fares data to support the functionality of their in-house fare quotation systems; (ii) airlines use the system for decision support, revenue accounting and internal paper tariff publication; and (iii) publishers use SITA fares data for their book production (in fact all major paper tariff publications use this system). Airlines may distribute their fares to GDSs world-wide free of charge via AIRFARE. SITA fares distribution provides airlines the unique capability to control their own fare updates via a PC connected directly to the AIRFARE data base through the SITA network.

TICKETING

The automatic printing of airline tickets is an important productivity aid to any travel agent with a high volume of business travel. It is important to understand the different types of ticket stock before we start looking at the different technologies used to produce them. To be able to issue airline tickets at all, the travel agent must possess an IATA licence. This licence stipulates the specific airlines that may be ticketed by the agency. With manually issued tickets, the agent is provided with a metal plate that is used to emboss the airline ticket, rather like a manually issued credit card voucher. The use of the *plate* terminology is carried over to automatically issued tickets as well. CRSs will only allow travel agents to issue automated tickets for those airlines for which it holds a *plate* issued by IATA.

At present, besides manual airline tickets, there are two types of tickets that can be printed by a computer system: off-premises transitional automated tickets (OPTATs) and ATB tickets. Each is explained in more detail below.

The OPTAT ticket

The OPTAT ticket (Fig. 3.4) has been in use now for at least the past five or six years. It is issued by BSP UK on behalf of IATA and is on continuous paper stock for use in computer impact printers. The stationery itself has several copies, each of which serves a different purpose. The image is carried from the top copy, by impact, onto the copies underneath. This is accomplished by means of the characteristic red 'carbon' that is coated on the reverse of all copies. The stationery is designed so that once printed and torn off the printer, it can be folded in two so that an IATA emblazoned logo (red on blue) appears as a kind of ticket cover. Travel agents invariably staple the

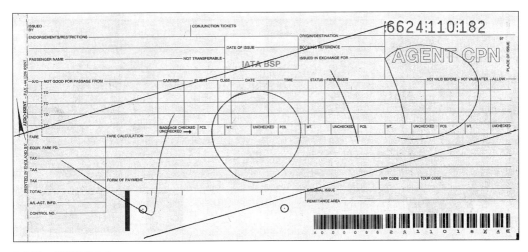

Figure 3.4 The OPTAT ticket

ticket into one of their own personalized ticket wallets. The stationery is formed in pairs of documents, the first being the ticket itself and the second being a credit card charge form voucher. The copies of the ticket are used as follows:

- **Airline ticket**
  1. Audit coupon (sent to BSP as part of settlement process).
  2. Travel agent coupon (filed in the office client file).
  3. Passenger flight coupon 1 (given to the passenger).
  4. Passenger flight coupon 2 (given to the passenger).
  5. Passenger flight coupon 3 (given to the passenger).
  6. Passenger flight coupon 4 (given to the passenger).
  7. Passenger copy coupon (given to the passenger).
- **Credit card charge form**
  1. Credit card company's copy.
  2. Travel agent's copy.
  3. Customer's copy.
  4. File copy.
  5. File copy.
  6. File copy.
  7. File copy.

The top copy of the OPTAT ticket used to be used for BSP settlement purposes (BSP is explained more fully in Chapter 7 – Agency management systems). In summary, it was Copy 1 of the OPTAT ticket that was used as the basis for the travel agent to pay the airlines, via an IATA clearing system called the BSP. Preparing these copies of the tickets issued is a laborious administrative chore but one that is crucial because prompt and accurate settlement of air ticket sales is a prime condition for retaining the travel agent's IATA licence; and without that licence, the agent cannot issue airline tickets.

ATB and ATB2 tickets

The use of ATBs is now widespread but has not yet fully replaced the OPTAT ticket. The ATB offers both airlines and travel agents some important productivity advantages and makes travelling easier for the customer as well. The ATB stock (Figs 3.5 and 3.6) is completely different in concept and appearance to the OPTAT ticket. The ATB is constructed from card and comprises only a single copy. It embodies a magnetic strip on the reverse side that is capable of recording data. Once data are recorded on the magnetic strip the data may be 'read' by special purpose reading machinery, which most major airlines are now installing at their airport check-in gates. The way in which the ATB is designed to work is generally as follows:

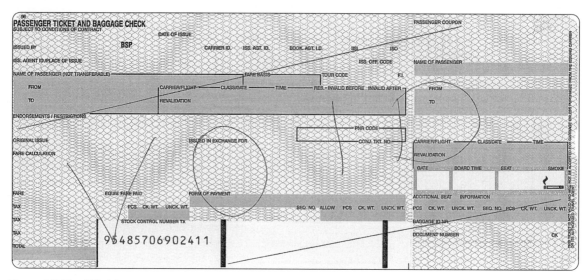

Figure 3.5 The ATB ticket

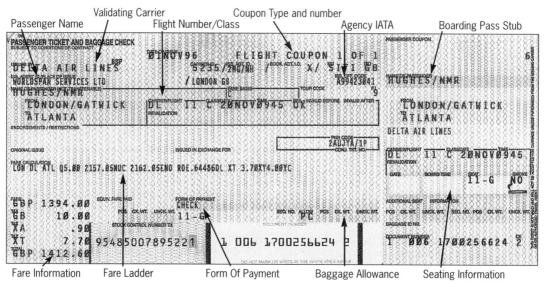

Figure 3.6 The ATB2 ticket

1. The ticket is printed by the travel agent using a special purpose printer. Besides actually printing the ticket, this printer also magnetically encodes the brown metal oxide strip on the ATB ticket stock with the details of the airline ticket purchased by the customer.
2. At the same time the printer also prints the other copies as required by the travel agent for the travel agent's files and the customer for supporting expense claims.
3. The traveller takes the ATB and presents it to the check-in gate. The check-in clerk inserts the ATB into a special reader that verifies that the passenger is destined for the correct flight, a reservation exists for the passenger, a seat is allocated and that the other ticket details are in order. The reader can also add data to the magnetic strip on the ATB denoting the check-in details.
4. When the passenger boards the flight, the ATB is presented to the flight welcoming steward

at the gate who inserts the ATB into another special reader that retains one part for the airline's revenue accounting purposes and returns the other portion to the passenger as the passenger's boarding card.
5. Meanwhile, the airline in whose system the reservation was made and that produced the passenger's ATB, reports the ticket details direct to BSP, electronically. This saves the agent from having to batch and submit physical ticket stubs manually.

The advantages to the airline of the ATB over the OPTAT ticket are fairly obvious. They allow the airline to mechanize its flight departure operations to a far greater extent than ever before. But there are important advantages to the travel agent as well. Avoiding the BSP batching and reporting task is perhaps the greatest advantage. Then there is the satellite ticketing advantages as described in the following section, as well. Finally, the customer even benefits because using an ATB will mean a faster transit through the airport and onto the flight.

Electronic ticketing

Electronic ticketing, or e-ticketing as it is commonly referred to, eliminates the need for paper tickets. Instead of a ticket being printed and the various copies processed by the agent, the passenger, the airline and BSP, an electronic image of the ticket is used to control flight boarding, settlement and revenue accounting. E-ticketing is possibly one of the most significant developments within the field of travel automation. Not simply because it makes life easier for the airlines but because it has far reaching implications for new electronic distribution systems like those based on the Internet. So, before we can explore these implications and consider the new systems that are beginning to surface, it is important that you grasp some of the fundamentals of e-ticketing.

E-ticketing has been in widespread use in the USA for some time. However, in Europe, it is in the early stages of roll-out. Scandinavian Airlines System (SAS) and Lufthansa were two of the first European airlines to introduce e-ticketing; and in the UK, ticketless travel is supported by British Airways and British Midland. It is important to realize that e-ticketing is a CRS supported function that can be used to book: (a) airline seats sold directly to customers by the carriers themselves, and (b) airline seats sold to customers via travel agents. Travel agents obtain e-ticketing functions via their GDSs, which are of course connected to the airline's CRS. So, for e-ticketing to be possible from a technical standpoint the airline's CRS and the travel agent's GDS must both support e-ticketing functions. Take, for example, the domestic air travel business of British Airways in its home market – the UK.

British Airways e-ticketing

British Airways piloted its e-ticketing system on the Gatwick to Aberdeen route for a period of six months. The pilot trial was judged to be very successful and as a result, BA together with most of its franchised carriers, introduced e-ticketing on all UK domestic routes on 12 March 1997. By early 1997 BA was carrying around 3,000 ticketless travellers per day. Also during this year certain travel agents could use Galileo or BA Link (British Airways' own viewdata system), to interactively book e-tickets. Agents using other GDSs could issue BA e-tickets, but only by queuing them to the BA system, just like a ticket on departure. However, all GDSs have plans to add e-ticket functionality within the UK very soon.

British Airways e-ticketing via Galileo travel agents

Travel agents in the UK can use the Galileo GDS to issue e-tickets on British Airways domestic routes. The principal way that this is accomplished is via the various Galileo PC workstation products. The following section therefore explores in more detail, how BA e-tickets are issued by travel agents using the Galileo GDS. Before we begin, there are certain conditions that must exist before an e-ticket can be issued:

- **E-ticket requirements** The standard ticketing process commences when three basic conditions have been met: (i) a reservation has been made for an airline seat, usually via a GDS; (ii) payment for the ticket has been received by a travel agent or other intermediary; and (iii) the departure time is imminent. For e-ticketing to take place, however, there are a couple of additional conditions. First of all, the ticketing carrier must support e-ticketing on the customer's route, as booked. This is usually denoted by an 'E' on a GDS availability display. Second, the customer must have some form of plastic card that will enable them to use the self-service check-in machines at the airport (more on this later). For British Airways this may be either an Executive Club membership card or an E-ticket Access card. However, not all customers of the travel agency will necessarily possess an Executive Club card. In such cases the travel agent simply issues an E-ticket Access card from a stock supplied by British Airways (supplies of E-ticket Access cards may be obtained by the travel agent by telephoning a special British Airways hot-line). This enables an E-ticket Access card to be issued to a customer prior to commencement of their journey. Finally, it is necessary for the customer's card number to be present within their booked PNR.

- **E-ticket issue** Once these conditions have been met, the travel agent makes a simple entry on the GDS terminal that initiates the issuance of an e-ticket (incidentally, if the travel agent does not use a GDS, the agent can still request the booking source – usually the agent's airline reservations office – to e-ticket on the agent's behalf). Instead of a physical ticket being printed within the travel agency, an electronic image of the ticket is created within the ticketing carrier's system. This is really no more than a data record containing all the items shown on a physical paper ticket, plus a few extra fields. The e-ticket image will even contain a ticket number, just like a standard paper ticket. This ticket number, which is prefixed with the airline's unique three digit code, is used for accounting and reporting purposes in the usual way (see BSP in Chapter 7).
- **Other documentation** The only remaining piece of paper that must still be handed to the customer is that stating the terms and conditions of carriage, otherwise known as an airline passenger notice (APN). These terms and conditions were originally agreed by all the airlines at a convention held in Warsaw many years ago. They must be given to the customer by the agent prior to departure. This is not normally a problem because there are usually several other items of paper also given to the customer at this time. For example, the itinerary, luggage tags, hotel vouchers and information leaflets provided by the travel agent. At least one of the information leaflets given to the customer will explain how they should check-in when they arrive at the airport.
- **Self-service check-in** Checking-in with e-ticket is quicker and easier for passengers when they arrive at the airport terminal, than with conventional paper tickets. The whole process takes only a few minutes and is controlled by self-service machines that resemble ATM cash dispensers but which use efficient touch screen technology. These machines are usually located near to both the drop-off zone and the terminal concourse, often within their own area designated as a self-service check-in lobby. They accept passengers holding either an ATB or an e-ticket. (For an ATB ticket holder, the customer inserts a cardboard ticket – the machine then prints a few additional fields and electronically converts the ATB into a boarding pass.)

E-ticket holders first insert their E-ticket Access card or Executive Club card into the self-service check-in machine. The traveller's card is validated by the British Airways system controlling the self-service machine. The system will check the number on the card against the number that has previously been entered into the PNR by the travel agent at the time of booking. Should a traveller have more than one e-ticket for travel on the day of check-in, the self-service machine will show a list from which the traveller selects the appropriate one. The flight details are then displayed along with the seat number originally allocated as part of the booking process. The traveller then has several options: either

1. Retrieve the card and a boarding pass, which is automatically printed by the machine. The passenger may then proceed straight to the departure gate and board the flight.

2. Change the pre-allocated seat, i.e. the one originally assigned to them by the travel agent. This is done by viewing a seat plan and choosing another available seat on the aircraft.
3. Choose to receive a boarding pass for both the outward and return journey, provided the passenger is returning within 24 hours of departure.
4. Print another copy of the e-ticket itinerary and retrieve this from the machine along with the boarding pass.

- **Departure** Having done this, the traveller proceeds to the departure gate, boards the flight and departs. This can be a very speedy process for passengers with just carry-on luggage. However, if the passenger has baggage that needs to be carried in the aircraft's hold, then the passenger must still queue up at the manual check-in counter, hand over the baggage and collect a baggage check.
- **Settlement and accounting** In terms of the settlement process, e-ticketing allows settlement to be effected by current BSP processes. These systems had to be modified slightly in order to support e-ticketing, although this work has now been completed as far as the UK's BSP is concerned. For example, the most obvious modification was that the ticket image passed by GDSs into the BSP systems had to possess a special indicator designating it as an electronic ticket for which there was no paper equivalent. A similar identification had also to be present for refunds. In fact, many GDSs have already built automated e-ticketing refund processing into their back-end systems. Otherwise, e-ticketing from a sales reporting and settlement perspective is fairly straightforward. The processes involved are not significantly different from those used for paper tickets (see Chapter 7 for a description of BSP).

Finally, let's briefly consider the situation where BA customers book their flights directly, for example, via a BA telephone sales unit. These customers have not had any contact with a travel agent. However, they can still use the self-service devices at the departure airport to: (a) collect their travel documents as described above, and (b) pay for their tickets using their plastic cards.

E-ticket benefits and the future

E-ticketing offers many potential benefits to travellers, travel agents and the airlines. It cuts down on paper processing, eliminates the 'I've lost my ticket' problem, allows last minute changes, i.e. right up to 30 minutes prior to check-in, eliminates the need for TODs, simplifies the refund process and even has indirect benefit to airports in terms of easing congestion at check-in desks. Looking to the future, e-ticketing holds the promise of delivering vast improvements in the quality and level of management information available to airline customers. However, e-ticketing does require a fair investment of resources of airlines. For a start there is the construction and operation of the self-service check-in machines and lobbies. But more importantly perhaps, there is the investment needed to enhance the airline and GDS systems to support e-ticketing.

While this may not be quite so challenging for an airline's domestic routes, a whole new dimension of complexity is introduced when international routes are considered for e-ticketing. The reason for this is the existence of inter-lining agreements that airlines have negotiated with each other. What happens, for example, when a passenger on an overseas journey 'holding' an e-ticket wishes to change their flight for another airline that does not support e-ticketing? The administrative complexities and opportunities for endless bureaucracy are substantial. So, e-ticketing will no doubt come to pass, certainly for many domestic routes and even for flights within the EC. However, its widespread use around the globe may have to wait a few years. This is why for the moment, e-ticketing is only possible in the UK for BA customers with less than four sectors, none of which may be international flights or flights involving other carriers. Some of the other major issues that e-ticketing faces are as follows:

- **Commission levels** Many airlines claim, quite justifiably, that e-ticketing reduces the amount of work that a travel agent needs to do, e.g. print the ticket and process the ensuing settlement. It has been estimated, for example, that issuing an e-ticket takes an agent 20 per cent less time, compared with issuing a conventional ticket. Airlines are therefore cutting, or at least discussing the possibility of cutting, agents' commission levels, e.g. SAS and Lufthansa have already reduced the commission on domestic e-tickets to 5 per cent.

- **Direct sales** Some travel agents suspect airlines of trying to use self-service ATM ticketing machines to build a customer prospect data base and then to use this to solicit these customers for new business. Although most airlines deny this, it is no doubt an issue that will only be resolved over the long term as new distribution channels evolve (see Chapter 5 for a discussion of disintermediation).
- **Legal issues** Will the terms and conditions of flight, as agreed at the Warsaw convention be communicated to passengers effectively? Because this will only be a separate piece of paper and not an integral part of the ticket as is currently the case, will travel agents always remember to give this to their customers and will passengers pay any attention to it?
- **Customer acceptance** Some customers will always feel more secure with a piece of paper and may in fact demand that their travel agent produces one for them.
- **Airport security** There remain many airports around the world where people are not allowed to pass into the 'air side' of the terminal unless they hold a valid ticket. The location of the self-service check-in lobby is therefore a critical issue for airlines. In most cases they will therefore need to be installed on the concourse before airport security.
- **Interlining** This refers to the flexibility currently enjoyed by passengers using airlines that have an Interline agreement. Such agreements allow a passenger to travel on another airline instead of the one shown on the ticket. The problem is that in order for this to be effected, the validating airline needs a copy of the ticket in order to prove that the passenger has the right to travel on their airline.
- **Shared check-in machines** At present each airline that supports e-ticketing in a country needs to install their own self-service device within the airline terminal. In the UK, British Airways has already done this and British Midland is thought to be planning to install similar equipment. How long will it be before airlines start sharing each other's self-service machines? This would make a lot of sense from a cost angle and from a logistics angle, i.e. in other words it could save a plethora of different and incompatible self-service devices from being installed in UK airports.

It is estimated that 20 airlines that participate in the UK's BSP will move to e-ticketing over the next seven to ten years. These airlines have the required level of in-house technology to make e-ticketing a practical reality; and because these airlines tend to be the larger ones, it is not surprising that they represent about 60 per cent of the UK BSP's ticket volume. Looking this far ahead, it is also possible to forecast e-ticketing for non-air carriers. However, despite these glimpses of a paper-less future, it is highly likely that travel agents will always need to keep a stock of hard-copy airline tickets for certain situations.

Satellite ticket printing

A type of printing of tickets, invoices and itineraries that is frequently used to service large business travel customers is known as satellite ticket printing (STP). It allows a small inplant location to produce airline travel documents without the need for travel agent workstations.

STP works like this. First, the central travel agency must have a fully functional reservations terminal and a data line connected by a communications controller to the remote inplant location. Then the inplant must have at least one printer that is connected via its own communications controller to the central travel agency site. It is usually best for the inplant to have three printers: one for airline tickets, one for itineraries and a third for hard copy. It is then possible for reservations made by the central travel agency location to be channelled to the inplant for document production, Because the inplant is usually situated right on the customer's premises a very speedy service can be provided.

It is important to remember, however, that the remote inplant location must obey all the relevant IATA rules for the holding of airline ticket stock. This includes, for example, the need for a safe on-site in which to store the blank tickets, fully trained staff and adequate security arrangements. In other words the inplant must hold a full IATA licence. If, however, the remote location is using an ATB printer and ticket stock then the process is a lot simpler. With ATBs, only the passenger's ticket need be printed remotely. The other parts of the ticket for the agent's files, for example, can be directed to a printer at the agency's main office. This does away with the need to have specially trained staff at the remote satellite location who know how to process the various copies of the traditional OPTAT stock. The net result is lower operating costs for the agent and a speedier service for the customer. STP is also available on a European-wide basis. For more information on Euro-STP, please refer to the section on BSP contained in Chapter 7.

Intelligent ticketing

There is little doubt that an industry standard for the electronic encoding of travel ticket media is long overdue. The Association of European Airlines (AEA) recognized this in 1996 and instituted a study of the possibilities and feasibility of establishing standards for a common ticket that could be used for several different forms of transport. Given the appropriate standards, a single multi-modal ticket could be used by a traveller, for example, to fly from Madrid to London, stay overnight in the city and then catch the Eurostar train to Paris. Or by a transatlantic passenger arriving at London Gatwick and using the shuttle train to Victoria station in London. There are even better examples of how a single multi-modal ticket could be used in Scandinavia to travel between bordering countries by air, rail and, especially, ferry. A multi-modal ticket with data recorded on it either in the form of a magnetic strip or an embedded chip, would enable the passenger to pass through machine-controlled gates and obtain boarding passes for travel.

Unfortunately, the AEA study was terminated, mainly because there was insufficient commitment on the part of some leading travel suppliers but also because the rate of change in the industry is so rapid. In such a turbulent environment, who is going to try and 'put a stake in the ground' and commit to a set of standards for a media that may no longer exist in the next few years? Then of course there are the inevitable competitive issues. This is especially true now that rail services compete so directly with airlines between, for example, London, Paris and Brussels. To illustrate this, the new cross channel rail services have taken between 20 per cent and 30 per cent of the market on these routes, from the airlines. In any event, many rail companies are embroiled in developing their own set of standards based on the ATB. Their objectives are to allow travel agents to produce rail tickets from the same printers that they currently use to produce airline tickets. Finally, there is e-ticketing, which threatens all forms of paper-based tickets in the longer term. If e-ticketing becomes truly widespread then physical airline tickets could become a thing of the past. Consequently, the intelligent ticketing project never really got off the ground.

Smart cards

Despite its demise, the goals of the intelligent ticketing project live on in the form of smart card technology. Indeed, many of the standardization objectives of intelligent ticketing are as relevant to this rapidly evolving area as before. A smart card is simply a plastic card in which a miniature computer processor chip with its own memory, is embedded. The smart card can be read and/or updated either: (i) by passing it through a swipe device just like a normal plastic card, (ii) by inserting it into a socket with pin connectors, or (iii) by its proximity to a remote reader/processor. Smart cards could solve the problems that intelligent ticketing sought to address. They could also help support the spread of e-ticketing and some of the new travel distribution technologies, many of which are explored in this book. Evidence of the potential opportunities that are offered by smart cards may be drawn from several industry wide experiments currently taking place. For example, the experiment that IBM, American Express, American Airlines and Hilton are undertaking. This aims to test the ability of a general purpose smart card that could, in the first instance, be used primarily for business travel.

It is interesting to speculate how travellers would be affected by smart card based services, in the future. One possible way is for the service to be based on an Internet site sponsored by either a GDS, a card issuer or even a joint venture between two or more travel and financial institutions. Customers of the service provider would be issued with smart cards that would be registered uniquely in their name. The customer wishing to embark on a trip could, for instance, connect to the Internet, sign-in to the site supporting the smart card application and then link to a GDS booking engine. A booking could be made in the GDS and the booking record downloaded for storage in the computer chip embedded in the customer's card. Downloading could be accomplished either by means of a specially adapted telephone or by a smart card reader device attached to the customer's PC. Then, with special smart card readers installed in airports, hotels, rail stations, retail outlets, ATMs and other locations, the smart card could be used to:

- Register at the airport terminal for e-ticket processing and boarding pass collection.
- Check-in at a hotel by using a self-service ATM type device to select a room and obtain an electronic room key.
- Check-out of the hotel via the self-service machine without needing to queue at the hotel's check-out desk.
- Swipe the card through rail ticket barriers to board a train.
- Obtain cash more securely from ATMs by using the intelligent authentication features of the smart card.
- Spend electronic cash by using the stored value features made possible by smart cards, at designated retail outlets adapted for this purpose, e.g. the Mondex experiment undertaken in Swindon, UK.
- Purchase goods, services, restaurant meals, fuel and other retail items, using the enhanced point-of-sale authentication and credit limit monitoring features of the smart card. This could use local card chip processing to authenticate items under a certain value with on-line authorization being required only for purchases exceeding an upper threshold, depending upon the card holder's personal profile as set by the card issuer. This could dramatically reduce the amount of on-line traffic, and therefore cost, that card issuers incur to run real-time authorization systems.
- Store personal health data in the card for use while travelling.
- Record loyalty reward points (this is already being done by a leading German airline).
- Store a full transaction profile of the card holder that would enable: (i) the card holder's travel and financial services to be finely tuned to meet their specific personal requirements, and (ii) allow suppliers to carry out some very highly targeted marketing aimed directly at the individual.

There are many more uses for smart cards but there are also a number of issues that will need resolution before much progress can be made in this area. One significant issue is the opportunities that smart cards offer to reduce the amount of fraud that is currently experienced with magnetic strip based plastic cards. It has been said that these cards have been severely compromised through counterfeiting. Interestingly, the introduction of smart cards in France has seen the level of fraud decrease substantially. This is achieved through improved authentication methods at the point-of-sale, built-in credit control and a higher barrier to counterfeiting. The ability of the smart card to be updated as it passes through the travel consumption chain is also a factor that helps reduce fraud and enhance service levels.

While it is true that the magnetic strip encoded onto ATBs can also be updated as travellers are processed through airline check-in desks, there are constraints on the future use of this technology, for example: (a) the amount of information stored and updated on an ATB is limited; and (b) the more widespread use of intelligent ATB processing outside of air, rail and ferry services in the future, is unlikely. One of the reasons for this is the cost of the equipment needed at the service point. In fact, the investment in point-of-service processing is probably the primary obstacle to more widespread use of smart cards. In the area of travel, there is no doubt that smart card readers are cheaper than ATB ticket printers. However, in the context of the global retail distribution chain, the travel agency sector is relatively insignificant. Even a comparatively small add-on cost affecting every point-of-sale till would require a very large investment of capital indeed; and who would pay? The obvious answer is the card issuers, because they stand to gain the most from the consequent reduction in fraud. However, there are many competing arguments on this topic and it is a game that is being played for very high stakes. The dual issues of cost and funding therefore represent the principal barriers facing smart card issuers as they consider the more widespread use of this technology.

Another fundamental issue is whether detailed information is actually stored in the smart card itself or whether the card simply contains a key to information that is held in an external computer system. The GDS booking record or PNR is a particularly good example. From a technical viewpoint, the PNR could be stored completely within the smart card. However, if this is done then a number of related issues arise. For example: 'Who owns the data stored on the smart card?' If it is owned by the airline (or GDS), then could the card issuer process the data, for example, to produce integrated management information for customers? Also, how would the data be synchronized between that stored within the smart card and that stored within the GDS? Airlines sometimes have to change the booking records of their customers to reflect a service change. How would this new information update the travellers booking record if it was stored on their card? An alternative approach is for the smart card to store just the PNR locator that could then be used to retrieve the full booking record from the GDS. However, holding only a key might seriously constrain the opportunities for fully exploiting smart card technology. These are just a few of the reasons why airlines and GDSs are reluctant to give up overall control of the booking information that they store on each of their customers.

There is also the fundamental issue of: 'Who owns the smart card service?' If this is to be the bank card issuers then it will be interesting to see how alliances with airlines, GDSs and other travel suppliers develop. Finally, there is the biggest issue of all – standards. Standards are pivotal to the success of smart cards in travel. Without standards, the uses that I have identified above would not be feasible. Standards that define how the data are recorded within the chip and what form of security and encryption techniques are used. The problem is that these standards need to be agreed throughout the travel and financial services industries and between competitors. A very challenging issue, but one that should not be ruled out as unattainable. After all, when the stakes are high in terms of financial pay-off, there is an enormous drive for standards to be agreed. There are anyway, plenty of examples of intra-industry, intra-competitor collaboration in the standards field, e.g. EDI and the current magnetic strip plastic cards. So, the signs appear to be quite encouraging; and a great deal of groundwork has in fact already been done in establishing basic standards for the multiple use of ATBs, electronic tickets and smart card data storage. Nevertheless, this whole area is one that is rife with issues although at present, i.e. mid 1997, it is also bereft of solutions and definite directions. Smart card technology in travel and tourism will be one to watch over the forthcoming few years.

Hotels

There is little standardization in the area of hotel systems. The kinds of technology used throughout the hospitality industry, while bang up-to-date, vary widely depending upon the size and type of hotel. However, there is a lot going on in the area of IT within the hotel industry. Some of it is in the high profile area of distribution and the Internet is a prime example (see Chapter 5). But hotel back-office systems are becoming increasingly sophisticated and are now widely recognized as being a key to improved profitability; and it doesn't stop there. Guest services are the current focus of attention and there are some really interesting experiments being conducted on a large scale. For example, Hilton, American Express and IBM are jointly exploring the opportunities offered by smart cards. Finally, in-room technology is rapidly becoming a very real competitive differentiator, especially at the top end of the hotel market. So, first of all, let me try and categorize the different ways in which hotels implement their systems. These really fall into three main areas: (a) in-house systems owned and operated by the hotels themselves, (b) packaged systems that are purchased from software companies, and (c) outsourced systems where a third party will run all or some of a hotel's application functions. Let's take a look at each of these in a little more detail:

- **In-house** Many of the larger hotel chains have managed to develop their own in-house systems over a period of several years. Some of these systems have been purchased from other hotels and even other suppliers in related industries. They have been modified to suit chains' individual needs and have developed into an important operations and distribution capability upon which their businesses have become highly dependent. However, not many hotels develop their own systems these days; it is considered far to risky and expensive a venture, which can detract a hotel from its core business.
- **Packages** Small to medium sized hotels may also develop their own systems but more usually, they purchase a pre-packaged system available from one of the many software companies specializing in the hotel sector. There are now a wide range of packaged systems available to hotels, each providing a specific function. I'll be taking a look at these functions and different types of packaged systems in more detail in a moment.
- **Outsourced** Some hotels find the outsourcing option very worthwhile. There is now a wide choice of different outsourcing options available to suit hotels seeking to automate specific areas of their operation. This can be attractive to smaller hotels that do not wish to invest in running systems themselves.

I am going to concentrate primarily on the packaged and outsourced solutions in the remainder of this section. The reason is that in-house systems are by their very nature unique and therefore not really representative of the industry. It is the packaged and outsourced systems supplied by third parties that are particularly relevant and interesting. The blanket term often used to describe the systems that support hospitality industry automation is a property management system (PMS). However, this terminology has been somewhat abused and there are in fact many more functions that can be automated by specialist application software. Let me put PMSs into context with other systems that are frequently used throughout the hospitality industry:

- **Property management system (PMS)** This is probably a hotel's core system. It maintains all details that describe the hotel, such as the size, type and number of rooms, the rates applicable for rooms and other facilities, and all details concerning guests. It supports guest check-in (also known as front desk support) and check-out functions, which produce guest bills and record payment. At any point in time, the PMS may be used to determine the status of a room, e.g. clean and vacant or not yet made-up. Hotels usually implement their PMS on an in-house computer. That doesn't mean to say they usually have an in-house IT department to support it. Most PMSs are turn-key systems that, as the name implies, are simply turned on and off by hotel staff.
- **Self-service kiosk systems** This is a new area for many hotels and is the subject of much experimentation at present. Self-service kiosks are an attractive way for a hotel to reduce guest check-in and check-out procedures while at the same time reducing the load on the front desk staff. It is, however, widely recognized that self-service kiosks will not be used by all guests. Nevertheless, business travellers who are becoming increasingly computer literate may wish to swipe their cards through a self-service machine, select the check-in menu and then choose their rooms from those currently available. It is quite possible for these self-service kiosks to issue an electronically encoded room 'key' and a printed set of personalized instructions for the guest. If this saves a tired business traveller 15 minutes queuing at the front desk and going through the old familiar check-in question-and-answer session, then the traveller may well prefer the self-service kiosk. Especially if the same kind of time savings can be enjoyed on check-out.
- **Reservation systems** These systems control the forward inventory of hotel rooms. The hotel maintains an inventory of the rooms available using its PMS. This is complemented with

additional information, such as periods when the room or rooms are blocked out as unavailable. The reservation system allows operators, or more generically, users, to allocate a room for a specific period of time to an individual. There are many instances where hotels outsource their reservations function. The reason for this is that reservation systems are expensive. They require non-stop computer systems, on-line applications and plenty of capacity in order to ensure busy times are covered. This all adds up to high operating costs. Sharing these resources with others via an outsourced operation can be attractive to small to medium sized hotels.

- **Point-of-sale systems** These sub-systems comprise payment handling devices that produce bills and accept various financial instruments from guests, e.g. credit cards, cash, guest cards and so on) Depending upon the hotel's size and level of sophistication, these point-of-sale devices are linked by communications lines to a central communications processor that in turn is linked to card authorization systems. The systems also feed transaction data directly into the hotel's accounting system, which is often integrated within the PMS.
- **Food and beverage systems** These systems control the food stock required for the hotel restaurant's various menus. It allows changes in suppliers' prices to be evaluated against projected customer demand for certain menu items and, as a result, enables the restaurant to optimize its order position with alternative food suppliers.
- **Back-office systems** This is the term given to a set of systems that records all transactions, and maintains the hotel's books of account. These systems receive transactions from the other sub-systems, as described here, and compile data on the hotel's expenditure and income. The systems support the purchase ledger, i.e. accounts payable, the sales ledger, i.e. accounts receivable, and the general ledger. A full set of financial accounting reports is usually included within this suite of systems.
- **Sales and marketing systems** These systems store a vast amount of information on virtually all the key parameters that are used to measure the accommodation and hospitality business. Of particular importance is a customer data base that can profile individual guests and is a powerful source for pull marketing (see Chapter 5 – The Internet). Another important function is yield management. This enables a hotel to review the past performance of its rooms, e.g. occupancy levels, average length of stay), and set the optimum rates that will maximize the property's yield management programme.
- **In-room systems** Successful hotels are constantly striving to adapt their services to meet the needs of their customer base. These days, hotel guests, particularly towards the upper end of the market, are becoming increasingly technologically oriented. Business people need to access the Internet, principally for e-mail purposes, during their business trips. They sometimes find video-conferencing useful, especially when a company holds a business meeting in a hotel; and there are many other examples. While not all hotels will cater for sophisticated IT services, there is a pressing need for basic resources to be made available to business guests, such as Internet plug-in points, power sockets for portable PCs, in-room PCs, faxes and overhead projection facilities.

As you will no doubt gather, there are a wide diversity of support systems that need to be used by hotels. The problem is that there are very few software suppliers that can deliver an all embracing and fully integrated solution to a hotel's automation needs. While one software company may be particularly strong on core PMS systems, its food and beverage package may be weak; and so, hotels are forced to go to different software suppliers for different applications. Naturally, there are several software company's that claim an all round automation capability; but, nevertheless, hoteliers are faced with the issue of: 'Whether to choose a reasonable all-round system from a single supplier or choose several specialist application packages from a number of software suppliers?' This is not an issue that is easy to resolve; and it leads on to the second problem faced by hotel management wishing to automate – standardization.

I have discussed standardization before in the context of several areas of travel and tourism automation (see Chapter 1 and TTI). Sadly, in the area of hotel automation, there is very little compatibility between sub-systems supplied by different software companies. Therefore, unless a hotel chain puts all its eggs in one basket and opts for a single software package providing all-round functionality, then it faces a serious obstacle to future growth. Even if a single package is chosen, the situation can still mitigate against the hotel's original choice. Say the hotel branches out into an area of the business that is completely new. Its single system chosen by the hotel in its formative years may not provide the required functions for the new service. So, another software supplier's system needs to be obtained and the hotel is straight-away faced with the old problem of incompatibility: 'How can the new system feed data to the old system so that management can enjoy integrated accounting tools and overall management control of their entire business?'

The problem of non-standardization seems to be just as bad all around the world. There seems to be virtually no consistency between different systems. This problem has often inhibited mergers and acquisitions between hotel groups. After all, how can a hotel buy another if its entire inventory and room management systems are entirely different. OK, they can be changed and one hotel's systems can be migrated to the others. However, this is a costly game to play and there have been instances where the migration effort has had to be abandoned at great cost. Only in the USA does there appear to be light at the end of the tunnel. The Hospitality Industry Technology Integration Standards (HITIS) project has been initiated by the American Hotel and Motel Association. This group has been charged with evaluating the possibility of establishing standard interfaces between various types of hotel systems, e.g. between PMSs and POS devices, between PMSs and central reservations systems, between PMSs and food and beverage systems). The only drawback is that there is no representation on this body by either Europe or any of the other major trading blocks around the world. However, the project represents a serious attempt at addressing the standards issue, as evidenced by its 23-strong committee including three members from Microsoft. Its mission statement is:

'To direct a non-proprietary, consensus based process to develop voluntary standards for the integration of evolving computerised system and sub-system transactions in the hospitality industry. The process will seek to synthesise and disseminate previous and current efforts to resolve such issues and allow hospitality operators and vendors alike to save on costly retrofitting solutions and enable faster technology adoption and evolution within the various segments of the hospitality industry.' More information can be found at http://www.hitis.org.

This has been a quick overview of the kinds of systems that hotels used to manage their operations and control their inventory. I won't be discussing distribution systems in any more detail here because this topic is covered by several subsequent chapters, i.e. GDSs, videotex and the Internet. However, there are a few other miscellaneous aspects to hotel automation that you need to know about before we embark on the alternative distribution channels. I have discussed each one below:

- **Room rates** There are many different types of room rates that hotels offer to various categories of guests. Business travellers often pay peak rates because their visits are booked at short notice, during peak times of the week and they require a higher standard of accommodation. However, partially offsetting this is the willingness of hotel chains to grant special rates to certain companies that deliver a high volume of bed-nights to their hotels. By contrast, holiday-makers book well in advance, stay at off-peak times, e.g. weekends, and do not always require such a high standard of room. Leisure travellers are often able to enjoy special bargain breaks and deals that involve complimentary or reduced price meals. Each type of guest is therefore allocated a different type of rate by the hotel, even if they stay in the same room. So, in order to understand hotel systems, it is worth spending a little time introducing some basic terminology in relation to room rates. It seems that each industry has its own jargon and buzz words and the hotel sector is no different in this regard. These may be summarised as follows:

- *Rack rate* This is the full rate as published by the hotel. It is the rate for a room that does not include any discount at all. In other words it is the rate that you or I would probably be quoted by the hotel if we were to walk in off the street and ask for a room.
- *Corporate rate* This is a special rate that is strictly only available to business or special users of the hotel. It is a rate that a hotel may give under its own discretion in exceptional circumstances.
- *Negotiated or preferred rate* This is a rate that is negotiated by a travel agent or a company directly with a specific hotel. It is usually dependent upon a minimum amount of business volume and is available only via the party that negotiates the rate. A travel agent with a large business house account can often negotiate a special rate on behalf of his/her corporate customer.
- *Promotional or weekend rate* These rates are only available via a package that the hotel arranges and provides for its guests. Packages often include a meal plan and some other activity organized by the hotel. Promotional packages are usually only available at weekends and are aimed at filling rooms unused by business guests that are the mainstay of a hotel's trade during the working week.
- *Net/net rate* This is the lowest possible rate and as such is one for which the hotel does not pay a commission. Some hotels will, however, supply certain special customers with rooms on this basis. Despite the fact that these bookings represent loss making transactions, they are nevertheless supported in order to provide an all round service. Many reservation systems do not support a net/net rate because there is no way for the hotel to recover the cost of making the booking.

- **Assured bookings** In the past it was sometimes the case that hotel reservations were subject to the dreaded 'lost booking' syndrome. This occurred when a travel agent made a reservation for a customer in a point-of-sale system that appeared to have worked perfectly. The trouble used to be that the travel agent's message to the hotel was never successfully delivered. Consequently, when the customer turned up, there was no reservation and everyone involved in the booking blamed one other. Once a problem like this is experienced, the word soon spreads that hotel systems are unreliable; and it takes a long time to re-build lost confidence. Since then, distribution systems have improved dramatically and hotel switches have virtually eliminated this problem.
- **Travel agents' commissions** As far as hotel commission payments are concerned, travel agents used to regard themselves as lucky if one in ten bookings resulted in a commission cheque being received from the hotel. From the hotels' perspective, they were faced with dealing with hundreds or perhaps thousands of very small commission cheques that were mailed to hundreds of travel agents all around the world. Travel agents had to deal with individual cheques worth only a small amount each and often written in a foreign currency. All in all, neither the travel agents nor the hotels were very impressed with the old hotel booking and administration process.

However, a lot has been done to solve these problems over the past few years. The hotels have realized that in order to obtain a greater proportion of their bookings from travel agents they need to make the process as simple and as accurate as possible; and there is substantial potential to be gained by hotels from travel agents. At present only about 28 per cent of hotel bookings are generated by travel agents. The remaining 72 per cent are booked directly by consumers or companies. One of the facilitators that will enable the travel agency distribution network to be a prime source of hotel business growth is the widespread use of GDS technology. This has tended to not only cure the problems I outlined above but has made it less costly and therefore more profitable to handle hotel reservations, provided they are done using computer terminals and not over the telephone. A telephone call takes a lot of selling time and can cost a lot of money.

The key issue for a hotel is yield management. This is a term used to cover the process of applying the most effective rate to a room in order to maximize overall revenue for the hotel. This is a balance between: (a) charging the highest possible room rate, but running the risk of fewer bookings; or (b) filling all hotel rooms, but at a lower rate. Obtaining the optimum balance is an exercise in pure economic theory. Too high a room rate and you will not attract sufficient numbers of guests to make a profit. Too low a rate and you will not receive sufficient income to cover your operational costs. Reservation systems have helped hoteliers with this problem. Systems have enabled hotels to shift the focus of their rates away from fixed room types each with an associated rate, and instead to move towards having individual rates for each room.

Take, for example, a hotel with 20 rooms, of which five are superior suites, ten are twin bedded rooms and five are single rooms. The classical approach taken by the hotel to market these rooms was to assign a rate to each of its three room types for each season of the year. Although this works OK in practice, it can be inflexible, particularly in times of either high occupancy levels or lulls when many rooms are empty. What many hotels are now doing is to designate all their rooms individually, each with its own rate that varies from time to time. In our example the hotel would describe each of its 20 rooms individually and assign a price to each one. Each room is described in terms of, for example, a room with two beds having a view of the sea from the top floor with a South facing balcony. The room would be assigned its own particular room rate that would vary not just by season but interactively depending upon the hotel's occupancy levels at any point in time. This approach is more flexible and allows the hotel to achieve a far higher degree of control over its rates depending upon current and forecast occupancy levels. So, if eight of its old style twin bedded rooms were taken, the remaining two could be priced slightly higher. Pure market economics.

Now this would have been rather difficult to achieve before the advent of computerized reservation systems. But with an on-line system that can be updated rapidly using computer terminals, an individual room can easily be assigned a rate that varies on, for example, a daily basis. Another supporting factor has been *seamless connectivity*. With *seamless connectivity*, end users accessing the hotel's reservation system via a GDS can see their displays exactly as they are formatted by the hotel's own system. In the past, GDSs have edited hotel displays because they could not show them in the same format as the screens that the GDSs use for airline availability. Most GDSs and other booking channels, such as the Internet, now support *seamless connectivity* to hotel systems. Hotels that have adopted flexible pricing for individual rooms have enjoyed a higher yield, which should mean increased profit. It does, however, place increased demands on hotel general managers to possess marketing and entrepreneurial skills in preference to classical abilities focused on running an efficient hotel operation.

This has been a brief look at how hotels use IT to automate their in-house operations. As I have mentioned before, it is important to understand these systems because many of them provide a platform for alternative distribution channels which are the subject of subsequent chapters. Finally, let's take a closer look at a particularly good example of a hotel chain that uses technology very effectively – Marriott Hotels.

MARRIOTT

Marriott International is a hospitality management company. Lodging products and services include International hotel brands such as Ritz Carlton, Marriott Hotels and Resorts, JW Marriott luxury hotels, Courtyard by Marriott, Fairfield Inn & Fairfield Suites by Marriott, Residence Inn and Townplace Suites by Marriott, Executive Residences by Marriott, Marriott's Vacation Club International, Renaissance International, New World, Ramada International and Marriott Conference Centres. One of the key success factors in Marriott's growth over the past few years has been their central hotel reservations system called MARSHA (Marriott's Automated Reservation System for Hotel Availability). MARSHA was developed in the 1970s and used airline reservations systems technology as its platform. The software which supports this platform is IBM's TPF (Transaction Processing Facility). TPF is a sophisticated piece of software designed to run (not surprisingly), on IBM main-frames. Its key strength is its processing efficiency which enables extremely fast response times to be delivered to large populations of reservation operators using main-frame linked terminals. One of the ways it accomplishes this is by the use of clever coding systems which reduce the size of a reservations message and its response to just a few characters. However, as we shall see later, this technology, although still very fast and efficient, does not lend itself quite so well to graphical based GUI systems and messages which comprise large amounts of data.

MARSHA is used by Marriott as the group's central reservations facility. As such it is connected to all of the world's major GDSs (Global Distribution Systems) which also operate on a TPF platform, and from these systems to PC terminals used by travel agents at the point-of-sale. They also connect to some lesser volume producing GDS's and the Internet via the Thisco switch (see Chapter 5: Distribution Systems) MARSHA is also accessed by Marriott's central telephone reservations operation which is housed in a purpose built centre in Omaha Nebraska in the US. This centre, chosen for its central geographic location within the US, houses around 1,600 telephone reservations operators who work shifts to provide round the clock service worldwide. Each operator uses a telephone headset to receive incoming telephone calls and is able to provide customers with live information on all of the chain's properties around the world using their MARSHA computer terminal.

MARSHA has been an excellent platform to support the growth and development of Marriott's hotel brands and its associated businesses. One of the basic reasons why Marriott have been so successful in exploiting technology to support their business is a clear information systems strategy. The keywords here were and still are, uniformity and standardisation. Marriott always made it a rule that any property or franchise which joined the Marriott chain must have a PMS (Property Management System), which is compatible with MARSHA and which can support a two way automated link between the two systems (i.e. between the property's PMS and Marriott's MARSHA). This is a fundamental axiom which guarantees last room availability to Marriott's customers. This is so important that it is worth explaining it in a little more detail. When a Marriott telephone reservationist uses MARSHA to inquire on the status of a room in a specific property, they may for example see that only one room is left for sale on a given date. If they are going to be sure that they can sell this room, they must be able to rely 100% on the accuracy of the MARSHA data base. This is possible only because the central MARSHA reservations system is in a continual two-way dialogue with the PMS used by the property concerned. So, if the PMS has one room available then Marriott's automated reservation system for hotel availability (MARSHA) will show that room as available for sale. Once the reservationist sells that room for the date specified, it is not only recorded as such in MARSHA but is also immediately communicated to the hotel property's PMS. The two systems are therefore always in synchronization, thus guaranteeing the quality of the information available to central reservations. Today, over 98 per cent of Marriott's properties use an on-line two-way link between MARSHA and their PMSs.

The benefits of Marriott's reservations systems standardization strategy have also paid off in the area of GDS interconnection. In the early days of the 1980s when GDSs were known as CRSs and they focused almost entirely on airline reservations, the inclusion of hotel and car services was a revolution. So, when Marriott established a direct connection to the System One CRS in 1987 it was breaking new ground. This was followed in 1989 with a fully automated link to Sabre in 1989. Marriott connected MARSHA to the Apollo, PARS and Abacus systems progressively over the ensuing two years. This GDS–CRS interconnection was no trivial matter for either parties. There were significant developments that the host CRS–GDS systems had to fund and there were similar modifications also required to MARSHA. After all, when you think about the basic products, i.e. airline seats and hotel rooms, they are fundamentally different in many ways. Airlines have fields such as departure–destination city pairs, and a number of seat classes: hotels have a range of room grades and special accommodation packages. It is because the two products are so different that these modifications had to be made to the systems involved. Eventually, Marriott, like many other hotel chains, decided to establish a single connection to a hotel industry switch called Thisco. Thisco could then take on the onerous job of interconnecting to the GDSs on behalf of their hotel members (see Chapter 4 for a full description of Thisco). Today, electronic reservations generated by all hotels via the GDS channel have grown from 25 million in 1995 to over 33 million in 1996.

Clearly then, Marriott recognizes the importance of its reservations system in supporting profitable, high quality, customer service led growth. The company continually strives to enhance and develop MARSHA further to provide its customers and users with a higher level of service. Over the past few years, the population of MARSHA's GDS travel agency users has grown significantly, especially in the USA where 80 per cent of all agency hotel reservations are now made via GDSs. The problem has been that this percentage is far lower outside the USA. In Europe, for example, during the early part of 1995, travel agents made only 2 per cent of their hotel bookings via GDSs. Although this has grown steadily to over 50 per cent by 1997, a continual programme of enhancements is needed to drive this growth further in the international arena.

One such enhancement goes under a number of terms, such as Galileo 'inside availability' and Sabre 'direct connect availability' (again, see Chapter 4 for more information on these GDS terms). However, they all refer to the 'seamless' functionality that allows a GDS user, i.e. a travel agent, to use their terminals just as though they were connected directly to MARSHA. There is no translation undertaken by the GDS and the travel agents see the displays just as MARSHA formats them. To support this, MARSHA was upgraded to MARSHA III in 1995. The new system now supports full room descriptions, rate details and many other important fields just as they would appear on a native MARSHA screen. This is particularly important when a user is discussing a customer's requirements and trying to decide on a room type. Does the code for superior twin-deluxe room in one system mean the same thing in another, for example. With the new comprehensive MARSHA descriptions, the review and decision process can be made in the comfort that it is based on higher quality information.

Despite the success of Marriott's automation programme, today's consumers expect the systems they use to support full multi-media technologies and GUIs. Although travel agents may be happy to continue using text based GDS systems for some years yet, even they may eventually come to expect a next generation reservation system. This is one of the reasons why Marriott has expanded MARSHA to provide a full information, reservation and payment facility using the Internet. This whole area is explained in more detail in Chapter 5 – The Internet.

Tour operators

Tour operators are in the business of combining travel products from several suppliers into unique holiday packages that are marketed to consumers. So, they are: (a) very much in the leisure business, and (b) highly dependent upon successful inventory control for their profitability. As with airlines, most of their sales originate via travel agents (although some direct sell tour operators are beginning to emerge). It is therefore vital that a tour operator's performance objectives include such factors as: the availability of systems that enable internal operations to run effectively, networks that enable their products to be booked easily by travel agents, and reporting systems that enable information on sales to be tracked on a current basis. IT is a key success factor in helping tour operators achieve these performance objectives.

Tour operators aim to save their customers the hassle of booking their own flights to their destinations, searching for the best and most affordable hotels and coping with all the ancillary chores that go hand in hand with foreign travel. Chores such as baggage handling, transportation between the airport and the hotel, tips, sightseeing tours and, last but not least, the language. The tour operators do this at the lowest possible cost by buying in bulk, packaging all the required services into an all inclusive holiday and selling this package to customers via travel agents.

The process starts a long way in advance of the date that the holiday will actually take place and it commences with the holiday design phase. The tour operator's marketing staff decide upon the type of holiday that they think: (a) will fit the operator's image and reputation, (b) will sell sufficiently well enough to make a decent profit, and (c) will appeal to the type of customer that the operator is targeting. Having decided upon the holiday design and specification, the operator will despatch its buyers to the destination resorts to negotiate the number of rooms needed to satisfy the expected demand for the holidays being sold. In most cases the hotel space is arranged one to one-and-a-half years before the season. Aircraft seats are also obtained a fair time in advance and these are obtained in blocks. Then there is the uniformed representatives to arrange and all the other ancillary services to purchase. While all this is going on, the brochure is being printed. To control this process the operator creates a large inventory of travel products that have been contracted with the various suppliers, e.g. charter airlines, hotels, local transportation companies, etc. This inventory is stored on a computer data base.

In these early stages of a packaged holiday's life, the data base provides the operator with a control mechanism to ensure that the required services are in fact contracted and that the rates and conditions are stored for pricing purposes. The operations staff work in conjunction with the marketing staff to link the services contracted with the tour package to be marketed. This is how the tour cost is calculated and the selling price determined, which in turn allows the inventory to be created. It is this tour inventory that is referenced each time a customer enquires about a booking. Now, let's look at an actual example of a tour company and how it uses IT to control its business operations.

COSMOS

Cosmos makes very effective use of IT to support the distribution of its package tour, coach tour and seat only businesses. But before we plunge into a review of this company's deployment of IT, it is worthwhile considering how Cosmos came into existence and understand a little more of the background that led to it becoming a leading supplier of package holidays. Cosmos is owned by the Globus Group. Globus itself is a private company, which was formed in 1928, and is run from its headquarters in Lugano, Switzerland. Globus focused initially on selling coach tours within Europe. Then in the early 1960s, Globus decided to enter the packaged tour business from the UK and created Cosmos. Soon after this, Cosmos formed Monarch, its own charter airline, to support its new package holiday business. Globus then moved into the USA market with the acquisition of Gateway Tours, which provided both air travel services into the USA linked with coach tours of the country and a growing market for the supply of coach tours of Europe including the British Isles. Globus and its TOURAMA coach programme, is now the largest coach tour operator in the world. It has general sales agents (GSAs) and associates in most of the major destination countries of the world. More recently, Globus acquired Avro, which is a leading seat-only air travel supplier, and in 1995 followed the general industry trend towards vertical integration with the acquisition of a leisure travel agency chain called Apollo Travel. Apollo has five retail outlets and a large telephone sales operation.

It is interesting to consider the way in which Cosmos has deployed its IT resources during the time of its fairly rapid growth. The two main drivers of Cosmos' automation programme have been: (a) communications networks that link tour operations' headquarters with destination service offices overseas, and (b) distribution systems that provide direct access to travel agents for sales support. Let's take a more detailed look at both these areas:

- **International communications** Cosmos has adopted a distributed approach to systems development but with a firm control on technology standards. This has dual benefits: (i) it allows destination areas to have the flexibility to develop the systems they need for their own local operations and (ii) it ensures that the overall business retains a high degree of compatibility across all its different systems. Cosmos uses Microsoft products for local applications development and office automation. This started with the simple need to transmit a rooming list from Cosmos' headquarters in the UK to a destination office in a resort area. This commonly used function has been gradually enhanced over the years using standard Microsoft Access products. It is currently an application that is initiated by the destination office by dialling into the company's computer system and downloading a rooming list file for local printing. The format of the transfer is EDI and this enables the local destination office to import the file into its local data base.

Globus and its Cosmos subsidiary have a growing need for international telecommunications services. The rooming list example I gave above is just one of the many applications that need to be distributed around the world. There are many others. Cosmos needs a telecommunications network that is: (a) available as and when needed, i.e. not necessarily permanently on-line; (b) capable of providing the high transmission speeds that its business applications require; and (c) is cost effective. As business volumes grow at Cosmos, this requirement for a global telecommunications service will increase in importance to the company.

- **Distribution system** Cosmos launched its viewdata booking system in 1988 when it had become clear that videotex was a technology that the majority of travel agents had committed to. The Cosmos main-frame, an ICL VME computer, was enhanced with multiple front-end communications processors. These each have a direct connection into the two leading networks that distribute videotex access services to travel agents throughout the UK (see Chapter 6 for more details on these communications network companies). The configuration used by Cosmos offers a high degree of reliability and integrity due to two main factors: (i) the host main-frame is insulated from the communications processing functions that pose a higher level of interference risk, and (ii) the multiple front-end configuration allows for either a single computer or communications link to fail without necessarily shutting down Cosmos' entire distribution system (see Chapter 6 for more information on Cosmos' videotex distribution system).

Cosmos decided some years ago that because its core business was tour operations, it would outsource its ICL main-frame and front-end computer operations. The overheads in terms of staff levels and skills required to run such an operation are considerable and Cosmos decided that it would be best to let a professional computer service company provide these services. These computers were therefore outsourced to the SEMA Group, which houses the computers and provides a full physical operational environment for them: although the Cosmos ICL main-frame and front-end computers are actually located and run off-site, they are nevertheless controlled and operated remotely by Cosmos' own operations staff. This enables Cosmos to concentrate on its local in-house office technology and overseas destination service support systems. So, the various departments, such as accounting, reservations, administration, contracting and aviation control, all run their own PC networks that are either based on Novell or Microsoft Windows NT network operating systems.

However, outsourcing your computer operations does not remove the responsibility for applications development and maintenance. Cosmos has decided that for sound strategic reasons, it will migrate its main-frame operating system environment and the associated applications to an open systems architecture. This means converting its current main-frame environment to a UNIX-based operating system. This allows for more 'openness' – in other words, a wider choice of hardware platforms and applications support software, e.g. data base management systems. Having reviewed its options for information management, Cosmos has opted for a standard flat file approach for core operational systems rather than using newer technology such as relational data base management systems. The reason for this is that speed and end-user response time are the critical factors that determine Cosmos' required service levels. While relational data base management systems are excellent for data manipulation, e.g. management information and enquiry systems, their inherent flexibility gives rise to a processing overhead. Flat files are better for straightforward fast processing of repetitive transactions in a real-time on-line environment.

The Cosmos tour reservation system, like many others, uses a technology called 'inside access' to review, obtain and confirm scheduled airline seats.

This is an interesting function and one that is worth examining in a little more detail because it will no doubt be used increasingly by tour operator systems in the future. When a Cosmos reservations operator receives a call from a travel agent wishing to book a package tour for a customer, the operator needs to provide instant confirmation of the booking. If the package tour happens to use a scheduled flight then, historically, the only way of providing an instant confirmation was to put the travel agent caller on-hold while a separate GDS terminal was used to check availability and make a booking. This is because airlines are reluctant to allow tour operators to hold allocations of their scheduled flight seats in the hope that they will all be sold. If the tour operator does not manage to sell their allocation, the airline is left with the job of selling these seats, usually at the last moment. This often involves selling the seats at a discount and consequently loses the airline valuable flown revenue. Instead, airlines prefer tour operators to use their GDSs to access their seat inventories and book those actually needed by their customers.

With 'inside access', this is done automatically by the tour operator's system. The reservations operator first uses his/her in-house tour system's reservations function to select the desired package holiday from the inventory. This presents the operator with one or more screens of information about the holiday. Then, when the travel agent enquires as to the availability of the holiday, the operator selects an availability-check function. In the background, the system links directly to the tour operator's chosen GDS, sends an availability request transaction, receives the response and displays this to the reservations operator. This exchange is accomplished entirely automatically using a set of machine-to-machine EDI messages without the operator being aware of the dialogue with the GDS. A booking is made in a similar way; the inventory is decremented and a booking message sent to the GDS to reserve the scheduled airline seat. Inside access is now also an integral part of the viewdata booking system and is an important new technology that has helped Cosmos improve service levels and increase the productivity of its reservations operations while also maximizing airline seat revenues.

Like most other tour operators, Cosmos is facing several important business issues, many of which relate to distribution technologies. Because these are closely linked to viewdata and the new Internet-related technologies, I have discussed them in more detail in Chapters 5 and 6. However, I hope this short analysis of Cosmos has given you an insight into the technologies used by tour operators that often also represent their platform for product sales distribution.

Rail companies

By its very nature, rail travel is run on different lines in different countries (you can't argue with that, can you?). This national culture of the railways is reflected in the systems and technology that are used to support the sale of rail tickets. Systems differ widely across different countries and there is no clear or consistent pattern to these systems. So, I shall pick the UK as an example of a country that has a thriving national rail network and that uses IT to sell its product.

Rail travel in the UK

Rail travel in the UK has undergone a radical shift in its core infrastructure during the 1990s as a result of the privatization programme initiated by the British Government. The old monolithic British Rail network has been broken up into two main components: (i) Railtrack – which operates the railway lines, switching points, bridges and stations; and (ii) the train operating companies – each of which operates the trains in various regions of the country for a contracted period of time. In addition to this there are several support organizations that deliver specialized services to the businesses. All of these individual entities are monitored by the rail regulator, an independent authority that is appointed by the Government to ensure fair play, appropriate pricing and the delivery of a consistent service to the public.

Rail tickets are sold to travellers by the train operating companies. Each of these train operating companies are required by the regulator to be a member of the Association of Train Operating

Companies (ATOC). This is strictly an association and is not a trading company, although it does appoint companies to perform certain support functions as part of a commercial arrangement. One of the main functions of ATOC is devolved to a group designated the Rail Settlement Plan (RSP). This group is principally responsible for ensuring that the revenue derived from the sale of rail tickets by whatever means, is fairly and equitably shared among the train operating companies. So, for example, a ticket from Brighton to Edinburgh will be purchased as a single ticket at a pre-set price. However, this ticket revenue must be apportioned between Connex South Central, the London Underground and Great North Eastern Railways. The ability to do this is provided by a complex set of inter-related systems that have been developed over many years by different parts of what used to be British Rail. Responsibility for the operation and development of these systems has been contracted-out by RSP to the SEMA Group, a large European facilities management company.

SEMA is therefore responsible for what used to be known as the British Rail Business Systems (BRBS). The portfolio of systems that are supported by this new group are many and varied. They include: (i) the large and powerful central main-frame computers located in Nottingham and Crewe; (ii) the point-of-sale systems used to distribute ticket sales to travellers; (iii) the information systems that support the retail sales activities, such as the telesales centres; and (iv) the revenue accounting systems that feed the train operating company's general ledgers. Let's take each of these types of systems in turn.

Central main-frames

There are two main-frame computers located in Nottingham and Crewe, each of which is capable of processing the other's work. This provides a fail-safe computing environment that safeguards the network against failure of a single facility. These computers hold the rail inventory that comprises timetables, reservations and fares:

- **Timetable** The operating dates and times of every train run by each train operating company is stored in a large data base called the computer aided timetable enquiry (CATE). This data base is updated daily by an interface from Railtrack's train planning systems. The CATE data base is the core of many other systems, some of which I will be explaining in more detail in a moment.

- **Reservations** This application, known as the central reservation system (CRS), supports train seat reservations and APEX yield management. It contains a great deal of information on stations, routes, restrictions and holds all retailing rules.

  Depending upon the response to a timetable enquiry, as described above, a reservation may be requested. This is made on a special purpose screen that asks for position (window or aisle), direction (facing or non-facing), dining seat, smoking/non-smoking or sleeper compartment. Only those facilities available on the train service requested are shown. Where a reservation is made for a customer, the rail main-frame computer in Nottingham creates a PNR, so that the reservation may be recalled at any future point in time for customer service purposes.

- **Fares** The fares data base is called the central prices file (CPF) and is enormous. It holds over 54 million individual fares. The reason it is so large is due to the commercially oriented pricing strategy established by British Rail many years ago. In essence, this strategy considers every single possible combination of origin station, destination station, route and class in the entire network as a possible individual fare. The rail main-frame computer system that runs CPF provides access to the fares data base and makes selecting the best fare deal for a customer very straightforward. Besides showing fare options, CPF can also show fares by ticket type. Most of the information that is needed to produce a ticket, including the fare, for example, would already have been entered as part of the reservations process described above. However, additional information may be added prior to ticket printing. This could include, for example, the form of payment. This may specify the usual means or a combination of payment methods.

  The fully automated customer enquiry terminal system (FACETS) enables an itinerary to be planned for a customer by entering the from and to cities (in fact either the full names of the cities or the three-letter codes for cities/towns may be entered), the time of travel and the type of fare requested. The system responds with a screen showing the suggested itinerary as specified. For each leg of the itinerary, i.e. the from/to stations involved in each train journey, the system shows the station name, the departure time, arrival time, the accommodation available on that train, whether it may be reserved and, finally, whether there is any available space to be reserved.

The central main-frame system represent the core around which both current and future UK rail automation systems are based. This core comprises a number of systems that are highly inter-related and some of which are being released in stages over the next few years. These systems, originally known collectively as the joint distribution system (JDS), also support the point-of-sale rail functions that I will describe in more detail in a moment. The JDS product was first demonstrated at the 1992 World Travel Market in London.

The JDS was so called because the intention was to support jointly the reservations and operations of both the UK's train operating companies and European Passenger Services Ltd. The latter, which became Eurostar .UK Limited (EUK), operates the new channel tunnel based rail service to the continent. These services that are branded 'Eurostar', operate principally between London and either Paris or Brussels, and started operations in 1994 when the tunnel opened. Cross-channel car and passenger services are provided by a separate train operating company called Le Shuttle. The original concept was that a single PC system would have access to both the UK rail main-frame computer in Nottingham and the EUK computer in Lille. As it has subsequently turned out, there are two systems in current use: (i) the Tribute system, which is aimed mainly at in-house rail servicing points; and (ii) the TSG system, which is the strategic travel agency point-of-sale system. Both systems derive most, if not all, of their data from the main-frame systems in Nottingham and Lille. I have explained each one in more detail in the following sections.

Having said this, two factors need to be recognized: (a) the central main-frame in Nottingham will continue to be the platform for all automated UK domestic rail services, and (b) the local train operating companies in the UK are perfectly free to develop their own systems, e.g. for marketing to travel agents and for internal use in local rail stations and booking offices. The latter point may become more relevant as time goes by: so, for the moment, I've chosen to concentrate solely on ATOC's main-frame systems in Nottingham and Crewe because they have the most significant potential for domestic rail travel in the UK.

Point-of-sale systems

Over the past two or three years, ATOC has been focusing on developing the next generation of technology to support sales from its own ticket offices as well as travel agents. In particular, it has recognized the trend among travel agents towards the more widespread use of GDS PCs and the necessity to produce automation products that are inexpensive and easy to use. This has to a large extent been driven by the need to support new products, such as the continental channel tunnel services, and the desire to direct more ticket sales via travel agents, especially for business travel. The expectation is that travel agents will be selling a higher volume of more profitable rail products in the future and that this will be accomplished to a large extent by using sophisticated point-of-sale technology. Looking longer term, there is the potential for increased overseas sales of rail products and services via GDS networks that extend to other countries.

Rail tickets may be issued to the travelling public from a variety of different machines. Each machine has been designed to serve a specific purpose, whether it be for train operator staff in ticket offices, by travel agents or via self-service consumer activated machines. Here is a description of the main types of point-of-sale systems used within each part of the rail distribution network:

- **Stations** Ticket offices in stations use an all purpose ticket issuing system (APTIS) to issue tickets and record sales transactions. APTIS has been used successfully by British Rail ticket offices since 1986. The machine is a purpose built piece of hardware and can only be used for issuing rail tickets. It also requires two separate power sockets and an on-line communications capability. The communications facilities needed are a British Telecom Rapide socket and an approved modem, which are required so that tariff data can be downloaded from the British Rail main-frame and ticket sales can be uploaded overnight. One of the few problems nowadays with APTIS is that few rail stations have devices that are capable of processing the bubble memories used in the original machines (bubble memory is a rather outdated storage technology).

Although it is possible for APTIS to be used by travel agents, they need to generate a high volume of rail business to justify it economically. This is because the machine must be leased on an annual basis and is quite costly (about £4,000 in the first year alone). A general guideline is that an agency needs to have British Rail ticket sales of at least £0.5 million per year in order to justify an APTIS machine. It is for this reason that only the largest of the multiple travel agency branches use APTIS machines. Often such branches house a central rail ticket issuing operation that services several regional branches (or even all UK branches). In total, there are currently 2,500 APTIS machines installed in the UK.

In the future, SEMA will be enhancing the software that runs on the APTIS hardware. The hardware is extremely robust, reliable and has a keyboard designed exclusively for rail ticket sales. This hardware can therefore be used as a firm base to develop on-line links to the main-frame timetable and fares data bases and also to card authorization systems.

- **Trains** Ticket inspectors on trains use the Super portable ticket issuing system (SPORTIS) to issue tickets and record sales transactions. SPORTIS is a portable ticket issuing machine used by roving ticket inspectors on trains, at departure gates and smaller stations. It is less sophisticated than APTIS and stores fewer fares and ticket details. Data are exchanged with SPORTIS by removing the memory and inserting the removed cartridge into a special device. This device reads the stored data on the tickets issued and writes updated tariff data into the memory. The memory is then re-inserted into the SPORTIS machine. There are 3,000 SPORTIS machines currently in use throughout the UK at present.

  The software that supports SPORTIS will be used as a base on which to enhance the system further in the future. A new hardware platform is to be developed with a capability to produce magnetically encoded card-sized tickets for use on the underground. The challenge for this development is to find batteries that are powerful enough to drive the encoding technology, which is rather power hungry.

- **Telesales centres** Telesales centres and other retail servicing points use the Tribute system to provide information support for incoming telephone calls from customers. Tribute is a software product that runs on a PC. The PC uses telecommunications technologies to communicate with the main-frame computers for fares, timetable and reservations support. The main user is the National Rail Enquiry Service (NRES), which was set up at the direction of the UK's rail regulator. This is a telephone service centre that receives calls from customers and is accessible via an (0345) local call number from anywhere in the UK. In total, there are 700 Tribute systems installed in the UK. The major features of Tribute are:
  – *Full function rail system* Tribute is used for a wide variety of functions related to the sale of rail tickets. Although the focus of the system is on the InterCity and Eurostar services, all UK train services are supported. The principal functions are ticketing, quick issuing and balances.
  – *Ticketing* When all information has been entered the user may request an ATB ticket to be printed with an encoded magnetic strip on the back. Cancellation of tickets can then easily be processed because the ticket to be cancelled is simply inserted into the ATB printer, which reads the ticket details encoded on it and automatically displays a cancellation screen for further processing. Credit card charge forms may also be printed automatically.

- *Quick issue* Each Tribute PC workstation has the capability to store up to 50 frequently used itineraries. Once stored, these template itineraries may be called up and used to shortcut the entire journey planning, reservations and ticketing process. Because each operator may have his/her own set of quick issue itineraries, the total scope for tailored processing is considerable.
- *Balance function* The user may request a status of the sales made to date and the amounts of moneys taken, at any time. This supports the control of telesales cash and sales processing functions.

- **Self-service** Self-service machines branded QUICKFARE are located on the concourse of 1,000 stations around the country. These machines issue tickets in exchange for cash, i.e. notes and coins. At present these machines do not accept plastic cards. The QUICKFARE machines are supplied from a Swiss company called Ascom Autelca and are extremely robust and reliable. In addition to QUICKFARE, some train operating companies have decided to use ATM style ticket issuing machines, which are marketed by a separate company called SHERE. There are now over 20 SHERE self-service ticketing machines installed in several stations. These machines issue ATB format rail tickets and accept plastic card payments from customers.

  SEMA are working with Ascom Autelca to evaluate further enhancements to the QUICKFARE machines in the medium term. One option is to extend the current payment method from cash to include plastic cards. This will in turn require an on-line link to the major card authorization systems from each QUICKFARE machine. Although the QUICKFARE machines are high quality, robust and reliable, the fact that they must handle cash makes them an expensive proposition for widespread roll-out to more stations.

  So, a more cost effective solution is being sought by SEMA. It is currently considering the development of its own self-service ticketing machine that would only support plastic card payments. This new machine would work in conjunction with the telesales centres. Customers could telephone a telesales centre to discuss options and book their journeys. They would make arrangements with the operator to collect their tickets upon departure at the stations nearest to them using one of the new self-service machines. The insertion of the customer's card would enable the ticket details to be retrieved from the main-frame computer, printed locally and dispensed from the machine. This would of course require on-line telecommunications from the self-service machine to the Nottingham main-frame computer and the major card authorization systems.

- **Travel agents** Travel agents use a number of systems to obtain information on rail travel and issue tickets to their customers. Although some very large agencies use APTIS, as described above, for most travel agents the two most commonly used systems are: (i) the reservations functions available via the GDSs, and (ii) the agent ticketing system (ATS). I'll describe each in more detail:
  - *TSG* ATOC has designated TSG as the travel agency system of the future for rail sales and servicing. Incidentally, the term 'TSG' is to be renamed soon and explaining its initials would only serve to confuse so, let's stick to TSG for the purposes of this explanation. Over 400 travel agents currently use TSG and the medium term target is to grow this to 2,000 or more. With TSG, travel agents are able to use their GDS PC terminals to link into the rail main-frame computer in Nottingham. This link is effected by means of a switching technology that is slightly different for each GDS (see Chapter 4 for a description of how each GDS implements non-air supplier access). So, TSG is a set of enabling technologies that distributes Nottingham's central main-frame functions, such as reservations, ticketing and servicing, to travel agents using their existing point-of-sale GDS PC terminals. In concept, this is similar to the way in which airlines distribute their sales functions via the GDSs. As such, it is *not* therefore a piece of software that runs in the agent's PC. TSG was originally developed by Eurostar to enable travel agents to gain access to their new channel tunnel train services via Galileo.

Using TSG, seats can be reserved on certain UK domestic rail journeys in both first class and standard coaches of all InterCity trains using GDS terminals installed in travel agencies. Reservations can also be made on most inter-urban rail services. First class single and standard twin berth sleeper compartments are reservable on all InterCity sleeper trains. The travel agent can choose certain reservations preferences for their customers such as, window or aisle seat, facing or back to the direction of travel, dining seat or non-dining, smoking or non-smoking seats.

Travel agents can use their existing GDS terminals to access the rail main-frame system using the 'BRL' entry. Agents without a GDS can use viewdata terminals using the Imminus or AT&T travel networks, again using 'BRL' as the access code (see Chapter 4). Agencies with the latest Galileo Focal Point UK terminals have the additional benefit of access to FACETS. This system runs on the rail main-frame computer and supports an integrated fares, timetable, availability and reservations facility that compares favourably in terms of functionality with most airline systems. At a certain point in time, a few hours before departure, the system prints reservations dockets for the departing train that a train operating company employee places in the appropriate headrest of each seat.

TSG can generate a rail machine interface record (MIR) for back-office accounting purposes. This main-frame created data record is therefore used to generate accounting transactions and is stored for future management information purposes (see Chapter 7 for more details on back-office or agency management systems). It remains the responsibility of the travel agent to ensure that their back-office system is capable of successfully processing the rail MIR.

ATB based ticketing is an important feature of TSG. This is supported by means of an ATB2 style ticket printer that is directly connected to the GDS PC. Incidentally, the ATB2 style ticket is the one with the magnetic strip on the reverse side (a plain old ATB ticket has no such magnetic strip). Only the ATB2 has the magnetic strip that can be read by devices at the ticket gate or *en route*. For continental travel, this will help speed the passenger through check-in formalities at the new Eurostar terminals instead of the old style travel document that will have to be exchanged for an ATB ticket before he/she can start his/her journeys. Shorter check-in times are of course especially essential to business customers who travel at peak times and usually pay full fares.

You may recall our earlier discussion of ATB type tickets and their associated printers in the section on airline reservation systems or GDSs. Well, the rail ATB (Fig. 3.7) complies with the IATA 722 encoding standards. This means that the data that are encoded on the magnetic strip on the reverse of the ATB, conform to a standard that has been set by IATA and used by all the airlines. The ATB printer is, however, different from most conventional printers in one important respect. In addition to printing an ATB and encoding it simultaneously, it can also read the magnetic strip on a previously printed ATB ticket. The potential is therefore in place for a travel agent to have just one ATB printer in the office that can print and process both air and rail tickets.

Clearly therefore, one of the long term objectives of ATOC is for travel agents to use a single printer for producing all UK rail tickets. This will enable a single ATB printer to be loaded with a set of airline ticket stock and a set of rail ticket stock. The GDS systems use special software to control contention between the various workstations that need to print a ticket. Contention occurs, for example, when Workstation 1 initiates an airline ticket printing command at the same time that Workstation 2 issues a rail ticket print command. This special contention handling software makes it possible to use just a single ATB printer at the point-of-sale. This is important because ATB printers can be quite costly. Looking even further ahead it may one day be possible

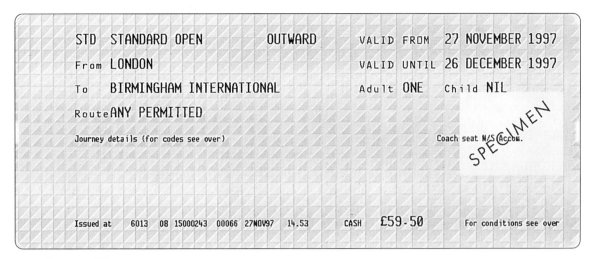

Figure 3.7 A rail ATB

for a single stock to be used for both airline tickets and rail tickets. However, in order for this to happen, there will need to be a lot more work done in the standards area. Because this is all organized by committees from leading airlines, rail companies and other travel suppliers, it may well take some time to agree such standards.

The ticket printed by the TSG system is for a maximum of three legs or sectors of an itinerary. More sectors can of course be ticketed but these will require more than one ATB. So, for example, a five-leg journey would have Legs 1, 2 and 3 printed on the first ATB and Legs 4 and 5 on the second. Standards are a critical issue for industry systems as I am sure you will have gathered from the section in Chapter 1. ATOC has adhered to the international union of railways (UIC) and rail combined ticket (RCT) standards for European ticket issue. This means that tickets produced on UK rail systems will be acceptable on the continent and can be read and processed by other non-UK systems. Incidentally, the rail printer will also be capable of printing credit and charge card forms. These will comprise a portion for use by the card company and a tear-off slip for the customer. This will no doubt be a valuable time saver for most travel agents.

In terms of GDS connectivity, the rail system is accessed via a special partition within the GDS multi-access capability. There is a GDS language entry that a user must enter into the GDS terminal in order to request seat and sleeper reservations. This is converted by the GDS system into the appropriate rail system entry (the rail system uses an alternative format to the airline systems), and transmitted via telecommunication lines into the main-frame computer in Nottingham. The rail system response on this return half of the dialogue is not converted but instead appears in native mode on the travel agent's GDS system display.

– *ATS* ATS is a 'stand alone' PC software package that has been provided to travel agents since 1991. ATS produces train tickets on continuous stationery and supports automated settlement of ticket sales to RSP. It is primarily for travel agents who have rail ticket sales of up to £0.5 million per year. The software is currently used by around 150 travel agencies. To use ATS the travel agent needs to have a dot matrix printer attached to their PC. This printer is loaded with continuous rail ticket stock that is very similar to OPTAT airline ticket stock in size and format. The main difference is that this ticket stock comprises just three parts: (i) a travel copy, (ii) a copy for the travel agent,

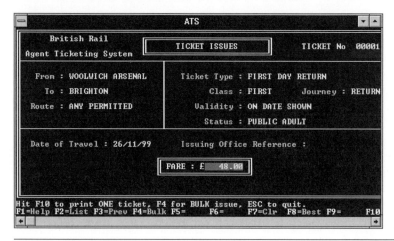

Figure 3.8 An ATS screen

and (iii) an audit or accounting coupon. The main menu offers the following set of functions: ticket issue, ticket cancellation, back-up sales data, daily sales report and best-seller input. There are also parameters that control ticket stock, accounting periods and VAT codes. The principal function is, however, the issue of rail tickets and this is therefore explained in more detail below.

The first question the ATS system (Fig. 3.8) asks the user, is to verify that the number of the next ticket to be printed tallies with the ATS PC stock records as displayed on the screen. Once this has been confirmed the ticket issuance process starts. The system works most effectively when it has been set up with details of the most frequently used itineraries. There may be up to 50 such itineraries and each is pre-set with: from city, to city, fare, class, etc. When a ticket is required from the best-seller list, the agent selects the route, adds the date and a reference and the ticket is printed automatically. If the journey is for a non-best-seller itinerary, the agent simply enters all the details that are required on the ticket, such as from city, to city, class, fare (from the rail tariff book), date and reference. Finally, when all data have been entered the ATS system may be instructed to print the ticket.

The details of all tickets issued for a day are added to a data base, which forms the basis of the end-of-month rail sales return. Each month the system produces a floppy disk that is mailed to RSP along with a copy of all the audit coupons of the tickets issued for the month. This saves a great deal of manual effort and totally eliminates the need for a hand-written rail sales return. It is also possible to print consolidated daily and monthly sales reports for all tickets sold by the agency.

The ATS system also interfaces with Galileo's President Agency Management System (PAMS). This is achieved using a connection from the serial port of the ATS PC to the serial port of the PAMS PC. This connection allows a MIR to be transferred from ATS into PAMS for every ticket issued. A MIR contains all the information needed by the PAMS back-office system to process the sales ledger functions (including the automated printing of customers' invoices) and other accounting tasks associated with rail ticket sales. This makes it possible to capture some basic accounting data at the point just prior to the ticket actually being printed. Information such as client account number, cost centre, product code, method of payment, credit card type and card number. This saves the travel agent's back-office accounting staff from having to re-key the ticket information already keyed at the point-of-sale into PAMS and then having to add other customer account information.

Retail servicing points

There are many different places where information on train services are required by customers. For example, there are the telephone sales centres scattered around the country. There used to be 45 of these telesales centres although they are being rationalized at the direction of the rail regulator and eventually there will only be around four large telesales centres and six smaller units. Rationalization has resulted in a new consolidated NRES. This is accessible via the telephone using a single national number charged at local call rates. But there are other retail servicing points besides NRES, such as travel agents that do not necessarily sell rail tickets, small stations and rail shops. The main systems used to provide train operating information to these retail servicing points are:

- **Tribute** This is a PC-based software product. It performs certain rail related functions locally and also uses ISDN telecommunications to connect into the rail main-frame computers in Nottingham. This provides its retail servicing users with the power of a local PC system but with the enormous resources of a main-frame just a phone call away. An important IT architectural feature of Tribute is co-operative processing.

  The term co-operative processing means that Tribute's system functions are provided by a team of computers working together and sharing the workload. Tribute uses co-operative processing techniques because it shares the total processing workload between its host PC and the remote rail main-frame computer. It therefore closely resembles an airline GDS. The screens that the user sees and interacts with are based on main-frame responses but these are enhanced locally by Tribute's special PC software. The resulting screens are extremely user friendly and are similar in appearance to the Microsoft Windows format that is by now so familiar to many PC users.

  SEMA plan to migrate parts of the data base held on the Nottingham main-frame to the local Tribute PC hard disk. Processing and storage functions within Tribute would then be increasingly shared between the Nottingham main-frame and the local PC. The fares data base, for example, will be split into two parts for storage and access purposes. Local fares will be stored on the user's PC and refreshed each night via the main-frame link, while all other fares will be stored only on the main-frame and accessed as needed. This should help take the load off the central main-frame and make Tribute more responsive. Another main-frame support functions accessed by Tribute include FACETS. FACETS combines the CPF fares system, the CATE timetable system and the British Rail CRS reservations system.

- **CATE and FACETS** These are the timetable and fares systems that run on the Nottingham main-frame. They are accessed from high volume rail servicing points by dedicated terminals connected by leased lines to Nottingham.

- **The customer information system (CIS)** This is located on station concourses. It shows departure details of trains leaving from a station within the next hour and in some cases also shows arrival information.

- **Rail planner** The rail train timetable has been computerized for some time. It is stored as a very large data base, comprising some 89 Mb of data storage, held on the main frame computer in Nottingham. Until 1992 this timetable was only available in the form of a large and somewhat complex book known as the *British Rail Great Britain Timetable*. The principal aim of Rail Planner is to simplify the planning of customers' journeys by providing local access to this timetable data base.

  Rail Planner is another 'stand alone' PC based software product (Fig. 3.9) that allows users to plan their journeys simply by specifying their origin/destination requirements. This is achieved by providing the entire rail timetable on a PC data base, which is compressed so that it occupies only 1.8 Mb of PC hard disk space. This has been done by using special software and a user friendly man/machine dialogue. Rail Planner is available in two options: (i) Rail Planner software plus a timetable data base supplied twice each year, or (ii) Rail Planner software with a timetable data base that must be updated each month with changes. The core element of Rail Planner is the timetable data base that, with its monthly refresh

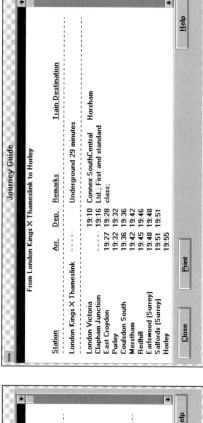
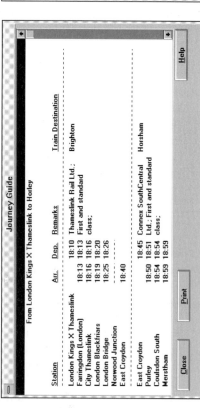
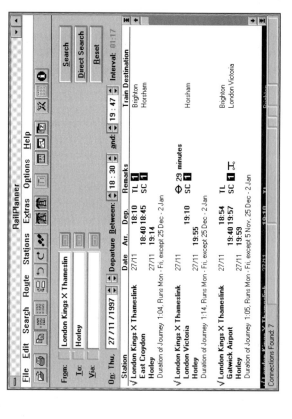

Figure 3.9 The Rail Planner screen

function, is close to being as accurate as the timetable data base stored on the main-frame.

Rail Planner is a Windows-based product and uses familiar GUI standards. The user simply enters the from and to cities, the date of travel and the approximate departure time. The system responds with a display of the itinerary showing the intermediate stops and train changes *en route*. Also displayed are several earlier and later options that may be of interest to the traveller. When the route has been finalized, a map option may be requested. This pictorially shows the chosen route with major stations and all interconnecting points. The map can be printed and handed to a traveller. Rail Planner has been available from British Rail since September 1992 (when it was branded Journey Planner). The software was originally supplied by a German company and has now been modified for use in the UK. The system is distributed on floppy disk and is mainly used by individual travellers and corporate travel planners.

Central accounting functions

One of the main functions of the RSP is to ensure that the revenue derived from rail ticket sales is distributed equitably between all the train operators and other third parties, e.g. the London Underground. The SEMA group undertakes this task on behalf of RSP using several large mainframe computer systems, the main one of which is called CAPRI. CAPRI is fed with ticket sales data from the various point-of-sale systems that I described above. However, the revenue allocation process is not as straightforward as you might think. Let's consider the logistics in a little more detail.

First of all, the details of the actual routes that the passengers travelled may not automatically be derived from the ticketed sales data generated from the points-of-sale. The UK rail network is sufficiently large and diverse to enable travellers to take alternative routes between their origin and destination points. So, it is not always possible for the revenue accounting system to calculate how a ticket's sales value should be split between two or more train operating companies on the basis of the route actually travelled. Instead, a set of rules have been laid down for each ticketed route that allows certain assumptions to be made regarding the allocation of ticket sales revenue. As you can imagine, with over one-quarter of a billion tickets sold each year, this is a job that is only feasible to perform using a large and powerful computer.

Once the ticket sales transaction data have been analysed, the CAPRI system determines the share attributable to each train operating company and also calculates the commission due. It then determines the amounts payable to third parties such as the London Underground and ferry companies. The end result of this process is a set of entries that are passed into the books of account of each train operating company.

Rail sales distribution

An enormous volume of rail tickets is sold in the UK each year, i.e. over one-quarter of a billion tickets annually, but compared with other travel industry suppliers, only a small proportion are sold via travel agents. One reason for this is that travellers tend to buy their tickets from the station as they depart. Another possible reason is that from the travel agents' perspective the revenue on a British Rail ticket is usually perceived as being rather low for the average journey, especially when the rather laborious monthly manual settlement procedure is taken into account. In fact, the average commission rate on rail tickets is some 9 per cent, which is actually quite competitive with domestic air travel.

Once again, business travel is a little different and most business travel agents will offer their corporate customers British Rail tickets. The reason for this is twofold: (a) it provides an all round service and gives a better and more caring image to the customer if a comprehensive range of products can be provided, not just the profitable ones; (b) business travellers often travel first class and the journeys are usually long distance, all of which means that the revenue earning potential of these type of British Rail tickets is in fact quite attractive. For this reason therefore, some travel agents do carry a stock of British Rail tickets.

The sale of rail tickets that are sold via travel agents is, however, set to increase significantly in the not too distant future. In fact, it is expected that the total volume of rail tickets booked through the UK travel trade will more than double by the end of this decade. This is primarily due to the channel tunnel, which opened in 1994, and the commencement of Eurostar and Le Shuttle, both of which provide new high speed surface links to the continent that challenge air and ferry services.

In summary, if travel agents are to be encouraged to sell UK domestic rail tickets and the new cross-channel products, it is imperative that they have access to efficient and cost effective sales and ticketing technology. Rail companies are therefore developing some very sophisticated supporting technology that should make the sale of most rail tickets highly profitable for travel agents and lower the cost of direct sales.

Information suppliers

Information is the raw material of the travel and tourism industries. There is an enormous number of travel suppliers and each one describes their products using a wealth of data; and besides travel suppliers there is a whole set of reference information that the industry and travellers themselves need if they are to travel the world safely and effectively. Unfortunately, these data and their assembly into useful information are scattered across many different companies and sources. Life would be a lot simpler if information on travel and tourism was stored in just a single place and easily accessible from anywhere in the world. This ideal is, however, unattainable at the present time, although the future holds the promise of new technologies that could rationalize data sources and standardize the way we access and use information. It is therefore essential that we explore companies that specialize in the accumulation, storage, distribution and provision of information to the travel and tourism industries. Probably the leading company in this field is the Reed Travel Group.

REED TRAVEL GROUP

Reed Travel Group is a member of the Reed Elsevier plc group, a world-leading publisher and information provider in the areas of scientific, professional, business and consumer publishing. With its principal operations in North America and Europe, Reed Elsevier has annual sales in excess of £3 billion and employs more than 25,000 people worldwide. Reed Travel Group is the world's largest independent provider of travel information products and services to: (i) business travellers, travel planners and the global travel industry; and (ii) GDS and airlines. The Reed Travel Group mission statement is:

> 'To be the preferred supplier to the global travel community of: (a) comprehensive, impartial information; (b) essential news and opinions; and (c) effective solutions and knowledge. Our corporate culture will inspire and empower employees to provide our customers and associates with products and services of superior value and to deliver a fair return to our shareholders. We will be good corporate citizens with respect to the communities and environments in which we work.'

With more than 2,500 employees in more than 50 offices around the world, Reed Travel Group's businesses serve all the major global travel markets including air transportation, hotel, cruise, meetings, leisure, cargo, rail and other travel sectors. Although I won't be covering all of these products and services in this book, there are several that are very relevant to the effective use of IT in travel and tourism, which I will be exploring in a lot more detail in this and other chapters.

Reed Travel Group's leading business portfolio includes: OAG, *Travel Weekly*, *Meetings & Conventions*, *Hotel Travel Index*, TravelNet, Weissmann Travel Reports, ABC Corporate Services, Reed Travel Training and Utell International. The company's products are distributed globally and the business operates from centres in the UK, USA and Singapore. Reed Travel Group's products and services serve many different markets: (a) travel principals, such as airlines, airports and hotel/car companies; (b) the corporate market, including travellers, travel arrangers and buyers; (c) travel intermediaries, such as travel agents, CRS, GDS and freight forwarders/cargo agents; and finally (d) USA Government departments, including the Federal Aviation Administration and Government travellers. The complete range of products may be categorized as follows:

- **OAG** This encompasses the print and electronic travel information products and services that are offered to travellers, corporations and the travel industry. These publications supply the information needed to make effective travel decisions. OAG and its electronic products are described in more detail below. The specialist GDS fares and schedule distribution service (known, respectively, as OAG Genesis and OAG Direct) are described earlier in this chapter (see the Fares Distribution section above).
- **Utell International** This is the world's largest hotel marketing, sales and reservations service promoting 6,500 hotel members in more than 180 countries. This service is described separately in Chapter 4 – Distribution Systems.
- **Weissmann Travel Reports** This is a leading provider of global electronic and hard copy destination information to the travel industry. Its data base includes destination information on more than 10,000 cities throughout the world and is available by country profile, state/province profile, city profile and cruise port-of-call profile. It is particularly relevant to IT in travel and tourism because it is a source of information for many Internet sites, some of which are described in Chapter 5 – The Internet.
- **Travel Weekly** This is the leading travel trade newspaper in the USA. Published twice weekly, it provides late-breaking industry news, features and practical information for travel agents and the travel industry as a whole. Several other travel related publications also are included in the portfolio. I have not, however, covered this aspect of Reed Travel Group's business in any more detail in this book.
- **TravelNet** A leading real-time travel booking tool, with advanced data base software, that enables companies to manage and reduce travel expenses. TravelNet, simplifies the travel planning, booking and management process by providing a company's employees with the ability to book air, hotel and car rental reservations from their personal computer. This product is explored in more detail in Chapter 5 – The Internet.
- **Reed Travel Training** Courses are offered for travel agents, travel arrangers and administrative assistants. Reed Travel Training is endorsed by IATA and the Universal Federation of Travel Agents' Association UFTAA. I do not cover these training products in this book.
- **Hotel & Travel Index** This is a leading hotel directory providing up-to-date booking information on 45,000 hotels, resorts and inns world-wide. Published quarterly, *Hotel & Travel Index* contains all pertinent hotel information including rates, accommodation details, contact names, addresses, toll-free phone and fax numbers, representatives and commission policies and GDS access codes. These products are not covered in more detail in this book.
- **ABC Corporate Services** This is the original provider of quality corporate services for independent travel agencies. The company's primary products include: (i) *Premier Hotel Plan*, a comprehensive programme including negotiated rates, value added amenities and block space; (ii) *Business Breaks*, a meetings' facilities guide for corporate travel planners; (iii) the travellers' emergency service system (TESS), a 24-hour emergency hot-line; (iv) an international rate desk and (v) Global Connect, a global travel management network. These products are not covered in more detail within this book.
- **EasyRes** A leading viewdata-based reservations system for leisure travel agents in the UK, offering free and easy access to a wide choice of airlines and fares, last seat availability, hotels and car rental booking facilities. The EasyRes service is presented in more detail in Chapter 6 – Communication Networks.

As I have mentioned before, information is the raw material of travel and tourism. Insofar as Reed Travel Group is concerned, this raw material is a prime company asset. It is an asset that is represented, at its lowest level, by collections of individual data items. These data items are stored in a way that enables them to be combined in different representations and this is what we call information. The core data base, which forms the basis for so many of Reed Travel Group's information products, is stored on a main-frame computer using relational data base management software. One of the simplest ways in which this data base is used is in the production of the OAG reference

Validity From	To	Days of Service	Dep	Arr	Flight No.	Acft	Class	Stops
FROM LONDON UNITED KINGDOM **LON** Continued								
-	3Oct	1-3-56-	**1915** LHR	2230 MAD	**IB** 3167	320	CMLWK	0
			*0105 MAD	***1020**	**IB** 6811	747	PJDYL	1
-	5Oct	------7	**1915** LHR	*0845 EZE	**AR** 1165	747	JYMLW	2
			*0940 EZE	***1055**	**AR** 1240	737	JYMLW	0
				***AR** 1165 Equipment change en route*				
-	6Oct	1-3-56-	**1915** LHR	2230 MAD	**IB** 3167	320	CMLWK	0
			2355 MAD	***0950**	**LA** 171	767	PJYBQ	1
-	7Oct	-2-----	**1915** LHR	*0845 EZE	**AR** 1115	747	JYMLW	2
			*0940 EZE	***1055**	**AR** 1240	737	JYMLW	0
				***AR** 1115 Equipment change en route*				
-	9Oct	---4---	**1915** LHR	*0845 EZE	**AR** 1135	747	JYMLW	2
			*0940 EZE	***1055**	**AR** 1240	737	JYMLW	0
				***AR** 1135 Equipment change en route*				
-	-	-2-----	**1915** LHR	2230 MAD	**IB** 3167	320	CMLWK	0
			*0050 MAD	***1045**	**LA** 175	767	PJYBQ	1
4Oct only		-----6-	**1915** LHR	2230 MAD	**IB** 3167	320	CMLWK	0
			*0025 MAD	***0955**	**IB** 6811	747	PJDYL	1
6Oct only		1------	**1915** LHR	2230 MAD	**IB** 3167	320	CMLWK	0
			*0105 MAD	***1030**	**IB** 6811	747	PJDYL	1
7Oct	10Oct	-23-5--	**1915** LHR	2230 MAD	**IB** 3167	320	CMLWK	0
			2355 MAD	***0950**	**LA** 171	767	PJYBQ	1
8Oct	10Oct	--3-5--	**1915** LHR	2230 MAD	**IB** 3167	320	CMLWK	0
			*0105 MAD	***1020**	**IB** 6811	747	PJDYL	1
11Oct	-	123-56-	**1915** LHR	2230 MAD	**IB** 3167	320	CMLWK	0
			2355 MAD	***1050**	**LA** 171	767	PJYBQ	1
11Oct	-	1-3-56-	**1915** LHR	2230 MAD	**IB** 3167	320	CMLWK	0
			*0100 MAD	*0730 GIG	**RG** 711	763	CYMHL	0
			*0930 GIG	***1400**	**RG** 920	767	CYMHL	1
				***RG** 920 Equipment change en route*				
11Oct	-	-----6-	**1915** LHR	2230 MAD	**IB** 3167	320	CMLWK	0
			*0105 MAD	***1130**	**IB** 6811	747	PJDYL	1
12Oct	-	------7	**1915** LHR	*0845 EZE	**AR** 1165	747	JYMLW	2
			*0940 EZE	***1155**	**AR** 1240	737	JYMLW	0
				***AR** 1165 Equipment change en route*				
13Oct	-	1-3-5--	**1915** LHR	2230 MAD	**IB** 3167	320	CMLWK	0
			*0105 MAD	***1120**	**IB** 6811	747	PJDYL	1
14Oct	-	-2-----	**1915** LHR	*0845 EZE	**AR** 1115	747	JYMLW	2
			*0940 EZE	***1155**	**AR** 1240	737	JYMLW	0
				***AR** 1115 Equipment change en route*				
16Oct	-	---4---	**1915** LHR	*0845 EZE	**AR** 1135	747	JYMLW	2
			*0940 EZE	***1155**	**AR** 1240	737	JYMLW	0
				***AR** 1135 Equipment change en route*				
-	29Sep	1------	**2000** LHR	2205 CDG	**AF** 821	321	CDYSK	0
			2320 CDG	[illegible]55 GIG	**RG** [illegible]	M11	FCYM[illegible]	2
			1525 GIG	[illegible]	[illegible]	737	JYP	
-	3Oct	1-3-5--	**1345** LGW	1620 FRA	**BA** 2716	757	CM	0
			2015 FRA	***0530** GRU	**LA** 171	767	PJYBQ	1
-	-	-2-----	**1345** LGW	1620 FRA	**BA** 2716	757	CM	0
			2100 FRA	***0625** GRU	**LA** 175	767	PJYBQ	1
6Oct	-	123-5--	**1345** LGW	1620 FRA	**BA** 2716	757	CM	0
			2015 FRA	***0630** GRU	**LA** 171	767	PJYBQ	1
-	27Sep	-----6-	**1445** LGW	1720 FRA	**BA** 2716	737	CM	0
			2015 FRA	***0530** GRU	**LA** 171	767	PJYBQ	1
4Oct only		-----6-	**1445** LGW	1720 FRA	**BA** 2716	737	CM	0
			2015 FRA	***0630** GRU	**LA** 171	767	PJYBQ	1
11Oct only		-----6-	**1445** LGW	1720 FRA	**BA** 2716	737	CM	0
			2015 FRA	***0530** GRU	**LA** 171	767	PJYBQ	1
18Oct	-	-----6-	**1445** LGW	1720 FRA	**BA** 2716	737	CM	0
			2015 FRA	***0630** GRU	**LA** 171	767	PJYBQ	1
-	3Oct	1-3-56-	**1455** LGW	1815 MAD	**BA** 2466	737	CM	0
			2355 MAD	***0530** GRU	**LA** 171	767	PJYBQ	0
4Oct	10Oct	123-56-	**1455** LGW	1815 MAD	**BA** 2466	737	CM	0
			2355 MAD	***0630** GRU	**LA** 171	767	PJYBQ	0
11Oct only		-----6-	**1455** LGW	1815 MAD	**BA** 2466	737	CM	0
			2355 MAD	***0530** GRU	**LA** 171	767	PJYBQ	0
13Oct	-	123-56-	**1455** LGW	1815 MAD	**BA** 2466	737	CM	0
			2355 MAD	***0630** GRU	**LA** 171	767	PJYBQ	0
-	3Oct	1-3-56-	**1700** LHR	1930 FRA	**LH** 4007	AB3	CDHBL	0
			2015 FRA	***0530** GRU	**LA** 171	767	PJYBQ	1
-	-	-2-----	**1700** LHR	1930 FRA	**LH** 4007	AB3	CDHBL	0
			2100 FRA	***0625** GRU	**LA** 175	767	PJYBQ	1
4Oct	10Oct	123-56-	**1700** LHR	1930 FRA	**LH** 4007	AB3	CDHBL	0
			2015 FRA	***0630** GRU	**LA** 171	767	PJYBQ	1
11Oct only		-----6-	**1700** LHR	1930 FRA	**LH** 4007	AB3	CDHBL	0
			2015 FRA	***0530** GRU	**LA** 171	767	PJYBQ	1
13Oct	-	123-56-	**1700** LHR	1930 FRA	**LH** 4007	AB3	CDHBL	0
			2015 FRA	***0630** GRU	**LA** 171	767	PJYBQ	1
4Oct	-	-----6-	**1730** LHR	1930 BRU	**BQ** 608	737	Y	0
			2130 BRU	***0530** GRU	**VP** 791	M11	FCYBM	0
-	3Oct	1234567	**1800** LHR	2100 JFK	**UA** 909	763	FCYBM	0
			2240 JFK	***0920** GRU	**UA** 987	763	FCYBM	0
4Oct	-	1234567	**1800** LHR	2100 JFK	**UA** 909	763	FCYBM	0
			2240 JFK	***1020** GRU	**UA** 987	763	FCYBM	0
-	3Oct	1-3-5--	**1900** LGW	2205 MAD	**IB** 3183	M87	CMLWK	0
			2355 MAD	***0530** GRU	**LA** 171	767	PJYBQ	0
-	3Oct	--3-5--	**1900** LGW	2205 MAD	**IB** 3183	M87	CMLWK	0
			2345 MAD	***0500** GRU	**IB** 7825	747	JDYLW	0
					/**AR** 7825			
-	9Oct	-2-4--7	**1900** LGW	2205 MAD	**IB** 3183	M87	CMLWK	0
			*0105 MAD	***0635** GRU	**IB** 6823	340	PJDYL	0
6Oct	-	123-5--	**1900** LGW	2205 MAD	**IB** 3183	M87	CMLWK	0
			23[illegible] MAD	***0630** [illegible]U	**LA** 171	7[illegible]	PJYBQ	0
8Oct		-3-5--	[illegible]GW	220[illegible]	**IB** 318[illegible]		CMLWK	0
			[illegible]D	***0**[illegible]	**IB** 78[illegible]		[illegible]DYLW	
9Oct	-	1234567	**1250** LHR	1550 IAD	**UA** 919	777	FCYBM	0
			1715 IAD	**1939** SEA	**UA** 171	767	FCYBM	0
-	8Sep	1234567	**1445** LHR	1730 SFO	**UA** 931	777	FCYBM	0
			1945 SFO	**2133** SEA	**UA** 2056	735	FYBMH	0
-	8Sep	12345-7	**1445** LHR	1730 SFO	**UA** 931	777	FCYBM	0
			1915 SFO	**2106** SEA	**UA** 2382	733	FYBMH	0
9Sep	-	1234567	**1445** LHR	1730 SFO	**UA** 931	777	FCYBM	0
			1935 SFO	**2126** SEA	**UA** 2058	733	FYBMH	0
-	2Sep	12-----	**1515** LGW	1840 STL	**TW** 723	767	FCYBQ	0
			2020 STL	**2244** SEA	**TW** 375	M80	FYBQK	0
3Sep	6Oct	1234567	**1515** LGW	1840 STL	**TW** 723	767	FCYBQ	0
			2017 STL	**2243** SEA	**TW** 375	M80	FYBQK	0
Seoul SEL								
-	14Sep	------7	**1150** LHR	***0715**	**BA** 041	744	FJM	0
-	-	1---5--	**1320** LHR	***0840**	**BA** 041	744	FJM	0
-	-	--3----	**1330** LHR	***0840**	**BA** 041	744	FJM	0
-	-	1-3-56-	**2110** LHR	***1605**	**KE** 908	744	FCYHV	0
-	-	------7	**2230** LHR	***1720**	**KE** 908	744	FCYHV	0
TRANSFER CONNECTIONS								
-	-	1--4---	**0920** LHR	1240 ARN	**SK** 526	M80	CSMBH	0
			1505 ARN	*0820 NRT	**SK** 985	763	DCMBV	0
			*1120 NRT	***1340**	**JD** 251	D10	CYBK	0
-	-	--3-56-	**1050** LHR	*0430 BKK	**TG** 913	744	PJYMQ	0
			*0720 BKK	***1635**	**TG** 634	777	JYMQB	1
-	-	----5--	**1050** LHR	*0430 BKK	**TG** 913	744	PJYMQ	0
			*0710 BKK	***1425**	**TG** 770	744	PJYMQ	0
-	-	-23-567	**1120** LHR	1405 CPH	**SK** 502	763	CSMBH	0
			1540 CPH	*0930 NRT	**SK** 983	763	DCMBV	0
			*1120 NRT	***1340**	**JD** 251	D10	CYBK	0
-	-	12-4--7	**1200** LHR	*0610 BKK	**TG** 911	744	PJYMQ	0
			*0720 BKK	***1635**	**TG** 634	777	JYMQB	1
-	-	-2-4--7	**1200** LHR	*0610 BKK	**TG** 911	744	PJYMQ	0
			*0710 BKK	***1425**	**TG** 770	744	PJYMQ	0
-	-	1234567	**1845** LHR	*1345 HKG	**CX** 250	744	FJCWY	0
			*1450 HKG	***2035**	**CX** 420	777	JCWYB	1
-	-	1------	**2200** LHR	*1745 KUL	**MH** 1	744	FRCWY	0
			*2120 KUL	**‡0830**	**MH** 64	330	FRCWY	2
-	-	--34---	**2200** LHR	*1745 KUL	**MH** 1	744	FRCWY	0
			‡0035 KUL	**‡1000**	**MH** 50	330	FRCWY	1
-	-	----56-	**2200** LHR	*1745 KUL	**MH** 1	744	FRCWY	0
			‡0055 KUL	**‡0815**	**MH** 66	330	FRCWY	0
Seville SVQ								
-	-	1234567	**1715** LHR	**2050**	**IB** 4143	320	CMLWQ	0
Shanghai SHA								
-	-	--3---7	**0740** LGW	0955 BRU	**BA** 2412	737	CM	0
			1540 BRU	***1120**	**MU** 552	340	FCYHL	1
-	22Sep	1--4---	**0830** LGW	1120 MUC	**B**[illegible] 4715	733	CM	0
			1815 M[illegible]	***1300**	[illegible]558	340	FCY[illegible]	1
[illegible]	-		**0830**	[illegible]120 MU[illegible]	[illegible]15	733		
			181[illegible]	[illegible]		340		

Figure 3.10 A sample page from the OAG *World Airways Guide*

books. This is, therefore, a good place from which to start our review of Reed Travel Group's IT products for travel and tourism.

OAG

The OAG publications have, for many years, served as the ultimate reference source for travellers, companies and travel agents. The core product is a book that contains the airline fares and schedules. The data base that is used to produce this book forms the basis of many other products and services marketed to the travel industry by the Reed Travel Group. So it is worthwhile spending a few moments on the OAG *World Airways Guide* before I explore the electronic products in more detail. After all, the OAG *World Airline Guide* is a very important reference source used by most travel agencies and many companies whose people travel extensively on business.

The OAG *World Airways Guide* is published monthly by Reed Travel Group. Its purpose is to provide a reference source for direct flights, transfers and connections between the world's airports. In other words, it shows you all the various ways in which you can get from A to B. The OAG *World Airways Guide* is published in two editions: (a) a travel trade edition, which – in addition to schedules – includes airline fares information; and (b) a corporate edition, which includes information on ground transport and destination information as well as health, visa requirements, public holidays, business hours, business hints and general travel information. It has become the standard reference for efficient travel planning with over 480,000 users spanning 180 countries of which 190,000 are travel agents and 200,000 are in the key corporate sector.

The OAG *World Airways Guide* contains details on all the 650,000 scheduled airline flight sectors in the world. Because there are more than 760 airlines and each has a route structure and timetable of its own, you can begin to appreciate the size and complexity of the publication. The OAG *World Airways Guide* shows every single flight sector operated by every airline, from each city airport to all other cities. Each of these sectors is described in detail, including, for example, the days of the week that the flight operates and the times of departure and arrival, the aircraft type and the class of seats available. An example of a page from the OAG *World Airways Guide* is shown in Fig. 3.10.

The OAG *World Airways Guide* also has different sections that describe the airports of the world and the cities and countries in which they are located. There are sections on all airlines and on travel by air in general. All in all, the OAG *World Airways Guide* is an essential reference source for a travel agent or traveller and, in fact, most agencies have at least one copy readily available. But the guide is used extensively also by business travel planners and by business travellers themselves. In fact, Reed Travel Group publish pocket editions of the OAG *World Airways Guide* that are known as the OAG *Pocket Flight Guide*. There are four pocket editions, which cover: (i) North America; (ii) Europe, Africa and the Middle East; (iii) Latin America and the Caribbean; and (iv) the Asia Pacific region. These guides have a total combined circulation of some 300,000 world-wide. There is also the *Africa Flight Guide*, which is published locally.

Historically, in the pre-CRS days, the OAG *World Airways Guide* was the first thing a travel agent used to arrange a customer's airline itinerary. The process went something like this: (a) the travel agent would take down details of where the customer wished to travel and the cities that needed to be included in the itinerary; (b) the OAG *World Airways Guide* would be consulted and the first city would be looked up to find the flights from that city to the next one on the customer's itinerary; (c) if there was no direct flight from that city to the customer's next city on the itinerary, then a combination of the atlas and the OAG *World Airways Guide* would be used to identify the next nearest city; (d) the process would be repeated until the flight details of each city pair had been ascertained and noted; and then finally (e) as far back as 1959, the agent had to give the details of all flights to the first airline on the itinerary, which would make all the reservations in the list. Quite a lengthy process as you will, I hope, by now appreciate.

So, when CRS terminals began to appear on travel agency desks, they changed the way in which the OAG *World Airways Guide* books were used. This is because the CRSs have most of the information built into their data bases and provide a 'look and book' facility. The OAG *World Airways Guide*, on the other hand, while not being able to offer this transactional facility, still has an important role to play in a travel agent's operation. It does this by continuing to offer a unique, reliable and fully comprehensive source of unbiased world-wide scheduled flights and transfer connection information that is updated and published monthly in chronological order. There are also many other sister publications that are too numerous to cover in detail here. For example, there are the following, although please bear in mind that this is not a comprehensive list, but simply a list of particularly relevant publications:

- **OAG *Guide to International Travel*** A hard-copy directory containing destination information on visa, health, passport, customs, currency and local business hints. Also includes information about major airports, cities and general travel related information.
- **OAG *Agents Gazetteer*** A set of six hard-copy reference books for UK travel agents containing unbiased information and reports on accommodation properties, key leisure destinations and resorts world-wide.
- **OAG *Air Travel Atlas*** Provides an easy to use guide on routes for all domestic and international scheduled flights world-wide. It also contains full coding structures for airlines, airports and cities. This publication is updated twice each year.
- **OAG *Rail Guide*** A portable hard copy directory of UK rail services in London and the South East plus Eurostar services. First published in 1853, it is the only guide with timetables and fares combined.
- **OAG *Travel Planner*** A desk-top reference guide listing hotel information, destinations and transportation information. It includes city maps and airport diagrams and is published quarterly.
- **OAG *Cruise & Ferry Guide*** A hard-copy directory that provides essential information on cruise itineraries and ferry schedules world-wide, including ocean, river and waterways, as well as cruise ship and ferry profiles. Updated quarterly.
- **OAG *Holiday Guides*** A set of four hard-copy reference books for UK travel agents containing details of programmes featured by all bonded, commission-paying UK tour operators.

Table 3.3 OAG FlightDisk functions

Function	Description
Airline schedules	Around 650,000 world-wide flight schedules including direct and transfer flights for domestic and international routes
Itinerary planner	Comprehensive itinerary planning and storage
Print	A facility for printing selected travel information and personalized itineraries
Notepad	An electronic jotter for temporarily storing flight details while planning complicated itineraries
Directory	Additional information covering aspects of travelling abroad such as health, currency and destination details
Files	A private store for the travel planner's own information relevant to a company and its travellers
Help	Comprehensive on-screen help

However, even in the era of automated systems like GDSs, the OAG *World Airline Guide* still has a sometimes crucial part to play in arranging travel itineraries. Take the following example. A friend of mine was planning to spend some time in St Malo and during his stay there, wanted to fly to another town in France for a short business trip. He visited the local travel agency to enquire about possible flights. The travel agent first used the GDS to make an enquiry but this drew a blank. There were no direct flights or connecting flights shown in the GDS data base, from St Malo. So, the agent pulled down the OAG *World Airways Guide* and together with a rather dog-eared atlas, began to search for the nearest city to the desired destination and an appropriate airline and flight. Eventually, this was successful and my friend was sent on his way with an appropriate airline ticket. So, I hope you can see from this little example that the OAG *World Airways Guide* book still has an important role to play in a travel agent's operations.

OAG's electronic products

In November 1991 Reed Travel Group decided to re-position its electronic products by using new optical disk and PC LAN technology. A brand new set of products was launched, the core of which is derived from the renowned OAG *World Airways Guide*. The new electronic product range is based on compact disk read-only memory (CD-ROM) technology; which has achieved substantial growth rates already, with increases forecast over the next few years. This technology was chosen by Reed Travel Group because the alternative – central storage and transmission of information – was considered too costly and time consuming. It would, for example, take a large number of floppy disks to store all of the OAG *World Airways Guide* data whereas just a single compact disk can store the same amount of information. CD technology, therefore, is characterized by the ability to store enormous volumes of data reliably that can be accessed quite quickly by relatively inexpensive PC devices. Nevertheless, if a user wishes to receive the product on diskette then this is also possible. The core electronic products are OAG FlightDisk and OAG HotelDisk.

Disk-based products

The disk-based products are designed for 'stand-alone' use and are implemented on a user's workstation or lap-top PC. To use either the FlightDisk or HotelDisk electronic products, users need a personal computer, i.e. 386SX or better, with a high resolution VGA colour monitor and a 3.5″ diskette drive *or* a compact disk device. It is the software in conjunction with the information on the disk that provides the interactive air and hotel displays. The information on flights and hotels is recorded on a floppy disk or on a compact disk, which looks just like an audio CD. Information

Table 3.4 OAG FlightDisk and HotelDisk LAN version hardware and software requirements

IT requirements	OAG FlightDisk	OAG HotelDisk
File server	Hard disk space: 12 Mb Microsoft Windows 3.11 or higher 3.5″ diskette drive or CD-ROM drive	Hard disk space: 12 Mb Europe 8 Mb Asia Pacific 20 Mb North America 40 Mb World-wide total Microsoft Windows 3.11 or higher 3.5″ diskette drive or CD-ROM drive
Workstation	PC with 386SX processor or better VGA monitor Microsoft Windows 3.11 or higher 4 Mb RAM Printer (optional)	PC with 486SX processor or better VGA monitor Microsoft Windows 3.11 or higher 4 Mb RAM Printer (optional)

cannot be erased or re-recorded on these CDs, so, when a new version of the disk is needed, a replacement disk is distributed by OAG. More information on these two core electronic products is given below:

- **OAG FlightDisk** OAG FlightDisk is updated monthly in the UK and twice each month in the USA (for LAN versions only – see below). It provides information on scheduled airline flights and transfer connections plus additional reference data on all aspects of travelling abroad. It also contains information on airports and ground transport facilities, destination information on more than 200 countries, vaccination and visa requirements, public holidays, banking and business hours, general business hints and local etiquette. When putting together a trip around a series of meetings the user simply keys in the departure and destination airports, date of travel, preferred timings and airlines. Then, based on these criteria, it will display the appropriate information and highlight the most suitable flights. The OAG FlightDisk product uses the familiar Windows GUI. The facilities available are as shown in Table 3.3.
- **OAG HotelDisk** A sister product also is available for hotels, which is called OAG HotelDisk. HotelDisk is updated quarterly and is aimed at travel arrangers within the corporate market. This works in a similar way to OAG FlightDisk but has the additional benefit of hotel maps that show hotel locations and enable users to see walking or driving distances at a glance. More than 46,000 hotels can be selected by name, location, quality rating, room rate or any combination of up to 30 different hotel amenities. All essential details that describe the hotel, its contact details and up to 30 amenities are shown on the HotelDisk display. Users can plot their homes and overseas offices on HotelDisk maps. This enables walking time and taxi distances between hotels and offices to be determined. Companies can store their negotiated hotel rates within HotelDisk. Although OAG FlightDisk and HotelDisk products are aimed principally at the corporate traveller, they could also be of great use to a travel agent.

LAN-based products

Both the OAG FlightDisk and OAG HotelDisk products are available in two basic versions: (i) a single-user version in CD-ROM and floppy disk formats, which is published each month and distributed to subscribers; and (ii) a LAN version, also published monthly, which allows a user with many PC workstations to share a single source of OAG FlightDisk information on their servers. The first of these, i.e. the single-user CD-ROM/floppy disk-based product, is aimed mainly at corporate travel arrangers, who can access the disk from their desk-top or lap-top PC. This version has grown in popularity by 87 per cent since 1995. The second, i.e. the LAN version, is aimed mainly at the corporate buyer market, which comprises well or-ganized companies with the following characteristics: (a) they have many travellers who need travel information for pre-trip planning purposes, and (b) they maintain a high level of control over their travel patterns. In order to use the OAG FlightDisk or OAG HotelDisk products, the user needs the hardware and software shown in Table 3.4.

The LAN version includes a Gateway option, which can be customized by the corporate user. This allows a company to design its own home page through which its employees access the OAG electronic products. The home page may contain company branding and shows a main menu that includes the following options: (i) flight information, (ii) hotel information, (iii) company travel policy, and (iv) latest travel news. This allows a company to record its travel policy on the corporate server for reference by travellers and travel arrangers. Other menu items, such as travel related news, can be added as needed by the company. The system also provides the ability for an itinerary to be stored as it is built up and then displayed and printed at the end of the session. In fact, any page may be printed at any time during the use of the system.

The market for electronic travel information

The OAG FlightDisk and OAG HotelDisk products are sold by subscription and are aimed at the business travel and corporate markets. They are PC based and are not priced on the frequency of access. These products have therefore been a commercial success over the longer term as both travel agents and companies migrate to PC technology as part of the general office automation movement.

Also there is no need for travel agents to be concerned by the availability of the OAG disk products to businesses. The chief benefit of this product from a travel agent's viewpoint is that it has the potential to make the travel agent's life easier. In theory, a business customer would consult the OAG travel disk in order to help plan the trip. When a fairly detailed itinerary has been constructed in this way, the customer can telephone the travel agent with a request to book the required seats. This takes a great deal of the workload away from an agent. If only all bookings could be that simple!

Looking to the future, it is possible that developments in hand-held computers – known as 'palm-tops' – might make it feasible to have a portable OAG *World Airways Guide*. Either the data could be loaded directly into the palm-top's memory or, the data could be inserted on a high-capacity disk. We shall have to await developments in this new technology to know whether or not there is a demand for such a service.

Product distribution

It is technically possible for suppliers to distribute their products to travel agents directly. In the past this approach has been taken by some airlines, international hotel chains and car rental companies. They have provided selected travel agents and other high volume users with terminals connected by communication lines to their corporate mainframe reservation systems. Bearing in mind the expense of doing this, a direct connection approach is usually only provided to extremely high volume users, such as central reservations units operated by some of the major UK multiple travel agencies. However, if suppliers could provide travel agents or even consumers with direct access to their data bases, without any other system being 'in the way', then sales would arguably be maximized. However, providing agents with dedicated terminals connected to a single supplier system does have some drawbacks:

- **Single access** Only the largest of hotel chains and car rental companies provide direct access to their inventories as the airlines used to do. This is partly because the reservation systems used by hotels and car rental companies are not as heavily regulated as airline CRSs and GDSs and partly because their volume in relation to airline ticket sales are relatively low. So there have historically not been many common networks or distribution methods for these companies to use that can easily reach the travel agent, other than the airline CRSs and the more recent GDSs (although this is changing with the advent of the Internet). Consequently, at present you won't find many single company terminals used for airline, hotel or car rental reservations installed in travel agents. Most agencies now use GDSs. While GDSs are usually considered fine for general purpose supplier reservations from the agent's viewpoint, they are nevertheless, not so ideal from the supplier's perspective. This is because they show competitors' products and allow the buyer to compare prices and service levels easily. Ideally, suppliers would rather a potential consumer or agent communicated with them directly during the sales process. However, this is not so because the GDSs have, over the past ten years or so, stolen the high ground and now dominate the agency distribution channels at present. The Internet is changing this situation and this is explored in more detail in Chapter 5 – The Internet.

- **Focus on business travel** Air, hotel and car rental customers tend to be business travellers. The leisure traveller usually purchases a holiday package in which the hotel and possibly the car, have been obtained in bulk and factored into a package price. Business travellers on the other hand need to purchase these products on an *ad hoc* basis often at short notice. The problem from the supplier's viewpoint is how to reach the business travel agent or the corporate business traveller other than through a GDS. In the past there has been no real answer to this question. Only viewdata offered a potential alternative although there are substantial barriers, for example: (a) viewdata is principally a leisure travel technology, not ideally suited to business travel with its demand for high speed response; and (b) there are a lack of viewdata PCs in business travel agencies and companies. Once again, however, the Internet offers a very real alternative method for suppliers to establish a direct communications channel with their customers.

- **Relatively low revenue source for non-air products** Although hotel and car businesses are profitable, they are regarded by many agents as very much secondary to air sales. One reason for this is that it is difficult to track the commission due from hotel sales (although see Chapter 8 for a description of the commission tracking systems that overcome many of these problems). Each transaction is usually for a relatively small amount of revenue and the overhead involved in keeping tabs on which hotels owe what amounts for large numbers of bookings is a chore most busy travel agents can do without. A back-office system can help here but there really is not the same demand from agents for dedicated hotel and car system terminals, for this reason. In other words, it has not been economically feasible for suppliers to install their own dedicated reservations terminals in agencies. Instead, they have used the GDS route to travel agents.

The net result is that there are not many dedicated airline, hotel or car reservation terminals in travel agency locations. The bulk of the automated reservations come from the GDSs that co-host or connect to most of the major car and hotel systems. Nevertheless these travel supplier companies do market their own dedicated systems to a small niche within the travel agency population. The thrust of their marketing effort tends to be the very large business travel agents, the headquarters of some of the large travel agency multiples, some very large companies and the specialist hotel reservation centres whose sole business is making hotel reservations for agents, companies and the general public.

It is for these reasons that I shall not be covering dedicated airline, hotel or car systems in any detail in this book, i.e. single company terminal systems. In my view, most of what you need to know on this subject can be found in the GDS chapter and in the description of hotel distribution systems like Thisco and Utell. Finally, any opportunities for suppliers to distribute their products to travel agents, companies and consumers directly, although problematical in the past as described above, are now becoming realities with the advent of the Internet and Intranet technologies, which I address in Chapter 5.

4
Distribution systems

Introduction

In Chapter 3, I discussed how suppliers have used IT to run their businesses and support the sales process. The next step is to consider one of the ways in which these suppliers can distribute their products and services to consumers. GDSs are presently the leading distribution channels for most major travel suppliers. This is certainly true for airlines, hotels and car rental companies, all of which participate in GDSs. There has been a rapid rate of change within the GDS sector over the past few years and no doubt this will continue as new electronic distribution channels open up. But before we explore newer distribution channels like the Internet, it is very important to understand clearly how the established distribution systems work because they have been channelling the vast majority of travel products and services via travel agents to consumers in both business and leisure sectors of the market, for some years now. Incidentally, let me expand upon what I perceive as the difference between a CRS, a GDS and a HDS. These are important distinctions because the terms seem to be used interchangeably by many people in the industry.

- **CRS** A CRS is a travel supplier's own computerized reservation system. It is owned and operated by the travel supplier, although some CRSs may provide co-hosting services to other suppliers – rather like a kind of outsourcing arrangement for smaller travel companies. The term CRS is mainly used to describe an airline's own computer reservation system, usually a large main-frame computer. In order to connect to a CRS, travel agents used to have a dedicated dumb terminal that was connected only to a single airline's CRS computer, or in a minority of cases, a dedicated hotel reservation system. All other communications with other airlines were done via the airline to which the travel agent was connected.
- **GDS** A GDS is a super switch connecting several CRSs. Each GDS is powered by a large main-frame computer that performs many of the end-user functions that are delivered to travel agents using PC-based terminals. GDSs use a co-operative processing architecture in which some functions are driven directly by the GDS computer and some are controlled by the airline CRS selected by the end user. Most GDS computers have their own large data bases which are used primarily for the indexing and control of booking records.
- **HDS** A HDS is rather like a GDS that has been designed exclusively for the hotel business. At its core is a large computer system that often comprises several super-servers. These large servers can provide certain operational functions to hotels that access the HDS via on-line terminals. In addition to this, the HDS is connected to other hotels that have their own in-house reservation systems and to the major GDSs. In summary then, HDSs distribute hotel systems to GDSs.

GDSs were formed from alliances of several CRSs, each of which had its own airline backer. So, their original formation was to some extent influenced by intra-airline relationships as well as the technical architecture of each airline's CRS. Once formed, there was a period of some consolidation and shake-out, after which four main GDSs emerged: Amadeus, Galileo, Sabre and Worldspan. In addition to these four major GDSs there are also four CRSs or smaller regional CRSs, each of which has its own particular niche. We are still witnessing the next stage in the GDS's life cycle. This is the gradual distancing of the parent airline owners from the GDSs they originally created. The degree to which GDSs become truly independent of their airline creators remains to be seen. But let's take the first step in our exploration of GDSs and discuss in a little more detail, exactly what a GDS is and is not.

WHAT IS A GDS?

The term GDS is used primarily to describe the systems that travel agents use to book airline seats for their customers. Most of the world's major GDSs are therefore owned by airlines and most travel agents are connected to one of the four major systems (more on that later). But in fact, GDS technology is also used within the hotel industry to distribute accommodation services to travel agents and consumers. The common technological thread running through both types of GDS is the so called *switch*.

In this context a *switch* is simply a computer that is connected on the one side to many different supplier systems and on the other side to many different end users. The systems connected to the supplier side of a switch are generically known as *host systems*. The end-user side of a *switch* is more varied. End users of a *switch* comprise travel agents, consumers and other distribution networks that themselves make the supplier services available at the point-of-sale. So, it is possible for these *switches* to be inter-linked and in fact, many of them are: it is this inter-linking that makes the whole subject of *switches* and GDSs so complicated.

For historical reasons, the airline GDSs currently 'own' the travel agent distribution channel. The reasons for this are as follows. It was the individual airline CRSs that started offering their reservation terminals to travel agents, many years ago. At this time, the hotel industry was only just getting established in the field of marketing automation. In those early days of travel and tourism technology, hotels were focusing on building computer systems to maintain and control their inventory of rooms and associated accommodation services. So, by the time hotels were sufficiently developed in the area of computerized room inventories and reservation systems, the airlines had already sewn up the market for terminals in travel agencies. Also, airline ticket sales have always been both: (a) high volume, and (b) high revenue earners for travel agents. So, it made sense for travel agents to invest in a technology that maximized their airline sales productivity.

Hotels therefore found themselves in the position of having to connect their reservation systems into the airline GDSs in order to reach the travel agent. But this threw up further challenges for the hotels. It meant that in order to reach a global spread of travel agencies, they were required to connect to several different GDS switches. The problem is that each GDS has different interconnection requirements. So, the cost incurred by a hotel in connecting its reservations system to multiple GDSs was quite significant. This is why hotels have banded together and created their own special type of GDS.

So, let's first start with a review of the airline GDSs. Although similar in concept, each one has its own individual characteristics. That is why I have included all four major GDSs in this section. A good understanding of each one of these is critical to several other distribution technologies that I cover in other sections of the book; then, once you have a good understanding of airline GDSs, we can go on to look at hotel distribution systems (HDSs), what their objectives are and how they work.

Airline GDSs

Airlines distribute their products, i.e. airline tickets, to customers via several channels. The cost of doing this represents one of the major portions of an airline's bottom line expenses and is therefore a major determinant of profitability levels. To illustrate this, airline distribution costs average around 18.2 per cent of an airline's total expenditure on international services (*source*: Pierre Jeanniot, IATA Director General, 3–6 November 1996): a great proportion of these costs are directed at electronic distribution systems or GDSs. An airline GDS is a *switch* that connects several airline CRSs to travel agents. However, the use of the term switch rather underplays the functions that are provided by airline GDSs. For a start they do a lot more than just switch message traffic between CRSs. Most GDSs also provide a great deal of functionality themselves, for instance, PNR consolidation, common language interfaces and the control of remote ticket printing; and the networks that some of the GDS switches use, are massive wide-area networks comprising thousands of telecommunication lines and dedicated switching computers that span the globe.

Historically, the principal GDS customer has always been the travel agent, with just a handful of large companies also using the system. However, with the advent of the Internet, the consumer has joined the ranks of the GDS customer base. What both sets of customers are looking for is access to supplier systems that provide travel products and services, and this is where the GDS world starts to become segmented. In the USA, 80 per cent of all travel bookings are accomplished perfectly satisfactorily using a GDS system. However, in Europe and other parts of the world, only 20 per cent of all bookings are done via a GDS. The reason? Well, in the USA, most people travel by air and sometimes stay in a hotel or rent a car. In Europe, people travel a lot more by other means, such as rail and ferry. Also, package holidays are far less important in the USA, whereas in Europe they are a large part of the travel market.

In the early days of CRSs the terminals were offered to travel agents by the airlines themselves. These were the bad old days of bias where airlines deliberately showed their own flights at the top of the list when an itinerary was being developed in a CRS by a travel agent. However, along with deregulation came a dictate that no CRS could bias its system either in favour of its own flights or indeed against those of its competitors on an identical itinerary. This took the sting out of the CRS as a principle marketing tool for the airline business itself.

Over the following years, airline systems became ever more sophisticated and more importantly, CRSs became almost indistinguishable in terms of the functions offered, i.e. the core airline booking functions, such as last seat availability. More recently, airlines are beginning to devolve themselves of GDS ownership. This has occurred principally because trading block bodies such as the EC consider it unfair for large airlines to influence their sales to consumers unduly as a direct result of their size. For example, mega-carriers can invest substantial sums in GDSs and thereby enjoy an advantage over their smaller competitors who cannot compete on resource grounds alone. All this has been done in the interest of the consumer.

However, there are still some carriers who participate in GDSs but block some of the functions of their systems when accessed within their home market. This is usually when the airline distributes its own CRS within its home market. In such cases, this action tends to boost the usage of the CRS and hence the sales of the airline that distributes it. Because this occurs in only the minor markets of the world, it is not a significant point. What is significant is that generally speaking, airline reservations functionality is virtually a level playing field for all airlines. This is especially true in terms of the way in which airlines participate in GDSs.

So, all of these factors have changed the way in which both the airlines as suppliers and the travel agents as intermediaries view the GDSs of the world. It is for these reasons that GDSs now compete for travel agency users far more, using factors such as geographic distribution, richness of functions, access to non-air suppliers, reliability, customization and so on. Airlines focus on increasing revenues by improving their core service and GDSs focus on increasing revenues by attracting more users and generating higher booking volumes for which they derive a booking fee from the recipient airline.

The initial focus of GDSs was to provide travel agents with a single reservation system to support the sale of airline seats and related travel products, such as hotel and car hire, via a single computer terminal, usually a PC. GDSs require a massive investment because they are extremely large computer systems that link several airlines and travel principals into a complex network of PCs, telecommunications and large main-frame computers. There are therefore very few GDSs in the world today. Some say that eventually there will only be two or three GDSs in the next decade as more

Table 4.1 The world's major global distribution systems

GDS	Terminals	Locations	Countries	Airlines	Hotels chains	Properties	Cars
Amadeus	168,000	39,000	117	440	268	35,000	55
Galileo	128,000	33,000	66	500	208	37,000	47
Sabre	130,000	30,000	64	400	215*	35,000	50
Worldspan	39,101	15,000	45	414	182	26,000	40

*Sabre's hotel chains exclude chains within chains

Figure 4.1 The Amadeus logo

mergers take place and more strategic alliances are formed. Table 4.1 shows some statistics that characterize the world's major GDSs.

In the following sections you will find a presentation of all four major GDSs: Amadeus, Galileo, Sabre and Worldspan (plus a summary of Infini, a Far East GDS). They are presented in alphabetical sequence and hopefully in a completely unbiased way.

AMADEUS

Amadeus (Fig. 4.1) is now the largest of the world's four GDSs. Its 168,000 terminals are used in over 39,000 travel agency branches and 8,500 airline sales offices around the world. It serves 117 countries and employs a total of almost 43,000 staff. Amadeus is connected to 440 airlines (of which 102 are directly connected), 268 hotel chains (representing over 35,000 properties) and 55 car rental companies.

Amadeus Global Travel Distribution was formed in 1987 by Air France, Lufthansa and Iberia. In April 1995, Amadeus Global Travel Distribution acquired the System One CRS from Continental Airlines of the USA. It is now a global GDS owned by its founders, with Lufthansa, Iberia and Air France each holding 29.2 per cent of the shares and Continental the remaining 12.4 per cent. In addition to these owners, Amadeus has over 32 partner airlines, each of which is directly connected to the Amadeus system. The company has a decentralized organizational structure, with its headquarters and marketing functions located in Madrid, Spain, its development activities in Sophia Antipolis, near Nice in France, and its operational centre in Erding, near Munich, Germany. National marketing companies (NMCs) are located in each major country. It is worthwhile exploring the components of this structure in a little more detail before we dive into the Amadeus system functions in more detail:

- **Amadeus Global Travel Distribution (Holdings)** Based in Madrid, this group supervises the group's activities and co-ordinates the subsidiary companies. It is responsible for the definition of corporate and financial strategy and goals as well as controlling the financial and legal obligations of the Amadeus group.
- **Amadeus Marketing** Responsible for all product marketing related functions, this group also co-ordinates the activities of all Amadeus NMCs in each country. It controls the flow and distribution of income between the service providers, the NMCs and Amadeus product marketing and management. Another set of key functions comprises the documentation, training and help-desk functions for NMCs, airline users and service providers.
- **Amadeus Development** This is the central systems development group and it is based in Sophia Antipolis. It designs, develops and tests

all Amadeus Central System software and provides a maintenance function for the group. Besides the central system, Amadeus Development is responsible for the AmadeusPro workstation products. In general, this group is charged with research and development activities for new Amadeus products.
- **Amadeus Data Processing** This group runs the Amadeus central computers and the global network (AMANET). It is located in Erding, just outside Munich in Germany and is home to a special purpose computer building that houses some of the world's largest main-frame computers and a whole host of smaller interlinked machines, e.g. six IBM main-frames, two Amdahl main-frames, four Unisys main-frames with 68 processors and massive amounts of magnetic data storage. The operating system environment is in several parts: an IBM transaction processing facility (TPF) for reservations applications, an IBM virtual machine (VS) and IBM multiple virtual storage (MVS) for software development and support. The standard Unisys operating system OS 1100 is used to run the Fare Quote system. These computers have a maximum capacity of 900 end-user transactions per second, and the current workload of approximately 30 million transactions per day is therefore well within this ceiling.
- **NMCs** There is one Amadeus NMC located in virtually every major country. Each NMC is responsible for the marketing of Amadeus within its own geographical area. This includes functions such as local customer service, help-desk support for travel agency users and running the national distribution system (NDS) network. Most NMCs own their local NDSs through which subscribers gain access to non-air services including rail, ferries, tour operators, event ticketing and back-office systems.
- **Partner airlines** Many Amadeus partner airlines own a share-holding in their local NMC (through which Amadeus is provided to subscribers). These airlines also use Amadeus for their own internal purposes.

The acquisition of System One from Continental Airlines in April 1995 was a significant move for Amadeus, and it is therefore worthwhile examining this in a little more detail. To be precise, the acquisition made by Amadeus was of all the CRS assets of Continental Airlines' System One subsidiary. As a by-product of this acquisition, Continental obtained a 12.4 per cent stake in Amadeus. Overnight, Amadeus became the largest single GDS in the world.

System One's travel agency customer base and information management software were transferred into a newly formed company called System One Information Management LLC. This new company is owned in equal parts by three stakeholders: Amadeus, Continental Airlines and Electronic Data Systems (EDSs). The resulting entity operates within the Amadeus organization as an NMC. It provides information management, marketing, distribution and customer support functions to travel agencies in the USA, Canada, Mexico, Central America, the Caribbean and markets in the Pacific area.

The System One Information Management LLC is the largest NMC in the Amadeus group with its products used by the top ten USA travel agency chains that themselves service a significant share of America's business travel market. This NMC has been charged by the parent company with the further development of the Amadeus product line, drawing on resources from EDS and Amadeus itself. The objective is to make all new products developed in this way available to all Amadeus subscribers around the world. This development programme has been assigned the name Operation Unison. This specifically includes the constant and gradual enhancement of System One functionality, culminating in the transition of all System One Information Management subscribers to the Amadeus Central System in the Erding data centre. In fact the first stages of this transition have already been completed with the linking of the two computer centres and the introduction of new functionality that enables subscribers to share bookings across both computers. The end-date for the complete transition was year-end 1997.

So, Amadeus operates its GDS through the decentralized organization described above. However, in addition to this, it also keeps close contacts with its primary subscribers through the Travel Agency Advisory Board (TAAB). This is a consultative group formed in 1989 with members from travel agencies in several different markets.

The prime objective of TAAB is to assist Amadeus in product planning and development activities.

The underlying IT architecture of Amadeus is subtly different from the other three major GDSs. In almost a kind of reflection of its organizational structure, Amadeus has adopted a more decentralized method of inter-connecting its systems components. At the core is the Amadeus Central System that controls access to the core supplier systems on a global scale, i.e. air, hotel and car rental. Then at the NMC and NDS level, i.e. the local market level, Amadeus provides interfaces to non-air suppliers, such as rail companies, ferry companies, tour operators and events providers. Finally, there is the AmadeusPro software, through which subscribers gain access to Amadeus and use it on a day-to-day basis. Let's take each of these components in turn.

The Amadeus Central System

The Amadeus Central System supports workstation access and host booking functions for global airlines, hotel chains and car rental companies. These central systems applications, which run on the Amadeus main-frames housed within the Erding data centre, fall into two main types: (i) host connectivity applications, and (ii) core end-user support functions. I'll cover each in a little more detail. First, the host connectivity applications for air, hotel and car.

Amadeus Air

The Amadeus Central System supports the flight schedules of more than 740 airlines of which 440 may be booked by subscribers. Of the 440 airlines, 220 participate with Amadeus in Standard Access, 140 in Direct Access, 150 in Amadeus Access and over 230 airlines show last seat availability displays. Incidentally, these individual figures do not add up to 440 because many airlines are connected to Amadeus so as to provide multiple levels of participation, i.e. they are counted more than once. So, it is important to understand the various levels of airline participation that Amadeus provides. The following are the main ones:

- **Amadeus Standard Access** This is the most basic level of participation. Flight schedules are loaded into the Amadeus data base from magnetic tape provided by participating airlines. All information is then shown on Amadeus principal displays, but seat availability and subscriber reservations are made using teletype messages that flow between Amadeus and the host airline's CRS.
- **Amadeus Direct Access** This enables users to be connected to the participating airline's CRS and therefore their seat inventories, on what is known as a secondary carrier specific display. This means that all messages are presented to the user in standard Amadeus format and all information on seats, schedules, fares and flight details is right up-to-date. Seat sales are reported to the participating airline by teletype and once confirmed, are guaranteed.
- **Amadeus Access** This is the highest level of participation and is achieved by means of an on-line real-time link between the Amadeus GDS and the participating airline's CRS. Three functions are supported within Amadeus Access:
  - *Full Amadeus Access* Participants at this level enjoy both the Amadeus Update and Amadeus Sell functions described below.
  - *Amadeus Update* This enables end users to view real-time schedule information including, for instance, flight irregularities and last minute changes, on their principal displays. When the system shows nine seats available, this actually means that at least nine seats are available for sale. When the system shows a number less than nine, then this represents the actual number of seats left on the flight.
  - *Amadeus Sell* With this function, when a user sells one or more seats, the sale is immediately confirmed by the participating airline's CRS. The airline's own record locator is sent to Amadeus and stored in the PNR. This is achieved via a special function invoked by the user, called Record Return.
  - *Record Return* This is a function by which a participating airline's CRS acknowledges the receipt of a sale made in Amadeus by sending its own record locator for display to the user. This may be requested by the end user who makes an RL entry on an Amadeus PNR. Record Return is automatically integrated within Amadeus Sell but is optional for airlines participating in Amadeus Direct Access.

I have mentioned the term Principal Display several times, so it is about time I defined it. First of all, Amadeus's Principal Display confirms to the neutrality rules of the EC Code of Conduct for CRSs (see Chapter 1). This means that all flights shown on a Principal Display are shown in sequence by: (i) direct flights by the departure time closest to that specified by the end-user's availability request entry, and (ii) connections by total elapsed time of journey with shorter times at the top of the list. The Principal Display is the display shown by Amadeus of its participating airline CRSs. Amadeus considers this to be accurate and constantly up-to-date thus obviating the need to view secondary displays that are produced directly from participating airline CRSs themselves. Amadeus Principal Displays may be customized by the end user in several ways. This is an important feature of the GDS, which I will therefore illustrate by way of some frequently used examples as set out below:

- **Neutral schedules** This is the basic display that shows the schedules of flights according to EC ranking guidelines as described above. It includes the schedules of all participating airlines, including those that are full and supports bookings on all participating airlines. Neutral schedules may be modified by users to rank flights by elapsed time, departure time or arrival time. They may also be restricted to the flights, or combination of flights, of up to three airlines.
- **Neutral availability** This type of Principal Display shows only those flights that have seats free for booking purposes. Possible modifications of this type of Principal Display are to show flights by connecting point, connecting time, time of departure or a list of flights up to one week ahead.
- **Dual city pair** This shows flights between the city of origin, i.e. the departure city, and two different destinations on the same screen. Alternatively, the display may show the outward and return flights on the same screen.
- **Carrier preferred** This type of Principal Display is only available for those airlines that participate in Amadeus at the Amadeus Access level. This shows only flights of a chosen carrier or other airlines that the chosen carrier wishes to display. All such flights are ranked by the carrier's chosen preference.
- **Direct access** This is identical to 'carrier preferred', as described above but with the added modification that displays may be ranked by the time of departure. Other modifications to the ranking are supported according to the participating airline's choice.
- **Timetable** Shows a timetable of all airline flights between a given city pair. The display may be ranked by date range or a specific day of the week.

The above Principal Display customization features are very important to a travel agent, or an airline sales office come to that. It enables the end user to tailor the GDS display to meet the needs of his/her own particular business or corporate customer: all this is provided by Amadeus under the umbrella of the EC Code of Conduct, which aims to limit the power of GDS systems to discriminate unfairly between airlines for commercial advantage.

A critical function of any GDS that is closely linked to flight availability is the fares that pertain to those flights. The Amadeus Fare Quote system supports this key function. It is a system based originally on the SITA and Air France fare quote systems. Amadeus Fare Quote now holds over 50 million specified fares and can build a virtually unlimited number of special fare combinations via its dynamic add-on processing capability. It obtains these fares from a number of sources, including SITA, ABC, ATPCO, IATA and approximately 90 airlines that maintain their fares on-line using the Amadeus Fare Quote system.

The Amadeus Fare Quote system can price up to 12 booked flight segments with 11 fare components. It can also support *Fare Driven Availability*, which allows the user to request a display of all flights on a specified itinerary that meet certain fare criteria. The *Best Buy* feature is also very useful in today's price sensitive markets. It automatically finds the lowest possible fare for an existing booking and allows the travel agent to re-book the itinerary using the lower fare. Finally, there is *Informative Pricing*, which allows an itinerary to be priced without having to create a PNR first.

Amadeus Cars

Amadeus has on-line links to most of the world's major car rental companies. This allows subscribers to book car rental services and either include them as part of an existing air PNR or to provide them as a separate stand-alone service. The computerized inventory control systems of car rental companies may be connected into Amadeus via two alternative methods:

- **Amadeus Standard Access** This method of participation is based on standard teletype messages that flow between the car rental company's computer and the Amadeus Central System computer. This allows booking messages to be sent to the car rental company, which responds by returning confirmations formatted as teletype messages.
- **Amadeus Complete Access** This is based on high speed telecommunications links between the car rental company's computer and the Amadeus Central System computer. When a booking is made, a booking request message is instantly sent from Amadeus to the car rental company's computer. This returns a message containing a confirmation number and other relevant data, within a period of between four and eight seconds.

Car rental displays fall into four main types: Car List, Car Inventory, Car Availability and Car Shopper's Guide. The Car List display shows all car rental offices and their proximity to local airports. Car Inventory shows car companies serving a specific location and displays the availability status of each car type, for each company. Car Availability provides availability and rates for a range of car types, including rate plans, rate categories and mileage charges. Finally, the Car Shopper's Guide is a multi-company display that searches for all car types and/or classes and gives an indication of the best car rental value in ascending order based on the period of rental and the location specified. All displays may be customized by the end users according to their own and their customers' preferences.

Amadeus Hotels

Over 35,000 properties around the world may be booked directly via the Amadeus Hotels facility: and in addition to the standard set of hotel rates available via participating chains, a facility to allow travel agents to store their own specially negotiated hotel rates was introduced in 1995. The Amadeus Hotels function was developed alongside the car rental system previously described. The advantage of this to the user, is that a common approach is used for both products. The levels of participation, for example, are Amadeus Standard Access and Amadeus Complete Access, which are virtually the same as described for car rental above.

Even the hotel display formats are similar in concept. These are Hotel List, Hotel Inventory, Hotel Availability and Hotel Features. Hotel List shows a comprehensive list of hotels for a specified city or country including the location of the hotel in relation to the city centre and the recommended form of transport from the airport to the hotel. Hotel Inventory shows room availability by type of room, over a 48-day period. Hotel Availability, as the name implies, shows the availability and rates for a combination of room types, including property codes, area identifiers, currency codes and hotel feature indicators. Hotel Features gives detailed information on the features and facilities offered at each property as well as booking policies and negotiated rate booking procedures. As for car rental, each display may be customized by the end-user according to their mix of business and the needs of their corporate customers.

Core functions

The core functions that support the Amadeus GDS are grouped under the banner Amadeus Service. These functions are generic and support nearly all the major activities of end users. They fall into two main groups: (a) functions that improve the productivity of end users by automating repetitive tasks, and (b) information that is commonly needed to service customers of travel agencies and the sales offices of partner airlines. Here are just a few of the main ones:

- **Central profiles** This enables travel agents to store personal information about their business travel companies and frequent travellers. Examples of this kind of information are company name and address, department codes, travel policy, traveller dietary needs, individual seating preferences, home address and contact telephone numbers. All of this information may be referenced during the booking process and certain fields may be automatically copied into the PNR. In fact a PNR can be created automatically from a Central Profile *and* a Central Profile can be automatically created from a PNR.

- **Card check and ticket check** This is a function that supports sales made by customers using credit and charge cards. It automatically verifies credit card sales by sending authorization messages to the computers of major card companies. The checks also extend to lost, stolen or blacklisted airline tickets.
- **Ticketing** Amadeus supports all three ticketing methods used around the world. These are: (i) the conventional transitional automated ticket (TAT), (ii) the ATB, and (iii) the new ATB2 (see Chapter 3 for a more detailed discussion of ticketing methods). The support provided by Amadeus for these different ticket types varies with the deployment of each type, on a market-by-market basis.
- **Amadeus' information system (AIS)** This provides users with access to on-line information on Amadeus products and participating suppliers. An up-front news page gives highlights of new additions.
- **Amadeus' instant marketing (AIM)** This is an important feature that suppliers can use to market their products and services to travel agents around the world. It allows suppliers to target promotional and information messages to subscribers via AIS pages, sign-on messages, broadcast messages and display messages (see the special section later in this Chapter).
- **Calculator** A self-explanatory feature but one that has been enhanced also to provide currency conversion functions and encoding/decoding of city, airport and provider codes.
- **TIMATIC** On-line access to passport, visa and health information for all countries.
- **On-line help** An information source that helps subscribers solve problems related to their use of the Amadeus system. This is probably the first step that a user would take in attempting to resolve a problem. If this was unsuccessful, the local Amadeus help desk would need to be contacted by telephone.
- **Practise training** This simulates live use of the system but does not create any live bookings and does not allow any entries to affect the live Amadeus system. It is extremely useful for first time users who wish to try out their recently acquired knowledge of Amadeus without affecting live work.
- **Scholar/teach** This is a self-learning facility that subscribers can use at their own pace to gain a solid grounding in how to use Amadeus. However, it is nevertheless secondary to attending a purpose-designed classroom training course specifically designed to teach people how to use Amadeus.

Finally, one of the most important functions in any GDS is the way in which PNRs are processed. The Amadeus PNR contains all the information related to a customer's travel plans. A travel agency may create and access its own PNRs and it may optionally grant access to its PNRs, to other affiliated agencies, which may be located in other countries. In addition to this, a travel agent may also authorize any one of 102 participating airlines to access and change a traveller's PNR. This can be a valuable customer servicing feature once the traveller has departed on their journey.

The 24 information fields that comprise an Amadeus PNR may be created by the end user in any convenient sequence, at the same time. Amadeus automatically places all such entries into logical or alphabetic sequence. PNRs may be modified by reference to the line number that contains the data to be changed. Flight classes and dates may be re-booked in a single transaction, thus saving time and maximizing an end user's productivity. All such changes are recorded in a PNR history file, which can be very useful when a dispute arises between a customer and their travel agent. Amadeus PNRs may be retrieved by record locator, name of traveller, flight number, frequent flyer number and a variety of other key fields.

Amadeus also provides special PNR features for groups of travellers. Groups, in this context, are more than ten passengers who all share the same itinerary. A specialized feature called Non-homogeneous PNR enables groups to be linked, but for their individuals to each have their own PNR entries. With this feature, the group view may nevertheless, still be displayed upon request. As mentioned above, a PNR can be automatically created from a Central Profile *and* a Central Profile can be automatically created from a PNR. Finally, the Replication facility allows a new PNR to be automatically generated from an existing one using a variety of rules, e.g. copy all segments but exclude elements not associated with the new PNR.

Local travel supplier systems

One of the keys to successfully connecting a number of non-air supplier systems to Amadeus at the local level is the back-bone telecommunications network, which is called AMANET. This is a high speed terrestrial network capable of transmitting 29 million bits per second via 14 data lines, each carrying 2 Mb/s. Incidentally, the word 'terrestrial' as used here, means that AMANET is based on a network of data lines rather than satellite links. The network comprises a total of 14 nodes located at strategic points around the world (in Europe, North America and Asia). In AMANET terms, a node is defined as a group of computers and modules that routes data in the proper direction; and data in this context can be anything from simple text messages to voice and video traffic.

Because Amadeus is totally dependent upon AMANET, Amadeus has taken many steps to protect the integrity and reliability of the network. All high speed main trunk lines, for example, are duplicated; and despite the fact that the core operational network is terrestrial, AMANET uses satellite communications technology to: (a) provide a back-up service in the event that land based lines become unavailable, and (b) transmit data to any point in Asia via its own satellite ground station. The network also has the capability to route messages via alternate paths to avoid areas of congestion or unavailability.

AmadeusPro

AmadeusPro is a PC-based travel agency management system that focuses on providing an easy to use Windows based GUI to the Amadeus Central System. The family of products that falls under the AmadeusPro banner is as follows:

- **AmadeusPro Res**   A software product designed for operation on a standard PC running the IBM OS/2 operating system. This provides a mouse and icon based interface to the Amadeus Central System as well as an interface to locally connected suppliers. It also allows users to work with standard OS/2 based integrated office applications, such as spreadsheets and word processors.
- **AmadeusPro Tempo**   This is almost identical to AmadeusPro Res but is based on a Microsoft Windows operating system environment. This product does not demand such a high level of PC specification, for the workstation platform.
- **AmadeusPro Base**   This is an extended version of AmadeusPro Res that supports locally stored client profiles and other booking reference data. All data stored locally on the PC in this way are used during the booking process in an entirely interactive way.
- **AmadeusPro Sale**   This is a mid-office product that provides support for the reservations functions, local client profiles, local storage of other data and integration with office productivity tools (see Chapter 7 for a definition of front-, mid- and back-office agency management systems).

These products all provide the user with three optional ways to use the software. The basic level is called the Guided Mode. It helps the inexperienced user through the booking process using pre-formatted screen and fill-in boxes. Speed Mode shows more native Amadeus Central System displays and supports special tool bars with colour coded push buttons that are activated by mouse clicks. Amadeus estimates that Speed Mode can reduce input effort by up to 68 per cent and thus help the user achieve significant productivity benefits. Finally, Expert Mode allows the user to work entirely with the native Amadeus Central System.

GALILEO

Galileo (Fig. 4.2) is one of the world's larger GDSs, with 27 per cent of the automated travel agency market. It required an initial investment of £200 million to set it up and was first introduced in 1991. Galileo International is the name of the

Figure 4.2 The Galileo logo

global distribution company that provides two core systems to countries around the world: Apollo in the USA and Galileo throughout the rest of the world. Within each country Galileo has a national distribution company (NDC) that is responsible for selling and supporting the Galileo service locally within that country. There are now 45 NDCs providing coverage across the Americas, Asia Pacific, Europe, Africa and the Middle East. In the UK the NDC is called Galileo UK, which is a part of Travel Automation Services – itself a wholly owned subsidiary of British Airways. But before we look at that, it is worth considering Galileo from a global perspective.

Galileo International

Probably the first major development in Galileo's history was the formation of the United Airlines Apollo system in 1971. In 1986 Apollo's owner was re-branded Covia, which became an independent affiliate of United Airlines. Galileo International was founded in 1987 by British Airways, Swissair, KLM and Covia. Originally, the headquarters were located in Swindon in the UK. More recently it was decided to relocate the headquarters to Chicago in the USA. Other key sites are Denver (where United Airline's Apollo system is based), Miami, Swindon and Hong Kong. In 1992 the European and North American owners of Galileo and Covia combined the two companies to form a major GDS. The combined group now has over 2,000 staff. Galileo International is jointly owned by eleven of the world's major airlines as shown in Fig. 4.3.

Additionally, Galileo International has two associate airlines: Ansett and Australian Airlines. The prime function of Galileo International is to provide the core reservation services for all of the NDCs. It is responsible for the day-to-day operation of the computers, data bases and telecommunications facilities that distribute 500 participating airlines (200 of which are linked directly to Galileo), 37,000 hotel properties and 47 car rental company systems, to NDCs in countries around the world. Between them, these two systems support some 33,000 travel agency locations using 128,000 terminals in 66 countries. The main categories of supplier systems include airlines, hotels, car hire companies, rail operators, ferry companies, sporting

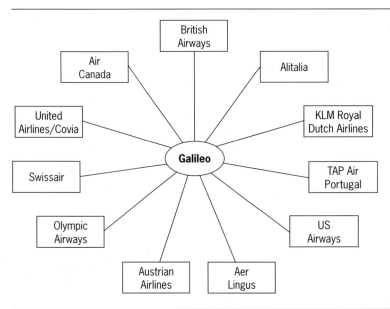

Figure 4.3 Galileo's owners

event promoters, theatre ticket agencies, general travel information providers and much more.

Before we evaluate the end-user aspects of Galileo, it is important to understand the core system. The core system used by areas of the world outside the USA is called the Galileo Central System (GCS). In fact, even USA users are gradually being migrated to this common core system.

GCS technology

The Galileo GDS is a global computer and telecommunications system that uses a central worldwide data base and high speed links to many airline CRSs. Users of Galileo interact with the system via a special airline oriented language that resembles a kind of special 'computer like' code. However, GUIs which run in Galileo's PC-based delivery systems are increasingly making this coded language easier to use for travel agents. This allows the central system to carry on using the highly efficient message format so ideal for computers but so disliked by human beings. Even with these coded message formats a considerable amount of processing power is needed just to run the network.

The central Apollo and Galileo systems are run on 15 IBM and Amdahl main-frame computers and other processors, all of which are housed in a large computer complex in Denver with a combined floor space of 21,924 m^2 (equivalent to 46 full-sized tennis courts). These computers have a processing power of 3,283 MIPS (million instructions per second) and handle over 66 million messages each day. The operating system environment is based on IBM architectures and uses a transaction processing facility (TPF), a virtual machine (VM) and a multiple virtual storage (MVS).

All computers on the global network are connected by a wide area network based on IBM's SNA protocol. Galileo distributes its systems around the world using a variety of telecommunications technologies. The company's sites in Denver and Swindon use a 'meshed' back-bone network to provide direct connections between participating supplier systems and travel agents. The term 'meshed' network means a system of leased high speed communication lines that are routed via alternative points. When viewed as a diagram, this appears to connect each location in a multitude of point-to-point lines that resemble a mesh. Network architectures such as this provide a high degree of resilience in the event of failure of a single line or node point.

In addition to the meshed back-bone network, Galileo makes extensive use of independent international network service providers to increase its reach into the global market. Where it is justified, Galileo has incorporated network hubs from strategic network providers into the high speed switching and routing systems in order to provide a fault tolerant front-end to the reservations systems. This reduces response times, i.e. makes the system faster, increases reliability and improves the quality of the managed service afforded to the customer.

So, the Galileo Central System is distributed around the world using a kind of super-highway. But at the local level in each country, Galileo changes onto 'B' roads that are more suitable to local market conditions. It is for these reasons that the NDCs in each major market use their own communications networks to distribute Galileo on a local basis. The UK NDC, for example, uses an X25 communications network with widely dispersed local nodes. The European gateway for this local X25 hub is located in Swindon in the UK.

Finally, because Galileo's operations are dispersed widely around the world, it is critical to the efficient running of the company that a fast and reliable internal communications network is available. The Galileo corporate back-bone network provides such a function. It supports LAN and WAN technologies between major sites, such as Swindon, Denver and Chicago. This internal network supports voice, data and video traffic.

Travel Automation Services (TAS)

British Airways owns a subsidiary in the UK called Travel Automation Services (TAS). This company started 'life' way back in 1977 as Travicom (see Chapter 3 – for the history of airline reservation systems). In 1997 TAS was re-structured to form three separate business units: Galileo UK, Chameleon and Icanos. Each of these companies focuses on a specific area of the travel automation market:

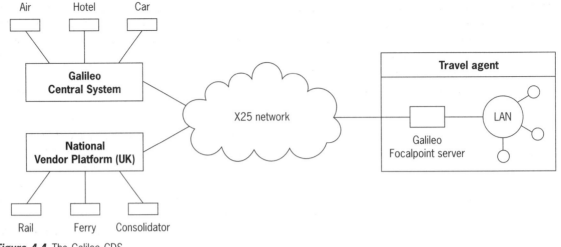

Figure 4.4 The Galileo GDS

- **Galileo UK** Galileo UK is the NDC that distributes Galileo International within the UK area. The Galileo system was first introduced to the UK in 1991 and from July of that year travel agents were gradually converted over from the old Travicom system to the new Galileo core. Galileo UK, as a business unit within TAS, is a wholly owned subsidiary of British Airways and has its headquarters in Maidenhead. Galileo UK currently has over 14,000 workstations installed (most of which are now PCs), in around 2,000 travel agents in the UK. This represents 68 per cent of the UK market.
- **Chameleon** This business focuses on training and consultancy. The Chameleon business was formed from the merger of both Speedwing and Galileo UK Training. Its primary role is to run courses for travel agents in how to use Galileo, its sister products and other GDSs.
- **Icanos** In its broadest sense, Icanos provides IT solutions to the travel industry. It will provide network services for individual customers, supply rail and ferry booking systems, market agency management systems and provide its customers with an IT development resource and centre of technical expertise.

Because this section of the book focuses on GDSs, it is the Galileo UK business that I am going to concentrate on here. I have tried to keep the description of Galileo's products as generic as possible and not solely restricted this analysis to a description of what is available only in the UK. However, some UK bias does nevertheless remain. The UK National Vendor Platform (NVP) is one such example, although it is probably mirrored in other NDCs around the world.

The UK NVP

The UK NVP is the technology that Galileo UK uses to connect its subscribers to non-air supplier systems. Galileo's general approach is a distributed one and uses X25 packet switching as its core communications methodology (Fig. 4.4). This enables Galileo users to link into non-core supplier systems, i.e. those that exclude air, hotel and car hire, without affecting the host switching systems of Galileo International. Examples of NVP host systems include train operators, ferry companies and air seat consolidators. Travel agents may book inventory on behalf of their customers from all NVP supplier systems although these entries are not at present integrated with Galileo's booking file (see later paragraph for a more in-depth presentation of booking files).

The Galileo NVP uses Sun Sparc Station technology to provide an X25 switching service to non-core supplier systems. Travel agents request a connection to their desired supplier system by entering a three character code on their Focalpoint terminal, e.g. RLY for rail, GAL for Galileo, PAO for P&O Ferries, STE for Stena Lines and GBT for

Guild Air Fares. This entry causes the Focalpoint LAN server or other Galileo terminal equipment to set up a communications session or path with the chosen host system via the NVP. Such a path supports all of the functions that the supplier's host system chooses to make available to travel agents via the Galileo route.

A particularly powerful feature of Galileo's UK NVP, and one that holds significant future potential, is the rail link. At present, when a user selects RLY on their Focalpoint PC or other Galileo equipment, they are connected to Eurostar's tribute sales guide (TSG) system. This system controls the inventory of Eurostar cross channel tunnel train seats. The Eurostar reservation system is located in a computer centre in Lille, France, and supports several important functions: such as the provision of timetable information, reservations on certain trains and ATB2 ticketing for Eurostar trains.

However, its true potential is illustrated by the fact that the Eurostar computer facility is co-hosted with France's SNCF computer, i.e. the computer of the French national railways; and this computer facility is linked back to the TSG system and the UK's ATOC computer in Nottingham, UK. There are also plans to provide an NVP machine interface record (NVMIR) that could be used to feed back-office systems connected to Galileo. It is therefore possible that given the right authorities and technical interfaces, Galileo users can be provided with reservations access and ATB2 ticketing functions for at least three of Europe's main rail services within the not too distant future.

Galileo's delivery products

A delivery product is a terminal that connects the user to the GCS. These days most delivery products are PCs that run special GUI-based applications within a standard PC operating system such as Microsoft Windows 95. This enables users to have reservations functions that coexist with standard office productivity tools such as word-processors and spreadsheets. The Galileo delivery products are:

- **Focalpoint** This is Galileo's main GDS access product and provides the primary interface to the GCS. In the past the hardware platform was Olivetti PCs with an in-built LAN capability. In future these will be replaced by Trigem PCs. The Galileo file server supports central storing of agency data, word processing, more sophisticated use of the PC's function keys, scripting and the connection of up to 24 workstations.
- **ET3000** This is the basic DOS-based Galileo airline reservations service for a small to medium sized travel agency, i.e. ET3000 is not Windows based. The ET3000 workstation (or travel agent terminal) is an Olivetti PC that is capable of running many of the packages available on the software market, such as word-processors, spreadsheets and data bases. The reservations services are described in more detail in the following section. ET3000 will gradually be replaced by Focalpoint products over the next few years.
- **Leisurelink** This provides a link to viewdata networks from a Galileo PC. It is designed for those travel agents who are principally in the air sales business but who need occasional access to tour operators and other principals that distribute their products via viewdata. The viewdata emulation is provided by the Travipad for ET sites and by the Galileo workstation file server (GWFS), for Focalpoint sites.

Besides these delivery system products, there are several agency management systems that are marketed principally in the UK by TAS's Icanos business unit (see above for a description of Icanos). One of these is Travel Manager and another is Travel Edge. Both are newer technologies that will no doubt gradually replace the President Agency Management System (PAMS). PAMS is Galileo's basic back-office system product that has by now been around for many years. All of these systems provide travel agency accounting and management information functions that are described in more detail in Agency Management Systems, Chapter 7.

Galileo's delivery systems technology

The above delivery system products describe what the user interacts with, in terms of functions and access methods. However, a great deal of co-operative processing now takes place on the end-user's workstation. Co-operative processing in this context means the sharing of processing between the GCS and the delivery system. The delivery systems are now capable of performing some significant functions, in addition to just being GUIs to the GCS. It is therefore worthwhile spending a short time reviewing the technological platforms upon which these products are based.

- **The workstation** The latest Galileo workstations are based on Trigem PCs. For the Focalpoint product there are two models of PC, one for the central file server (GWFS) and the other for the workstation. The central file server is of course a key component of the Focalpoint product. This is a powerful Trigem 3560 PC known as the GWFS. It is a Pentium P100 processor with 32 Mb RAM, a 1.2 Gb hard disk, a 1.44 Mb 3.5″ floppy, a VGA colour monitor and Ethernet interfaces. Each of the workstations is a Trigem P100 with 16 Mb RAM a 1.2 Gb hard disk, a 1.44 Mb 3.5″ floppy, a VGA colour monitor and Ethernet interfaces. Travel agents can use their own PCs with Galileo Focalpoint Special Edition (but not with ET products). In such cases, Galileo stipulates that the minimum PC requirement is a 486/66 MHz with 16 Mb RAM.

    The GCS can support up to five sessions at any one time, i.e. five work areas per agent sign-on. This is rather like having five separate terminals on your desk, with each one carrying on its own dialogue with the GCS. But using a Galileo PC, you don't need all five terminals because the combination of the GCS and the Windows operating system under which Galileo runs, supports several activities at once. This can be useful for carrying on a reservation for client A while also checking availability for client B. It is also possible to use this facility to construct several alternative itineraries for a client and to offer the client the option of which one to actually book. Galileo supports up to nine windows for displaying information simultaneously.

- **The Travipad** In order to work effectively the Galileo workstations must be connected to the Galileo core system via a communications controller called a 'Travipad'. This supports up to 48 workstations and handles all telecommunications tasks using X25 telecommunications technology. The Travipad is a black box as far as the travel agent is concerned and is virtually never used except when something drastic goes wrong. In such cases the Travipad may be used to test the status of the data line connecting the travel agent to the Galileo communications node. In terms of physical location the Travipad therefore needs to be positioned as close as possible to the data line terminating point in the office.

- **The printers** There are several different types of printers used as part of the Galileo system. These comprise the ticket printer, the invoice/itinerary printer and a general purpose printer. It must, however, be noted that if back-office accounting functions are required, an agency management system will be needed by the travel agency (see Chapter 7). Also, if word processing is required then not only will a general purpose printer be needed but a word processing software package that runs on the Galileo PC must also be installed.
    - The ticket printer may be either: (a) a dot matrix printer that is capable of using the OPTAT airline ticket stock available on continuous stationery from IATA, or (b) an ATB printer with several hoppers capable of supporting different ticket types. In some cases, this printer can also be used to print customer itineraries and invoices. In so far as dot matrix printers are concerned, to save continually changing the paper stock loaded in this printer it will usually be necessary to have at least two ticket printers of this type; one loaded with continuous airline ticket stock and the other loaded with pre-printed paper stock bearing the travel agent's logo, which can be used for invoices and itineraries.
    - The hard copy printer is usually an inexpensive and low quality device used to produce a printed copy of a Galileo screen for archive purposes. It may also be used to print queue messages. Generally speaking, although a hard copy printer is a necessity in a travel agency, its regular use is to be discouraged. This is because paper clutters up the office and the booking files. The travel agent user, once fully confident with the reliability of the Galileo system, will find little need to

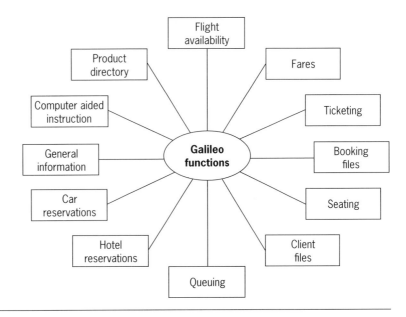

Figure 4.5 Galileo functions

print everything and will gradually regard a client file as an electronic image only.
- The general purpose printer may be either a dot matrix or a laser. A dot matrix is cheaper but the quality of the typeface is not considered particularly professional these days and it has the added disadvantage that it is noisy. The laser printer is expensive but produces an extremely high quality output and is virtually silent in operation. Both options are available from Galileo.
- **The software** The Galileo software is the critical ingredient that makes everything work. It exists in two places: (a) in the main-frame that runs the Galileo core system; and (b) in the Galileo workstation, which is a powerful PC. It is the workstation software that is of prime interest to the travel agent. The underlying operating system used to control this workstation is Microsoft Windows.

  Besides using the Galileo workstation for the core business, the travel agent may elect to use the PC for other purposes when it is not being used for reservations. In order to do this the PC will need to have some additional software loaded. This software can be stored on the PC's hard disk until it is needed. Software such as spreadsheets, word processors and data base management systems can all be used.

Galileo's core functions

There are many functions provided by Galileo and it would be impossible to cover them all in detail in a book such as this. The only way of learning more about Galileo and of becoming proficient in using the system, is to attend one of the training courses run by Chameleon. The intention here is therefore to give you an overview of the main functions and to help you understand the structure of the system. The functions that Galileo provides and which I am going to describe in this section are as shown in Fig. 4.5.

Flight availability

This is one of the most frequently used functions in the Galileo system and represents the first step in making a booking. Before looking at an availability display, the agent will normally have discussed the customer's travel requirements and established the basic routing of the journey. As an integral part of doing this, the agent will usually access the timetable display within Galileo.

- **Timetable** This shows the connecting flights between any two airport cities in the world, for 28 days from the specified date of travel. This is far more efficient than using other paper-based sources and reference books. It shows the days of operation of each flight and the dates that

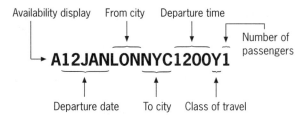

Figure 4.6 Galileo availability display

services start and finish. A range of qualifiers, such as specific times or actual airports, may be requested and particular carriers can be either included or excluded from the display.

- **Flight specific timetable** This is a more targeted and detailed display of flights that is constructed from the data base of Galileo's participating airlines. It shows total flying time, time spent on the ground, the classes of travel available, the terminal numbers and any stopovers involved. Other relevant information is also shown, such as non-smoking flights and code-share flight data.

Once the basic itinerary has been ascertained from the timetable function, the agent then enters the first sector of the itinerary as demanded by the customer's travel plans. This will usually comprise the departure date, a city pair between which travel is required, a preferred departure time, the class of travel and the number of seats required. An example of a Galileo availability display between London and New York on 12 January, leaving around 12.0 noon for one passenger in economy class is shown in Fig. 4.6.

The system will then respond with an unbiased display of the flight details that most closely match the customer's requirements, for up to seven days ahead. Up to 26 classes can be shown on the display and specific connection points and carriers may be included or excluded as required by the agent. The sequence in which flights are shown on the availability display is as follows:

1. The non-stop flights of participating airlines are shown in a sequence determined by how close the available flight's departure time is to the customer's preferred time of leaving.
2. Then, the flights that stop during the flight are shown. This may be for refuelling purposes or to pick up other passengers from connecting flights along the way. In any event the passenger does not have to change flights and often does not have to leave his/her seat.
3. Finally, the flights that require the passenger to change aeroplanes are shown.

The agent may try various alternative displays until one is found that meets the customer's requirements. This whole process is repeated for each sector of the client's itinerary including the return leg. These availability displays allow the agent to build a proposed itinerary for the customer. The most basic display is known as a free-sale display and it can be used to sell seats from all of the airlines participating in Galileo. An acknowledgement of a free sale booking can sometimes take several hours to be received. So, if the agent wants to be sure that those seats are actually reserved for the customer then an on-line seat availability display is necessary. To do this, the agent enters a special code to request Galileo to send messages to the other airlines involved in the itinerary to request an up-to-the-second status of the seats requested. Each airline responds to Galileo, which displays the status codes on the availability screen. There are currently about 211 airlines in Galileo that respond on-line to availability requests like this. The various availability options are as follows:

- **Instant availability** The Galileo system dynamically constructs connections for any requested journeys that are not already stored in the system as city pairs. This means that for a given city pair, Galileo will first attempt to show the direct flights available. Then, if no direct flight is stored, it will find the best set of flight segments involving other cities, i.e. airports, that most effectively join the requested from and to cities specified.
- **Inside availability** Galileo retrieves the availability status from an airline's own system and integrates it into the Galileo availability display. This is done automatically by the system without the need for the user to make any further entries.
- **Carrier specific display** Users of Galileo can request an availability display direct from a preferred airline's own system. This provides a real-time direct link into an airline's own

database, which provides the very latest information available.
- **Last seat availability** Called numeric availability in Galileo, this display shows the exact number of seats left for sale on a particular flight, up to the quota sale level of the airline involved.

At this stage of the booking, the travel agent will have viewed the alternative flight options and discussed them with the customer. The next stage is to confirm the actual flights required by making a firm booking.

Booking files

A firm booking means that a file must be created within the Galileo core system so that the booking can be referenced in the future. The Galileo booking file is an integrated collection of all reservations made for a customer, whether it be for a number of different airlines, hotels, car rental companies or other special services. All such information about the booking is held in this booking file. So, the first step is to sell the seats desired by the customer. Invariably, these seats will be displayed on the screen as a result of the availability request described above. Seats may be sold directly, i.e. a feature known as Direct Booking, or from an availability display, in a number of different ways:

- **Secured sell (interactive)** This is the highest level of booking available in Galileo. The booking request is sent directly to the airline's own reservation system via a direct link and a seat is immediately reserved in the airline's own data base. At the end of the transaction the agent receives a positive acknowledgement record locator direct from the airline's system that is automatically stored in the customer's Galileo booking file. This guarantees that the seat is held for the customer.
- **Super guaranteed sell (positive acknowledgement)** The booking is processed through Galileo's constantly updated data base. At the end of the transaction Galileo instantly sends the airline a booking request message that causes a prompt response to be initiated and returned to Galileo for storing in the customer's booking file.
- **Guaranteed sell** At end transaction, a message is sent directly to the airline on whose flight a seat is required and a guarantee indicator is displayed in the booking file. Every flight booked from a numeric display in Galileo is guaranteed by the airline.
- **Standard sell** At end transaction, Galileo sends the sell request to the chosen airline, which promptly processes the booking. This method is available for each of Galileo's 500 participating airlines.

In order to create a complete booking file, several critical fields must be entered. One good way to remember which fields must be entered is to think of the word PRINT. After all, sooner or later a ticket will be required and this will need to be PRINTed. Each letter of the word PRINT can be associated with one of the key fields required for a booking file to be created that will eventually enable a ticket to be printed successfully. These are:

- **Phone field**, i.e. the telephone number of the travel agent.
- **Received from**, i.e. the identity of the agency making the booking.
- **Itinerary**, which is the end result of the availability and reservation process.
- **Name of the passenger(s)**, an important entry that is difficult to change once entered.
- **Ticketing information**, i.e. information required to print the ticket and any other supplementary information.

All of the PRINT fields must be entered in order to complete a booking. Having completed the booking entries in this way, the final step is to 'end transaction'. When this is done, Galileo responds with a booking locator that is supplemented by the initials of the agent making the booking. At this stage the booking has actually been made and is stored in the Galileo central system. Additionally, each of the other airlines providing sectors for the itinerary has also stored a record of the booking for their sector only, within their own CRS system.

To retrieve a booking all that is needed is the booking locator. It doesn't matter whether or not British Airways or any of the other Galileo partner airlines are included in the itinerary or not, the locator will always enable the Galileo core system to retrieve the booking. Should the booking locator be lost or unavailable for some reason, it is possible for Galileo to be requested to display all passenger bookings with a specific surname. The appropriate one may then be selected from the list and the booking retrieved. For those airlines not supporting a locator, the agent must know the PNR and the host CRS before the booking can be retrieved.

Finally, there are the special services that customers often require as part of their trips. Examples include wheelchair requests and tickets that are to be collected at the airport terminal on departure. Each airline system requires slightly different commands to be entered in order to request these special services. Galileo provides a purpose-built function called enhanced booking file servicing (EBFS) that eliminates most of the complexities involved in this task. Galileo provides users with a standard command entry for special requests, which is the only one that a travel agent need learn. Galileo automatically translates this into the individual formats required by each airline system. All such details are stored in the customer's booking file.

Fares

Galileo can construct over two billion fares from its data base, which can be accessed by travel agency users. These fares are set by the world's airlines who keep each other updated on a regular and frequent basis. In fact Galileo's global fares data base is updated three times each day. This is a complex and fast changing aspect of the airline business, especially when you consider that there are, for example, over 30 different fares in force on the average transatlantic jumbo flight.

A fare quote is the automated construction of an airline fare for a booking that has been made. The fare quoted is guaranteed, provided that the agent produces an automated ticket within a period of seven days and has adhered to the rules associated with the fare. A useful facility is the ability to display all fares on a particular route for a specific date. This display shows the cheapest first, then the next cheapest and so on, up to the most expensive. Another useful facility is the ability to display a fare in local currency, i.e. pounds sterling. Because not all airlines quote their fares in pounds on their displays, this can be of great help to the travel agent. The exchange rate used is also displayed.

Ticketing

Galileo supports the production of airline tickets via two broad approaches: (i) by means of a set of ticketing control and printing functions, which are integrated within the GCS; and (ii) by passing an MIR to a separate back-office system, e.g. PAMS, Travel Edge, etc. It is the former method that is presented here. Ticket production via back-office systems is covered in more detail in Chapter 7 (although the related Galileo MIR is explained later in this section).

The GCS produces tickets by combining information from various sources, which are primarily: the availability data base, the booking file, the data base of filed fares and the agency accounting table (AAT). Other ticket modifier parameters may be specified by the travel agent, such as specific carrier selection, form of payment, commission percentage and endorsement data. All of this information is consolidated and used by Galileo's ticket control program to produce the following types of airline ticket:

- **OPTAT and TAT tickets** Galileo supports the printing of airline tickets on continuous paper stock supplied by the local BSP. These may be OPTAT or TAT. Approved ticket printers include the Datasouth A3300, the Texas Instruments TI810 and the MT5200, which is also a multi-purpose printer.
- **ATB** Galileo's ATB ticketing system has been gradually rolled out within the UK since early 1994 (see Chapter 3 for a fuller description of ATB ticketing in general). This ticketing system is supported by a number of models of special purpose ticket printers including those manufactured by IER, Texas Instruments, the MT5200 and Unimark Mark1. These printers produce the required ATB copies, i.e. for flight coupons, audit coupons, agent coupons, credit card coupons and the passenger receipt, by printing each one individually. It does not take much longer to print these copies than it does to print one of the older style OPTAT tickets, i.e. 10–11 seconds. There is also the added advantage that each copy of an ATB ticket may

be printed at a different location. This supports satellite ticket printing in large corporate locations without the need to have trained staff on site to process the ticket copies as is the case with OPTAT.

- **Electronic ticketing** This method does not of course result in a printed ticket at all (see Chapter 3 for a more in-depth description of electronic ticketing). It can only be used in countries where the BSP has granted the necessary authority. Electronic ticketing is designated by the presence of an 'E' in the 'Sell Response' field, which is a part of the booking file. When a ticketing instruction is issued, the validating carrier's system authorizes the request and flags the booking file. As a by-product of this process, a facsimile of the ticket is stored in the validating carrier's ticket data base.

Galileo checks to ensure that all required data are present in the booking file and that the information is correct. If credit card details have been entered as the payment method, then Galileo first carries out a card authorization check. This check is performed via Galileo's link to SITA, which itself has connections to the credit card companies' computer systems. Once authorized, the ticket is then produced by one of the above methods. From this point onwards, the ticket information is available via the ticket invoice numbering (TIN) report. This report shows details of all tickets that have been issued or voided over the past 30 days. It is a valuable report for a travel agent because it allows a reconciliation to be made with the BSP billing analysis.

Seating

Galileo's advanced seat reservation function can display information on the computer screen that represents a seating plan of the flight being considered. This plan is known as a seat map. Seat maps show the actual position of every seat on the aircraft and whether it is near a window or an aisle. The seat map also denotes seats that are either smoking or non-smoking. Several different types of seat map displays are supported by Galileo:

- **General seat map** This function enables a seat map to be displayed for a particular flight, class and segment. The display shows all of the important features within an aircraft, such as toilets, emergency exits, aisles and galleys.
- **Seat availability map** This display shows a seat map that includes only the available seats for a specified flight, class, date and segment. This can show the whole cabin or, by entering preferred seat characteristics or row number, it can show only those seats that match the customer's personal criteria.
- **Specific seat characteristic display** This allows the agent to determine the exact characteristics of a specified seat number in plain text. The seat to be shown is selected from a previously displayed general seat map or seat availability map.

A seat can be reserved in one of two ways: either (a) a specific seat can be requested for the customer by entering the seat number, or seat map co-ordinates, into the system; or (b) a seat type can be requested, such as aisle non-smoking. Once the entry is made, Galileo responds with a confirmation of the seat reserved or a message to indicate that the seat requested is unavailable.

Client files

This is a powerful customer servicing capability that is especially useful for business travel, where travel agents service a large number of frequent travellers. A client file stores information about a client that can be called up instantly on the screen and inserted into a booking file whenever necessary. So it not only enables a travel agent to give a more personal service by knowing customers individually, but it saves a lot of time and keying effort. There are three parts to a client file:

- **The agency file** This is the top level record, which contains details about the travel agency making the reservation. It is a part of the profile that does not often change.
- **The business file** This is the corporate profile, which contains information on a company for which the travel agent provides a business travel service. It might include, for example, some information on the company's travel policy, the departments involved and the account settlement details. In cases where an individual is a frequent traveller but does not work for a

company, this profile is used to store information about that individual.
- **The personal file**   This record contains information about each of the individual travellers who work for the company as defined within the corporate profile. It contains information such as the traveller's name, phone number, address, department and personal preferences, e.g. smoker or non-smoker, window or aisle seat, etc.

In addition to these structured profiles there is another very useful function provided by Galileo called the frequently flown itinerary feature. This is designed for frequent travellers who often have the same or very similar itinerary details. A record of these can be set up in skeleton form and can then be copied into a booking file with only the date and number of seats needing to be added to make the booking complete. This is particularly useful to a travel agent for servicing employees of a company with say a head office in Paris where a lot of the itineraries will be virtually identical, i.e. flight London/Paris, standard business class, company corporate hotel, etc.

Queuing

The queuing facility is a powerful tool for controlling work flows and increasing productivity. Queuing is all about the passing of messages: (a) between travel agency staff within the office; (b) between offices using Galileo; and (c) with airlines, hotels and car rental companies. Not all bookings run like clockwork! There are usually complications of one sort or another that require communication between the various parties involved. For example, the airline changes the departure time of a flight after a booking has been made for a client. Or the hotel can only supply a room at the back of the hotel without the sea view requested. Or the travel agent in office A has made a reservation for a client that needs follow-up by an agent in office B. In these cases the airline, hotel, car rental company or agent from office A would send a message to the travel agent informing them of the changes to the client's itinerary details.

The message is placed in a special area of the system that is available for travel agent control purposes, known as a 'Q'. This special area is known as the Queue. It is called a queue because there may be several messages for an agent and these messages are stacked up in the order in which they were sent, for the agent to read and process in sequence. The queue area is divided up into separate sections and each of these is identifiable as a different type of queue. There is usually a general queue area for the agency, a ticketing queue and a queue for each travel agency staff involved in the servicing of customers. You should note that the onus is on the agent to read the queues regularly. If the queues are not accessed and processed in a timely manner then vital itinerary information may be missed with dire consequences for the traveller.

Hotel reservations

Galileo provides its users with the ability to reserve a hotel room for their clients at over 37,000 hotel properties from over 200 hotel chains worldwide. The service is called 'RoomMaster' and it is used by a simple process of filling in an electronic form that is displayed on the Galileo screen. The ways in which hotels are connected to Galileo are very similar to the different ways in which airlines are connected. Some are connected indirectly and provide free-sale room reservations. Others are on-line and provide instant confirmations.

Indirectly connected hotels have a computer system, but it is one that is not capable of being connected directly to the Galileo computer and cannot therefore undertake a two way dialogue with the booking agent. This means that when a reservation is made, it is on a free-sale basis. In these cases the reservation message is passed by a communications network to the hotel's computer, which puts it in a kind of queue for subsequent processing. While this is happening the travel agent will note the reservations request and proceed with other business. At some future time the hotel will process the reservation request and send a message back to the travel agent either confirming that the booking has been made successfully or advising that the room requested is not available.

Directly connected hotels (of which there are 45 in Galileo at the time of writing), have a computer that is on-line to Galileo and can hold an intelligent two-way dialogue with the travel agent making the booking. Galileo often refers to these hotels as having 'inside links'. In these cases when

Table 4.2 Galileo RoomMaster entries

Entry	Meaning
HOA	Standard hotel availability display for a specific city, airport or reference point
HOC	Complete hotel availability, which displays additional room rates and/or room types for one property
HOI	Hotel index, which displays a list of properties for a specific city, airport or reference point
HOU	Hotel update for previous availability or index request
HOR	Hotel reference point list, which displays reference point list associated with a metropolitan area
HOD	Hotel description which displays a keyword menu to access chain or property policies/information
HOM	Hotel modification: modifies or deletes fields in a booked hotel segment

the booking screen is completed it is sent to the hotel computer, which immediately responds with either a confirmation or a 'not available' message. This allows the travel agent to advise the client of a confirmed booking while they are on the phone and saves the agent from having to call the client back as is the case with an indirectly connected hotel.

RoomMaster offers a wide range of functionality for travel agents who wish to book hotel rooms for their customers. The main entries that are outlined in Table 4.2 will, I hope, give you some idea of the range of functions available.

The Galileo Spectrum product (Fig. 4.7) runs within Focalpoint and is fully integrated with Room-Master. It is a CD-ROM-based hotel mapping system covering over 30,000 hotel properties (1,400 of which are mapped) and hundreds of world-wide, regional, metropolitan and city maps that enable users to zoom in and pin-point a hotel's precise location. The maps show places of interest, convention centres and other reference points and may be printed for passing on to customers as part of their travel documentation. Spectrum also allows users to customize the maps to show offices they intend to visit, the hotels where they will be staying and the nearest affiliated travel agency. Distances can also be calculated by Spectrum. So, if a customer has to walk from his/her hotel to a convention centre across town, Spectrum can calculate the exact walking distance.

Spectrum also contains a great deal of detailed information about each hotel, the facilities offered and the rates. Each participating hotel has six displays on which to show details such as amenities, sports and recreation facilities, dining and entertainment, and business services. These details can be annotated with additional details entered by the travel agent. To ensure that the information is kept constantly up-to-date, an updated CD-ROM is sent to the travel agency user every three months.

Once a travel agent has viewed the information on Spectrum with a customer, the RoomMaster function within Galileo is then used to look at the chosen hotel's availability and room rate. Finally, RoomMaster is used to make a firm booking for the customer. Spectrum is a Windows-based product that runs on the whole range of Focalpoint 3.0 platforms, provided of course that the PC is connected to a CD-ROM drive.

Car reservations

Car rental reservations are handled by Galileo's CarMaster system. This operates in a very similar way to that described for hotels. There are currently 47 car hire companies available in CarMaster and most of these have a world-wide network of servicing locations that number over 15,000, where cars can be collected and dropped off. Bookings with vendors, such as Hertz, Avis, Budget, British Car Rental, Eurodollar, Alamo, Europcar and Thrifty, can all be made in Galileo and fully integrated with airline reservations in the booking file. The main entries for CarMaster are shown in Table 4.3.

A great deal of standardization has been accomplished in the car hire area. There is, for example, a common four character coding system that identifies whether the car is an economy or compact size, an estate car or saloon, automatic or manual

Table 4.3 Galileo CarMaster entries

Entry	Meaning
CAA	Standard car availability
CAL	Availability from low to high rates
CAQ	Availability qualified by vendor, car type, rate
CAI	Index of vendors for a specified city
CAU	Update previous information
CAM	Modify dates, car types after sale

gearbox and air-conditioned or non-air-conditioned. The actual codes are as shown in Table 4.4.

General information

The Galileo Information System (GIS) provides a number of information pages that may be regarded as an electronic book for travel agents to use. The information is structured into chapters and pages, which are indexed for easy reference. The following are the main headings of information that are provided by GIS:

- Airline information
- Consulate information
- Country information
- Customer services
- Galileo and its services
- Galileo's product directory
- Help information
- Product overviews

One useful facility is the TIMATIC information service. This is operated by a Swiss-based company that assembles and maintains a large data base of essential travel related information. This is available to Galileo users via an on-line computer interconnection with the GCS. TIMATIC provides travel information on countries throughout the world and covers a wide variety of topics that are all kept current. It is organized rather like a book with chapters on different subjects. The chapters of TIMATIC are: health, visa, airport taxes, passport, customs, news and country (geography, customs and currency).

Finally, GIS provides an on-line encode/decode function to translate automatically the various codes used throughout the Galileo GDS. There are a multitude of these codes, many of which cannot be carried in a person's head. The main codes are for airports, airlines, aircraft types, countries and cities. The user simply keys in the text and GIS translates this either into a code or, if a code was entered originally, into the full textual name.

Computer-aided instruction

There are several pages of text stored in Galileo that comprise a self-teach course for travel agents. These pages enable an untrained user to undertake a self-paced learning exercise that can give them a good overview and a basic grounding in how to use Galileo. However, computer-aided instruction is designed to complement and support a programme of classroom training. It is not intended

Table 4.4 Galileo CarMaster types

Car size		Car type		Transmission		Air-conditioned	
M	Mini	B	Two-door	M	Manual	N	No
E	Economy	C	Two- or four-door	A	Automatic	Y	Yes
C	Compact	D	Four-door				
I	Intermediate	S	Sport				
S	Standard	T	Convertible				
F	Full size	X	Special				
P	Premium	W	Estate				
L	Luxury	V	Van				
X	Special	F	Four-wheel drive				
*	All sizes	R	Recreation				

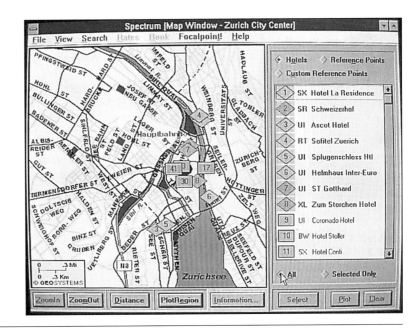

Figure 4.7 Galileo spectrum

to replace attendance at a course lead by a competent trainer who can impart many technical and non-technical aspects on how to use the Galileo GDS in a travel agency environment.

Product directory

Galileo's product directory is an ordered set of information concerning other non-air travel products. Some examples of services available in the product directory are theatre tickets, citicorp bank drafts, limousine services, UK chauffeured parking services and Columbus Insurance. The product information is entered and maintained by each participating supplier using a dedicate maintenance terminal supplied by Galileo. These services are available by entering 'PD' into any Galileo screen.

Once in the product directory, information is processed in a fashion that is very similar to a viewdata system (although the product directory is not a viewdata system itself). The product directory main menu is a directory that offers choices of: (a) a list of participating suppliers, (b) a list of suppliers by product category, and (c) a list of participants offering products in a particular city. Descriptions of the different products are stored in logically arranged sections that are accessed using sub-menus. Pages of information can be viewed and other pages requested for display. To book one of the products displayed, an electronic form is completed that is printed out at Galileo's operations headquarters. The printed copy of the electronic booking form is then sent to the supplier by facsimile transmission. From that point onwards the dialogue between the supplier and the travel agent may be either via the Galileo system or, as is often the case, handled off the system, usually by telephone or mail.

Other Galileo productivity functions

In addition to the core reservations functions described above, Galileo offers its travel agency customers some additional productivity enhancing tools. These fall into two main areas: (i) functions available on the Galileo Central System, and (ii) functions that are designed to run on the Focalpoint PCs that deliver Galileo's Central System functions to the user.

GCS productivity functions

A set of optional functions is available on the GCS that enables users to maximize their productivity. These functions are computer applications that actually run on the Galileo host computer system and are therefore available on a global basis:

- **Galileo's past date quick (PDQ)** This provides travel agents with access to booking files which may be up to 13 months old. The requested historical information is provided within a response time determined by the age of the file. Booking files that are from one to 60 days old are displayed immediately. Items that are more than 60 days old are displayed within 24 hours. The information retrieved may be displayed, put in a queue or printed. Search and selection criteria are also available to streamline the process.
- **Galileo's MIR** A Galileo MIR is a special message that contains all information about a booking. It is passed from Galileo into an agent's back-office system where it can be used for a variety of purposes, such as ticket printing, itinerary printing, accounting and management information reporting. Each Galileo MIR comprises over 700 data elements pre-formatted as an ASCII flat file, which may be transmitted securely to the back-office system in real-time. MIRs may be sent to a variety of locations, depending upon the needs of the individual agency. They may, for example, be sent into the back-office system in the office where the reservation was made or they may be sent to a central computer in cases where the agent operates a central group accounting and MIS system. The choice is the agent's. The MIR specification is available to any software company wishing to develop an interface to Galileo. At present there are over 300 software houses and travel agents currently working with Galileo International and its NDCs to integrate accounting solutions with the Galileo system for MIR data acquisition.
- **Galileo's selective access** Galileo allows travel agents to share customer information across different locations, sometimes across different countries of the world. Groups of agents known as affiliates may be formed and selective access partnerships established between them. These affiliations may be formed from within a travel agency's multi-branch network or between separate agencies that have formed a strategic partnership. The information that may be shared includes booking files, client files, documentation and queues. Access may be controlled at several pre-determined levels thus allowing only certain staff within an agency to access shared information about their customers.

Focalpoint productivity functions

Galileo has developed a number of applications that run within the Focalpoint PC system. These are in most cases applications that run in what I have described earlier as a co-operative processing environment. In other words, the applications use data from the GCS and add local processing functions to deliver enhanced services to users via a simple Windows based GUI. These functions are as follows:

- **Galileo's relay productivity tools** This is a set of tools that combines the features of Focalpoint with the flexibility of Microsoft Windows to bring productivity support to a travel agency user. It is available on any Focalpoint Version 3.0 or higher product and comprises the following functions:
  - *Queue Manager* Supports the handling of queues according to pre-determined agency rules. This allows booking files to be scheduled for action automatically at different times of the day and by different staff within the agency.
  - *Booking File to Client File* This allows a client file to be created automatically from a booking file, according to pre-determined rules set within the travel agency. The client files created in this way will all conform to a format determined by the agency and saves the repetitive entry of data.
  - *Response Capture* This is an advanced 'cut and paste' feature that allows information to be copied onto a clipboard from one or more of Galileo's windows and pasted into any other application such as, for example, Microsoft Word or Excel.
- **Focalpoint's Scriptwriter Plus** This is a Windows based Focalpoint product that allows a travel agent to automate the entries used by his/her staff to make a booking. A set of commands similar to a program code, called scripts, are used to define exactly how a booking is to be transacted. Each of these scripted functions is stored within the PC and represented on the screen as an icon. When a user clicks

on the icon, the script is executed and the user is prompted for certain entries. Scripts can be written to detect and store certain responses received from the GCS and to exchange data with applications using dynamic data exchange (DDE). This is a powerful way for a user to carry out automatically a complex booking transaction very quickly and with the minimum of key-strokes. It also is a way of reducing errors by using quality control checks on data before they are sent to Galileo.

There are three sub-products that comprise Scriptwriter Plus: (i) Scriptwriter Plus Build – which is used to create customized scripts for users, (ii) Scriptwriter Plus Run – an execution utility that enables scripts to be initiated from icons or the tool bar, and (iii) Scriptwriter Plus Convert – a conversion utility that allows scripts to be upgraded from one version to another. Scripts are fully configurable on installation thus allowing, for example, scripts to be written by head office and used by branches that do not have the authority to modify them.

- **Galileo's Premier**   This is a Microsoft Windows based product designed for Focalpoint that allows travel agents to search and book hotels and car rental services. It is fully integrated with the RoomMaster and CarMaster central system functions. There are three major components:
  - *Itinerary View*   the current booking file is retrieved from the GCS and re-formatted into an easy-to-read style. To book additional hotel or car products the user points and clicks on icons and the tool bar instead of entering more complex airline type commands.
  - *Premier Cars*   an easy-to-use interface to CarMaster. This function retrieves previously stored information, such as the customer's corporate policy on car rental, frequent traveller details, licence requirements and insurance. Search criteria are automatically constructed from the itinerary information and CarMaster is used to find the services required.
  - *Premier Hotels*   similar in concept to Premier Cars, this function provides a simple GUI to RoomMaster. This allows users to access specially negotiated room rates stored on their local Focalpoint systems and embed these into the reservations and booking process.

Galileo is developing an enhancement to its Premier product that will extend its use to support airline bookings also. This will work in a similar way to the above – it will act as a GUI to the GCS and allow infrequent users to quickly master the system for all airline booking functions.

- **Galileo Travelpoint**   This is a software product that is used by a travel agent's business house customers, who are usually sophisticated and frequent travellers. It provides travellers with access to Galileo via a simplified Windows-based GUI, but channels all reservation and ticket requests via the travel agent. It can be a useful tool for frequent travellers and can be implemented on their lap-top computers while allowing the travel agent to retain control of ticketing, MIS and quality control.

SABRE'S TRAVEL INFORMATION NETWORK

Sabre (Fig. 4.8) now represents the largest privately owned computer system in the world (the largest being the US Government's SAGE defence system). Sabre is one of the world's leading GDSs and is used in over 29,000 travel agency locations around

Figure 4.8 Sabre Europe logo

the globe. It is a system that enables its subscribers to check the availability and make reservations for seats on airline flights, hotel rooms, rental cars and many other travel-related products.

The Sabre system drives over 147,000 computer terminals of which 122,000 are located in travel agencies. The migration away from dumb terminals to PCs is well under way and at present over 90 per cent of the terminals used in these agency locations are now PCs. Most of these travel agency workstations are distributed throughout 64 countries in North America, Europe, South America, the Caribbean, Australia, Africa, the Middle East and Asia. In the UK there are over 750 locations using more than 4,000 Sabre terminals.

The history of Sabre

American Airlines' first automation program started in the 1930s with the 'request and reply' system. A travel agent would telephone a regional American Airlines central control point where inventory was maintained, to enquire about space available on a flight. A response would be returned via teletype. Through the mid-1940s, reservations were recorded manually using a pencil to mark up different coloured index cards known as 'Tiffany cards'. These cards were named after the famous lamps with the attractive coloured glass shades. These cards were arranged in a 'Lazy Susan', and flights were controlled by half a dozen employees sitting around a table spinning the 'Lazy Susan' for index cards that would correspond to particular flights. By counting the pencil marks on each card a clerk at the reservations centre could give a 'yes' or 'no' to a request for a seat.

In the larger reservations centres a wall-sized status board was installed to display seat space available on each flight. The board summarized much of the information on the index cards in the 'Lazy Susan'. As new reservations came in, workers at the table passed the information to 'board workers' who removed seats from a particular flight's inventory status until no seats remained. Using the Tiffany system to complete a booking for a round-trip reservation from New York City to Buffalo required 12 different people performing more than a dozen separate steps during a three hour period – longer than the flight itself took!

In 1946 American Airlines developed the Availability Reservisor, the industry's first electrical/mechanical device for controlling seat inventory. The Reservisor applied basic computer file technology to the task of keeping track of American's seats and flights. Even though it could not sell the seat or cancel a reservation, the system represented a milestone in adapting electronics to airline reservations.

By 1952 the airline had introduced basic magnetic storage technology – a random access memory drum and arithmetic capabilities – to the Reservisor. With the Magnetronic Reservisor a reservations agent could check seat availability and automatically sell or cancel seats on the electronic drum. As advanced as the Magnetronic Reservisor was for its time, the reservations process was still intensely manual. All passenger information was handwritten onto record cards and was kept separate from the seat inventory and reservations information. Clearly, the airline needed a better method for handling reservations and managing its inventory.

In 1953, a chance meeting of two Mr Smiths on the same American Airlines flight from Los Angeles to New York sparked off a series of technological innovations that eventually led to the development of Sabre, as we know it today. The first step along this evolutionary path was the development of a data processing system that would create a complete passenger record and make all the data available to any location throughout American's system. The end result of the conversation between C. R. Smith, American's president, and R. Blair-Smith, a senior sales representative for IBM, was an announcement on 5 November 1959 of a semi-automated business research environment, i.e. Sabre (Fig. 4.9).

Sabre enabled American for the first time to link (in a single electronic unit), a passenger name to a specific seat sold on an aeroplane. Sabre also made possible a link to passenger inventories in other CRSs thus laying the groundwork for the way airlines handle interline reservations today. In 1960 the initial computer centre was installed in Briarcliff Manor, New York, and Hartford became the first American Airlines office to use Sabre in 1963.

In its first year of operation Sabre could process 85,000 phone calls each day, 30,000 requests for fare quotations, 40,000 confirmed passenger

Figure 4.9 An early American Airlines reservations terminal

reservations, 30,000 queries to and from other airlines regarding seat space and 20,000 ticket sales. American's initial investment in the research, development and installation of Sabre was almost US$40 million – the price of four Boeing 707s at the time! By 1964 the USA telecommunications network of the Sabre system expanded from the east coast to the west coast and from Canada in the north to Mexico in the south. At that time, it was the largest real-time data processing system in the world.

Ten years later in 1974, American initiated a joint carrier feasibility study to explore the prospects for a CRS that could be owned and run by a joint venture of several leading USA airlines. This project received antitrust immunity from the Civil Aeronautics Board and included several other airlines. In 1975 a study group deemed the system economically practical but United Airlines withdrew from the project and announced its intention to provide travel agents with its own CRS. Because this system had schedule displays and connections biased in favour of United Airlines, the installation of United's CRS terminals posed a competitive threat to other airlines. Consequently American announced its intention to market its own reservation system. In May 1976 American Airlines installed its first Sabre unit in a travel agency. Such has been the success of the system that there are now more than 110,000 Sabre terminals installed in over 25,000 locations throughout 64 countries.

American Airlines' commitment to new technology extends well beyond the flight reservation systems. The systems operations centre (SOC) in Dallas is a prime example of the innovative use of technology. The SOC is home to a unique blend of human skills and IT systems that together enable the world's largest airline to operate all its flights smoothly and efficiently. The SOC is the nerve centre from which every single American Airlines flight is controlled. From the pre-flight planning of crew, food and fuel, through the load and balance calculations for take-off, to the weather charting and flight planning process. All this is undertaken by skilled and experienced airline and meteorological staff supported by an array of sophisticated technologies.

These technologies include over 400 Apple Macintosh workstations each with between 4 and 16 Mb or RAM with 100 Mb of hard disk connected into a fault tolerant Ethernet LAN with ten servers. Each of these workstations can display several windows showing weather maps and other graphical information as well as providing access to the flight operations system (FOS) that runs on a large main-frame computer. Knowledge-based systems are used in the Automated Load and Balance routines and the Hub Slashing application. This uses artificial intelligence to assist with decisions on which flights to cancel in complex situations arising from things as diverse as weather bound airports to defective equipment. Unfortunately it is outside the scope of this book to delve any more deeply into this fascinating side of airline flight operations and control. What we need to concentrate on, is that part of Sabre's technology that is of prime interest to travel and tourism practitioners; and this is the Sabre GDS.

Sabre

Sabre's corporate entity has undergone a substantial restructuring over the past few years in order to: (a) position it to compete effectively on a global scale; and (b) distance itself from its founding parent, American Airlines, which is a separate business in its own right. The Sabre Group is therefore now a separate legal entity, listed on the New York stock exchange. It devolved from the AMR Corporation, the owner of American Airlines and its original founder, over the period 1992 to 1995. Following the flotation, the AMR Corporation retained a majority stake in The Sabre Group. This devolution to an independent publicly owned corporation gives The Sabre Group several significant strategic advantages including: (i) the flexibility to grow under its own control, (ii) the ability to explore a wider range of marketing opportunities, and (iii) access to a source of investment capital for growth and development. The Sabre Group itself comprises several operating divisions. These are:

- **The Sabre Travel Information Network (STIN)** This is the marketing arm of Sabre which sells the system to travel agents and travel suppliers around the world. More than 25,000 subscribers in over 64 countries world-wide access travel information through Sabre. STIN includes Sabre Europe, a subsidiary that controls several distribution companies in Europe, the Middle East and Africa. STIN employs 2,200 staff.
- **Sabre Interactive** This division encompasses emerging technologies. It provides a full range of information management services including: systems development, network design and management, telemarketing, reservations services and systems, data management and technical training. It employees some 5,000 staff.
- **Sabre Decision Technologies** This is a software business aimed mainly at supporting other airlines. It includes: (a) software applications development and project management; (b) application marketing of products, like flight scheduling software and yield management; and (c) decision support technologies that, for example, simulate airport terminal passenger movements for planning and evaluation purposes. This organization is also responsible for the development and application of Sabre functions. It employs 2,000 staff.
- **Sabre Computer Services (SCS)** This is the part of the business that actually runs the Sabre main-frame computers and is responsible for the co-hosting of over 40 airlines that use the Sabre system. It is also the data processing activity that develops and maintains Sabre. SCS is responsible for the assembly, custody, administration and protection of corporate telecommunications and computer-based information resources. It employs around 2,000 people.

Like most of the other three GDSs, gone are the days when the CRS was a sales tool for the airline business. Legislation has done away with all that. Nowadays, GDSs must be so unbiased that they are no longer regarded as a key part of the airline's sales and marketing function. Sabre's primary mission is therefore to increase the sales of its automation products. Hence it is important to emphasize that following the corporate restructuring exercise described above, Sabre's activities are no longer driven by the need to support the sales of American Airlines' seats. In fact only a few years ago, when airlines were experiencing one of the industry's worst recessions, Sabre regularly generated more profit than the American Airlines business itself.

Decentralized processing

In the early days of IT, back in the 1960s and 1970s, Sabre was accessed by travel agents who predominantly used dumb terminals for everyday access to Sabre. These terminals were connected to an office based communications controller that itself was connected via a dedicated communications line to the Sabre host main-frame computer. Terminals such as these possessed no inherent processing power of their own and relied solely on the Sabre main-frame system for all functions initiated by the user. In fact, there are still quite a few such agents around the world who use these devices for their day-to-day reservations purposes. However, for most travel agents, the new PC-based Sabre products are more productive, more user-friendly and are richer in functionality than the 'old' dumb terminal approach. Before we dive into an analysis of Sabre's new products, it is worth considering how this shift in GDS IT evolved.

One of the clear trends in IT over the past ten or even 20 years has been the decentralization of computer processing power. The major reasons for this are as follows: (i) the power of processor chips has increased exponentially, (ii) the price of these chips has fallen in almost an inverse price/performance curve, and (iii) the physical size of circuitry and chips has shrunk to microscopic proportions. Evidence of these factors has been the widespread use of the ubiquitous PC. All of this has meant that more processing can be carried out closer to the end user, usually on their local PC.

So, the legacy main-frame computers on which the core airline reservation systems run, have seen some of their processing functions distributed to outlying PCs via global telecommunications networks. This phenomenon is called co-operative processing and it uses distributed PCs to make it work. Even if the main-frame systems do not shed any functionality, the distributed processors can do a lot more local checking and validating of users' entries long before transactions actually hit the host system. This local validation processing generates several important benefits, such as: (i) lower transaction volumes because many erroneous entries are trapped locally and not even sent to the main-frame; (ii) reduced processing load on the main-frame because much of the validation checks and error detection/reporting/correction functions are undertaken locally on the PC; and (iii) faster response times for the user because data entry and validation transactions do not have to be transmitted across a network and are instead performed locally, which supports an almost instantaneous response.

Then there are the added user interface benefits. Local PC processors allow the unfriendly face of the main-frame to be replaced by a GUI. The problem with a lot of legacy main-frame systems is that they are usually very user un-friendly. They often require complex entries to be made by users, which often bear a remarkable resemblance to computer programming instructions. These coded entries are difficult for users to remember, are susceptible to mis-keying errors and require lengthy training courses. However, local PCs can act as interpreters for the main-frame and therefore offer the user many benefits including: (i) a far easier interface that is quicker to use and generates fewer mis-keying entries, (ii) the automatic generation of certain fields that historically had to be laboriously keyed by the user, and (iii) less training before the user can become a proficient operator.

So, the trend is clearly away from dumb terminals and towards a co-operative processing architecture that uses a combination of legacy main-frame systems and local PCs. Evidence of this trend can be clearly seen in virtually any travel agent's office. Although dumb terminals were the norm right up to the 1980s, in the 1990s travel agents gradually replaced them with PCs linked to GDSs. Nevertheless, the Sabre central system is a massive computer facility. The main site is based in Tulsa, Oklahoma, and a major portion of this facility has been designed to withstand both natural and man-made catastrophes. It uses 17 main-frame computers to deliver 4,000 MIPS of processing power with 15.3 Tb of storage. These computers are housed in a high security data centre with 11 km^2 of floor space. The networking capabilities are equally impressive with 180 communications processors and numerous mid-range UNIX-based computers. Over 200,000 PCs are managed and a voice network comprising over 45,000 telephone numbers and 10,000 voice mail boxes generate over 115 million calls each year.

Now, it is time to consider the range of products offered by Sabre to its subscribers around the world. There are several products that I describe here and all of them use a combination of: (a) PCs as the primary travel agent's workstation, and (b) functions provided by the Sabre main-frame core system. But before we can begin to understand Sabre's end-user products, it is essential first to understand something about the core reservation engine that drives the Sabre GDS. Understanding Sabre's core functionality provides a solid foundation for exploring the many front-end products marketed to the travel industry by this major GDS.

Sabre functionality

The Sabre core system runs on an extremely large main-frame computer that is situated in an underground bunker in Tulsa, Oklahoma. This system has two types of functions. First there are the back-end functions that are concerned with Sabre's connections to other airline reservation systems. Then there are the front-end functions that provide access to the Sabre system by travel agents and other users, such as regional American Airlines reservation offices. Let's review the back-end functions first.

Sabre's back-end functions

Sabre is connected to a large number of other CRSs via direct computer-to-computer links. Each link is supported by a set of complex computer functions that themselves have various degrees of sophistication. The degree of sophistication depends upon several factors, such as: (a) the functional capabilities of the participating CRS, (b) the price that the participating airline wishes to pay Sabre for participating in their GDS, and (c) the communications network used to connect into Sabre. The terminology that covers this aspect of Sabre's intra-CRS connections goes under the banner of 'levels of participation'. The reason it is so important to understand the various levels of Sabre's CRS participation, is because it is the dominant factor affecting the displays and functions that subscribers experience when they use Sabre.

Sabre therefore provides its airline customers with several options regarding the precise way in which they participate in the Sabre system. The most basic level is the basic booking request (BBR), which is a relatively new service designed for smaller airlines requiring only the simplest of booking functions at a low cost. It uses on-line two-way communications for reservation request messages, but the timing of responses is not immediate. The standard level of participation is called Full Availability and this is the service used by the majority of airlines connected to Sabre. In addition to this there is a set of optional services, which falls under the umbrella of what Sabre calls Total Access. Total Access therefore comprises several different sub-levels of participation, which are described as follows:

- **Direct Access** With Direct Access, Sabre communicates directly with the other airline CRSs and 'takes' the required seat(s) from the airlines' inventories. This provides last seat availability.
- **Multi-Access** This is a form of participation that is similar in many ways to Direct Access. The difference here is that the user must by-pass Sabre and use the airline system that is being accessed. Although the Sabre common language can be used for outbound entries, the inbound displays received from the multi-access host will be displayed in native format, i.e. non-Sabre format. Multi-Access provides last seat availability.
- **Answer Back** This is a kind of enhanced messaging system. In this case, the participating airline sends a two-character answer-back code to Sabre via a teletype message, i.e. something like a telex message, confirming that the reservation has been successfully made within its own system. Sabre updates its PNR with this answer-back code. This gives the travel-agency user (not to mention the traveller), a degree of comfort that the reservation has actually been made in the other airline's system. However, it is possible for a reservation to be withdrawn even after confirmation and so Answer Back is not the optimum level of participation.
- **Direct Connect** This is the premier level of participation and is sometimes known as 'Seamless Connectivity'. Instead of the participating airline sending just a two-character confirmation code, it actually sends its PNR locator reference. This is automatically added to the Sabre PNR. With the other airline's locator embedded within the Sabre PNR, the booking is not only as secure as it could ever be, but there is a speedy reference point into the other airline's system should enquiries ever need to be made, or a ticket needs to be collected on departure. With Direct Connect it is possible to sell a seat from an availability display that actually shows zero seats available. This is because although the availability display sent from the participating airline shows no seats available, Sabre can 'go into' the airline's system and if there is a seat left, it will take it.
- **Direct Connect Availability** Otherwise known as Seamless Availability, this is the ultimate level of participation that another airline may have with Sabre. It provides the user with a real-time availability display using information obtained directly from the participating airline's reservation system.

When a reservation is made in Sabre, a PNR is created and this is given a PNR locator code. This may be used to retrieve the PNR on subsequent occasions, although the PNR may also be retrieved on entry of the travellers name. A Sabre PNR generally contains details of all products booked via Sabre, even though some of them may in fact be non-air segments delivered by non-Sabre host systems, e.g. hotel reservations, rail reservations and car bookings. That brings us nicely on to non-air host systems that are connected to Sabre.

Non-air hosts

The term 'Sabre central hosting' summarizes the architectural principle that is the cornerstone of Sabre's non-air host development program. This principle is based upon connecting all non-air supplier computer systems, no matter where they are located, directly to the Sabre host computer in Tulsa. Centralizing these connections allows Sabre to distribute all non-air host systems to its population of users located around the world. Consequently, Sabre has developed a range of new non-air products for each marketing region of the world. In Europe, for example, the project name is European local vendor access (ELVA). The objective of this product is to make Sabre more of an attractive utility to travel agents who book a fair amount of leisure travel business.

Some GDSs connect local non-air systems to their nearest national node. While this is fine for local access from within a country, it is not always suitable for global access. Take the following hypothetical situation. Say that a French non-air supplier's booking system needed to be connected to Sabre. One might think that this could be done via Sabre's node in Paris. Such a solution would seem at first glance to be perfectly satisfactory because it would appear to provide access for all Sabre users within France. But this is not how Sabre works. If, for example, access were required to the supplier's system from Sabre users in say Italy, then the switching technology required in France would be complex and costly for Sabre to implement. So, the Sabre national node in France does not have a switching capability and instead the French supplier's system would be directly connected to Sabre's GDS central switch in the USA. This illustrates the arguments that Sabre uses for its policy of central hosting for non-air suppliers.

It is for reasons such as these, that Sabre has adopted a policy of central hosting for all non-air suppliers. This means that non-air supplier booking computers are always connected to Sabre's central switch in the USA. An architecture such as this allows any Sabre user, no matter where they are located, to have access to the central non-air host system, provided of course that they have the appropriate authorizations and access privileges to allow this.

In the USA, travel agents are highly focused on three core travel products: airline seats, hotel accommodation and car rental; with cruise lines a close fourth. Sabre has, for many years, had a range of these non-air suppliers connected to its central system. However, in Europe, the travel pattern is different and many travel agents also require access to additional travel suppliers. Examples include: rail companies, tour operators and ferry companies. The ELVA project has now widened the span of Sabre in Europe by adding many new non-air suppliers including, for instance, SNCF, the French rail operator (and thereby the Eurostar cross-channel service), tour operators and ferry companies. New suppliers are being added to Sabre all the time. For example, there are one or two leading ferry operators who have already agreed to connect their systems to Sabre under the banner of Sabre Navigator, e.g. P&O Ferries. Also, several leading European tour operators will be joining the list of Sabre leisure travel host systems under the banner of Sabre TourGuide. There are even plans afoot to connect other rail services into Sabre.

Non-air suppliers are invariably connected into Sabre using the Multi-Access level of participation. Just to refresh your memory, this means that the user 'talks' to the non-air supplier in the native host 'language' of their booking computer system. In other words, Sabre does not undertake any translation in an attempt to communicate with the user in a common 'Sabre type' language. This has the benefit of allowing the user access to the full range of functions and commands that are available in the non-air host system. However, Sabre does apply some intelligence to the booking process by combining and integrating all booking segments into a single PNR for the passenger. So, if the passenger's trip involved a flight, a car rental, a train journey and a ferry crossing then all of these services would be stored in a single PNR for the trip. In the future it should also be possible to request availability between two cities, say London and Paris, and see a Sabre display showing rail and channel tunnel services alongside airline flights.

More recently, Sabre has launched a new booking service called Direct Request for Hotels, aimed principally at the smaller hotel and bed and breakfast establishments (SMEs). These properties do not usually have on-line reservation systems of their own. So, Sabre has adapted straightforward fax technology to communicate with these SMEs. This opens up sales opportunities from over 29,000 travel agencies around the world that are Sabre subscribers. Sabre includes participating SMEs in its hotel availability displays. Then, when a booking is made by a Sabre subscriber, the central host system automatically generates a fax message that is sent directly to the SME. Once received, the SME fills in the appropriate fields and faxes the fax back to Sabre. The Sabre system then electronically reads the incoming fax using optical character recognition (OCR) technology. The decoded information is then used to update the PNR booking record, which may be viewed and acted upon by the originating travel agent.

Finally, there is a wealth of information stored within Sabre that is available to its users. The following are just some of the topics available via Sabre: (a) theatre information and seat bookings, (b) tourist information on countries and regions, and (c) information on national governments and health/visa requirements of their countries.

Well, that completes our brief review of Sabre's back-end functionality. Now let's see how these host systems, including the vast information resource stored on Sabre's central data base, are distributed to subscribers around the world. Distribution of Sabre is effected by several inter-linked front-end systems, each with its own set of associated products.

Sabre's front-end functions

Sabre's front-end core functions are the driving force for its many travel industry products. Only by understanding these functions can we expect to appreciate these new 'look and feel' products in detail. It is therefore necessary first of all to examine the basic reservation functions at Sabre's core, so that we can understand how it is that they can be made to look so simple within a distributed processing environment.

In many ways, the front-end core system doesn't need to know what kind of workstation is hanging on the other end of the line. It may be a dumb terminal or a PC running any one of Sabre's new products, such as Sabre For Windows, Turbo Sabre, Planet Sabre or Travelocity (Sabre's Internet product – see Chapter 5). As far as the front-end core system is concerned, the entries and the responses it processes are identical. The functions that comprise this front-end core system go under the banner of Professional Sabre. It doesn't matter what front-end product you are using, it is Professional Sabre (or a sub-set thereof), that you will actually be interacting with on the Sabre host main-frame. This is the reason why I have chosen to use a review of Professional Sabre as our springboard for understanding Sabre and its full range of GDS products.

Professional Sabre

Professional Sabre has the most comprehensive set of functions available to non-AMR users who are of course in the main, travel agents. Let's start by considering the use of Professional Sabre as accessed from a dumb computer terminal. If we can grasp how these core functions are performed at this level then it will be far easier to understand how the more sophisticated products work. Most of the basic system functions in Sabre are initiated by depressing a number on the keyboard. Each number has a certain function and these are described in more detail as follows:

- **1 Availability** This is probably the most frequently used function of all. As the name implies it allows a user to enquire about the availability of a specific flight or to review the various services between two cities. Participating Total Access hosts allow last seat availability and guaranteed bookings. All that is needed for a minimum availability request entry is the date of travel and the city pairs. The system responds with a list of flights in the sequence recommended by the ECAC CRS Code of Conduct (see Chapter 1), i.e.

- *First*: Direct flights in departure time order.
- *Second*: Flights with stops but same flight numbers.
- *Third*: Connecting flights in sequence by the most direct route.

- **2 FliFlo** This entry is used to obtain flight information for participating carriers. The information is presented in real-time, which means that it is up-to-date. It shows such information as the latest estimated (and actual) arrival and departure times, the flying time between two cities, the weather *en route* and so on.
- **3 GFAX** Otherwise known as other service information (OSI) and/or special service request (SSR). OSI messages do not require a response, whereas SSR entries do. This is how you request special services that are needed on a non-American Airlines flight by your customer. Services such as special meals, seat requests, meet and assist, and so on. Incidentally American Airlines offers over 30 different types of meals that if ordered 24 hours in advance can be served on the flight, from low lactose meals to an all-American burger!
- **4 AFAX** This entry is made when a response from American Airlines is needed. AFAX is used to request SSRs from American Airlines or send OSI information. Replies to GFAX SSRs are sent to AFAX. When the reply is received, Sabre stores it as part of the PNR. It may subsequently be recalled and displayed by the AFAX entry.
- **5 Remarks** Besides the obvious use for this entry, coded remarks are supported. This means that the user need only enter a shorthand code in order to make a special remark appear on a PNR, a ticket, an itinerary or an invoice. Examples of some common entries are the passenger's address and the form of payment.
- **6 Received field** Each time the PNR is accessed and changed by the travel agent, an entry is made in the history file. The Received field is used to store data that are added to indicate who authorized the booking, or who requested a change, etc. It is a mandatory field for the original booking and also for certain changes. This is useful when a passenger says 'I never asked for that service to be provided'. With a PNR history, the travel agent can always display the precise sequence of events that led to the current status of the booking. In fact a '6' field must be entered before a PNR can be finalized, e.g. an 'E' after the '6' field entry, ends the transaction.
- **7 Ticketing field** This field contains ticket instructions that must be entered prior to ticketing, and after ticketing has taken place it is updated to contain certain ticket-related data. Besides the basic reservations information held in the PNR, there are certain key fields that are needed for ticketing and these are: the airline code that designates the ticketed carrier, the commission rate (per cent), the form of payment and the baggage allowance. The ticketing field is a sophisticated function that is impossible to cover fully in just a short section of this book. Suffice it to say that Sabre maintains a diary of ticketing actions and updates this part of the PNR automatically.
- **8 Ticketing time limit field** This field is used to record a time limit beyond which the PNR is cancelled if a ticket has not been printed. When ticketing does occur, the ticket field (see above) is updated to reflect the new status and this ticketing time limit field is eliminated.
- **9 Phone/contact field** This field contains details of how the traveller may be contacted. It may be entered automatically by the Sabre host main-frame system from some previously recorded customer profile information held in a storage area called special traveller account record (STAR), which I will be covering in more detail later. Alternatively it may be entered at the time of booking by the travel agent.
- **0 Sell or reserve a product** Finally, the field that allows a travel agent actually to sell an airline seat or other product. The field entries are simply the class of travel and the line number of the availability display that contains the details of the flight (or other product).

There are a few other keys that have special significance and I will cover these briefly here. First, to cancel a segment the character 'X' is used, followed by the segment that is to be cancelled. So, for example, entering X3 would cause the third segment of the itinerary to be cancelled; and entering XI would cause the entire itinerary to be cancelled. A '/', i.e. a slash, allows a segment to be inserted into the itinerary; a '.', i.e. a full stop, changes the segment's status; and a ',', i.e. a comma, allows a new number of passengers to be entered, e.g. four were originally booked, one is to drop out and so the new number of passengers is three but the itinerary details remain the same.

This sounds quite complicated and one may ask why is it necessary to use all these peculiar combinations of characters to mean so many different things. The reason is partly historical and partly practical. For historical reasons, the original Sabre system was designed to use a single character to denote a specific function and from a practical view, the use of a single character is fast. It just takes a bit of getting used to really. But there are more functions available via the keyboard:

- **Programmable function keys (PF keys)** PF keys are a powerful productivity tool. PF keys are the set of keys on a keyboard that are usually located across the very top of the keyboard. There are 12 PF keys on most modern keyboards, which together with the shift key enable 24 PF keys to be available to workstation users. These keys are available to be used by the application that happens to be running in the terminal controller at the time. Each PF key may be used to store a set of functions equivalent to several key depressions. Each PF key may be programmed a different way by its user. So, for example, a travel agent sales person may decide to set their PF keys up one way, whereas a colleague in the same agency may set their PF keys up another way. It is, however, usually the case that a standard is decided upon in an agency of any size, so that people can move from one workstation to another without having to re-designate the PF keys.
- **Special traveller account record system (STARS)** STARS is known in other systems by the term *profiles*. Sabre's STARS is widely regarded in the travel agency industry as one of the most powerful features of the system. Each STARS record supports the storage of up to 200 lines of information. By using the STARS facility wisely, a travel agent can gain substantial productivity advantages and greatly enhance the level of personal service provided to customers. In essence STARS is lines of information that may be moved automatically into a traveller's PNR by the travel agent making the reservation. The data are maintained under the complete control of the travel agent. STARS is usually used to keep a file of reference information on each company and frequent flyers, with whom a fair amount of repeat business is transacted. This allows information to be copied into PNRs effortlessly and for the agent to demonstrate effectively to the customer that they really do know a lot about them. There are three levels of STARS:
  - *Level 0: The travel agent profile* This contains the basic static information that describes the travel agency. Items of information such as the agency's name, address, telephone number and other items of reference information. There is therefore only one Level 0 STAR for each travel agency location.
  - *Level 1: Company STAR* This is a collection of information that describes a business travel account customer. It would normally contain details about the head office of the company to which the travel agent provided business travel services; name, address, telephone, etc. This Level 1 STAR may also be used to store the details of an individual who is a frequent customer of the travel agency but who does not work for a particular company.
  - *Level 2: Individual traveller STAR* All the employees of a company have a single STAR record each. This is associated with the Level 1 STAR for their company and describes the traveller in detail. Identification information is stored along with the preferences of the individual concerned. Examples of individual preferences include: vegan, smoker, aisle sitter and window gazer. Other STAR fields are the form of payment and the traveller's department code.
- **Sabre Scribe** Another powerful facility to tailor Sabre for an individual user or a travel agency is the script facility. This allows a 'fill in the blanks' format to be defined for a

particular job. An example of when this would be useful is for capturing a number of repetitions of a similar transaction. Or for capturing a standard set of management information fields for a specific company booking. A script can be set up so that when recalled, a kind of electronic form is presented on the screen to the user. The cursor is positioned next to the field to be entered and when this has been input, it moves on to the next field and so on until all data have been entered. Nine pre-programmed scripts are provided with the basic Sabre Scribe product; others can be set up to meet specific user requirements.

- **User defined interface data (UDID)** Also available as an important tool to help tailor Sabre for a travel agent's own particular needs is the UDID facility. A UDID is a special kind of remarks field in Sabre. It can, however, be interpreted by a back-office computer system that is programmed to recognize certain UDID fields. A UDID entry can be used to record special information within Sabre and format a record or part of a record that will automatically be fed into a back-office system. Each UDID is assigned a label that, when entered, allows the data that follows to be associated with that UDID for later processing.

  A good example of UDID usage in practice, is the entry of a corporate customer's travel requisition request number as part of a booking. A travel requisition incidentally, is a form used by many companies to enable a traveller to obtain the necessary in-house approvals from management, in order to make the trip. This is usually a paper-based form although with the increasing popularity of e-mail, this is fast becoming an electronic 'document'. When the reservation is made, the traveller specifies the travel requisition number that is entered by the agent, for example, as the UDID entry '5.U1.-AT1476'. This causes the travel requisition code AT1476 to be stored in the PNR. At ticketing time, the ticket interface record is transmitted into the back-office system which is programmed to search each incoming record for a field identifier of 'U1'. When it finds this, it knows that the following data is the travel requisition number. This number can then be extracted from the ticketing record and stored on the back-office system's MIS data base for later reporting.

  It is also possible to get even more sophisticated and combine Sabre Scribe with the UDID facility. This enables the UDID field to be edited by Sabre at the time of entry. In our example above, the UDID entry could be checked by Sabre Scribe to ensure that it was six characters long and that the last four characters were numeric only. Using this approach Sabre can be virtually pre-programmed to capture the precise information needed by the travel agent's business travel customers. This illustrates the labour saving and customization facilities that Sabre's UDID and script features provide a travel agent.

- **Ticketing** The control over Sabre ticketing is provided by the host main-frame computer. In a dumb terminal environment, such as the one we are considering here, the ticket printer is connected to the Sabre communications controller installed in the travel agency. At the time a ticket is printed the user can also request an invoice and/or an itinerary to be printed. The ticket printed by Sabre in this environment is called the Phase 3 continuous stationery ticket.

- **Back-office interface** Whenever a transaction is invoiced or a ticket is printed by Sabre, the data can be stored for transmission into a back-office system. The repository for this information is called DWLIST. This is really a special kind of file that is stored on the Sabre host and that can be downloaded into the back-office system via the agency gateway on the file server. In fact this download process can either be interactive, i.e. it is done as soon as a ticket is printed, or it can be done in batch at certain times of the day. Sabre can interface to a wide variety of industry back-office systems including, for example, Galileo's PAMS!

- **Leisure travel support** The Sabretext product supports access to leisure travel supplier's reservation systems that are based on videotex technology, from a Sabre PC workstation. This is effected via the X25 link from the travel agency to the Sabre gateway that controls the type of access. The access from the Sabre gateway may be either direct into Sabre or in the case

Table 4.5 A summary of Sabre products

Market segment	Segment classification	Sabre product
Travel agents	Agents with an average mix of leisure and business travel bookings	Professional Sabre Dial Sabre Sabre With Windows Sabre For Windows 95 Planet Sabre
	Agents with a large volume of business travel bookings	Turbo Sabre
	Agents with their own in-house booking systems	Sabre EDI Gateway

Corporates	–	Commercial Sabre (USA only)
		Business Travel Solutions
		    SabreExpress
		    SabreExpress Fax
		    SabreExpress Mail

Consumers	–	EasySabre (USA only) Travelocity (USA only)

of Sabretext, directly into the Istel network. AT&T Istel supports a number of viewdata leisure travel reservation systems that are connected to it. In Sabretext mode, the workstation acts just like a viewdata terminal.

- **Concierge** This is Sabre's general information system. It is supplied from a variety of sources, as well as Sabre itself. A virtual travel related encyclopaedia of useful facts can be accessed on a wide range of subjects including, for instance, health, visa, weather, geography and tourist information.

Sabre's front-end products

As mentioned earlier, the core Sabre front-end system is capable of being accessed in a number of different ways by different types of users; and the profile of the average travel-agency user is changing quite radically. It used to be the case that the focus of required skills needed to make a good travel sales person (or travel counsellor as they are sometimes known), were largely technically oriented. For example, many advertisements for travel sales positions stated the need for applicants to be skilled in a particular GDS system. This is not quite as important a requirement as it used to be, chiefly because: (a) systems have become, or at least are becoming, far easier to use; (b) the time required for training a new user has decreased substantially with the newer GUI and intuitive systems now widely used in travel agencies; and (c) there is a more pressing need for travel sales staff to be experts in the field of travel itself (or at least to know where to look for travel information), rather than computer experts. In fact it is widely assumed that, as a result of today's modern education, young recruits are anyway sufficiently computer literate. However, the users of Sabre are not restricted to travel agents.

Following some in-depth market research, Sabre has segmented its users into three main types: (i) travel agents; (ii) companies, otherwise known as corporates in Sabre terminology; and (iii) individual consumers. A complete range of products has been developed for each of these market segments. This product segmentation is usually achieved by means of: (a) local processing carried out in the user's Sabre PC, and (b) 'filters' that have been built into Sabre's front-end system. As a result of this, there are therefore several Sabre reservation products, each of which has its own brand name and is aimed at a particular section of the market. Some of these products have been around for several years while others are only just

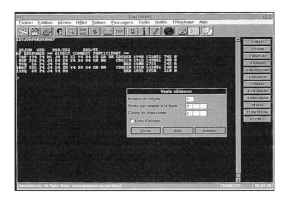

Figure 4.10 Dial Sabre

being introduced. However, Sabre has a policy of supporting all established products for the foreseeable future, i.e. in IT terms. The main Sabre reservation products are summarized in Table 4.5.

The reason for the growth in new and innovative Sabre front-end products is the benefits that PC GUIs bring to end users. The core Professional Sabre commands are concise and efficient in terms of system resource usage, but at the same time they are complex, difficult to remember and error prone from the user's viewpoint. Also, some hotel commands entered via Sabre can be over 40 characters in length; and with each character having a specific meaning, this represents a source of errors, mis-understandings and repetitive training needs. As a result, most users only remember the formats of between 30 and 40 per cent of Sabre's available functions. The other less frequently used but nevertheless very useful functions, are just not memorized. So, Sabre has introduced a series of evolutionary GUI-based products that are summarised in the above figure and that are presented below in a little more detail.

Travel agents

Sabre has designed a set of products especially for its main subscriber base, which is the travel agents of the world. There is virtually a customized product for each of the main types of agency (see Chapter 7). These products range from those providing simple, easy to use features that can be slower to use, up to the faster versions that require more training to use effectively. The travel agency products are:

- **Professional Sabre** This is the name of the product that is used by travel agents who do not have PCs. Professional Sabre is covered in the preceding section in greater detail. The functionality of this core system is, however, enhanced by co-operative PC processing that allows the following products to be supported.
- **Dial Sabre** This product is designed for the infrequent Sabre user, often smaller travel agencies, who cannot justify a dedicated reservation terminal. As such it is designed mainly for small independent travel agents, especially those concentrating on the leisure market. It supports only a limited set of the entire range of Sabre functions (Fig. 4.10).
- **Sabre for DOS** The advent of cheap and widely available PC technology has provided Sabre with an ideal opportunity to make its reservations product even easier to use and more sophisticated than was possible using dumb terminals. Sabre for DOS has been in use for several years now although it has recently been superseded by the Microsoft Windows family of operating systems. The product is more precisely identified as Sabre System 5.0 Enhanced Software for DOS Version 3.3 or higher (in actual fact, Sabre Revision 5.1 is now available). Any IBM compatible PC may be used to run Sabre for DOS and the product may also be used on a LAN. The screen is subdivided into several parts, which are generally as follows:
  - *Memo area* This is a coloured rectangular area (usually blue) located at the very top of the screen, which is a kind of electronic memo pad. Its purpose is to allow the travel-agency user to display a message on the screen. This can be useful when the workstation is unattended for a period of time and the operator needs to leave a message for a colleague.
  - *Function key labels* Function keys provide some powerful pre-programmed functions that may be specified by the workstation user. Whenever these functions need to be performed, a single key depression will execute them. A tall thin rectangular area on the right-hand side of the screen is reserved for a display of the function keys and their meaning.

- *Work area* This is a large almost square area in the middle of the screen, which is the main work area. This is the area where the user keys the Sabre entries and sees the displays that are sent from the host system.
- *Mode* There is a small block at the bottom of the screen that shows which mode is being used to access Sabre. This could show, for example, Professional Sabre, Sabre Scribe, etc.
- *Clock and date* These two useful reference fields are displayed in the bottom right-hand corner of the screen and are self-explanatory. Additionally, the depression of certain keys causes a calendar and a calculator to be displayed. The calendar is a very useful feature because it is automatically updated by the PC's internal clock powered by a small long-life battery that is continuously running. The calculator displays an image of what a typical calculator looks like with buttons for numbers and the common arithmetic functions. These buttons are 'pressed' by selecting them on the keyboard.

- **Sabre With Windows** This is a product that was launched world-wide in April 1993. It was rolled out in the UK and the rest of Europe by mid-1993. Sabre With Windows really does make the PC more than just a dumb terminal emulator. Besides providing an extremely user-friendly front-end to the Sabre host system, the PC can run any Microsoft Windows Version 3.1 compatible software at the same time. Any PC that is IBM compatible can run Microsoft Windows Version 3.1 or higher as supplied by Sabre. An industry standard version of Microsoft Windows is provided as part of the Sabre With Windows product. The plain vanilla version of Windows 3.1 has been enhanced as the result of a joint effort between Sabre and Microsoft. This has produced a version of Windows that enables the user to get the most out of Sabre and yet be able to run industry standard Windows 3.1 compatible software without any special modification. Sabre is simply just another software product that can be loaded and accessed via standard Windows.

A common question often asked by travel agents is: What benefits do I get from Sabre With Windows? This is a good question but there is an equally good answer to it – a lot of extra functionality that can generate a higher level of productivity! The following are just some of the advantages of using a Windows environment:

- *Multi-tasking* First, there is the multi-tasking capability. Multi-tasking means that you can run several programs at one time. It is possible, for example, to have Sabre available in one window, a client file from your back-office system in another window and a word processor in a third window. Your PC might be printing a report as you are accessing Sabre. The overall reason you might want to take advantage of multi-tasking, for example, is so that you can build a personalized itinerary for a customer and produce a quotation for them.

  Multi-tasking allows you to add 'boiler-plate' text to the document and put a few finishing touches to it before storing and printing it. With Windows, you just start the printer off and then you are free to do something else. The Windows multi-tasking software takes care of controlling the physical printing while you get on with another customer proposal or make another reservation.

- *Clipboard* This is where the second powerful Windows facility comes into its own; the Clipboard. The clipboard is a standard feature of Microsoft Windows and it allows you to select and copy information from one window and insert or paste it into another. So, you could, for example use the Sabre window to build an itinerary for the customer using the availability displays and other features that I have described above. Then you could copy the relevant pieces of the itinerary, swap into the word processing window and paste the itinerary into a document. Then you could activate the back-office system window and display the relevant client file that contains all the static information about your customer. Once again you could copy the appropriate information on say name, address, telephone number and special preferences, swap windows back to the word-processor and paste the information into the document you are building.

The Microsoft Windows Program Manager is the front screen for Windows. Besides the standard Microsoft windows containing various general program functions, there is a Sabre window. This shows several icons, each of which when double clicked with the mouse, offers some important services to the user. It would be impossible to cover the richness and depth of these functions in this book. It must therefore be borne in mind that the following description covers these areas in very general terms only:

- *Sabre utilities* The first of these utilities is network security, an often overlooked yet very important subject. Sabre network security is provided by Novell LAN software. This controls which groups of functions are accessible by the user of the workstation, the rights granted to users in terms of whether they can read or write to certain sensitive files and a directory of users. LAN diagnostics are also presented pictorially. An actual image of the LAN in the agency is shown, which is a powerful tool for use by the supervisor within the agency.
- *Sabre applications* A calculator and a calendar, just like with Sabre for DOS as described above. The only difference here is that the user is free to use a mouse instead of (or in addition to), the keyboard.
- *Gateway to Sabre With Windows* Besides providing a gateway to the core Sabre system as described in the main body of this section, several other peripheral functions are offered in this window. A revision history of the Sabre software used by the travel agent is available. This is especially useful in ensuring that the travel agent is using the latest version and has loaded the latest set of updates. It is surprising how often this is overlooked and it can cause some horrendous problems if not done properly. User information is provided that shows who is logged in and what peripheral devices, such as printers, are doing – a sort of 'big brother is watching you', kind of a function that is so loved by the agency supervisor! Then there are the Sabre With Windows tutorials and help functions: (i) About Sabre With Windows – a general introduction and overview of the product and the functions available at a general level; (ii) Getting Started – an overview of the basic entries that are, for instance, needed to obtain an availability display and make a simple reservation; and (iii) Practising Sabre With Windows – an extremely useful facility that enables a new user to practise most of the Sabre entries without any danger of impacting the live system.
- *Sabre Help* This, as the name implies, provides information on certain subjects that are needed to accomplish a task that the user is unsure about or needs a quick refresher. It is a powerful support facility that can be accessed in a number of ways. First, it is like a book in that there are certain headings and sub-headings that can be accessed via a kind of table of contents. Then there is context sensitive help, which is probably one of the product's strongest benefits. When a user gets to a point in the transaction at which further information is needed to complete it, the depression of the 'Ctrl' and 'H' keys initiates Sabre's context sensitive help function. The system looks at the current entries and extracts what it considers to be the most relevant section of the help text that is most likely to address the user's problem. A diskette is periodically distributed by Sabre with updated help information recorded on it. This is loaded into the system and is then available to users.

The Windows approach enables Sabre to provide not just an easy to use interface but also one that can be tailored to each country of the world in which it operates. This functionality is one of the in-built functions of Microsoft Windows. So, although the core Sabre host system is the same, no matter where it is accessed, the PC running Windows can be set up to display standard workstation responses in the national language, e.g. Help Tutorials, Windows Help, Sabre With Windows Help and Sabre Help; and this includes Japanese Kanji symbols too. It is probably fair to say that Sabre With Windows is a strategic product that will be around for many years to come and will continue to be enhanced and refined.

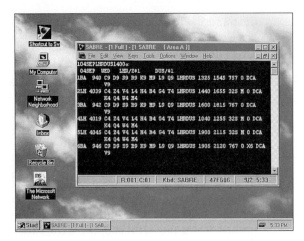

Figure 4.11 Sabre for Windows 95

Figure 4.12 Planet Sabre (logo)

- **Sabre For Windows 95** As the name implies, this is a version that runs under the Microsoft Windows 95 PC operating system. It is, however, generally the same as Sabre With Windows from a functional viewpoint.

  One of the important optional features with this product is graphical ticketing. Graphical ticketing addresses the minefield of complexities surrounding the automatic printing of travel tickets for customers, whether they be for air, rail or ferry. Instead of the old set of one character commands required by Professional Sabre to drive out a ticket successfully, graphical ticketing presents the user with a virtual image of the actual ticket as it will be printed. The image is constructed from the information already keyed by the user, following the reservations process. The supporting software is also a lot more intelligent and works out for itself the likely contents of various fields that are to appear on the ticket. The travel-agency user can then view the ticket image on the screen, make any necessary adjustments and when it looks OK, release it for ticket printing. Because humans are more adept at understanding and manipulating images rather than text and numbers, graphical ticketing can reduce errors and accelerate booking procedures. With the increasing complexity of ticketing rules and procedures, graphical ticketing is therefore an extremely useful and productive feature of Sabre For Windows 95 (Fig. 4.11).

- **Planet Sabre** This product, which was launched in the USA during 1996, replaces Sabre With Windows, i.e. Microsoft Windows Version 3, and also Sabre For Windows 95. It is characterized by an attractive welcome page showing the 'travel planet', which also serves as a top level menu of functions (Fig. 4.12). Simply clicking on the appropriate image shown of the travel planet will link the user to the associated set of Sabre functions. Planet Sabre was launched in Europe in two phases during 1997:
  - *Phase 1* This focused on two key areas: (a) formatting all non-air commands using a new GUI, and (b) incorporating the graphical ticketing sub-system. The reason for the emphasis on these two areas for Phase 1 was that they represent the areas where most errors are made by users. These are the kind of errors that arise from the occasional use of complex functions *vis-à-vis* commonly used simple 'bread and butter' functions, e.g. airline availability and reservations.
  - *Phase 2* This includes the automatic formatting of air booking functions. In essence, this replaces the use of a single character code with selectable menu options for all booking functions.

  Planet Sabre uses a full-screen GUI within a Windows 95 operating environment. It has a toolbar situated at the top of the screen that scrolls horizontally. The body of the screen

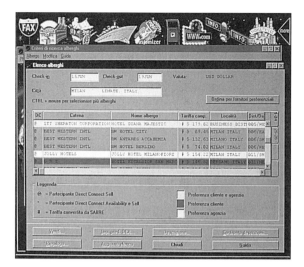

Figure 4.13 Planet Sabre (screen 1)

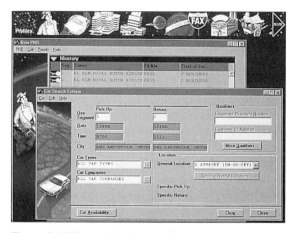

Figure 4.14 Planet Sabre (screen 2)

(Fig. 4.13) shows Sabre options and drop-down menus, all of which are selected by point and click commands using the PC mouse. All user initiated entries are cross-referenced and checked for validity prior to being sent to the Sabre host. Many entries are automatically generated using intelligent retrieval of contextual information, e.g. by using STARS, by using previously entered data and by populating hotel booking fields from basic air data already entered. Finally, Planet Sabre includes an improved version of graphical ticketing as a standard feature of the product (for a fuller description of graphical ticketing, see under Sabre for Windows above).

Besides all this, Internet access is fully supported along with a web browser (see Chapter 5 for more details of Sabre's Internet site – Travelocity). Planet Sabre requires a PC with a Pentium processor chip, a CD-ROM device and Windows 95.

- **Turbo Sabre** This is a Microsoft Windows product, which was launched in early 1996 and designed specifically for the high volume business travel agency. Its focus is on travel agent sales staff who need to achieve high levels of productivity in all business travel products. Thus, the design of Turbo Sabre is based on providing functions that maximize the speed of booking.

It comprises a window that is sub-divided into four mini Sabre displays. The lower right-hand quarter contains 16 Windows push buttons that represent the 16 most commonly used Sabre commands. Each push button is associated with a key on the keyboard. The keys are arranged so that the ones that are easiest to hit are the ones associated with the most frequently used commands. When a key is depressed, the window in the lower right-hand quarter shows the next lowest level of commands. This process may be repeated until eventually, once the lowest level is reached, the screen shows either: (a) the actual booking data that must be keyed, or (b) the relevant display as requested in one of the other free quarters of the screen.

The main advantages of Turbo Sabre are: (i) the speed of operation; (ii) the reduced amount of training time required for a user to become proficient; and (iii) the display of a large amount of information on a single screen, e.g. an availability display in the top right-hand quarter, a customer's STARS display in the top left quarter and a hotel availability display in the lower left quarter. Using Turbo Sabre, training time can be reduced from the one week required for most Windows products to just two days. However, users nevertheless still need basic training in the airline business.

Newer and faster versions of Turbo Sabre continue to be released. The latest version supports the re-display of the last 20 entries and provides more sophisticated customization features. This means that users are almost totally insulated from the cryptic CRS formats that are so difficult to learn and remember. This has been found to reduce the number of keystrokes required to create a PNR by 50 per cent thus reducing overall booking time by 25 per cent. It simultaneously reduces errors by between 20 and 50 per cent, while increasing preferred vendor sales by 20 per cent.

- **Sabre EDI Gateway** This product has been designed for those very large travel agents who have developed their own sophisticated in-house computer systems and branch network. Such agencies already have their own dedicated links to certain supplier systems that they then distribute from their head-office computer centre to their own terminal devices used in remote branches by means of a proprietary telecommunications network. Rather than have a variety of different interface programs (one for each large travel agency), Sabre has instead chosen to make its interface available using the world-wide standard known as Electronic Data Interchange (EDI). This approach has several benefits: (a) it is a computer-to-computer standard that is supported globally, (b) it means that Sabre need only maintain and support a single interface program for all such travel agency links, and (c) travel agents wishing to interconnect in this way can become readily familiar with the interconnection method. An example of one such large travel agency group that has connected its central system to Sabre is Club Med.

Corporates

Sabre has developed a set of end-user products that are specifically designed for large companies that have a high number of staff who fall into the category of frequent business travellers. This sector of the travel market, i.e. business travel, has its own particular requirements. For instance: (a) the travellers are usually individuals who are very knowledgeable in the field of travel, and (b) the companies usually have well defined travel policies that are carefully monitored by dedicated staff. Sabre has developed some relatively new and sophisticated products for this niche market:

- **Commercial Sabre** This is designed for use by corporate customers themselves. It is a product that has been around for at least four years, but has only a limited future because it will soon be replaced by Business Travel Solutions (BTS, see below). Despite its direct use by customers, it doesn't cut the travel agent out of the action at all. The way it works is generally as follows. The corporate customer (or business house account), uses a PC to access a sub-set of Sabre, i.e. a restricted set of Sabre's functions. The PC is connected to Sabre via a publicly available dial-up computer network called 'Compuserve'. The users of Commercial Sabre can request an availability display and select seats on flights that they wish to be booked. The bookings are then routed to the travel agent by the Sabre main-frame. The travel agent uses a dedicated reservations processor PC installed in the agency and supplied by Sabre to: (a) perform a quality control check on the bookings, (b) enter the more technical details into the PNRs, (c) make the necessary entries for ticketing, (d) check with the customers' established travel policies, and finally (e) confirm the bookings. Once the travel agent has confirmed the bookings, copies of the itineraries are sent to the customers' administrative offices and tickets are printed for the travellers. It is even possible to use Satellite Ticket Printing actually to print the ticket in the corporate customers' own offices, all under the control of the travel agent.
- **SabreExpress Fax** This is a product available under DOS or Windows and is now incorporated within BTS (see below). Roll-out started in 1993 on a global basis and was available to anyone who wished to have an airline reservations capability of their own. Like EasySabre, the corporate traveller completes a simple screen on their PC that specifies the booking required. The difference with SabreExpress Fax is that the booking is sent to a local travel agent via a fax board installed in the PC. (A fax board is a printed circuit board that contains a fax modem and is installed in one of the PC's expansion slots inside the case; it usually has some special communications software associated with it that runs in the PC itself.)

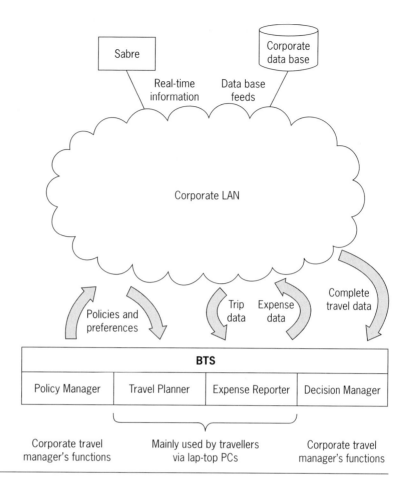

Figure 4.15 Sabre BTS

The travel agent receives the fax message and checks the booking adding any necessary fields before confirmation and ticketing.
- **SabreExpress Mail** Again this is a product available under DOS or Windows and its functions are now incorporated within BTS (see below). It is identical to SabreExpress Fax except that instead of using the fax service, an electronic mail service is used to deliver the booking request to the travel agent.
- **Business Travel Solutions (BTS)** This is one of Sabre's new products that is still in the development stage (the product was in Beta testing during 1997). It replaces SabreExpress and is designed for the corporate travel environment. BTS (Fig. 4.15) is a family of PC software packages that supports: (i) travel planning, (ii) expense management, and (iii) travel management functions. It is a product that is generally sold to large companies by Sabre itself as well as travel agents and other airlines who have an established business relationship with one of their large business travel customers.

BTS is used principally by corporate travellers who often implement the software on their lap-tops and take it with them when they travel on business. It is also implemented on the company's own network of PCs, which are invariably linked by a LAN. The products are as follows:
- *Travel Planner* This software provides access to Sabre for real-time bookings and the BTS core for administration. It runs on the traveller's own lap-top computer and accesses the Sabre network via a dial-up telecommunications link. It allows the traveller to view the availability of all travel products and services supported by the system.

Travellers can book airline seats and other products according to their own pre-set preferences and within the context of their companiest travel policies. Frequent trips can be stored as a template, which is then used to create a new trip of the same type. This accelerates the booking process considerably. Travel Planner also incorporates other easy-to-use features such as 'drag and book' – the appropriate icon is simply dragged across the screen and placed on the correct date in the graphical calendar.

- *Policy Manager* The company's travel policy is pre-programmed into the Travel Planning module. In this context, the policy is expressed as a set of rules governing such options as the class of travel, suppliers to be used, routes flown for certain destinations, category of room, type of car to be rented and other major factors that influence the cost of travel. Although the software can detect bookings that contravene the company's travel policy, it nevertheless allows the traveller to override the warning and make a booking outside policy. However, such bookings are flagged for eventual reporting to the corporate travel manager on an exception basis.
- *Expense Reporter* As expenses are incurred by the traveller they are logged and categorized by BTS. Some expense items are recorded automatically as a by-product of the booking process. Others are keyed by the traveller as they are incurred during a trip. At any point in time, the traveller may use their lap-top to view the expenses recorded to-date. Once the trip has been completed, the Expense Management sub-system provides the traveller with an electronic trip expense exhibit as a screen image that can be viewed, amended and eventually finalized before being electronically submitted for approval by a corporate executive. Although physical documentary supports, i.e. receipts, are required for some categories of expense, BTS comes as close as possible to a paper free expense management system – something most business travellers dream about!
- *Decision Manager* This is the software used at the company's headquarters to centrally monitor actual and planned travel expenses for all employees. As previously mentioned it produces exception reports highlighting instances where the company's travel policy has been contravened and keeps records of travel expenditure by various categories. An important function provided by this feature of BTS is management information. This is available not simply via voluminous reports but by report generator software and enquiry tools that feature a user-friendly GUI. Decision Manager can help lower costs by improving negotiated rates, enhancing pre-trip authorizations and stream-lining policy compliance.

Consumers

For many years, Sabre has recognized the importance of supporting its customers directly, while at the same time protecting its main distribution channel – the travel agent. This is not as much of a balancing act as it might at first appear. While it is all well and good to provide customers with information that helps them to plan their itineraries, making a reservation firm and printing a ticket require specialized skills that only a travel agent can provide. The following products have all been designed to support consumers, yet maintain the travel agent's key role in the loop:

- **EasySabre** This is a product that has been available in the USA since 1985, although it is not marketed in Europe at present. It is a version of Sabre that runs on a consumer's own PC. A modem is required to access Sabre via on-line services such as the Compuserve network and a simple 'fill in the blanks' screen is used to capture the travel requirements. Once this is captured by Sabre, the reservation is channelled via a local travel agency for ticketing and follow-up action. The user does not need to be familiar with the Sabre reservations language.
- **Travelocity** This is Sabre's Internet product offering and it is described in more detail in Chapter 5 – The Internet.

Finally, Sabre supports electronic ticketing (see Chapter 3 for a fuller description of electronic ticketing). This is currently in widespread use by

Sabre users throughout the USA and is planned for release in Europe during 1997.

Sabre pricing

Sabre pricing, like many GDSs, comprises two main elements: (i) equipment rental and service charges, and (ii) volume related credits. A separate pricing schedule is agreed with each user, which is designed to encourage bookings made via Sabre. There are two inter-linking components. First, there is the cost of the equipment and associated network usage: second, there is the booking fee credits that Sabre rebates to the subscriber via a productivity based agreement (PBA). These components are described in more detail as follows:

- **Costs** The equipment rental and service charges vary according to several factors such as: the number of computer terminals or PCs required by the user, the operating system running in the user's PC, the number of remote locations that require access to Sabre, the type of ticketing used, the level of functionality required, the storage space within Sabre allocated to the user, the software products implemented and the amount of training needed. These factors are all taken into account and a monthly fee calculated.
- **PBA** Against this monthly fee is set the PBA. This is a booking target that is set by Sabre in conjunction with the subscriber and is reviewed every three months. The PBA is the vehicle used to share the booking fee income derived by Sabre from other airlines, for bookings originating in Sabre. The greater the number of segments booked of all products, i.e. air, hotel, car, etc., the greater is the credit given by Sabre to the subscriber. There are, as one would expect, a few ground rules that the subscriber is expected to observe. A good example is the non-allowance of passive segment. Subscribers may not create a booking in an airline system using some other form of communication, e.g. telephone or other GDS, and then duplicate the booking in Sabre. Such passive segments will not count towards the PBA target. Nevertheless, it is quite possible for the cost of the entire system to be off-set completely by the PBA credits.

Sabre's integration technology

Sabre runs within a surprisingly varied number of different technical environments. The software is available, not only in the IBM and IBM-compatible world, but also in the Apple Mac world. Sabre has a technical team whose job it is to make the system work in whatever environment the customer wants (within reason of course!). In Spain, for example, there is a customer who accesses Sabre via an in-house DEC VAX minicomputer using two different terminal networks: (i) dumb DEC terminals, and (ii) Sabre PC terminals. In this configuration, each type of terminal has access to both the DEC minicomputer and also the Sabre system.

In the straightforward IBM-compatible world, Sabre runs in an environment with a file server and one or more workstations. The type of PC needed to run Sabre will vary with the product used. Sabre for DOS is less demanding on the workstation, which can be a 286 or higher. Sabre for Windows requires a 386 or 486 workstation PC with at least 4 Mb of RAM and a mouse. The server in each case needs to be a 386 or 486 with at least 4 Mb of RAM and a minimum of 80 Mb of hard disk space running under Novell Netware Version 2.15.

Sabre Europe has introduced Pentium PCs as the primary hardware platform for Windows 95 users. The hardware is assembled and provided by the Dell Computer Corporation. A range of models is available from the minimum – a Pentium P100 processor with 8 Mb RAM, 1.2 Gb hard disk, quad speed CD-ROM and PCI video with 1 Mb of RAM, to the maximum – Pentium P133 processor with 32 Mb RAM and other devices.

WORLDSPAN

This leading GDS (Fig. 4.16) has its origins in two of the world's most important CRSs, namely Delta Airline's Datas II and TWA/Northwest's PARS. These two reservation systems combined their resources and skills in 1990 to form Worldspan Global Travel Information Services. The resulting company is now owned jointly by Delta Air Lines, Northwest Airlines, Trans World Airlines and ABACUS Distribution Systems PTE Ltd. Incidentally, ABACUS is one of the largest GDSs

WORLDSPAN

Figure 4.16 Worldspan logo

in the Far East and its owners include Singapore Airlines, Cathay Pacific and Dragon Air. ABACUS and Worldspan each have a cross share-holding in each other's companies. Worldspan's world headquarters and host computer are both co-located in Atlanta, USA. This computer handles 1.2 billion messages globally in a peak month (an average of 1,377/s), stores an average of 7.8 million international PNRs and has a data base of over 85 million fares. The organization has two systems development centres in Kansas City and Fort Lauderdale.

With operations in 45 countries, more than 15,000 sites around the world, 8,000 of which are in the international arena, Worldspan is truly a global GDS. Its Europe/Middle-East/Africa (EMEA) region alone comprises 33 countries and is home to 29 independent Worldspan offices. The central market development activities, along with other support activities, such as finance, sales and marketing, the region's central help desk and technical support functions, are located in the international division at London's Heathrow airport. Over 6,600 travel agencies in the EMEA area can book from a total of 414 airlines, 40 car rental companies, 165 hotel chains, 29,000 hotels and 38 special travel service suppliers including cruise lines, railways, ferries and tour operators, all of which are available via Worldspan.

Since its inception in 1990, Worldspan has grown by over 550 per cent. This is especially significant, bearing in mind the company's niche marketing strategy. This strategy may be summarized as comprising two major elements: (i) the provision of customized systems, rather than fixed products; and (ii) an emphasis on the mixed leisure/business travel agency market, rather than, for example, having a primarily business travel focus as do many other GDSs.

Worldspan has always focused on providing its customers with bespoke systems, tailored to individual and specific business needs. This has resulted in the development of over 83 unique customer systems, outside of the central host GDS application. Now that Worldspan is firmly established as one of the world's leading GDSs and the niche marketing programmes of the past few years have been successfully completed, other new distribution channels are being considered. Prime examples include the corporate market and on-line services. Worldspan is in a strong position to consolidate product offerings and exploit the emerging technologies of these new distribution channels, mainly because of: (a) its wide exposure to so many different customer requirements within the global travel business; (b) the unique customer solutions that have been successfully developed and delivered; and (c) its global network, which supports many different communication protocols including the Internet (see Chapter 5).

So, there is an opportunity for Worldspan to pick some of the common functions that run through all of its bespoke customer developments and construct key generic products that have a wide appeal to the travel and tourism market. In the short term, therefore, Worldspan remains a substantial transaction processor; however, in the medium to longer term, the company's strategic focus is likely to shift so that it becomes even more of a technical partner for its customers. As a technical partner, Worldspan is in a strong position to provide consultancy and expertise in areas such as: helping customers to choose the best PC for their in-house departments, setting up an Internet site, distributing head office functions throughout an enterprise or establishing a strategic direction for a customer's technical environment.

Supplier connectivity

A good starting point for a walk-through of any GDS functionality is to consider the way in which suppliers are connected into their global network. This really breaks down into two main areas: (i) there are the suppliers that have been connected to the central host system for many years and provide the global dimension to the Worldspan service, and then (ii) there are the local suppliers that are connected in their countries of origin. Because both types of supplier are key to Worldspan's utility within several key user markets, it is important that you understand the supplier side of this major GDS before we consider the end-user functions.

Central supplier systems

In terms of central host connectivity, it is widely recognized that all the four main GDSs provide virtually the same levels of connectivity. Although some differences remain, generally speaking, all of the four major GDSs offer similar supplier interface functions. Some of the smaller airlines favour the particular GDS with which they are formally associated and deny some marginal functionality to other GDSs. However, airlines are becoming much more clearly focused on their core products and key competencies, which are the provision of airline seats to a globally distributed customer base. Airlines, and consequently their associated CRSs, are becoming much more relaxed about providing full functionality to any major system that can sell more seats on their aircrafts, thereby maximizing group revenues.

This levelling of the GDS playing field also extends into the area of host participation. Besides the basic level of participation entitled AccessPlus, which provides last seat availability on over 134 airlines, Worldspan offers its host airline CRSs the following levels of participation:

- **Airline Source** This is the highest level of participation that is available to airline customers. It features a real-time communications link between the participating airline's CRS and the Worldspan host computer. Each time a user, e.g. a travel agent, requests an availability display that involves a segment from the participating airline, an interactive on-line dialogue takes place. In such cases Worldspan provides the user with a transparent response from the participating airline's CRS. This enables the user to view availability as though they were connected directly to the participating airline.
- **Direct Sell** This allows Worldspan users to access a participating airline's CRS directly and hence its inventory. It therefore allows users to view inventory status interactively in real-time, just as though they were themselves connected to the carrier's CRS. This means that they may directly decrement the airline's seat inventory prior to ending the transaction. The participating CRS generates an acknowledgement record including the PNR locator, which is stored within the Worldspan PNR. This is known as Positive Acknowledgement.
- **Direct Access** This is a real-time link between Worldspan and the participating airline's CRS, which is slightly more sophisticated than Direct Response. This provides the end user with true last seat availability.
- **Direct Response** This is the most basic level of participation and provides the other airline system with the ability to return an acknowledgement message, including a PNR locator. This message may be generated either: (a) from a manual teletype entry, or (b) from an automated computer response. When the locator reference is received, it is placed into the Worldspan PNR.

If you compare these levels and types of participation with other GDSs included in this chapter, you should see that they are all very similar. This illustrates the comment I made earlier emphasizing the increasing levels of co-operation among airline CRSs.

In terms of hotel and car systems, Worldspan also provides the kinds of linkages and connectivity one would expect of a global GDS. The Worldspan host computer is both: (a) connected directly to the computer systems of the major hotel chains; and (b) connected to hotel industry switches, such as Thisco and Sahara. This web of inter-connectivity is masked from the end user by the Worldspan back-end system. This decides the optimum routing for reservation messages and also standardizes the responses from each of the different supplier systems used.

All the user does, for example, is to select a hotel property from a list of those available on the system and enter a service request, e.g. an availability display for a particular room type. The back-end system then decides whether to: (a) route this message via a direct connection to the hotel system, or (b) route the message via the relevant hotel switch. Once a response is received it is displayed to the user in a standard format.

Local supplier systems

What really differentiates one GDS from another are two critical success factors: (i) the range of local supplier systems available to customers, and

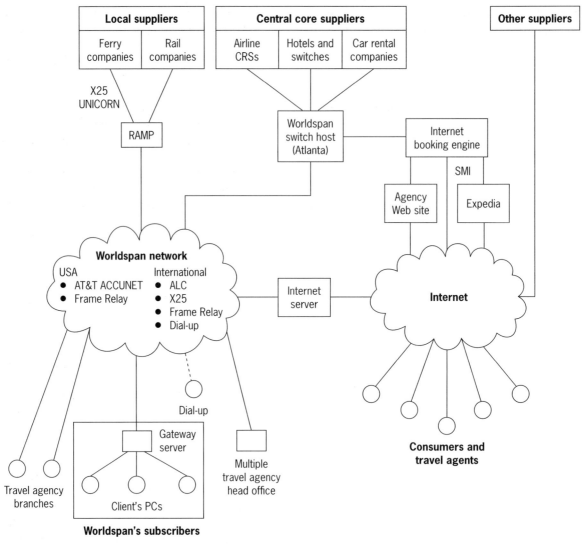

Figure 4.17 The Worldspan GDS network

(ii) the distribution network that the GDS uses to reach its customers. It is these two areas that have been the principal focus of Worldspan's attention over the past six or seven years. Let's take a closer look at the first of these two critical success factors – the way in which Worldspan connects into what are often called non-core local supplier systems.

As previously mentioned above, the core supplier systems are those that provide Worldspan users with access to airline bookings, hotel reservations and car rental services. All of which are connected into the main host computer located in Atlanta. This large and powerful main-frame computer provides end users around the world with all of the functions normally provided by a GDS and is described above in the section on central supplier systems.

Worldspan's approach to connecting into non-core supplier systems is a decentralized one. This means that non-core suppliers are connected into the Worldspan network (Fig. 4.17) in the countries in which their systems are located. End users may then access these supplier systems via the Worldspan communications network without having to be routed via the host computer in Atlanta. The two principal areas where many non-core supplier systems have been interconnected into Worldspan are ferries and rail companies. Worldspan's relevant GDS product for each is as follows:

- **Ferry Source** Ferries are a good example of locally connected supplier systems. Most leading ferry company systems are connected into Worldspan's X25 network. Worldspan's ferry booking function provides users with direct access to the leading ferry companies' own host systems. Ferry Source is available to any Worldspan DOS Alliance customer on the X25 network who is also an account holder with the ferry company accessed.

    These ferry systems use the UNICORN standard for information and booking messages to communicate with Worldspan (see TTI in Chapter 1 for more information on UNICORN). This means that travel agents may use their PCs to connect into any one of the ferry systems available using a common language, i.e. a common GUI. Participants include Stena Line, P&O European Ferries, Hoverspeed, Brittany Ferries, P&O North Sea Ferries and Moby Lines.

- **Rail Access** Worldspan has several links to rail service computers around the world. In many cases access is limited to users within the country in which the rail company operates. However, increasing use of overseas rail services is being planned and developed by Worldspan. Rail host connections are therefore effected in two ways: (i) a direct connection to the Worldspan host main-frame computer in Atlanta, or (ii) a local connection into the Worldspan network in the country of origin. Centrally connected rail suppliers support the integration of non-air reservations into a single Worldspan PNR, along with other air segments that may be part of the same booking. However, bookings made via locally connected non-air hosts must be filed separately from airline PNRs within Worldspan.

    Because access to rail systems is core to Worldspan's business in Europe, discussions are being held with most major rail operators throughout the area. The current situation, as at early 1997, is as follows:

    – *Belgium* For the past two years, Worldspan has been providing access to Societe Nationale des Chemins de Fer Belges SNCB – Belgian railways, for Belgian users only. This enables agents to check timetables, book rail journeys and issue rail tickets.
    – *France* Subscribers will soon be able to access Societe Nationale de Chemins de Fer Francais (SNCF) and enjoy similar functionality to that available in Belgium.
    – *USA* All Worldspan users have access to Amtrak, the USA national rail network, which is directly connected to the Worldspan host system in Atlanta. Worldspan is currently the only approved GDS in the UK that can issue Amtrak BSP tickets. Amtrak tickets sold by IATA licensed travel agents in the UK may therefore be settled via the UK's BSP process (see Chapter 7 for more information on BSP).
    – *Canada* Also connected to the Worldspan host computer in Atlanta, is Canada's Via Rail network. Reservations are effected using standard airline entries that can also support Via ticketing.
    – *Germany* The Fly Rail service (a German domestic service), is available to German users of Worldspan only.
    – *UK* In the UK, users may access the European passenger service (EPS) Tribute system for trains that use the channel tunnel for travel to Paris, Brussels and other European destinations. The link to EPS is effected via a special terminal connected to the Worldspan network (although this is planned for upgrade in 1997). Access to the UK's domestic train services is available via Worldspan's link to the FACETS computer in Nottingham (see Rail in Chapter 3 for more information).

Looking to the future, Worldspan is investing substantial resources in enhancing its ability to interconnect with even more local suppliers. The vehicle for this strategy is a general purpose communications interface system that has been developed by Worldspan under the code name of 'Project RAMP', i.e. the Regional Applications and Messaging Platform. RAMP is an important new development that forms the infrastructure for

Worldspan's future supplier distribution strategy. It does the same kind of job as Sabre's ELVA, Galileo's NVP and Amadeus's START/SMART/Estoril products. However, RAMP has one very important feature – it is based on the Internet's communications protocols. This is one of the key reasons why Microsoft decided to use Worldspan as the booking engine for its Expedia web site. RAMP is discussed in more detail in Chapter 5, which also includes an in-depth discussion of Microsoft's Expedia.

Worldspan's host functions

The Worldspan GDS provides its users with a rich set of information and booking functions. These are distributed by a global network that links travel agents' PCs with the Worldspan host computer and other supplier systems. However, before we consider the distribution network, it is critical that you gain a sound understanding of Worldspan's end-user functions. After all, it is these core functions that are distributed across the various end-user networks:

- **World File (client profiles)** A client profile consists of those details that describe a travel agent's customer in terms of flight preferences, personal contact details and corporate information. The storage of these client profiles may be either at the local workstation level or on the Worldspan host main-frame. The advantage of the host option is that the profile is available from any authorized user around the world. World File profiles may be used by travel agents to create and populate PNR fields automatically that can save a great deal of time during the customer booking process.
- **Airline schedules and availability** This shows flights for all participating airlines in an unbiased display that conforms to the regulations set by the UK's Department of Trade and the European ECAC (see Chapter 1). The flights are shown in the order of 'best trip', i.e. least flying time first.
- **Airline fares** Worldspan's international fares data base comprises 85 million fares of which 50 million are for the European area alone. Each day Worldspan processes an average of 750,000 fare changes. Also shown for each fare is the text describing each rule and its associated routings that have been filed with the authorities in each case. The Worldspan fare products that are available on the system are:
    - *MoneySaver* This product automatically displays fares in low to high sequence.
    - *Low Fare Finder* Identifies and books the lowest fare applicable to a booked itinerary.
    - *Ultimate Fare Search* Instantly displays fares for the travel dates specified.
    - *Power Quote* With this tool, the travel agent does not need to have prepared an itinerary as part of a PNR. With simply the from/to city pairs specified, the system will find the lowest fare.
    - *Power Pricing* A key component of Power Quote is known as Power Pricing – Worldspan's low fare finder. Given an itinerary, created as part of a PNR for a passenger, it will find three alternative lower priced options.
    - *SecuRate Air* This is a product that offers participating carriers and subscribing travel agents an electronic means of creating, managing and distributing a wide range of negotiated fares. These fares are proprietary to a specific travel agent and may not be viewed by others.
- **Hotels** Worldspan provides its users with access to large and sophisticated hotel information and booking systems. This is supported by high speed links to hotel switches and hotel computer systems themselves. Because of its interdependence with the hotel industry, Worldspan is a member of HEDNA (see Chapter 1 for more information on HEDNA). The relevant Worldspan hotel-related GDS products are:
    - *Worldspan Hotel Select* This feature allows the travel agent to view detailed rate information, availability displays, amenities information and confirmed bookings for over 182 hotel chains and 26,000 properties. Access Plus links users directly into the reservation systems of 67 hotel companies, thus allowing instant confirmation numbers to be obtained. Other features include: (i) the hotel default record – this allows each travel agency user to tailor his/her own hotel reservation screen so that certain pre-set defaults are always shown at the outset of a booking (examples

of defaults are the number of nights, distance from airport and the rate plan code); (ii) Worldpoint – a geo-locating product that provides accurate distance and direction parameters to and from hotels and reference points such as airports, railway stations and local attractions (the locators are based on longitude and latitude grid references); (iii) Electronic Rate Update – allows hotel associates electronically to update hotel property rates dynamically thus ensuring the accuracy and availability of all rates offered by the hotel associate, within the Hotel Select product; and (iv) Negotiated/SecuRate – special hotel rates negotiated by travel agents can be entered into the Hotel Select display by the hotel concerned. This information is of course agent specific and saves additional phone calls by the agent that would normally be required to confirm rates for certain customers during the booking process.
  – *Worldspan Hotel Source* This provides an interactive, real-time seamless connection to the databases of participating hotel associates. The display shows up to the minute room and rate availability, rate rules, reservation displays, services and other information. This enables the travel agent to make hotel bookings directly in the system that is used to maintain the property's inventory of rooms.
- **Cars** The Worldspan main-frame host computer also links directly into 40 of the world's major car rental company systems. The services offered are:
  – *Worldspan Car Select* This feature supports user-friendly fill-in masks that facilitate the entry of fields, such as vehicle code options, rate variation, rate categories, car selection by hire company, price or vehicle type and the MoneySaver function. Access Plus provides last car availability for reservations with more than 40 leading car companies world-wide, i.e. 90 per cent of the car rental market, including the Association of Car Rental Industry Systems Standards (ACRISS) members. ACRISS recently elected Worldspan an honorary member. Access Plus provides a direct link into the internal reservation systems of the participating car rental companies, thus supporting rate verification and instant confirmation numbers.
  – *Car Point-Of-Sale* This feature enables car companies to load rates according to agent identity or geographic location. Rates are tailored according to the identity or physical location of the subscriber, thereby preventing the offer of un-saleable rates, i.e. rates that are available in the USA but not in, say, Holland.
- **Airline reservations** Single entries made by the agent at the Worldspan terminal PC allow up to 12 air segments to be booked. Seats that have already been booked may be cancelled as necessary and subsequently re-reserved. Air, non-air and combined air/non-air PNRs can be booked. PNRs may be retrieved by passenger name, PNR file address, flight, departure airport name or departure time.
- **Queues** Users may access Worldspan's automated queue control system to schedule time dependent actions that need to be carried out on customer PNRs. The Queue Record Search facility, for example, allows all PNRs for a specific airline, date, flight or other determinant, to be accessed using a single entry.
- **Tickets and travel documents** Tickets, boarding passes, complete itineraries and invoices may all be produced using the Worldspan system. These are requested by specific client name or flight segment and can be customized with, for example, specific PNR data and important remarks to clients. Additionally, the following are also supported:
  – *ATB2* Worldspan has recently completed Beta testing of its automated support for the printing of ATB2 tickets, i.e. a combined airline ticket and boarding pass. This will be rolled out in the UK and other markets as required. The ATB2 control software, usually known as a driver, will support two print hoppers and therefore two types of ticket: (i) an airline ATB, or (ii) a rail or ferry ticket.
  – *Electronic ticketing* Worldspan has been supporting e-ticketing for some time in the USA. This makes it straightforward for e-ticketing to be implemented in international

markets on an 'as needed' basis, i.e. as needed by the airlines in each market. Worldspan simply records those segments within a PNR that are available for e-ticketing. However, although a physical ticket is not printed in advance for the customer, a pass is produced at the airport check-in that allows the traveller to go through security and customs. That just leaves the issue of how to provide the traveller with the 'conditions of carriage' as agreed by the world's airlines at the Warsaw Convention held several years ago. An issue that has yet to be resolved satisfactorily.
- *Satellite ticket printing (STP) and WorldSTP* This allows reservations made in one location, e.g. a travel agency, to be queued for ticketing at another, e.g. the customer's own office. In fact this can be done across international geographical boundaries, which in Europe is particularly important. At the remote location, e.g. the customer's office, the only actions required are to: (a) dial-into the Worldspan network, and (b) enter a couple of command entries to verify the IATA licence number. No in-depth knowledge of how to use Worldspan is required at the ticketing location. The ticket printers used are the TI810 or TI830.
- **Other information systems** There are many other information related functions supported by the Worldspan GDS. Examples include HELP, which covers Worldspan entries, functions and current formats. The INFO topic provides users with explanations of Worldspan functions in a clear and easy to understand language. Finally, there is the Global Reference System (GRS), which provides a virtual encyclopaedia of information topics, including:
  - *Worldspan Travel Suppliers (WTS)* This provides product information and educational services related to the travel industry. Examples include theatre tickets, travel insurance, rail information, cruises and tours.
  - *Vacation Source by Travel File* This allows users to interrogate the Worldspan data base using simple fill-in screen templates and thereby retrieve information on a variety of subjects.
  - *TIMATIC* Worldspan's electronic version of the renowned travel information manual providing details on subjects ranging from health to visas as well as many other important facts essential to international travel.
  - *Travel Guides* This provides the user with tourist information on specific countries.
  - *Taxes* A complete list of taxes that are applicable to airline travel may be retrieved and displayed by country.
  - *Computer-Based Instruction (CBI)* A comprehensive self-tutorial program for all Worldspan functionality.
  - *Worldspan Indexing* An indexing system that allows users to access any topic in the Worldspan system rapidly, e.g. GET-SHOW provides the user with details of theatre services world-wide.

Worldspan's client functions

The users of Worldspan are invariably travel agents. They use their workstations (Fig. 4.18), usually called client PCs, to connect into the Worldspan network by means of several different gateway protocols. But, more on the gateway later. Let's first of all consider the client PCs, each of which is connected into the Worldspan distribution network by several gateway software products. These client PCs run on a variety of platforms, i.e. different operating systems, and use special Worldspan control software to deliver customized end-user functions.

The first and most basic of these platforms was Microsoft DOS. Whereas the functions discussed in the previous section are supported on plain old dumb terminals connected to the Worldspan host main-frame computer, with DOS, clients functions could begin to be decentralized. Some functions were therefore added to the Worldspan DOS client when the PC began to replace dumb terminals in travel agencies. For example, there is the FareDeal capability. With FareDeal, fares negotiated by the travel agent may be stored within the Worldspan PC as a separate data base under direct user control. When a fare is highlighted, it may be converted easily into a booking by FareDeal with just a few key strokes required of the user. The booking is then ready for processing

Figure 4.18 A Worldspan workstation

by the Worldspan host system, which can then create a PNR. Worldspan FareDeal is, however, solely a DOS application.

It wasn't long before Microsoft Windows began to replace DOS as the most common GUI used by travel agents and other PC users around the world. Worldspan therefore developed a family of product enhancements that capitalized on Windows technology. The all embracing title for this suite of software is Worldspan for Windows and it operates within the Microsoft Windows 95 environment.

- **Worldspan for Windows** This version of the Worldspan software is fully compliant with all Microsoft standards and, for example, allows up to ten reservation screens to be accessible at any one time on a workstation. Worldspan for Windows also supports:
  - *An open application program interface (API)* This allows other companies' software products to interface with the Worldspan main-frame. Examples include robotic programs that perform routine quality assurance checks, automatic low fare scanning systems, applications that make a set of different GDS screens look identical and interfaces to corporate computers.
  - *ScriptPro* This is virtually a programming language that may be used to develop customized applications for end-user travel agents. The use of scripting to automate repetitive functions and keyboard entries can reduce errors and increase the speed of service. Travel agents can: (i) write the scripts themselves, (ii) receive consultancy advice from Worldspan on how to write scripts, or (iii) instruct Worldspan to develop customized scripts especially for them. A library of commonly used scripts now exists within Worldspan and users can adapt many of these for their own purposes.
  - *Optional functions* Several added-value products are also available within the Worldspan for Windows 95 environment. The following are just a few examples of some optional applications: (i) E-mail – Worldspan for Windows includes licensed copies of Microsoft Mail (MS Mail) or intra-office e-mail (this, together with the scripting facility, makes a powerful tool-set for creating customized applications); and (ii) Compass – this is a management information function (Fig. 4.19) that provides travel agency executives with productivity statistics, charts and diagrams reflecting their business profiles on an immediate daily basis. Data can be viewed on screen, printed locally or exported to another software package for further analysis.

The PC that supports Worldspan for Windows can make use of several additional program function keys, known as Ready Keys. Once set up, the use of these keys can avoid repetitive entries and can invoke pre-defined scripts.

The client hardware that runs the Worldspan user application software, is supplied by IBM and comprises a Pentium 133 MHz processor with 16 Mb of RAM, a 1.2 Gb hard disk and a 1.44 Mb 3.5″ floppy diskette drive. Other devices, such as CD-ROM drives and multimedia components, may be obtained as optional

Figure 4.19 Worldspan Compass

extras. Worldspan recognizes that it needs to allow its users to keep pace with the rapid rate of technological developments. The specification of the Worldspan client workstations will therefore evolve as appropriate, so as to take full advantage of cost/performance developments in the world's future hardware and software markets.

Viewed from a regional perspective, there are over 17 different screen presentations that form part of the Worldspan for Windows client interface. However, many of these screen formats are customized for a particular market or country. This is a good illustration of the way in which Worldspan has customized its product offering to meet specific customer needs. Other additional products that are available for the Worldspan client environment are:

- **Commercial World** This is a product that has been around for some time and is aimed solely at corporate customers that have a relationship with a Worldspan travel agent. It provides corporate travellers who have their own lap-top computers with a user-friendly interface, albeit with limited functionality, to the Worldspan system. All bookings are queued to the travel agent who must then check the bookings and create the entries necessary to drive out tickets and related documentation.
- **ETravel** Worldspan is considering launching a new product in the UK called ETravel, which again is aimed at the corporate travel market. This product, which will replace Commercial World, comprises a set of software that runs on a business traveller's own lap-top computer. This can then be used by the traveller to dial-into the Worldspan network and make direct bookings. ETravel automatically ensures that these bookings comply with the company's travel policy and quality control requirements. The bookings, which automatically include all ticketing and related operational information, are then queued to a travel agent for processing. The significant difference with ETravel, however, is that virtually all the data entries and booking creation work have already been done by ETravel. Etravel may well also incorporate an expense management system for use by companies. This is a new product enhancement that is currently under consideration, i.e. as at the second quarter of 1997.
- **Quality Assurance** ScriptPro provides a high level programming capability that allows a travel agent to carry out quality assurance checks automatically on customer bookings. The quality control checks are developed using ScriptPro's high level scripting language, which is similar to Visual Basic and supports an open API. In this way, the travel agent's particular needs may be programmed into their Worldspan PC workstation. This product is also e-mail enabled; which means that when certain trigger events occur, an e-mail message may be generated automatically to alert the agent or customer to an unusual event requiring human intervention.

Worldspan's distribution network

As a result of several years of strategic growth, Worldspan now has a far reaching network that supports many leading communications protocols and is therefore able to support many different supplier systems. It is therefore important to review the Worldspan global telecommunications network in some detail because it is fundamental to so many of its current and future products. Figure 4.17 shows an overview of the Worldspan network, which I will explain in terms of its constituent parts, reserving any discussion of the Internet related features for Chapter 5:

- **Branch network access** Many of Worldspan's customers operate their own LANs. In such cases, the LAN will use a PC, designated as the Worldspan gateway server, which connects into Worldspan's global telecommunications network using a variety of communication protocols. This server may connect into several Worldspan networks, each of which is specific to a certain type of application. Most major server operating systems are supported including DOS, Windows 3.11, Windows 95, Unix, Apple's MAC and AS/400's OS:
  - *Gateway Plus* The Worldspan Gateway software spans a very wide range of telecommunications interfaces, which allows the agent's server PC to talk to systems using protocols such as X25, IBM3270, IBM5250, VT100, Tandem and the Internet. Protocols that themselves include E-mail, Intranet, access to the World Wide Web and FTP. Also supported are videotex (see Chapter 6) and Ferry Source, a ferry information and reservations function.
  - *DialLink* For smaller users, who are ABTA and/or IATA travel agents, yet who cannot justify a leased data line, Worldspan offers the World DialLink service. This provides virtually all the major functions of Worldspan but without some of the more costly telecommunications overheads that are required for a high volume, fast response on-line system. All that is required is a PC and a suitable modem. Two modes of operation are available: (a) a user-friendly interface that can be used without prior specialized training; or (b) by-pass, which is faster but requires the user to have a higher level of knowledge of Worldspan command entries. Finally, a ticketing version of World DialLink is available to IATA licensed agents. This requires a dot-matrix ticket printer connected to the travel agent's PC.
- **Customer head-quarter services** Worldspan has been particularly successful in signing up several large multiple travel agencies in most major markets. These multiples often have a requirement for their branch outlets to talk, not just to the GDS for booking services, but also to the company's in-house host computer at headquarters. This is particularly important for multiples that operate several branches across geographical boundaries. Worldspan's products for customers of this type are:
  - *World Solutions* Under the banner World Solutions, Worldspan provides a pre-sales consultancy service to its customers. Some customers, particularly multiples and specialist non-air suppliers, often require secure links to many different types of host systems and branch networks. Such capabilities allow head offices to communicate with branches and non-air suppliers to connect into travel distribution companies. This has resulted in many unique solutions including Worldspan links to IBM 3090 main-frames, Amdahl, IBM RS/6000, IBM System 36/38 IBM AS/400 minicomputers, Apple Macintosh, Bull, Data General, Digital, Hitachi, NCR, Texas Instruments, Unisys, Zenith and open platforms running the Unix operating system. In many cases, the end results of these consultancy assignments are standard products that can be re-used by other Worldspan customers.
  - *Worldspan Alliance* This allows a DOS PC using any one of several different communications protocols, to be supported over Worldspan's various networks. These include: the old airline telecommunications protocol (ALC), X25 (packet switching), IBM 3270, Frame Relay, Videotex and the ubiquitous dial-up. All protocols provide full access to the Worldspan host computer. In the USA, travel agents are connected into Worldspan via the AT&T ACCUNET service, which uses the high speed Frame Relay technology.

Each of these local networks connects into the Worldspan host system in Atlanta via high speed links provided by the world's major telecommunications companies. Worldspan uses AT&T, SITA, Compuserve and IBM to connect its international networks into the Atlanta host computer.

THE INFINI GDS

INFINI is the leading distribution system in the Japanese market. It is branded a CRS; but, theoretically, because it provides access to more than one airline system, it may be regarded as a GDS. It was established in 1990 as a joint venture between All Nippon Airways, which holds 60 per cent of the joint company, and the Singapore-based ABACUS distribution system, with 40 per cent. The resulting company is called INFINI Travel Information Inc. and it has almost 9,300 users in 6,300 agency locations. It was set up to provide the Japanese travel industry with an impartial, user-friendly computerized reservation system to link airlines with travel agents and the wider world of travel.

HDSs

The primary purpose of most HDS companies is to provide reservations and information services to travel agents via the 500,000 airline GDS terminals used in travel agencies throughout the world. However, this situation is changing rapidly at the present time, particularly as a result of the Internet. I'll discuss the Internet in more detail in Chapter 5 but for the moment, it is vital that first you understand the hotel distribution systems, the way in which they relate to the airline GDSs and how they work.

HDSs may be categorized as follows: (i) computer switches connecting a hotel's own in-house reservation system with the major GDSs for distribution purposes – a prime example of this type of industry switch is The Hotel Industry Switch Company (Thisco), and (ii) service companies providing smaller hotels with an outsourced marketing, reservations and distribution service, also with connections to airline GDSs – examples include Utell and Sahara. Both types of HDSs work closely with the world's GDSs to provide the hotel industry with automated sales and booking services. Each of these is explained in more detail below, starting with one of the leading hotel industry switches that is marketed by Pegasus Systems of the USA.

PEGASUS SYSTEMS

Pegasus Systems Inc. was created in 1995 as the parent of three important high-tech companies all of which are heavily involved in the world's hotel industry. The three companies are: (i) THISCO, (ii) the Hotel Clearing Corporation (HCC), and (iii) TravelWeb Inc. The parent company, which is located in Dallas, Texas, and which has an international head office in London, is owned by 15 of the world's largest hotel and travel companies (Fig. 4.20). In 1996 Pegasus sold US$7.5 million in stock to an investment company called Trident Capital whose largest investor is Dun & Bradstreet. The proceeds from this private placement will be used to build and enhance the TravelWeb interactive travel reservations web site, develop new travel industry product lines/software and continue growing the Thisco and HCC businesses. All these companies have experienced a compound annual growth rate of 52 per cent. Not bad for a company with only 100 employees! It is therefore important that we consider the main products and services of Pegasus Systems, because they play such a key role in today's international travel industry. First of all, let's take what is perhaps the core business – Thisco.

Thisco

The principle aim of Thisco is to provide its hotel customers with a single and standardized interface to the world's GDSs. More specifically, the objectives of Thisco are to reduce the operating costs and increase the efficiency of the hotel reservations business, which is distributed via airline GDSs to travel agents and to consumers via TravelWeb – Thisco's Internet site (see TravelWeb in Chapter 5). But before we dive into a more detailed analysis of Thisco, it is worth reviewing the background to this leading global hotel distribution company.

Before the formation of Thisco, hotels that distributed their products to travel agents were required to connect their hotel inventory systems to each airline CRS (GDSs came much later). Most of the major hotel chains were connected to more than one of the world's leading CRSs (usually to at least four). This meant that the same messages containing the status of hotel rooms and other reservations information had to be formatted

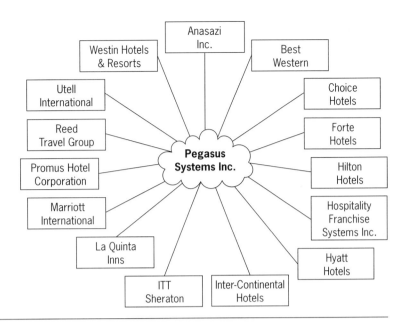

Figure 4.20 Shareholders of Pegasus Systems Inc. (as at 1997)

separately and sent to every CRS to which they were connected. Now because each CRS generally used a slightly different format for its computer interface, the hotels were having to convert their messages into the format used by each and every CRS to which they were connected, prior to transmission. This placed a heavy burden on the hotels' systems, both from a development and an ongoing operational viewpoint.

It also gave rise to inefficiencies in the booking process. At this early stage in the hotel distribution story, there were only two types of hotel to CRS connections: Type A and Type B. With the older Type B connections, booking messages were queued by CRSs before being delivered to travel agents. This meant that reservations could not be confirmed for several hours and this gave rise to multiple bookings for the same passenger followed by a high number of cancelled bookings. When Type A connections were developed, they helped the situation somewhat by using a more streamlined process that resulted in confirmations being received within 7 seconds. Although this went a long way to solving the early problems, it was not until much later when Seamless Connectivity was developed, that the hotel reservation process via the GDSs became a truly workable and efficient distribution method. Seamless connectivity provides the travel agent with a virtual direct channel to the hotel's own reservation system computer. This enables the travel agent to use the hotel's computerized reservation system just as though the agent was connected to it by one of the hotel system's terminals. With seamless connectivity, the other switching systems, e.g. the airline CRSs or GDSs, are simply transparent communications channels that serve only to support the dialogue between the travel agent and the hotel's computer system.

In the late 1980s, many of the world's largest hotel companies in the USA realized that they needed to use advanced IT to provide accurate and rapid information on room availability, rates and confirmation numbers. In 1988 several leading hotel chains joined forces to form the Hotel and Booking Research Association. The association's first objective was to evaluate possible solutions to electronic hotel distribution opportunities. The outcome of this review was agreement that a common need existed among members for a communications switch that would link each hotel company's inventory control computer to point-of-sale distribution systems of the world. The required solution would support the development of a single interface linking each hotel system to the common switch. On the distribution side of the switch, a standard GDS link would be developed that could be shared by all hotel participants. Thisco was formed in 1988 when 15 major hotel companies, including Utell International, Hyatt, Forte, Marriott, as well as most of the other leading hotel companies in the USA together with Murdoch Electronic Publishing (which later became the Reed Travel Group), agreed to invest in a hotel industry switching company.

The newly formed Thisco developed a computerized switch called Ultraswitch. The way this switch works is very much like a transparent link between the travel agent and the participating hotel. The switch has a supply and demand side. On the supply side, it connects to hotel inventory systems and translates their messages and commands into a standard Thisco format that is used for all processing within Ultraswitch. On the demand side, the Thisco switch communicates with all major GDSs using the proprietary message format of each one. In many ways the Ultraswitch computer acts as a sort of super-translator between the various hotel systems and the major GDSs. It provides full support for all GDS hotel functionality, including bookings, status messages, rate updates and seamless availability.

Ultraswitch now enables 70 hotel chains and 25,000 hotel properties to distribute their lodging products to more than 350,000 point-of-sale reservations screens in travel agencies throughout the world. Over the period 1990 to 1994 the volume of GDS bookings handled by Ultraswitch grew from one million to over eight million and by 1996 had reached 14 million. The switch now handles 40 million messages each month and the rate of growth is a compound 29 per cent. The system that makes all this possible is located in Phoenix, Arizona. Ultraswitch uses technology based on a scalable client/server computer running the UNIX operating system and a relational data base management system (RDMS). It uses high speed digital data circuits carrying between 56 and 64 kb/s of data and supports many different telecommunications protocols including SNA, X25, SLC and TCP/IP. Ultraswitch offers its users several products:

- **Ultraconnect** This is the basic reservations product that makes hotel reservations functions available to travel agents 24 hours a day, seven days per week. Ultraconnect uses Type A connection technology (see above), to link the travel agency user with the hotel computer via the GDS and Ultraswitch networks. It enables a travel agent to complete a hotel booking via a GDS screen in less than 7 s. With Ultraconnect travel agents may book, cancel or modify reservations and obtain immediate confirmation and cancellation numbers direct from a hotel reservations system.
  - *Stage 1* Travel agents make reservations via Ultraconnect in two stages. The first stage entails viewing static information screens that are displayed from information stored within the GDS used by the travel agent. This information is in fact created and loaded into the GDS by the hotel and then updated periodically to reflect changes. It includes data such as available properties, general information, available rooms and rates [see also Hotel Systems Support Services (HSSS) Limited later in this chapter].
  - *Stage 2* Once the travel agent has selected a property for a customer, stage two commences. This involves the creation of a booking request entry by the travel agent, using the GDS terminal. The resulting reservations message is transmitted from the GDS, via the Ultraswitch to the hotel system. When the reservation message is received by the hotel system, it checks the required availability and sends a response back to the travel agent via the Ultraswitch and the GDS. This two-way message flow continues until either a booking is made or the travel agent signs-off, i.e. the transaction is ended. In those cases where a booking is made, the hotel system sends the travel agent a confirmation number that may be used to guarantee the room for his/her customer.

To connect to Ultraswitch a hotel must of course have its own in-house room inventory system. This must be able to support both on-line and teletype messages as well as being able to generate reservation responses automatically to GDS messages. The connection to Ultraswitch is made via one or more high speed telecommunication lines and a customized hotel interface that is developed by Thisco to run on its computer system.

- **UltraSelect** This provides travel agents with seamless connectivity to a hotel's own reservations computer, just as though the agent was connected directly to it. This is a critical differentiating factor from the approach used by Ultraconnect. With Ultraconnect, the dialogue is formatted by the GDS, which simply translates the hotel system's responses into its own on-screen display. These displays are constrained by the GDS's airline oriented technology. With UltraSelect, however, the GDS and Ultraswitch act simply as a communications channel connecting the travel agent directly to the hotel's own system. The display that the agent sees on his/her GDS screen is therefore exactly that which is formatted by the hotel system. In essence, UltraSelect replaces Stage 1, as described above.
  - *Stage 1* So, instead of the travel agent searching the GDS information for a desired hotel property, room type and rate, the agent searches the UltraSelect data base that is stored within the Ultraswitch. This property information data base, which is maintained by the hotel systems in real-time, contains details on 70 chains with 25,000 properties and 2.3 million rooms. Information is recorded in two categories: (i) static, and (ii) status. Static information changes infrequently and includes the hotel's name, address, number of rooms, amenities, public facilities, transportation to/from the nearest airport and geographical referencing co-ordinates that pin-point the hotel on any map. Status information includes rates for the coming year and the availability of each rate during different periods.
  - *Stage 2* Stage 2 involves the travel agent connecting to the hotel's system to make a reservation. The travel agent selects a hotel property in Ultraswitch and then directly accesses the hotel's computer system to obtain even more detailed information. For authorized travel agents, this includes access to specially negotiated rates that are stored within the hotel's own computer system. Finally, a booking can be made by the travel agent when a property, room and rate are found that match the customer's requirements.

  One of the main advantages of UltraSelect, from a hotel's viewpoint, is that it gives the hotel direct control over the independent electronic marketing of its products. For instance, besides being able to display customized sales and marketing information about their properties and rooms, the hotels can also differentiate their products by means of full textual narrative instead of non-descriptive GDS codes. More complicated marketing opportunities are also possible, such as the selling of 'denials'. This involves a hotel system selling another hotel's rooms if the customer has denied the initial offering.
- **UltraRate** This product enables hotel reservation systems to deliver room rate information to GDSs electronically via the Ultraswitch. Prior to this product being available, staff within the hotel company manually entered rate changes directly into GDSs using computer terminals. This is a labour intensive task often requiring a dedicated member of staff whose only job it is to key this information into each and every GDS. Using UltraRate eliminates this manual effort, speeds up the entry of information and therefore increases the accuracy of hotel room rates shown by GDSs.
- **EasyView** This product is similar in concept to UltraRate. However, whereas UltraRate addresses the problem of updating room rate information, EasyView addresses the static information problem. Thisco's participating hotels must update their static information, which is repetitively stored in several GDSs, on a periodic basis. This information is in many cases manually keyed by hotel staff directly into each GDS. With EasyView the hotel can use its own Windows-based PC to update static information on all GDSs. The hotel's PC stores the static information in a standard format on a local data base. EasyView allows the hotel to interface to each GDS and to then re-format the static data as required.
- **UltraRes** This is a product that supports the processing of large blocks of rooms for hotels. Conventions, visitors bureaus and wholesale tour operators traditionally communicate their bookings to hotels via fax, mail or telephone.

This can often lead to inaccurate and delayed information being received by the hotel. With UltraRes the booking source can transmit block booking requests to the hotel via special entries made using a GDS.

Thisco provides a complete service to its hotel participants. This includes project management during the interconnection stages of a hotel becoming connected to the Ultraswitch and ongoing account management. Thisco also continually reviews alternative distribution systems for its hotel customers. One good example of such an alternative is the Internet. Thisco's tailor-made product for this purpose is called TravelWeb and this is described in more detail in Chapter 5. Another is the commission administration system called HCC, which is included under the heading of Financial Services in Chapter 7.

UTELL

Utell's services are aimed primarily at hotels that do not have their own large sales and marketing organizations or an internal central reservations computer. So, while a hotel may have its own PMS for internal operational purposes, this may not be sufficiently large or sophisticated enough to link directly into the Thisco switch (see above). These types of hotels need the support of a company that can market their services to travel agents and other sellers around the world while also providing an automated booking function. In effect, a company to which they can outsource their marketing and booking functions. Utell provides these hotels with just such a service.

Utell therefore provides smaller hotel chains and hotels with reservations facilities using Utell computers. This can be a great advantage to a small 50 room hotel in a resort area, especially when one considers that the Utell screen displays are all neutral. The sequence of the hotels displayed in response to an availability enquiry is purely rate-descending order within a city; and via the GDSs, the sequence of display is entirely random so as to eliminate any possible bias. A small hotel is therefore competing on a level playing field along with the giants of the industry.

Utell is now the world's largest hotel marketing and reservations company. It represents more than 6,500 hotels offering 1.25 million rooms in more than 180 countries. It also represents a wide range of hotels of many types, including for instance: budget properties, deluxe resorts, city centre locations, hotels in holiday areas, major international brands and independent hotels. It handled 3.5 million hotel reservations in 1996, which generated over US$1.4 billion in room reservations revenue. Utell's strategic objective is to grow this level of business to 22 million reservations by the year 2000. In order to achieve this high rate of growth, Utell will rely heavily on enhancing the company's underlying customer servicing infrastructure. This means delivering improved technology, encouraging marketing innovation, developing new systems and introducing new products. Quite a challenge. So, let's take a closer look at the company, the technology it currently uses and the future developments it plans to initiate. First of all, it is important to know something about the origins of the company and how it is structured.

Utell's origins and company structure

Utell International is a part of the Reed Travel Group and a member of the Reed Elsevier plc group. However, the company has its origins back in 1930 when Henry Utell, a travel journalist living in the USA, formed the company. Through his experience of international travel, Henry Utell soon became aware of the need for a global hotel sales and reservations service. In order to help fellow travellers, Henry created Utell International in 1930 with just two hotels. The business thrived and grew.

Utell Inc., as it was known by 1972, was bought out by Grand Metropolitan Hotels and in 1974 Utell International (UK) Ltd was formed. In 1976 a computerized reservations system was developed and implemented. The company rapidly expanded to represent 1,800 hotels in 1978. Then in October 1982 five directors of Utell International acquired the whole of the issued share capital from Grand Metropolitan plc making Utell International a private company again after having been a wholly owned subsidiary of Grand

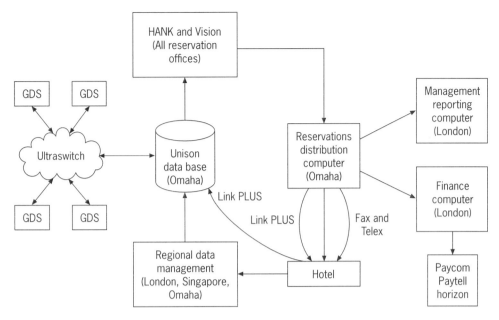

Figure 4.21 The pre-April 1997 Utell network

Metropolitan for ten years. The company continued to expand and besides enhancing its computer system still further, in April 1986 it opened a new US$3 million reservations centre in Omaha, Nebraska. Then in April 1987, Murdoch Magazines, a subsidiary of News America Inc. bought Utell International. Finally, in June 1989, Reed International (now renamed Reed Elsevier), the UK's largest publishing and information company, acquired Utell International.

Today, Utell International (Fig. 4.21) is an integrated part of the Reed Travel Group, which also owns OAG and ABC Corporate Services. Utell International has its corporate headquarters in London and is organized into three regional divisions: (i) North and Latin America, with a regional headquarters in Omaha; (ii) Europe, the Middle East and Africa, with a regional headquarters in London; and (iii) Asia Pacific with a regional office in Singapore. These divisions look after all aspects of Utell International's business in each region, which in total comprises 6,500 member hotels in over 180 countries.

It can clearly be seen that over the last 31 years Utell has been building a strong relationship with both hoteliers and travel agents. The company has increased reservations for member hotels steadily and has worked towards providing a reliable and comprehensive service to travel agents throughout the world. It has a strategic business plan that encompasses challenging growth objective, as already stated. The success of this growth will depend to a large extent upon the technology used by Utell to operate and distribute hotel marketing and booking services to sellers around the world.

Utell's current core system

The Utell reservations system has been developed over a number of years. It comprises an IT architecture that uses a central data base called Unison, a network of reservations offices supported by the Hotels Automated Network Know-how (HANK) system and a telecommunications network linking Unison to hotels, the Thisco switch and all four of the world's major GDSs. Let's take each one of these components in turn.

Unison – the central data base
This is the hub of Utell's reservation system. It is a large computer data base facility located in Omaha, Nebraska. It stores all property information on member hotels and is constantly updated

by regional computers in London and Singapore. Between them, these computers support all 52 Utell reservations centres around the world.

The Unison system supports two main functions: (a) hotel bookings and international yield management, and (b) full management information allowing hotels to control and direct their sales efforts. The system also incorporates features such as multi-level availability and rates, information and availability on packages, ability to sell different rates in different countries (or different GDSs) and to different travel agents, full management information (including full analysis on actual business, plus business denied for whatever reason).

The Unison system is connected to three types of users: (i) participating hotel customers, (ii) Utell International reservations offices around the world, and (iii) each of the world's four major airline backed GDSs. Let's explore each of these players in a little more detail:

- **Participating hotel customers** In order to provide fast and accurate hotel reservation system functions to the hotel industry, good clean data are essential. When new hotels join Utell, they need to provide information on their precise location, classification, size, facilities, rooms, rates and availability. This data is supplied to the nearest Utell regional computer centre by participating hotels via either manual or automatic methods. Once received and checked, the data are used to update the central Unison data base that resides in Omaha (see above). Many participating hotels use computer terminals with direct access to Utell in order to enter their update data themselves. This facility is known as LinkPLUS.

  Utell LinkPLUS was launched in 1986 to provide direct access between participating hoteliers and Utell's computer network. The link allows hoteliers to control their own rates and availability from their own front offices or central head office. Hotels can update availability and rates information directly and receive reservation messages direct from Utell electronically. It comprises two parts:
  1. Hotel PC and special software – the PC is installed within the hotel, e.g. either on the front desk of the hotel or, in the case of multiple property chains, in the chain's head office. This PC is linked to Unison by a data line that allows it to support two-way communications. It can, for example, download reservations messages and other e-mail from Unison and upload hotel rates and availability status responses to Utell from the hotel.
  2. Utell interface program – This is a piece of specially developed software that runs on each regional computer. Its primary purpose is to support the link to the hotel's LinkPLUS PC and allow it to update the Unison data base.

  In the absence of a PMS, LinkPLUS can provide a hotel with an enhanced degree of integrity over reservation messages while also enhancing the degree of control over its inventory as recorded in Utell's central Unison data base.

  In order to participate in the Utell system, hotels must first supply a great deal of detailed information about their properties, services, amenities, rates and capacity. In addition to supplying this information to Utell, a hotel wishing also to participate in the GDS distribution service must additionally build a presence within each GDS. This is a somewhat lengthy process although it is co-ordinated fully by Utell, and only needs to be done once during initial set-up.

- **Reservations offices** Utell International operates a network of reservations offices that has been created with the sole purpose of receiving hotel bookings via the telephone. Reservations operators in each of the 52 world-wide offices use a computer system called HANK, to support voice reservations functions and link to the central Unison system. While HANK handles the front-end incoming reservations functions, the link to Unison provides full access to the central data base of hotel information. Because the HANK system is, however, limited in the information that it can display, an enhancement called UtellVision is available to participating hotels:
  - *UtellVision* This product was launched in 1988 and provided additional support for Utell's reservations staff incorporating enhanced electronic colour pictures, mapping

and search criteria. Its aim is to increase the product knowledge of Utell reservations agents. Each reservations operator uses two VDU screens as part of the reservations process. The top screen displays television quality pictures showing detailed city maps, and the bottom screen provides on-line hotel data and booking facilities.

When an operator selects a particular hotel during the booking process, they use the bottom screen to access the Unison data base via HANK. When the hotel's details have been retrieved from Unison, HANK transmits a command to the PC controlling the upper terminal. This enables the UtellVision PC to retrieve graphical data automatically on the requested hotel. These images are stored on laser disks and are updated once each quarter. Participation in UtellVision is an optional extra cost for hotels although the selling opportunities are significant.

- **Airline backed GDSs** The Utell system is distributed to travel agents using GDS terminals via the THISCO Ultraswitch (see above for more information on Ultraswitch), in which Utell has a share-holding. Utell International connects 3,500 of its hotel members to GDSs that distribute their products to a total of 500,000 travel agency terminals world-wide. Utell is available via Amadeus (System One), Galileo (Apollo), Sabre and Worldspan. Travel agents use these GDSs primarily to book hotel accommodation for their business travel customers.

The Utell International hotel reservations system is also connected directly into every other major national airline and third-party hotel reservation system, including JAL Axess and Transnet in Japan, BookHotel in Scandinavia, HRS in Germany, HotelSpace in the UK and Sita Sahara throughout the rest of the world. Additionally, Utell is the only non-hotel operator to be a member of Ultraswitch, which is located in the USA (see Thisco above).

Some of these GDSs have their own branded name for their hotel reservations service. Take Galileo for example, into which Utell is itself connected as a host system. Galileo's own RoomMaster hotel reservation capability offers users some of the larger hotel chains as directly connected hosts. This means that a travel agent can opt to use a specific hotel chain's computer reservation system to make a booking. Now some of these hotels are also included in the Utell data base. To avoid confusion, those hotels that have a direct connection from their reservations computer directly into Galileo are not accessible via the Utell gateway on Galileo (this used to be known as the Utell partition within the old Travicom system, which has now been superseded by Galileo UK). A hotel can therefore only be accessed on Galileo UK through a single channel only. This explains why it is that Utell on Galileo offers just over 4,000 hotels out of the global total of 6,500. It is because the other 2,500 are available on Galileo UK as directly connected host systems. A similar situation exists on Sabre, which has its own SHAARP PLUS hotel reservation capability that is comparable to Galileo UK's RoomMaster as outlined above. However, it must be stressed that HotelSpace on Galileo UK is purely a UK product.

The GDS language that travel agents need to use to access Utell's HotelSpace tends to be somewhat complex. Now a GDS language is not the most user-friendly dialogue at the best of times, so you might wonder why it is made more difficult for hotel reservation purposes. The reason is that GDSs were designed initially for airlines and are beginning to run out of codes for non-airline systems. Most of the easy to remember codes have been used up in the core airline reservation language of the GDS. So, the codes left over to use for hotel reservations are limited and result in entries that are only just a bit better than Egyptian hieroglyphics. Still, they are quick to use once committed to memory.

Other Utell International services

The preceding section provides an overview of the core technology that is used to provide Utell International's services to customers and users around the world. The full range of these services is too great for inclusion in this book in detail. However, there are certain services and functions that are worth exploring in a little more depth because they have a direct bearing on the subject of IT in travel and tourism.

Private Label

What is not quite so well known is that Utell also provides hotel booking services to hotels under the hotels own name. This is called the Private Label service. Take, for example, Thistle Hotels. Utell International provides a reservations service that is badged or branded Thistle Hotels. The service covers voice support for reservations as well as distribution via CRSs and other systems. When the travel agent calls the Thistle Hotels' telephone number, they hear something like 'Hello, Thistle Hotels, how may I help?' In terms of connections to CRS systems, there will be a dedicated Thistle Hotels' data line that will in fact come from the nearest Utell network node. In other words the service appears just as though it was being run by Thistle Hotels itself. This is becoming more and more attractive to hotel groups as they strive to concentrate on their core business, which is after all running a profitable hotel business and not operating complex booking computers and manning reservation centres at all hours. There are many other similar examples, including the Private Label Voice service operated on behalf of Summit International Hotels.

GDS participation

Utell's interface with Thicso means that all bookings are transferred automatically from the airline systems directly into Utell's Unison database for immediate distribution to hotels 24 hours a day 365 days a year. This not only virtually eliminates any chance of human error but also contributes to the speed of distribution between Utell and its member hotels.

Utell handles reservations on a kind of sophisticated sell and report basis. The way it works is something like this. A travel agent uses the Utell system via one of its many distribution networks, be it via a GDS, videotex or voice. A reservation is made in the Utell system and a booking record is created. Then within 7 seconds, the travel-agent user will be provided with a confirmation of the reservation and a confirmation number, just like an airline PNR locator. As a result of this, the Utell system sends a message to the hotel concerned. In the case of hotel reservations, the communication of the reservation message from Utell to the hotel is sent by one of several methods, which may be either: (a) telex or fax, (b) directly into a PMS without human intervention, or (c) via LinkPLUS directly into a hotel's PC. Because the delivery of this message is so critical, let me explain each of these methods in more detail.

- **Telex or fax** These methods are well known but suffer from the vagaries of human intervention in the form of a person who must take a piece of paper and ensure the hotel records are aware of the booking. Inevitably, some of these pieces of paper are mislaid and the old 'no reservation on arrival' problem rears its ugly head. This is not a problem with Utell. This is because in order to participate in the Utell system, the hotel is required contractually to honour every Utell reservation, even if fully booked. So, provided the hotel has not previously sent a 'full up' message to Utell, then a reservation made via the system must be honoured. If the hotel does not have the room, it is responsible for finding a hotel of a comparable standard that can accommodate the customer.
- **PMS link** The reservations messages are sent via a communications link between Utell's nearest regional computer and the participating hotel. With this kind of direct link there is little room for messages to go astray and it is by far the best method that a hotel can use to communicate with Utell. However, not all hotels have a PMS and even if they do, they are not always capable of connecting to an external system such as Utell.
- **LinkPLUS** This is a dedicated system to system link, which is explained in more detail above under the topic of Unison's participating hotels and their interconnection methods.

Utell monitors the participating hotels very tightly. First, it has the means to trace and prove delivery of a reservation message to a hotel; and second, it keeps a record of all such misdemeanours made by a hotel and can thus formulate a list of repeat offenders. These delinquents are penalized at first, but if they continue to abuse the system then they are asked to cease using the service by Utell (in other words they are forcibly removed from the Utell system).

Marketing support programmes

There are many other services that are provided to participating hotels by Utell International. Besides reservations and training services, several important GDS marketing products are available to Utell hotel subscribers. These include products like: (a) Sign-ins – this allows a hotel's promotional message to be displayed on the travel agent's GDS terminal when it is activated; (b) Point-of-Sale Message – these messages appear on a travel agent's GDS terminal when a transaction relevant to a certain pre-set destination information is entered, e.g. when an agent books a flight to Paris the GDS screen will show a promotional message on Paris hotels; and (c) Electronic Mailshot – a message of up to 18 lines that may be sent to Amadeus terminal users whenever a queue is accessed.

Financial services

One of the historical problems experienced by travel agents has been the collection of commissions on hotel bookings. In the past this has been a major deterrent that has suppressed hotel sales by travel agents. However, Utell International has a solution to this problem in the form of two products: Paytell and Paycom. These services are described in more detail in Chapter 7 in the section on financial services.

Travel agency services

If one considers all hotel bookings made from all sources around the globe, only between 28 and 30 per cent are generated by travel agents. The remainder come from direct bookings from consumers and companies. Of all reservations handled by Utell, 98 per cent are generated by travel agents. The remaining 2 per cent come from customers directly, although Utell has not solicited business actively from end-user customers to-date. In the UK, 40 per cent of all bookings are obtained from automated systems. This is considered quite a high level in comparison with other countries, excluding the USA, which has a far higher volume of automated hotel bookings than most other countries. Automated systems in this context include videotex systems as well as the GDSs.

We can draw two conclusions from these statistics: (a) there is a significant opportunity for the travel agency sector to increase the volume of their hotel bookings and the associated revenue stream significantly, and (b) there is an untapped opportunity for Utell to market its hotel services direct to consumers via the Internet or some other direct marketing channel. Well, I discuss the Internet in more detail in Chapter 5. So, for the moment, let's focus on the opportunities for travel agents to increase their volume of automated hotel bookings via Utell International.

First, an important point; the Utell hotel reservations service is provided free of charge to travel agents. Utell International is able to do this because it is paid by member hotels to represent and market them world-wide. This payment is made as both a fixed cost membership charge and a volume related transaction fee that is similar to the way GDSs derive their revenue. The hotels are quite happy with this arrangement because for them to do all their own distribution and operate their own reservations departments would cost them a lot of money. Of course travel agents are pleased to have access to such a sophisticated hotel booking facility at no cost to themselves.

- **Travel agent terminals** As a general rule, Utell does not support the direct connection of individual travel agents to its own reservations computer. In other words, individual travel agents cannot approach Utell and ask for one of their terminals to be installed in their travel agency. The reason for this is that Utell does not wish to have the overhead and the ongoing support functions that go hand in hand with running a large network of directly connected travel agency terminals. Far better to use existing terminal networks, such as the airline GDSs. Also, from the travel agent's viewpoint, desk space is a premium. So if yet another terminal was installed in the travel agency alongside the GDS and videotex ones already in place, the agent would begin to run out of counter area. So, instead of running its own network, Utell International is distributed via other travel industry distribution channels. These include airline CRSs, videotex networks, the Internet, other national airline systems and third-party reservations systems world-wide.

- **Central reservations units (CRUs)** However, despite the above, Utell is nevertheless also connected into many of the high volume major travel companies' own automated systems, i.e. third-party business. These third-party companies comprise several multiple travel agencies and some hotel booking specialists. For example, many of the larger multiple travel agents in Europe and Asia Pacific have created their own specialized internal hotel units called central reservations units (CRUs). These units handle hotel reservations on behalf of their branch network centrally on a service basis. Utell allows these CRUs to have dedicated terminals installed for direct access to Unison.
- **Rates** Utell supports the many different types of rates that travel agents need to access. Preferred and negotiated rates are assigned by a hotel either: (a) to a large multiple travel agency chain, or (b) to a specific company whose business travel is handled by a certain travel agent. This company or travel agent may access a specially segregated part of the Unison data base that contains the customer's preferred hotel rates. Similar rates may be held for government departments and other categories of customer.

    This rather sophisticated capability is called Defined Viewership (also known as Multi-Rate Access). It provides personalized views of hotel rates for each travel-agency user. With Defined Viewership travel agents have access to the preferred and negotiated rates specified by the agent. So, when a particular hotel is being viewed for reservation purposes by a travel agent and the rates are requested to be displayed, they will show the negotiated or preferred rates. Other travel agents would not, however, be able to view these rates. They are held on Unison in something like a closed user group (CUG). The only type of rate not displayed is a net/net rate for which no commission is payable.

The hotel reservation services provided by Utell International and the related products that are designed for travel agents are sophisticated tools that if used properly can significantly increase revenues. There is a substantial untapped source of extra revenue for travel agents in the area of hotel bookings for their customers.

Utell International's new systems

Utell International has experienced a growth of more than 100 per cent in reservations volume over the past four years. Looking to the future, its strategic plan calls for the generation of over US$2 billion in annual revenues for its hotel members by the year 2000. It is for these primary reasons that Utell has recently decided to invest over US$24 million in enhancing and upgrading its IT infrastructure. As part of this overhaul, in October 1996 Utell International signed a software licensing and enhancement agreement with Anasazi Inc. This enables Utell to build a new state-of-the-art 'GDS-like' reservations and support system for its hotel customers. The basis for this is the Anasazi lodging enterprise system architecture (ALESA) product line. ALESA offers hotel companies a fully integrated technology solution including such functions as PMSs, guest history and recognition programmes, i.e. support for loyalty schemes, revenue management and an Internet capability. The resulting new systems will be implemented in two phases:

- **Phase 1** The creation of a new highly structured hotel information data base providing enhanced management of rates and availability information. This new information structure will be enhanced further by a new LinkPLUS system with improved functionality for hotel users. Also included in Phase 1 is the implementation of more flexible and efficient connections to GDSs. Migration of existing hotel customers to this new systems environment commenced in 1997.
- **Phase 2** The replacement of HANK with a new voice reservations processing system, including a new integrated version of UtellVision. This will be a Windows-based application that will be installed in all 52 reservations offices around the world. The new LinkPLUS system will be made available via the Internet. Further enhancements to Utell's GDS links will be made, which will feature a seamless availability function and better management information reporting via data warehousing techniques.

The new software will be capable of handling the complex data involved in marketing hotels in an increasingly competitive environment and providing far greater flexibility and responsiveness to customers. This ambitious IT programme will involve Utell upgrading many of its computer systems around the world. Communications network resources will also be upgraded and expanded to cope with new technology and higher transaction volumes. Utell is well aware of the potential for disruption during this major IT enhancement project. A progressive and phased implementation plan is therefore being carefully followed in every region.

Marketing on the GDSs

Much of what I am going to discuss here relates to the 200 or so hotel chains that use the GDSs to market their properties to travel agents. The GDSs are very powerful marketing vehicles because they reach one of the most underdeveloped areas for hotel sales. Yes, I am talking about travel agents. Travel agents generate less than 30 per cent of all hotel bookings and most of these are channelled via the GDSs; and because the GDSs distribute their terminals to these very same travel agents all around the world, any marketing opportunities supported by them offer great potential for increased hotel sales. By the way, most of these remarks also apply to car rental companies, which can also use the GDSs to market their products effectively to travel agents.

There are two important factors to understand within the GDS marketing world: (a) promotional or marketing messages can be sent by hotels to the point-of-sale screens of GDS users who are nearly all travel agents, and (b) hotels can use GDSs to aim their promotional efforts at a highly targeted section of the world's travel agency population. These are very attractive capabilities, which a hotel's marketing team can use to great effect. However, it is by no means a simple task. Each GDS stores its own set of its marketing information fields, which may be used in many different ways. The two most important of these are: (i) distributing static information that describes the hotel chain, its leading properties and features; and (ii) pushing dynamic information out to selected travel agency groups.

STATIC INFORMATION

The first of these, distributing static information, i.e. information which does not change very often, is a simple but effective means for hotels to communicate with travel agents at the point-of-sale. It involves storing information about the chain and its hotels in one or more GDSs:

- **Chain descriptions** This information can be quite descriptive because a number of pages may be used, depending upon how many pages the hotel wishes to rent from the GDS (each GDS makes a different number of pages available). Travel-agency GDS users may then access these pages by entering the appropriate hotel code. Most GDSs do not charge a fee to hotels that use this service, provided the number of pages used is reasonable.
- **Hotel descriptions** A limited number of lines of descriptive information may be created by the hotel in the GDS and stored in the data base under an index that uniquely identifies the hotel or its chain. As there is only a finite number of lines and a limited degree of flexibility, it is important that the hotel uses this facility carefully and wisely if it is to maximize its marketing exposure.

Most GDSs provide 'super indexes' which assist travel agents in finding the appropriate keyword used to identify the hotel. Very often the only way in which a hotel can publicize its static pages in a GDS is by direct mail, and other paper-based communications with the travel agent and by the judicial use of dynamic methods, as follows.

DYNAMIC INFORMATION

Distributing dynamic information is a more sophisticated technique. It capitalizes on the GDS's ability to profile its travel-agency users. This may be done by parameters that include whether the agency is part of a group, the geographical location of the agency and the type of booking that the agency makes. Consequently, there are a number of different ways in which dynamic marketing information can be communicated direct to selected travel agents:

- **Sign-on messages** Whenever specified groups of travel agents sign on to the GDS, they automatically receive a kind of 'welcome' screen. This welcome screen can display a marketing message that the hotel constructs. This is a useful facility but it only allows the hotel to get its marketing message across once or twice each day, depending upon how often the travel agent uses their GDS terminal. (The more technologically sophisticated travel agents do not, however, always see the GDS sign-on messages because their log-on functions are often automatically processed by front-end computer systems and communications networks.)
- **Bulletin boards** The hotel can create its own bulletin boards within the GDS for communication with travel agents. The hotel can post items on the board that may be current for the week. Travel agents may peruse the information on the hotel's bulletin board; which may, for example, include details of special offers at certain properties.
- **Headlines** This is the most sophisticated of the dynamic GDS marketing techniques. Whenever selected groups of travel agents enter a certain pre-determined city pair or a certain destination city, e.g. London/Paris, the GDS will automatically display a hotel's special promotion for the destination city. The travel agent can then book straight from the headline marketing message, which links the agent into the hotel's reservations function on the GDS. This can be a powerful way for a hotel to segment its market and aim products at specific sectors of its potential customer base.
- **Broadcast** With a broadcast message, the hotel has the opportunity to send a message to virtually every GDS terminal around the world during a fixed 15 minute time slot. This is particularly effective for creating an impact with travel agents in relation to a special announcement or the start of a major marketing campaign. This is akin to TV advertising when a specific time slot can be purchased by the hotel or car rental company. The message appears on every GDS terminal screen, whenever it is refreshed from the host GDS computer, which is usually every few seconds, as entries are made by users.

Dynamic information has been used to great effect by hotels and car rental companies that participate in the GDSs. It is particularly suitable for marketing purposes because: (a) the facility can be used over certain time periods that tie into promotional campaigns, and (b) the message gets straight through to the point-of-sale. Space is rented on the GDS data base for a fee that is usually time based, i.e. a cost per week, and also relates to the distribution coverage and size of message. Table 4.6 summarizes some of the main types of information display options (unless otherwise indicated, there are no additional costs for these entries).

There is no doubt that marketing via the GDSs is an excellent way for hotels and car rental companies to increase the sales of their products. Historically, there have been very few alternative distribution systems that allow a travel company's marketing message to be sent directly to the point-of-sale in travel agencies all around the world: the potential is there for a large increase in sales from travel agents. After all, if less than 30 per cent of hotel bookings originate from travel agents at present, it is highly likely that this could be grown to around the 80 per cent level. These are some of the reasons why GDS marketing is a relatively new and growing medium for the travel industry.

Hotel chains that participate in GDSs do, however, need to commit to keeping their information up-to-date. This commitment can involve a substantial amount of ongoing effort on the part of hotels. This is especially true if the chain has decided to participate in more than one GDS or even all four (individual hotels are not allowed by the GDSs to participate unless they are represented as part of a corporate group or brand). Each GDS must be updated with information such as property description, rates and promotions for every hotel property in the chain. Table 4.6 shows the kinds of marketing information that each of the four main GDSs allow hotels to store and distribute to travel agents; it is important to realize that this information is formatted in quite different ways by each GDS. While some of the larger hotel chains have on-line links between their CRSs and the GDS hotel data base, this is not always the case. It is not therefore surprising that some hotels and their chain parents, choose to outsource their GDS

Table 4.6 GDS marketing data summary

Amadeus	Galileo	Sabre	Worldspan
Sign-in message Up to two lines of 60 characters each. Ordered by day. Chargeable (statistics available)	**TD/News Page** Introductory line of 32 characters then five lines of 55 characters. Displays for one week, Monday to Sunday. Agent has to request a TD/News Display (no statistics)	**System Hot** Up to three lines each of 56 characters. Displays for three days. Agent has to request SYSHOT display (no statistics provided)	**Associate Market Place** Comprises three lines of 58 characters plus one extra line for page reference. Displays one to five days on request and is published Monday through Friday. The agent must request a GAMP display
Broadcast messages Up to 18 lines of 60 characters each. Ordered daily. Queued to terminal in one city, one country or a particular travel agency chain in one city or country. Chargeable (targeted receipt)	**Front page sign-in** Up to three lines of 55 characters each then either direct to GIS page which can be up to 999 lines or to a HOD page. Ordered by day. Chargeable (statistics available)	**Sign-in advertising** Up to two lines of 56 characters. Ordered daily. Chargeable (no statistics provided)	**Associate of the week** A week long promotion reserved on a first come first served basis. Receives top billing in AMP pages. One complimentary Prime SINE guaranteed
Display messages Up to two lines of 60 characters tied to airport pairs of hotel's choice. Chargeable (statistics available)	**Brochure line in chain display** Up to 63 characters. Displays each time a chain HOD requested (no statistics available)	**Sabre DRS pages** Up to 99 pages each with 99 lines of free format text (statistics available)	**Prime Sign** Maximum of two lines of 60 characters. Accessed by agents when they sign-in. Chargeable
AIS pages Unlimited pages, each of 999 lines in free format (statistics available)	**Chain information pages** Many pages of 99 lines each available (no statistics available)		**GRS pages** Up to 250 lines per page, each of 60 characters. Unlimited number of pages set up by Worldspan on an as-needed basis following usage of initial allocation
	Hotel information pages Many pages of 99 lines each available (no statistics available)		
	Apollo headlines A new promotional opportunity where a hotel chain can purchase a two-line display that is brought up on any given day any time an agent requests an airline display for a certain city pairing, e.g. JFK–LHR		

data maintenance functions to third-party service companies.

HOTEL SYSTEMS SUPPORT SERVICES LIMITED (HSSS)

One such company is Hotel Systems Support Services Limited based in Wokingham. HSSS provides technology consultancy services to the hospitality industry and an ongoing GDS maintenance service for hotel chains, i.e. HSSS does not have any direct contact with individual hotel properties. Although similar services are offered by some hotel representation companies, these usually involve the hotel outsourcing both its reservations and data base management functions. However, HSSS provides hotels with a greater degree of flexibility. A hotel chain can, for example, outsource just those GDS data base management functions that involve the updating of hotel and chain description pages and the maintenance of rate information. Or a hotel can use HSSS on an *ad hoc* basis as and when the need arises, for example, to make a large scale update to its GDS data base.

Hotel chains send their GDS data base updates to HSSS in Wokingham from around the globe, usually by fax but increasingly by electronic mail via internal e-mail systems. Each GDS requires its descriptive information to be presented in a slightly different format and there are even some fields that are peculiar to specific GDS systems. HSSS provides hotels with a ten page generic data collection form that is then used by operators in the Wokingham office to update each GDS data base. These forms are nearly always used to create information for a new hotel but they may also be used for regular update purposes.

HSSS employs a number of staff, each of whom is expert in the way in which GDSs store and use hotel related information. The HSSS operator logs onto the first GDS and signs into the data maintenance function within the hotel data base area, using a password. Then the operator updates the GDS data base with the information provided by the hotel. This is an onerous task because the GDSs use only simple character based update and editing techniques that require each data line to be changed individually. Once this has been done, the operator signs-on to the next GDS in which the hotel participates and repeats the data base update process. There can be dozens of rate changes for a hotel each day. HSSS has access to each of the four main GDSs and regularly maintains the data for 2,000 hotel/GDS combinations. However, many of HSSS's hotel customers use its services on a one-off or *ad hoc* basis. The *ad hoc* maintenance work can include such tasks as: (i) updating hotel descriptions when the accuracy has deteriorated over time; (ii) adding information when a GDS upgrades its system to allow more data fields to be stored; or (iii) loading new negotiated rates for use by travel agents and companies, which usually occurs at the end of each year, ready for access by GDS users the following year.

As already mentioned, information on rates and room types can be sent from the hotel to HSSS by fax or e-mail. However, it is now common practise to download this data direct from the hotel chain's CRS. These systems usually store the complete range of hotel rates on a central data base. Downloading them electronically via a dialled telecommunications link enables HSSS to carry out quality control checks on the updates prior to keying them into the GDSs. It allows HSSS to check that it is interpreting the updates in a way that is consistent with the chain's CRS. The same can apply to conversion tables for fields such as Room Types. A conversion table allows a hotel's room type code to be mapped to the GDS's room type code. For example, a hotel may designate a double room using the two letter code DB, whereas a GDS may use the code A1D to mean the same thing. HSSS can use these conversion tables to simplify the GDS update process.

The HSSS service can substantially reduce the amount of routine administrative functions that hotels face if they are to market and sell their products via the GDSs. This is especially true when you consider that the GDSs are enhancing their systems and adding new fields to their data bases almost continuously. Just to keep abreast of these format and field changes can be quite an administrative chore – certainly not one that should be at the heart of the hotel's core business.

5
The Internet

Introduction

This chapter deals with the Internet – possibly one of the most exciting developments in travel and tourism since the industry was invented. The omission of the Internet from the previous edition of this book illustrates how quickly it has become a major factor in travel. Only two or three years ago it was a fledgling technology used by a few scientists and some USA students for very specialized applications. Now, not only is it widely used within the travel and tourism fields to reach consumers, but more importantly it is perceived as one of the major influences affecting the travel industry of the future. Recent studies, for example, predict that 20 per cent of total bookings will be via the Internet within five years (*source*: Jose Tazon, Amadeus – at the Association of Corporate Travel Executives conference in Madrid). From a more general perspective, the US Government estimates that 20 per cent of all consumption will be transacted on the Internet within 20 years. There are currently over 50 million Internet users world-wide, over half of which are in the USA. The statistics for Europe are shown in Table 5.1.

I'm afraid that I do not include here any description of the Internet or the technologies that make it work. This would be an entire subject in itself and one that I could not possibly hope to even skim in this book. I therefore assume that you, the reader, understand the basic terminology and that you already know what an Intranet is, what a hyper-text mark-up language (HTML) is, what browsers are and basically how Internet telecommunications work. My analysis of the Internet in this chapter is very much viewed from the perspective of how it is *used* within the field of travel and tourism. I therefore do not explore the more esoteric technological aspects in any detail at all. After all, it's how the Internet is used that I think is most germane to this book's audience.

The chapter starts off with an analysis of the marketing aspects of the Internet. Then goes on to discuss one of the biggest single issues facing the industry at present – disintermediation. Following this, I analyse the various ways that some companies are using the Internet at present. Included here are descriptions of several leading Internet sites that have already established themselves in the global travel and tourism industries. Finally, I have included several examples of some particularly interesting Internet pages within each section. But please note that these pages are in

Table 5.1 Internet registrations in Europe (millions)

European Country	Actual 1996	Forecast 2000
Britain	2.40	10.00
Germany	2.00	6.90
France	0.30	1.20
Italy	0.20	1.00
Netherlands	0.20	1.10
Sweden	0.16	1.10
Denmark	0.08	0.80
Norway	0.10	0.40
Finland	0.14	0.40
Belgium	0.30	0.40

(*Source*: IHBRP, Inteco Corp, 1997)

fact 'screen shots' and that they do not show a complete Internet page. Most Internet pages are in fact too large to fit on a single screen and rely on vertical scrolling functions supported by most browser software. Nevertheless, I hope they give you a flavour of what functions and information are available on travel and tourism via the World Wide Web.

Marketing on the Internet

In my view, the Internet is an almost pure manifestation of marketing principles and practices. It is a marketing person's dream because: (a) it levels the playing fields, (b) it enables companies of different sizes to compete on more equal terms, and (c) it allows a company to open up a direct channel of communication with its customers. What's more, the success of an Internet site is not always directly proportional to the amount of money spent on designing it. We are all no doubt aware that the success of an advertising or promotional campaign depends very much on the amount of money spent on media advertising. This is because the company must broadcast its message to everyone, or at least a very large proportion of the population, in order to reach its desired target market. The amounts spent by large companies on television advertising, bill boards and the press are enormous. By contrast, smaller companies cannot afford such massive exposure and consequently their products do not become so well known.

However, with the Internet this is not necessarily the case. Companies of all sizes are much more equal in their competition for the consumer's attention. The main reason for this is that the pages that comprise one company's Internet site can be available to the same population of consumers as another company's site, yet without any significant additional amounts of expenditure. It is not quite so easy for a large company to throw money at their Internet site and as a result, expect it to be visited by vastly increased numbers of consumers. What is happening in the new electronic marketing world of the Internet is much more subtle. A whole new approach to sales and marketing is evolving. It remains to be seen precisely how this will crystallize into a tried and proven methodology, because the technology is so young and consumer reactions have yet to be measured accurately. So, everyone is learning the hard way – lots of experimentation mixed with liberal amounts of trial and error.

But first of all, let's get the relative size and importance of this new channel into perspective. Because the Internet is the focus of my marketing analysis, let's first of all consider what types of people use it and what its potential is. The current profile of a typical Internet user is remarkably consistent with that of a profitable potential travel customer. They tend to have a high level of disposable income and are in the 25–35 age group. Research shows that many Internet users are affluent and experienced travellers and this sounds just like the target market of many travel agents and suppliers. At the moment, over three million European homes have Internet access or subscriptions to on-line services, e.g. Compuserve and America On-Line (AOL). This figure is estimated to double over the next two years. Forrester Research believes that Web generated global sales will rise from US$2 billion in 1996 to US$61 billion in the year 2000. Forrester also predicts that the top three sectors for on-line shopping will be: (i) computer products, (ii) travel, and (iii) entertainment. (Forrester Research is a major research and consultancy organization that has carried out extensive Internet analyses).

Clearly, therefore, the Internet represents a significant new opportunity for a company to distribute its products and services direct to consumers. But in order to do this well, a good marketing campaign will be needed. The question is: 'How should a company's products and services be marketed to consumers via the Internet?' The problem is, there appears to be a lack of any established methodologies for successful Internet marketing. But despite this, it does seem that a set of critical success factors are beginning to be distilled. They are probably best described in terms of an evolutionary approach that several companies have taken towards the development of their marketing strategy for the Internet. The first of these, not surprisingly, is to establish a corporate presence on the World Wide Web.

WEB SITE PRESENCE

The first and most basic commitment that a company can make to the Internet is the establishment of its own Web site. Although many companies have taken this first step, it can be more complex than it seems at first glance. To start with, there are some fundamental issues to be resolved, such as: should the site be created and maintained in-house using the company's own computer or should it be outsourced to a local computer service bureau? Companies sometimes start by establishing a Web presence on a bureau basis and then, depending upon its success, move the Web site operation in-house. Having said this, there are a lot of companies that are perfectly happy with an outsourced solution to their Web presence. After all, unless the company has a cost effective in-house IT department, the expense of creating a Web site and coping with ever-changing Internet technology can be significant.

Another important decision is the establishment of a memorable name for the site. Site names are important because they need to be memorized easily by the consumer and they obviously need to represent a natural link to the name of the supplier company. Once established, they cannot be easily copied, or for that matter changed. Having established a site name or unique reference locator (URL), the next step is to design a home page. Again, this is no simple task. A home page needs to be attractive and must provide links to other parts of the site and to other related sites. Incidentally, it is quite possible that advertising fee income can be generated if a company's Web site incorporates links to other sites. It usually incorporates some form of main menu, but not in the sense of the old classical computer application. An Internet main menu is much more intuitive and user-friendly. It often involves graphics, sound and animation, as well as text. The problem is: 'How should a home page be designed so that it supports today's site visits, yet allows the remainder of the site and its other pages to grow and develop over time?' Well, the answer is that of course the remainder of the site must be designed at least in concept before the home page can be completed. However, this is easier said than done, especially when the site is expected to evolve and therefore change over the short term as more content and new sections are added.

Design is a critical aspect of any Web site; and design is not just about information content and layout. It is very much about the visual effect of Web site pages as they appear on computer screens. The Internet Web site design process is certainly not confined to computer programmers. It is a new skill that is best undertaken by graphic designers and creative artists who work in the advertising, publishing, marketing and corporate communications businesses. After all, if the pages are to be effective and have impact they need to be produced by the kind of people who design brochures, magazines, logos and advertisements. This is a specialist field and is not one in which either travel companies, tourism organizations or IT departments are known to excel. To get the best Web site design, an outside agency is probably the best approach. There are now many companies that provide these services, including the Internet providers themselves, and the only issue is how to decide the best and most appropriate one.

An important feature of the home page is the site owner's e-mail address. This is vital in order to begin the development of a rapport with the consumers that comprise the company's target market. It is here that the company will need to make its second major commitment to the Internet. If it is to publish its e-mail address then it must expect to answer incoming e-mail from site visitors. Again, this is easier said than done. As the Internet is a global medium, e-mail can be expected from virtually anywhere in the world; and they will come from a very wide range of people too. Students, casual browsers and serious customers are all potential sources of e-mail. The challenge is to weed out the serious customers yet maintain a reputation for all round good customer service. So, to develop a site successfully, it is very important that all e-mail is filtered and answered within a certain period of time. This is often implemented via a quality control measure that companies with successful Web sites embed within their employee work practices. This brings us to one of the golden rules of Internet marketing; develop a good communications channel that establishes a dialogue between your company and its consumers. The next steps are: (a) increase the number of site visitors, and (b) turn site visitors into buyers of the company's products and services; in other words, increase sales.

An essential way of increasing Web site visitors is to advertise the site. This can of course be accomplished by means of standard paper-based advertising and promotion, e.g. specialist Internet magazines and 'Whats On' publications. But there is an alternative electronic way to achieve the same thing. This is by establishing hyper-links from other Web sites to your own. Again, its all pure marketing. The company needs to research other companies that have a Web site and select those with whom a strategic relationship exists. In fact, no such relationship might exist at present because the other company is in an entirely different field of business. However, new relationships can be established by finding new inter-relationships between a company's products and those offered by other companies. Establishing hyper-links from other successful sites to your own is absolutely essential if your site visits are to be maximized.

Another way to increase site visitors and attract new customers is to make use of the Internet Newsgroup functions. There are many prime examples of individuals and companies who have used the Newsgroup facility to create new businesses. They establish a Newsgroup on a particular subject. Then they post open letters into the Newsgroup that describe a particular business opportunity, a new product or an innovative service. Internet users can log onto the Newsgroup index and if they are interested, pick the company's Newsgroup item. From here they can post open-electronic-news items of their own within the Newsgroup that other participants can also see. Using this kind of open communications channel, a company can build up a pretty good base of interested potential customers. All the company has to do is ensure that somehow or another it captures the Newsgroup user's e-mail address. The company can venture into the world of direct e-mailing. This whole area is a subject in itself and there are many books that specifically address the topic. All I have attempted to do here is make the reader aware that these electronic marketing opportunities exist.

Another approach to turning a Web site presence into actual sales is by providing site visitors with access to a booking engine. This can be via a supplier's own booking engine interface to its corporate computer or by linking from the supplier's site to another site that provides a booking service. Let's examine this in a little more detail.

BOOKING ENGINE

Once a company has established a Web site, the next thing that it will need to consider is selling its product directly to consumers. This is a significant step for any company and one that is obviously not taken lightly. To sell products and services direct to consumers via the Internet, companies really need some form of computerized inventory system. Most companies will already have such a system that they use to control stock positions and support the sales process. Airlines have their CRSs, hotels have their room inventory systems, tour companies have their booking systems and so on. At present many of these companies use their booking systems as platforms from which to distribute their products via existing channels such as the GDSs and videotex. To distribute products via the Internet, a new interface is required. This interface will allow the company to make an Internet booking engine available to its site visitors.

Even with an in-house computerized inventory control system, developing an Internet booking engine is a non-trivial task. However, it need not be a major obstacle for a company. This is because there are specialist IT service companies that have already sprung up to support just this type of Internet application. In fact I have given an example of one such company in the section below entitled 'Interfacing supplier systems to the Internet'. These companies have developed the technical infrastructure that enables an existing booking system to be interfaced to the Internet. This infrastructure makes the interfacing task relatively simple and straightforward. It allows all kinds of systems to be adapted for the Internet. Even old legacy main-frame computer systems can be presented to consumers as dynamic new Web sites using this approach.

Companies that have established their own Web sites and have complemented these with booking engines are in powerful positions to generate significant additional revenues from the Internet. This is especially true for companies that can: (a) sell their products or services to consumers all around the world; (b) sell their products or services without needing to deliver paper documents, e.g. an air ticket; and (c) accept payments from consumers via plastic card mechanisms. For these reasons, hotels are in a particularly strong position to exploit the Internet, and there are many instances that illustrate that hotels are in fact doing just that. The examples I have given in the following section include THISCO's TravelWeb, Utell's HotelBook and Marriott's own site. So, a Web site linked to a booking engine would appear to be the ultimate position for a company to strive for in the world of the Internet. But it really is just the beginning. It is at this point that highly targeted relationship marketing becomes a possibility.

TARGETED MARKETING

Companies with established Web sites and booking engines are in a position to undertake some highly productive marketing activities that have not been practical with older technologies. These all revolve around a customer data base and an activity known as 'push marketing'. First of all, let me explain this terminology. There are two types of marketing campaigns which are known in the industry as *push* and *pull*, respectively:

- **Push marketing** Push marketing is where a company's products are advertised widely to many people. The audience that is targeted may be very large and it is probably the case that only a small percentage of the audience will be attracted to buy the company's products or even simply to enquire about them. However, without the ability to know each one of their prospective customers individually, companies are faced with having to push the product at them in a kind of shotgun approach. The ultimate hope is that sufficient numbers will buy the product and thereby justify the high cost of the associated advertising campaign. Push marketing is what we are all used to and it will no doubt continue for many years, if not, forever. However, 'pull' marketing can be more cost effective and highly productive. It also happens to be a marketing technique that is ideally supported by the Internet.
- **Pull marketing** Pull marketing is much more consumer specific than push marketing. It relies on establishing a relationship with a customer or consumer. The best kind of relationship is that which flows from a customer's purchase of the company's products or services. When this happens, the company is in an ideal position to learn a great deal about its customer. If customer information such as this can be categorized, indexed and stored on a data base then it can form the platform for highly effective 'pull' marketing campaigns. A pull marketing campaign is one where specific products are aimed at precisely those consumers that have either made prior purchases or whose profiles exactly match the product being promoted. The concept is to pull these specific customers towards the company and encourage them into purchasing those products or services that are of particular interest to them.

Successful 'pull' marketing campaigns are highly dependent upon IT for their effectiveness. However, pull marketing is not a new concept. There are many cases, for example, where a single site hotel can afford to keep a handwritten card file on all their guests. Each guest's card would show their personal preferences and the kinds of services they have enjoyed on previous visits. Then, when the hotel decides to hold a particular event, it scans the card file for previous guests whose profiles would seem likely to fit that of the planned event. Those guests selected would receive personalized letters from the manager reminding them of the previous event and introducing them to the planned new one. The problem is that this approach is not really feasible on a national scale and is totally impractical globally without some degree of automation. This is where the new and emerging technologies can play a vital part in travel-related marketing programmes.

It is now possible to use a similar approach to the old card file system across entire multi-national corporations that have customer bases of several hundred thousand people. With new IT it is perfectly feasible to process millions of electronic 'card files' within a matter of seconds. In fact this capability is a combination of two new technologies: (a) the Internet, which provides the communications channel with the consumer and acts as a front-end for data collection; and (b) a good relational data base management system, which can index and organize the information gathered. Together, these two technologies enable companies to develop highly effective pull marketing campaigns. However, to be successful, a company needs to be highly disciplined in the way it deploys its IT on a global basis. Consider for a moment the key principals that a successful Internet-based pull marketing campaign should embody:

- **Internet Web site** If a company is to establish an interactive communications channel with its customers, it will almost certainly need to have a Web site of its own. To be effective, this needs to be highly interactive and responsive. It will probably use e-mail to exchange messages with existing or prospective customers. Ideally, the site should incorporate a booking engine and be capable of receiving post-booking feedback from the customer.
- **Customer data base** This is the core of any marketing effort. But for pull marketing to be effective, a customer needs to be identified individually. This is not so much a technical challenge as it is a logistical one. A method must be found that encourages a person to identify themselves to the Web site whenever they visit it. One commonly used approach is to request the user to enter their own user name and password whenever they visit the company's site. Once the consumer is registered other more detailed profile information, including their e-mail address, may be captured and stored within the data base.
- **Transaction history** While the presence of an individual's profile on the customer data base is critical, so are the transactions which that customer undertakes with the company. It is essential that all relevant details of each and every transaction is captured and stored so that it is linked to the profile recorded in the customer data base. The trick is to link what appear to be separate transactions, to a single individual on the customer data base.
- **Query tools** As the data base of profiles and transactions grows, so it becomes ever more important for the company's marketing team to be able to analyse the data and try to identify trends and patterns. This is the first step that a company can take towards understanding its customers. Only by doing this well can new products, services and special promotions be designed in the knowledge that a market exists for them.
- **Selection tools** Sometimes called profiling, this is a technique for selecting all customers from a data base that meet certain pre-defined criteria. For example, a hotel may select all customers who stayed in a certain room type as part of a weekend break anywhere in Western Europe over a particular holiday weekend (and who also booked using the Internet). Selection tools can be quite sophisticated and can specify very detailed parameters indeed.
- **Direct e-mailing** As more consumers use the Internet, so the number with registered e-mail addresses will grow. Because this is almost certain to be one of the data elements recorded within the customer profile, it can be used to communicate with those customers that have been selected. This is very similar to classical paper-based direct mail but with some important differences: (i) the degree of targeting is extremely high; (ii) the cost of an e-mail is virtually zero; and (iii) people are more inclined to reply to an e-mail than a letter, chiefly because it is hassle-free.

These are all very challenging principals for a company to implement successfully. More significantly, they all involve substantial amounts of expenditure in terms of both cash and people's time. However, there is clear evidence that most, if not all of these pull marketing principals are in fact being implemented by many companies right now. This, to a large extent, illustrates the faith that these companies are placing in today's fledgling Internet. So, as the world-wide population of Internet users grows in volume and Internet commerce grows with it, I think pull marketing will become a critical success factor for many businesses, particularly those in the field of travel and tourism. Companies that have started to experiment with electronic marketing in the early days will be well positioned and sufficiently experienced to capitalize on these critical business survival skills in the future.

INTERNET MARKETING RELATED ISSUES

The Internet is such a new distribution channel that there are many issues that both suppliers and intermediaries are faced with. In this section I am going to focus on some of the major issues that influence the way in which companies market their products and services on the Internet. Each of these issues is explored only briefly because they nearly all could consume chapters in themselves. However, the following encapsulation of these issues should provide fertile ground for further debate.

Search engines

When consumers first start surfing the net in search for holiday planning and booking sites, they often start by using a search engine, e.g. Yahoo. There are several popular search engines and they each work in similar ways although there are important differences in the way in which they catalogue and find sites for users of the Internet. Web site owners register their sites with the major search engines and provide them with a great deal of information about the site and its contents. Besides providing Web site search functions, the search engine companies also award their own prizes to what they consider to be the best sites of the week or month; and they obtain much of their income from advertising other companies' products and services on their Web search page.

Now, the issue is: 'How is the sequence of a search engine's Web Sites Found display determined?' Let's say the consumer enters search criteria keywords of 'air travel booking'. The search engine will identify several Web sites that provide air travel bookings, but how will the sequence in which they are displayed be determined? Often, this is on the basis of the number of site hits recorded, but the criteria vary. Isn't this rather like the old CRS biased display situation that was judged as unfair and discriminatory by various regulatory bodies in the USA and the EC a few years ago? Couldn't the big airlines, for example, pay vast sums in advertising revenues to the search engines to ensure their sites always came at the top of the list? If they did so, would this be judged to be unfair competition? It's an interesting issue, which to my mind has not yet been sufficiently debated within the industry.

The legal issue

This issue relates to the contractual position between the consumer and the supplier when a travel product is booked through an Internet site. If the Internet site is a GDS, for example, then a contract will exist for the provision of travel products from the supplier company to a travel agent. However, what is the legal position when a dispute arises between the consumer and the supplier? No such contract exists. Would it be possible for the travel supplier to claim that they did not formally approve the distribution of their products direct to consumers? In which case they might argue that because the consumer purchased the product directly from the GDS, then it is the GDS that should accept responsibility. After all, if a travel agent had been involved then the advice given might have been correct and no problem would have arisen. This issue is complicated further in situations where a product is purchased via the Internet by a consumer in a country in which the supplier and possibly also the GDS are not represented.

Booking fees

At present it is unclear how booking fees and commissions will be apportioned for travel sales made via Internet sites. Take, for example, one of the so-called supermarket sites (probably better described as one of the new intermediaries). Many of these new intermediary sites use a link to a GDS as their booking engines for air, hotel and car rental products. When a travel agent makes a booking via the Internet, what commissions must be paid by the supplier? Table 5.2 shows the various possibilities.

Assume for the moment that the supplier is a hotel. Should the hotel pay a GDS booking fee – after all the hotel's system is connected into the GDS and the hotel would normally expect to pay a booking fee if the travel agent booked via

Table 5.2 Booking fee possibilities

Booking fee analysis		Supplier pays booking fees or commission to:		
Booked by:	Booked via:	Travel agent	GDS	New intermediary
Consumer	Supplier's own Web site	No	No	No
	New intermediary and link to supplier's own system	No	No	Possibly (see note)
	New intermediary and link via GDS to supplier's system	No	Yes	Possibly (see note)
Travel agent	Supplier's own Web site	Yes	No	No
	New intermediary and link to supplier's own system	Yes	No	Possibly (see note)
	New intermediary and link via GDS to supplier's system	Yes	Yes	Possibly (see note)

Note: New intermediaries may collect a commission if they are, for example, registered travel agents as are Expedia. However, this is not always the case and many new intermediaries do not collect a booking fee from all suppliers, e.g. TravelWeb is not paid a fee for airline bookings that it handles for its customers.

their GDS terminal? If so, should the hotel also pay a travel agency commission as well as an intermediary booking fee? These new intermediaries will also need to keep their booking fees competitive with the GDSs. They must make sure it is cheaper for a supplier to sell a product to a consumer via the Internet than via the GDS/travel agent route. There are many related issues here – certainly sufficient to keep a class discussion going for quite some time.

Supplier interconnection strategies

With the expansion of new electronic distribution channels, suppliers without their own booking engines are now faced with a new problem: 'Which GDSs and Web sites should they connect to?' It would appear at first glance that a supplier should connect to as many GDSs and sites as possible in order to obtain the widest exposure. However, for a supplier without its own internal booking engine, there is a substantial overhead involved in connecting to a large number of third-party systems. Like so many issues within the area of IT in travel and tourism, the root of many of these problems is a lack of standardization. The problem is that for each system a supplier connects to, the supplier must support the following: (i) a channel through which it can receive reservation requests; (ii) a method of providing confirmations of reservations; (iii) a method for updating the inventory and product details held within the site's computer; and (iv) a translation of its internally used data standards into the format and standards used by the distribution system, whether it be a GDS or a Web site. The short answer to this problem is for a supplier to obtain its own on-line booking engine. However, this is expensive and not economically feasible for all but the largest of companies. Most small to medium sized suppliers will instead look carefully at the alternative distribution systems and make a value judgment on just one or two that are most relevant to their businesses.

Advertising policy

The publishers of newspapers and magazines know only too well that there are rules and regulations that govern how they take advertisements from other companies for inclusion in their publication. It would, for example, be regarded as unfair competition if one newspaper refused to take an advertisement for one of its rival publications. The issue is: 'Does this apply to the Internet?' Could, for example, a site owner refuse to advertise a competitor's Web site on its own, all other things being equal, e.g. space was available, other companies were advertised, etc. Would such refusal be regarded as unfair exploitation of the Internet as a public media and if so, which body could bring a prosecution and in which country?

Hotel Intranets

Internet technology allows hotel 'brochures' to be created electronically, complete with pictures, diagrams and a full set of room rates. What's more, individual versions of these electronic brochures can be created especially for corporate customers of hotels. These tailor-made versions can only be accessed by the client company via a special password and are not accessible by other general Internet users. These domains of private customer information that can exist within a hotel's Web site, are called Intranets. While most Intranets involve private networks owned by companies, hotels can distribute theirs via the World Wide Web. However, if a large hotel customer were to have their own networking capability, they could access the hotel Intranet via more secure means, e.g. via private leased lines or secure dial-up via ISDN services, both of which could use their own firewall for security and protection against unauthorized access. Once this begins to happen on a wider scale, hotels will have established a very powerful customer relationship that can be used to each organization's overall benefit. The hotels can then achieve increased sales with higher levels of profitability while the corporate customer can enjoy lower rates and provide a better service to their employees in terms of information availability and accommodation services.

Some of these issues begin to raise the question of what role intermediaries will play in the future world of the Internet and other electronic distribution channels. This topic has become known as 'disintermediation', which is a term I personally do not favour, particularly because it appears to be a misuse of the word. However, it is the term that is used throughout the industry to mean the possible stripping away of travel and tourism intermediaries. So, let's put the syntax to one side for the moment and consider exactly what the future role of travel and tourism intermediaries will be in the future.

Disintermediation

I thought this Internet chapter might be an appropriate place for a discussion on the future role of intermediaries in travel and tourism. After all, the Internet is one of the prime forces that could bring about disintermediation. The driving force for this is the cost incurred by suppliers in receiving a customer booking. It has been estimated, for example, that the cost of obtaining a booking via a telephone service centre is around US$10, to receive a booking via a GDS costs around US$3.50, but to capture that same booking via the Internet costs only 25 cents. These are broad brush figures but the message is nevertheless clear – intermediaries represent a substantial element of supplier distribution costs. It is not surprising therefore that disintermediation has already started and the only really interesting issue is the extent to which it will progress as time goes by. I hope the following preliminary discussion of the issues surrounding disintermediation will set the scene for the remaining sections of this chapter, which describe some of the more interesting travel and tourism Web sites that existed as at mid-1997. I just hope they are as relevant to you at the time you are reading this book as they were when I wrote it!

Travel intermediaries cover a wide range of organizations. Although travel agents are usually singled out as the primary intermediaries, there are many others that we need to consider. For example, the GDSs are intermediaries, principally between the airlines and travel agents. Then there are tourist offices, which are intermediaries between tourist organizations and consumers or tourists.

TRAVEL AGENTS

Let's take travel agents first. Travel agents are intermediaries between travel suppliers and consumers. They sell suppliers' products and services to their customers and derive a commission for doing so. A travel agent's added value to the customer is their expertise in travel and their knowledge of the relative strengths and weaknesses of various travel suppliers. A travel agent's added value to a supplier is their customer servicing role, one which is time consuming and costly for suppliers to handle themselves. These are pretty compelling reasons for the existence of travel agents as intermediaries. However, things are changing. But, what are the fundamental reasons for this change? There are three catalysts for change: (i) the spread of automation from suppliers via distribution systems to agents and consumers; (ii) the supplier's rising cost of distribution, much of which is paid to intermediaries such as travel agents and GDSs; and (iii) the customers' impatience with the slow pace of change among travel agents, who they often perceive as adding very little additional value to their transactions. Let's examine each in a little more detail:

- **Automation** It used to be said that travel agents were the custodians of four key abilities: (1) they had the ability to print airline tickets, (2) they understood the complex airline reservations and booking language used by the GDSs of the world, (3) they were licensed to print airline tickets, and (4) they had the expertise to know how to arrange travel for their customers. But how much of this is still true now that: (a) the Internet is distributing travel related information and booking functions around the world using simple GUIs, which can be used by people who are not trained in IT or travel; and (b) airlines are introducing electronic ticketing, which does away with the need for airline tickets and related ticket stock licensing issues? It could therefore be argued that many other organizations and individuals now have access to at least three of the above four key abilities. If travel agents do not focus on changing their core competencies to the proactive provision of added value travel management expertise, then they may well find that their traditional reactive services are no longer in sufficient demand to support their businesses.
- **Suppliers' distribution costs** With deregulation and increased competition, suppliers are increasingly focusing their attention on overheads. One of the most significant overhead items is distribution costs. These are the costs borne by suppliers in selling their products to customers through distribution channels.

Historically, the primary distribution channels for most suppliers has been the travel agency network; and it probably will continue to be for some time. However, there is no doubt that this situation is changing with the spread of new technology. In any event, at present travel agents sell the vast majority of suppliers' output. This is a double-edged sword from the suppliers' perspective. On the one hand it removes the overheads of dealing with customers from the suppliers. They do not need to worry quite so much about the time-consuming and often non-productive tasks that are an important part of the selling process. Tasks such as pre-trip planning, giving advice on areas of the world, helping to decide the best time for the trip, advising on health and visa requirements and much more. All this is handled for them by the travel agent. The suppliers can therefore devote as much attention as possible to marketing their products and operating them. On the other hand, paying travel agents commission is a costly exercise. One that represents a large chunk of the suppliers' distribution costs.

- **Travel agents' added value** Many customers, particularly in the corporate environment, feel that travel agents are simply reactive and not sufficiently proactive. Agents react reasonably well to customer requests for bookings but they are perceived as not proactively offering customers added-value information that either reduces their costs or improves their service levels. While agents are striving to address this issue by appointing dedicated account managers to business travel customers, those very same customers are being constantly exposed to technological tools that allow them to add value without the overheads of an intermediary.

So, suppliers are constantly searching for ways to leverage their investment in automated systems and thereby reduce their distribution costs. Travel agency commissions are therefore being constantly squeezed. There are many examples of this including 'commission capping', which is commonplace in the USA. Airlines stipulate that for certain types of air ticket, usually the ones on common point-to-point routes, they will only pay commission up to a certain fixed amount, regardless of the value of the ticket and the percentage commission that is usually paid. Then there is electronic-ticketing. The industry is rife with talk of the airlines restricting the commission paid on flights that are ticketed electronically. The argument being that travel agents have far less work to do for these sales and should therefore receive a lower level of commission. Finally, there are the smaller airlines that cannot afford the overheads of what they regard as a costly distribution channel serviced by travel agents. There are examples of airlines who are turning to direct sales to consumers and this had caused a backlash from travel agents who in some cases have refused to sell those airlines' tickets. Nevertheless, it is a strategy that appears to be working for certain airlines. So, what are the alternative distribution methods for suppliers wishing to sell their products to consumers? Here are some of the main ones:

- **Tele-sales centres** Suppliers can re-engineer their telephone customer service offices into fully fledged tele-sales centres based on new telecommunications.
- **The Internet** The Internet offers suppliers an opportunity to sell direct to consumers without having to pay sales commission to intermediaries. Also, Internet technology allows much of the travel advice and pre-trip consultancy to be given to consumers electronically.
- **Interactive television** This is a technology that is in its infancy and is way behind the Internet at present. However, it offers substantial potential for direct sales to consumers because nearly everyone has a television set, even if not that many currently have access to the Internet.
- **Self-service kiosks** These are intelligent ATM-style machines that are activated by consumers. They have links to suppliers' electronic distribution systems and sometimes include voice links and even video-conferencing.

I'll be examining each of these new distribution methods in more detail in a moment. But first, let's examine a key question: Why not use this new technology to by-pass travel agents and sell directly to consumers? This is really the heart of the disintermediation debate. However, there is no easy answer to this question. The push for suppliers to sell direct to consumers is driven by a powerful force – increased profitability. However, this is partially offset at present by some substantial barriers to change, even though they may be of a transitory nature; and as we all know, change is one of the most challenging issues for management to tackle. Let's consider some of the key barriers to change:

- **The threat to sales** Suppliers are in the position of being highly dependent upon travel agents for the vast majority of their sales. Most airlines, for example, derive around 80 per cent of their ticketed sales revenue from travel agents. Travel agents are therefore their primary distribution channel. So, although there may be new ways for suppliers to circumvent travel agents as their primary distribution channel and substitute them for something less costly, in the short term this is dangerous. It is obviously a dangerous course of action for suppliers to attempt to bypass a distribution channel that delivers the vast majority of their sales volume. The danger is that if they start pushing an alternative channel that threatens travel agents, then travel agents will retaliate by switching sales to other competitors. So, deadlock. Suppliers would like to change to a less costly and more direct channel but they do not wish to upset the apple-cart and disenfranchise their primary distribution channel and thus jeopardize sales.
- **Ticketing** At present, consumers who book directly with suppliers need to collect their tickets before they depart on their journeys. The only practical ways to deliver tickets to customers right now are:
    – *Ticket delivery using mail and courier services* This is perfectly practical but poses some problems. First of all there is the time taken to deliver tickets by mail. For someone departing soon after making a reservation there is always the danger that the tickets may get delayed and not reach the customer before they have to leave on their journey. Then there is the security issue. Tickets can get lost in the post or even stolen during transit, which can cause serious problems for both the customer and the supplier.

- *Ticket on departure* Customers can collect their travel documentation at the airport, immediately prior to departure. Again though, there are potential problems. First of all, customers have to queue at the airport at a service desk to collect their tickets. This can be a problem if insufficient time is left for this task and the ticket desk is busy with long queues.

In so far as the airlines are concerned, electronic ticketing holds the long term answer to the ticketing problem (see Chapter 3 for a description of electronic ticketing). The clear trend is for air travellers to use electronic ticketing increasingly. Whether they buy their tickets from a travel agent or directly, they will in the future use electronic tickets. So, if consumers book their travel via one of the new electronic channels, they will not need to receive printed tickets at all. They will simply receive boarding passes from a self-service ATM-type machine when they arrive at the airport.

- **Payment** Receiving payment from customers remotely always introduces some degree of risk. While consumers feel safer giving their card number to a customer service representative over the telephone when contacting a tele-sales centre, they feel less inclined to do so over the Internet. Although the issue of commerce on the Internet is being addressed at present, it has yet to be resolved finally. Consumers therefore still feel disinclined to enter their card information into an Internet page, no matter what guarantees are given by suppliers. However, this situation is changing and if the USA is anything to go by, consumers are becoming more comfortable with paying over the Internet using secure encryption technologies.

I think that despite these obstacles there is a clear trend for suppliers to sell an increasing volume of their products to consumers using some form of direct channel that bypasses travel agents. One only has to review some of the travel Web sites that I have reviewed in the next section of this chapter to see that this is true. The question is: 'How quickly will this direct selling channel expand and to what extent will it grow?' Clearly, the rate of expansion won't be any kind of a big bang but instead will be a more gradual process that will build its momentum over time. To explore how quickly and to what extent it will happen, let's take a look at the spectrum of travellers and the kinds of journeys they undertake:

- **Frequent travellers with simple itineraries** On one side of the spectrum are those frequent travellers who regularly travel between just a few destinations. These are relatively sophisticated travellers who know their destinations quite well and who are familiar with the alternative types of travel and competitive suppliers on their routes. Often, they are business travellers who work for smaller to medium size companies, but not exclusively so. People who have friends and family overseas also fall into this category. Such people make several trips each year to the same destination, which they get to know very well. This class of traveller derives little added value from a travel agent. All they really want is the lowest price ticket at a level of service for which they are willing to pay. There is little reason why they should not use a direct channel to obtain their travel products and services.
- **Independent travellers** These people do not buy pre-packaged tours and instead like to construct their own personalized itinerary. They include people who either know many areas of the world and simply wish to make their own arrangements to get there, or people who want to go exploring to more exotic locations. They usually find that the average travel agent will not know a lot about the kind of trip they wish to take because it is so specialized. What they want is to select the best air transportation, often the cheapest, add a car rental option, perhaps book the occasional hotel but usually make their own arrangements for accommodation when they are travelling. Again, these types of consumers often enjoy the process of researching their intended trip, reviewing alternative supplier options and building their own itineraries. Again, these types of consumers could well be attracted to a direct Internet channel, especially one that is rich in information content on far-flung destinations.

- **Packaged holiday-makers** A growing proportion of holiday-makers know their preferred destinations and are looking for simple packages at the lowest possible cost to one of the popular holiday resort areas. Good example of products in this category are fly-drive holidays to the USA, and either flight-only or flight plus accommodation packages to the beach resorts of southern Europe. There is clear evidence that many of these holiday-makers use television-based teletext information to research and book a suitable package. Once again, if these types of consumers have the opportunity, there is no reason why they could not book directly with a tour operator or consolidator via a direct channel.
- **Business travellers with complex itineraries** Many business travellers make extensive trips to a number of destinations on behalf of their companies. They tend to use a number of different airlines, hotels and car rental companies to meet their more complex travel requirements, which are often quite demanding in terms of travel time and pre-determined dates. Such trips really do require the services of a knowledgeable travel consultancy that specializes in route deals, corporate rates and can provide a high level of customer service. It is unlikely that these types of travellers will be inclined to make their own travel arrangements via a direct channel. So, this is an area where business travel agencies could develop their skills to offer a more specialized and proactive consultancy service to their customers.
- **Infrequent travellers** This category of travellers is relatively unsophisticated in terms of their knowledge of the world's travel destinations and need face-to-face contact in order to discuss their travel requirements. They would probably not feel sufficiently confident to choose a supplier or a destination without first having received some consultancy advice from a travel agent. They are therefore unlikely to simply book a package with an operator directly or arrange their own transport with a single supplier.

This brings us onto the issue of whether these consumers, who are eligible for direct sales, have the opportunity or the propensity to do so. I identified the main direct sale channels at the beginning of this section as being the Internet, supplier hosted tele-sales centres, the interactive television and customer activated self-service kiosks. However, I am going to concentrate my analysis of disintermediation on the Internet. But before I elaborate on this, I feel I should really say why I am not going to pursue the other direct sales channels in more detail:

- **Tele-sales** Take tele-sales centres – there is no doubt that supplier tele-sales centres have significant potential for handling a far greater volume of direct sales. The principal technologies that will enable them to accomplish this are: (a) third-party offerings that enable call answering tele-sales activities to be outsourced to companies in the telephone service business, and (b) re-engineered in-house supplier systems that support tele-sales operators. However, the issues governing the rate of change in this area are not as complex as those in other areas such as the Internet.
- **Self-service kiosks** Self-service kiosks that are activated by consumers will no doubt grow, but are unlikely to replace any of the other direct distribution methods that I have outlined. These kiosks will I think provide more of a customer servicing function. In terms of direct selling, they may well allow consumers to browse travel alternatives and obtain information for trip planning purposes. However, when it comes to booking, the approach being used by many of the current schemes is to put the consumer into contact with a remote sales assistant either by telephone or in the more sophisticated implementations, by video-conferencing methods. So, while the use of these kiosks will no doubt grow, they are unlikely to cause a paradigm shift in consumer buying patterns across the industry.
- **Interactive television** Interactive television is a different matter – this is a technology that really does offer some significant potential for direct sales of travel products and services. The issue here is the mechanism that will be used to support the interactive dialogue with the home television consumer. On the one hand

this could be a new technology and a new network that allows television users to connect into different supplier systems and information sources. But is this likely to be something entirely separate from the Internet? Televisions are already being manufactured with Internet access capabilities. Despite the fact that there are technical difficulties to be overcome, it seems unlikely to me that with the investment many companies are making in the Internet that a completely separate technical infrastructure will be built just to support interactive television. So, my argument is that while I believe that interactive television will no doubt grow and become widespread, the interactive part of it will be based on the Internet.

It is the Internet that I propose to focus on for the remainder of this section. My reasons for this are the projections of Internet growth that I quoted in the introduction to the marketing section at the beginning of this chapter and also some other very relevant market research. First of all, the growth rate in numbers of people who are able to access the Internet is very high. It doubled in 1995 to 26 million and almost doubled again in 1996 to 50 million. It seems that this rate of growth is set to continue or even increase as new technologies, such as interactive Web-enabled televisions, arrive on the consumer home market. This end-user growth has a related impact on the number of Web-originated travel bookings. Analysts predict, for example, that travel bookings on the World Wide Web, which currently stand at some US$400 million per year, will rise to US$4 billion by the year 2000 (source: Jupiter Communications, New York). Despite the hype surrounding electronic commerce, the estimated fraud rate involving Internet transactions is low, at around US$1 for every US$1,000 billed. This compares, for instance, with US$19.83 for every US$1,000 billed using cellular telephones (source: Forester Research 8/96).

Having analysed the issues that are most likely to affect disintermediation, the bottom line question is: What will be the likely impact of new distribution channels, such as the Internet, on travel agents? Well, I hope from the preceding discussion you will have gathered that, in my opinion, it is likely to be significant. That's not to say that it will be the end of travel agents. Far from it. Certain types of travel agents will thrive. But to do so they will need to change:

- They will need to focus on developing their true added value so that they can begin to offer quality advice, both to travellers and to corporate administrators. This should include the development of expertise on how people can travel most efficiently to different areas of the world with optimum use of supplier deals. It is difficult to see how any currently available electronic method can beat the all-round expertise of a travel expert in a one-to-one discussion. This is especially true for complex itineraries involving many countries and demanding travel schedules.
- Many of the simple straightforward transactions will be handled directly using new technology, such as the Internet. These represent the vast majority of business travel transactions that are often point-to-point return air tickets, possibly with a hotel.
- They will need to have access to some sophisticated business travel support technologies that will help them compete with suppliers, especially the airlines and GDSs. Many GDSs have developed business travel support systems that enable travellers to take care of their own travel arrangements, but consolidate information and control at the companies head offices. Although these systems currently keep the travel agent firmly in the loop, there is no practical reason why this should continue, especially with the advent of electronic ticketing. Unless travel agents have their own capability to do this, they could well lose their business travel accounts to either the airlines or the GDSs.

Now, to help illustrate some of the points that I have made above, let's take a peek into this future world in order to explore a few of the issues in more depth. Take a hypothetical company whose management employees travel a fair amount as part of its business. Assume that this company has decided to use a travel management software package that performs all the functions that the company needs to run its own travel arrangements. Such packages are available right now in any event.

So, in this future world, the company's sales director, for example, can use their lap-top computer to check availability via access to one of the GDSs that has a Web site. They enter their travel requirements and from an availability listing chooses a flight. The system checks that the fare and class are within the company's travel policy and that all required fields have been entered for future management information purposes. Their personal travel preferences are stored in the system on their profile and the system uses this to make a seat reservation. Now the fun starts.

There are clear rules that the airlines have agreed regarding the choice of ticketed carrier. The ticketed carrier is of course the airline who will issue the ticket and collect the fare amount via the BSP (see Chapter 7 for more details on BSP). Even though the ticket may be issued electronically, it needs to have a designated ticketing carrier. All right, let's assume that our GDS chooses the correct ticketed carrier. The next decision to be made by one of the systems involved in this future world scenario is who will collect the funds for the ticket. Airlines do not usually collect funds for ticket sales direct from passengers. This is usually done via the BSP. So, whereas in today's world the ticket would usually be allotted to a travel agent's IATA number, in our scenario, this would not be available because no travel agent is involved. So, now we come to the first issue: 'Who will collect the funds for direct ticket sales when no travel agent is involved?'

You might think this is simple – it should be the ticketed carrier. Well, if it is to be the ticketed carrier then consider this. Depending upon the route flown and the first carrier on the ticket, the ticketed carrier could potentially be any of the world's airlines. So, assuming the company's air travel is quite extensive, it will need to expect payment requests from a large number of airlines, i.e. each ticketed airline flown by the company's employees. From the airlines' viewpoint, each airline will need to send out payment requests to many different companies with all the associated payment processing functions that this will involve, e.g. sending out reminders, reconciling payments received versus payments due, controlling cash flows and outstanding receivables, vetting the credit worthiness of companies and, finally, coping with company liquidations and bad debts. In other words an airline's worst nightmare.

OK, so let's assume that instead of the ticketed carrier having to collect the funds from the company, each company will negotiate with a single airline to produce all its tickets and collect all funds. This airline would then be burdened with quite a substantial administrative task. First of all, it would still have a number of company customers with whom it would have to deal direct. The airline would therefore be burdened with the same kinds of problems outlined above. Also, for the tickets that it issued on behalf of other airlines, it would enjoy a positive cash flow. However, would those other airlines be so happy. They would be carrying the passenger but would probably not receive payment until some time later. In other words they would be out of pocket for longer than at present. So, this scenario is unlikely to be acceptable by the airlines either.

Well, that just leaves us with the option of having some third party involved who will collect funds from the company and use the BSP system to settle ticketed carrier funds on a consolidated basis to each airline in IATA. Sounds familiar? The travel agent rears its head again. But what about the BSP organization itself? Couldn't it extend its clearing house role to include collecting payments from companies? Well, it's just possible but I don't think this is very probable. After all, BSP is owned by IATA, which is itself an airline association. Once again, the issue here is: 'Will the airlines want to get involved in payment collection from their customers?' I think that BSP has enough of a job collecting funds from a limited number of travel agents. Collecting funds from hundreds of thousands of companies would be a nightmare of even greater proportions.

This may sound like I have argued that disintermediation will not happen, at least not in the business travel air segment. However, that is not the real point. Although it seems there may continue to be a need for a travel agent, the role that the agent plays in the future will be quite different. In our future world scenario, the airlines will almost certainly not wish to pay the travel agent the current levels of commission just to act as a third party for BSP settlements. After all, in this future world virtually all the routine tasks are undertaken by software. What added value has the travel agent contributed? Answer – very little; just the settlement function. Certainly nothing that would justify a percentage of the ticket value.

So, the travel agents of the future will have to derive their incomes from some other source. This comes back to the question of added value. The travel agents' added value is their consultancy advice. This expert advice is not always needed for every trip. In the case of our fictitious company, the sales director did not need any advice – they simply booked their trip using their lap-top computer. However, there will no doubt be instances where they will need to ask an expert what the best airline and route would be for a more complex trip. This is where the travel agents come into the picture and is an area where they can develop a niche for themselves. The agents should be able to apply expertise to help the traveller plan the trip and select the most appropriate airline, route, departure timing, departure airport and other travel arrangements. For this consultancy advice, the travel agents can expect to be paid. The problem for the travel agents is that they claim to have been doing this for some time and at no apparent charge to the customer (indeed, in most cases the customer has actually had money back from a share of the agents' commission). Travel agents will therefore need to work very hard to develop true consultancy expertise. This will need to be delivered to such a high standard that the customer will be convinced that it is worth paying for.

But value can be added in other ways. It can even be added by semi-intelligent machine-based processes. Some Internet applications already use a special piece of sophisticated software called an 'Intelligent Agent' (incidentally, the word 'Agent', as used here, has nothing to do with travel agents – rather it is an entity that acts for the user's own interests). An Intelligent Agent falls into that class of computing known as software robots. These are clever computer programs that understand user's requirements and search the Web for items that appear to match what the user is looking for. It is quite possible that Intelligent Agents will form an integral part of new Web sites operated by the new travel intermediaries. Intelligent Agents should be able to understand what a consumer is looking for; for example, a holiday to Indonesia costing less than a certain amount, selected from four or five preferred airlines with departures from London Gatwick. Many other more detailed requirements and preferences could be included. The Intelligent Agent should then be able to search the Web for sites that contain the kinds of holidays that match these requirements and present them to the user. In other words, they do all the hard graft of signing on to relevant Web sites, searching them, recording the responses, signing off, going to the next site via a search engine and so on. However, despite the distinct possibility that they may find a niche in the travel industry of the future, I think it will be a long time before Intelligent Agents begin to replace travel agents.

So, not the end of the travel agents, but a radical shift in their role. Similar parallels can be drawn within the leisure side of the business. Straightforward holidays can be booked directly, possibly using one of the new distribution channels, such as the Internet. However, some people and some more complex holiday requirements will demand more specialist advice. Here, once more, there is a role for the travel agent. However, it remains to be seen how the travel agents will derive their income from this situation. Will holidaymakers expect to pay for expert advice from their travel agent? Will tour companies pay travel agents to offer advice on their products only? It appears possible that the environment could develop along similar lines to the financial services industry where agents are either tied to a company or offer independent advice on all companies. Although this appears to be getting away from the subject of IT in travel and tourism, these potential shifts in the underlying structure of the industry are being driven by rapid technological change.

TOUR OPERATORS

Tour operators are intermediaries between suppliers and either travel agents or consumers. They purchase products and services from travel suppliers and package them into a product that they market to consumers. So, what opportunities are there for using the Internet to provide electronic packaging mechanisms that could bring about the demise of tour operators as intermediaries? Well,

I guess like many of the other disintermediation issues, it is not quite as black and white as all that.

Undoubtedly, there are some consumers who are adventurous enough to use the Internet to construct their own packages. In fact, there are several software products around that support this very function. It is only a matter of time before they are available on the Internet. Say, for instance, that an Internet site was available that enabled consumers to: (a) browse an inventory of cheap hotel deals in a particular resort area; (b) browse a data base of associated seat-only air services; and, finally, (c) add a few optional sightseeing trips to their itineraries. At the end of such a process, the consumers would have assembled their very own personalized packaged tours (also known as an, Independent Tour (IT)). It would only remain for them to print the itineraries, pay for the services and receive their documentation either through the mail, at the airport or electronically. All without purchasing a packaged tour from a tour operator – or is this really the case?

Why couldn't this kind of Internet site be run by tour operators? After all, they are the ones that have the relationships with the hotels and other services in the destination areas; and they often have their own charter airlines to these same destinations. So, maybe the only function that is at risk due to electronic commerce, is the packaging of these individual components for a consumer. Well, when you think about it, this is the very area that gave rise to most of the current problems for tour operators. Problems such as the decision process required to guess what arrangement of components will make a package that appeals to the widest number of consumers. The package holiday companies would like nothing better than for everyone to select their own combination of travel products from their inventories. Think of the massive reductions in brochure printing costs, advertising and agency commissions that this could bring.

However, I think it will be a long time before sufficient numbers of consumers become this sophisticated and confident to have a real impact on tour operators. However, it will undoubtedly happen, the only question is: 'When will it happen?' So, tour operators need to consider their strategic options and start experimenting with this new technology if they are to be capable of adapting to the new electronic business world of tomorrow. In fact, a very good book that examines this issue in more detail, as well as several others in the area of tour operations in the UK and Germany, is published by DeutscherUniversitats Verlag by Karsten Karcher entitled *Reinventing the Package Holiday Business*.

DISTRIBUTION SYSTEMS

GDSs and HDSs are intermediaries between travel suppliers and travel agents. The GDSs have their origins in the airline CRSs that were themselves originally designed to enable airline sales staff to sell seats on their flights. Over the course of time they were first distributed to travel agents, then enhanced to include access to hotels and car rental companies and, finally, consolidated with multiple CRSs to form what we now call GDSs. Finally, the interconnection technology that linked GDSs to hotels was vastly improved by means of specialist industry switches called HDSs. What is the next stage in their evolution? As you will see from the remainder of this chapter, many of them have developed an Internet interface of some form or another. Some of the HDSs have broken new ground by turning the tables on GDSs and offering consumers and travel agents their own hotel-based Web booking services that also include GDS access. Generally speaking, access paths to the consumer via the World Wide Web at present keep the travel agent firmly in the loop – but for how long? It seems quite possible that new intermediaries can offer a whole range of booking services to consumers without using GDS technology or travel agents. But, first of all, let's consider the future of GDSs from an airline's viewpoint.

GDSs

An airline's CRS is quite capable of handling the bookings of seats not just for its own flights but also the flights of virtually every other airline. The precise functionality of how CRSs handle reservations involving other airlines is governed by their respective levels of participation (see Chapter 4 for more details on this). Airlines must pay a fee for their participation in GDSs and this is usually levied by means of a booking fee. Again, this is one of the major components of their distribution costs that I analysed in more detail in the preceding section: and because distribution costs have a direct and substantial impact on profitability levels, any opportunity to reduce them needs to be carefully considered by airlines.

The Internet offers airlines a direct sales channel to consumers. Many airlines have developed their own sites, some of which also support booking and payment functions. The key question is: What effect will this have on their participation in GDSs? It could well be that as time goes on, a substantial proportion of their bookings could be derived from their own Internet sites or indeed from the new intermediaries (see next section for more details on the new intermediaries). Handling bookings directly via this channel has the dual benefit of: (a) eliminating GDS booking fees, and (b) eliminating travel agent commissions. This is a very sensitive subject for airlines and one on which they are unlikely to be very forthcoming. The reason for this coyness is that dangerous talk costs revenue. If airlines were thought to be considering this path they would disenfranchise their GDS as well as their travel agency relationships.

However, it is nevertheless the case that a direct Web site offers significant benefits that cannot afford to be ignored by the airlines. This explains why these sites are nearly all currently described as being quite separate from the main distribution channel and in many cases require the consumers to collect their tickets from their nearest travel agencies. But not all such sites require the consumer to do this. Some offer full payment processing with ticket collection on departure. The point is, it is rather like an insurance policy. Having an Internet site allows airlines to become familiar with the technology, to build a loyal client base (albeit a small one initially) and to establish some small degree of independence from both the GDSs and travel agents.

Now, let's consider the situation sometime in the future when most airlines have developed their own Web sites for information and booking purposes. Let's also further assume that many more people have access to the Internet and are using it heavily. Consider the situation from a consumer's viewpoint. Take someone who wants to fly to some foreign destination. Which airline Web site will they access? One might start with the national airline of the destination country. However, with competition and deregulation, national airlines are rapidly becoming a thing of the past. Even if they weren't, they do not necessarily always offer the cheapest or the best deals. The poor old consumer could, in this scenario, spend a great deal of time visiting one airline site after the other, looking for a suitable deal.

Far better surely, to have a special kind of airline search engine into which you enter your basic requirements and it finds several airlines that have deals to suit your needs. Again, doesn't this sound familiar? The old GDS concept rears its head once again. However, the guise is somewhat different. Instead of this new generation GDS being the main switching point between the airlines and other travel service companies, it is much more akin to an Internet search engine. It would need all the functionality provided by a search engine but with more sophisticated links to other sites, principally airline sites. These links would enable it to collect, disseminate and present options to consumers that would allow it to direct them to the airline best suited to their needs.

But this is not a scenario that the airlines particularly relish. It takes away the consumer influencing part of the buying decision process and vests it in a separate company over which the airlines have little or no control. Then there is the bias rules and regulations to be considered. Who would police these new Internet-based airline search engines? Enforcing rules on Internet service providers is a tricky business that so far has not been tackled successfully. How, for instance, could the EU enforce its unbiased rules for GDSs on an airline search engine located in say, Malaysia?

However, the stakes are high in this game. If an airline can develop an excellent Web site that proves highly successful and popular with consumers then it is going to generate a substantial amount of revenue: and this revenue is potentially free from GDS booking fees and travel agent's commission. Once this begins to happen, the writing is on the wall for the GDSs. But don't let's forget that most of the GDSs are currently owned by airlines. Having said this, one can't help but notice the gradual divesting of GDS ownership by airlines. American Airlines' parent company still

```
Room rate charged by hotel ..................................................................  100.00
  Less:              Travel agent commission at 10 %      −10.00
                                    GDS booking fee       −3.55
                          Hotel switch processing fee     −0.50
                             Booking service provider
     (e.g. representation company or hotel chain headquarters)  −9.00
                              Credit card service fee    −3.50
                          Corporate rate discount on room −10.00
  Total deductions ...........................................................  −36.50
  Hotel income ................................................................................  63.50
```

Figure 5.1 The economics of hotel bookings

owns over 50 per cent of Sabre, but this is a lot less than its total ownership situation as of a few years ago; and there are several other examples where airlines can be seen to be reducing or selling their equity investments in GDSs. So, quite frankly, who knows what will happen? I think it all depends simply upon how successful the new airline Web sites are. Only time will tell.

HDSs

Now, what about the view of GDSs from the hotel industry's viewpoint; and in particular, the view of HDSs and their hotel owners. At present around 28 per cent of all hotel bookings are generated by travel agents. In the USA, 80 per cent of these travel agency hotel bookings are made using GDSs. In Europe the figure is far lower at 35 per cent and in Asia Pacific it is just 15 per cent. The other 72 per cent of hotel bookings are generated by consumers themselves either via toll free telephone calls to specialist reservation centres or by direct contact with the hotel. To illustrate the pressures for disintermediation from the hotel industry's point of view, let's take a somewhat extreme example. Take a hotel booking that is worth US$100. Let's first of all assume that the booking was made by a business traveller who used a travel agent. The agent booked the room via a GDS and the customer paid using their credit card. The economics look something like those shown in Fig. 5.1.

At 36.5 per cent, the overheads of this booking channel appear excessively high from the hotel's viewpoint. Even if we consider direct bookings received via the toll free telephone service channel, the hotel is still looking at some horrendous costs of sale. It is estimated that voice calls made by consumers to toll free telephone booking centres average between US$10 and 15 with a frequently reached upper level of US$30. Clearly, there are enormous pressures on hotels to seek alternative distribution channels for their products. The Internet is one such channel and companies like TravelWeb and Thisco offer a far cheaper route to market than the classical GDS/travel agent combination that has been the established way of doing things for so long. Many hotels already participate in HDSs like Thisco and to use this as a platform for bypassing the GDSs and ultimately, the travel agent, is an attractive scenario. If we take a hotel with 100,000 bookings per year and assume that it could save US$13.50 per booking then this could generate US$1.35 million each year. Now, I accept that a hotel is unlikely to be able to realize quite such a large saving, at least not in the early years of this new distribution scenario. But the important point is – this is the target that seems to be attainable by hotels, and it helps explain the rationale and pressures that are the principal driving forces behind GDS disintermediation.

TOURISM

Tourist offices, often also known as destination service organizations, are intermediaries as well.

They are intermediaries between national tourism organizations, which are often sponsored by governments or at least local governments, and remote tourist offices in overseas locations. The general pattern here is that the central government tourism organizations are charged with developing and executing marketing plans that promote their country or region overseas. This usually involves: (a) building a data base of national information and supplier details, and (b) distributing this to overseas tourist offices where information is made available to consumers and travel companies in a pre-defined area. These overseas offices receive local enquiries either by telephone, mail or from walk-in clients. Enquiries are serviced by access to the reference data and by distributing booklets and pamphlets as required (see Chapter 2 for more details on how IT is used to support tourism in this way).

It is the Internet that poses disintermediation in tourism. This arises from the growing number of Web sites devoted to tourist information. These sites are becoming quite sophisticated and many contain all the information that potential inbound visitors and travel organizations would want to know. Those sites that also offer on-line booking of accommodation services and events are particularly attractive to end users in other countries. The key question here is: To what degree will these Web sites impact local tourist offices? It is highly unlikely that these sites will cause the ultimate demise of overseas tourist offices, but it could have a major bearing on the size and distribution of offices.

The new intermediaries

I have used the term 'new intermediaries' to encompass any Internet site that offers a full range of travel services directly to consumers. In some cases these new intermediaries are backed by an existing distributor of one or more major travel products. However, what makes them a new intermediary in my terminology is that they offer a range of other travel products, not all of which are provided by the site's main sponsor. In other words, they may be viewed as an electronic travel agent offering a wide range of travel services and travel-related information.

It is also the case that some of these new intermediary sites use travel agents for post-sales customer servicing. The fact that they use travel agents in this way does not dilute their potential for affecting disintermediation, it does not make them any less important to the direct distribution of travel and tourism, nor does it mean that they will not have a significant impact on the classical travel agency. The kind of travel agent that has formed an alliance with these new intermediaries is just the type of agent that I think we will see more of in the future. Those agents that stick rigidly to so-called tried and tested methods based purely on face-to-face high street sales are the ones most likely to be affected by these new intermediaries.

EXPEDIA

Not many people know that Microsoft is a travel agent – but it very definitely is. Its Internet site, branded Expedia (Fig. 5.2), is one of the most important examples of the new generation of travel intermediaries. So, I would encourage any travel agents who do not think the Internet will have an impact on their businesses to take a good look at Expedia. It represents what is arguably the first real electronic travel agency aimed directly at consumers. It is a Web site that was launched in the USA on 22 October 1996 and is already highly successful. In the early months of its launch it sold an average of 1,000 air tickets each day generating over US$1.25 million worth of air travel turnover per week. Along with this substantial volume of electronic air sales goes a significant amount of related hotel and car bookings. In fact the proportion of non-air sales made via Expedia is higher on average than the typical business profile of USA travel agents; and with a 20 per cent growth rate, Microsoft's business is already beginning to make serious inroads into the USA travel industry. At the time of writing this book, Expedia was only distributed to domestic consumers in North America. So, although anyone with an Internet connection could access Microsoft's USA site, only consumers actually resident in the USA and Canada were allowed to participate in the transactional booking functions of Expedia. However, Microsoft is now implementing its Expedia service outside the USA with other major countries including the UK, Germany and Australia.

So, it is evident that Microsoft has entered the travel business in a very serious way. Its Web site, branded Expedia, incorporates a vast amount of travel-related information that is available in both HTML pages of text and graphical images recorded in full colour. This information is stored in several relational data bases that are indexed and accessible via powerful search engines. Expedia is also linked to the Worldspan GDS via a booking engine interface that provides consumers with access to the full range of published scheduled air flights, hotels and car rental services. All these travel products and services are available via a very user-friendly front-end interface that may be accessed using most secure Web browser software products including of course, Microsoft Explorer.

Microsoft's commitment to its travel business is characterized by the 120 staff that it dedicated to Expedia in 1997 and by its possession of an IATA licence. Microsoft is therefore a fully fledged travel agency in its own right and makes regular payments for air sales via the USA equivalent of IATA's BSP, just like any other USA travel agency. At present, for purely logistical reasons, Microsoft has outsourced its USA travel servicing functions to World Travel Partners (WTP), a USA travel group based in Atlanta, Georgia. WTP provides Microsoft with services that include the issuance of travel documents for Expedia customers, including air tickets. These are mailed to customers' home addresses using the regular USA Mail postal service or special courier delivery services as necessary, e.g. Federal Express. However, with the increasing use of electronic ticketing (see Chapter 3), this aspect of WTP's service may well become less important as paper tickets decline in use. WTP also provides an after sales service, or post-reservations support function, that provides customers with classical travel agency services delivered via the telephone and electronic mail.

Travelling with Expedia

Microsoft's strategy on post-reservations support for international markets seems to be based very much on the USA model. In each country or region, a travel company is selected as a customer service partner. In the UK, for example, the travel partner is A. T. Mays. A. T. Mays has worked with Microsoft to develop a travel support function that includes several interesting facets (Fig. 5.3). Besides providing post-reservations support and fulfilment operations, A. T. Mays has built a data base of consolidator air fares and other travel-related information on a Web server that is located on the Microsoft network in Redmond Washington where Microsoft houses its headquarters and operations centre. It is these kinds of partnerships that are behind the real power of Expedia. Let me illustrate this by walking you through how a consumer in an international area (I've used the UK as an example here), interacts with Expedia to make their own travel arrangements.

Registration

To use Expedia for booking travel products, a consumer must first register themselves on the site. It is not compulsory to enter plastic card information, although this may be recorded and helps speed the booking process. A consumer may also elect to record their travel preferences within their own personal profile as part of the registration process. This enables the traveller's likes, dislikes and preferences to be entered automatically into booking fields at the appropriate time – a good example of Expedia's labour saving features.

General trip planning

Once registered, a consumer may browse the information stored within Expedia. This is an enormous data base of travel-related information that is maintained by Microsoft staff. Besides maintaining up-to-date information on destinations and all kinds of travel opportunities, Expedia also features chat sessions where a consumer can log-on to an electronic meeting place hosted by one or more experts in certain travel subject areas. The venue for these chat shows is published on Expedia and allows the consumer to choose when they wish to log-on and participate in the session. During a chat session, each participant's questions and observations are put to the host via a Forum Manager and are also distributed to all other consumers participating in the session. Microsoft uses full-time Forum Managers to provide its Expedia customers with expert travel consultancy on many

Figure 5.2 The Expedia home page (above)

Figure 5.3 The Travel Agent page (above right)

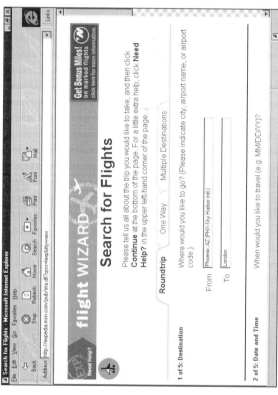

Figure 5.4 Flight Wizard

198 INFORMATION TECHNOLOGY FOR TRAVEL AND TOURISM

subjects and destinations. Much of the subsequent feedback and information distribution for these sessions is handled via electronic mail.

There are many general trip planning functions that are either part of Expedia or that may be found elsewhere within the Microsoft Web site. One of these is The World Guide. This presents the consumer with a simplified map of the world, divided into major regions. If a region is clicked, Expedia shows a more detailed map of the chosen region. Again, this shows a number of areas, each of which may again be clicked to show a lower level of detail. At the lowest level, textual information and pictures of famous places are shown. More information may be obtained by linking to another Microsoft site – the Encarta World Atlas On-line.

Another interesting tool available within Expedia's trip planning portfolio, is Mungo Park. This is a diverse collection of travel stories and information on the more far flung places of the world, which is branded by Microsoft as an adventure travel magazine. It even includes information on current and past expeditions to certain remote regions undertaken by specialist teams of explorers. Updates and reports on these expeditions can be viewed at any time. These often include dispatches transmitted from an expedition member's lap-top PC via a satellite link to the World Wide Web. Most of these dispatches are archived within the Mungo Park site for all to see.

Expedia's on-line data base pages make information-rich content directly available to the consumer. This can be a powerful way for a person to learn about a destination and plan their own itinerary. In fact, it is precisely the kind of information that people visit travel agents for. However, not only is it now freely available to anyone with an Internet connection but it can be obtained without the hassle associated with high street shopping; and what's more, it goes further than the average travel agent's capabilities. It can, for example, provide the more intrepid would-be holiday-maker with the kind of specialized information they invariably need to plan an adventure holiday in some far flung place; and adventure holidays are a growing sector of the travel market.

Once the consumer has decided on the kind of trip they would like to make, they select the Expedia booking function in order to plan their trip in greater detail. The following sections describe how the major travel products are booked using Expedia. Each product selected by the consumer, whether it is booked or not, may be added to a personal itinerary file. The itinerary may be built-up over one or many Expedia booking sessions and is always available for viewing by the consumer. At the end of the booking process, it represents a detailed itinerary that may be printed using the consumer's own printer linked to their Internet browser PC.

Let's look at each of the main booking functions and products in more detail, starting with air travel which is supported using Expedia's Flight Wizard.

The Flight Wizard
Having decided upon an outline itinerary, the next step is for the consumer to do some detailed trip planning, obtain some prices and availability and then start to build a more detailed trip itinerary. Let's begin with the air travel options that are supported by Expedia's Flight Wizard (Fig. 5.4). In order to deter non-serious users and check a consumer's details versus their registration, Expedia requires the consumer's zip code, i.e. postal code, to be entered prior to processing a reservation. Next, the consumer enters the destination of the first leg of their trip. This may be expressed either in terms of a full city name, an abbreviated city name or an airport code. Expedia assumes that the consumer is travelling from their nearest home airport, although this may of course be changed. Next the date and desired departure (or arrival time) of the flight is entered. Finally, the consumer may choose the sequence in which Expedia will show their availability display. This may be either: (a) all flights in ascending sequence on price, or (b) all flights in sequence on the desired departure time and minimum flight time. All fields are presented to the consumer in the well-known Windows style that makes abundant use of drop-down lists, check boxes and radio buttons. This makes the reservation requirements easy for an untrained consumer to define accurately.

The next part of the process is, to my mind, one of the most powerful of all Web-based flight booking functions currently available on the

Internet. I'll therefore explain the steps that Expedia takes in order to show an availability display in a little more detail:

- **Build flight requirements** The consumer's flight requirements are checked and stored by Expedia within Microsoft's Redmond based travel Web server. Once the consumer requests an availability display of their stated itinerary, Expedia formulates a data base query that it sends to the travel Web server housing the air fares information built by A. T. Mays.
- **Assemble consolidated fare options** The travel Web server receives the request for availability and, first of all, queries the data base of consolidated air fares. It tries to find all fares on the data base that match the consumer's preferences for city pairs, dates, times and other details. All matches are assembled within the travel Web server. For each of the selected flights, an availability message is constructed and sent to the Worldspan booking engine (see also the separate section on Worldspan in this chapter).
- **Obtain flight availability** The Worldspan booking engine is used to obtain the availability of the specified flights. These will be very specific availability requests that specify precise classes of seat reflecting the consolidated fare contracts. All such flight details are returned to Microsoft's travel Web server along with associated flight operating details.
- **Build available flight display** The travel Web server then merges the information provided by the Worldspan booking engine into a list of flight details that match the consumer's stated requirements. The result of this query is a mini-data base of flight information built specifically for the consumer. It contains both contracted fares, i.e. consolidated fares, as well as scheduled fares.

  This is an important feature that at present, is unique to Expedia. Most other booking engines show only the scheduled air fares for flights available from GDSs. However, Expedia also includes specially negotiated lower priced fares and their availability.
- **Show flight options** A summary of the available flights that match the consumer's stated requirements is then presented on a Web page with a scrollable list in the sequence requested. In the case of a listing by fare price, it shows the cheapest flights first, which are usually those featuring one of the consolidated fares specially contracted by A. T. Mays. Any stop-overs or connecting flights are clearly shown. These flights are designated Expedia Special Fare. Then further down the list will appear the scheduled flights that may be more direct and convenient, but are often more expensive. Scheduled flights are designated by means of a small graphical image of the airline's logo.

All of the above is undertaken in a matter of seconds, without the consumer being aware of the detailed processing steps involved. The consumer simply sees the results in the form of an easy to understand Web page listing the flights that match their requirements in the sequence requested by the consumer. In most cases, a number of flight options will be shown on this Web page, which is fully scrollable.

Each flight shown on the summary page may be viewed in detail by just clicking on a Web page 'button'. When this is done, the particulars of the selected flight are shown on a separate page of its own. Each leg of the flight is shown in detail including: aircraft type, flying time, check-in time, meal options and many other key items of information. In addition to this, the conditions of the selected fare are also shown. This is very important and the conditions are shown in full detail, including: applicable fare rules, usage restrictions, implications of post-booking itinerary changes, lost ticket conditions and so on.

Booking

When a flight has been chosen, one of the first things Expedia requires the consumer to do is to accept the conditions of the selected fare chosen for the desired flight. This is accomplished by requiring the consumer to enter a check in a box marked 'signifies acceptance of conditions'. At this point Expedia offers the consumer three options regarding payment:

1. The flight details may be saved in the itinerary but not booked. This action does not reserve a

seat on the chosen flight but records all the details in the consumer's itinerary, which is stored in Expedia. All such stored itineraries may be retrieved at any point in the future and either cancelled or booked by one of the other two methods described below.

2. An option on the chosen flight may be taken. This option is recorded by Worldspan with an associated time limit. The option is automatically cancelled by Worldspan if not confirmed by midnight on the following day. To take an option in this way, the consumer must enter the last four digits of their card number. Although payment is not actually taken at this point, the entry of card information denotes a serious intention on the part of the consumer to eventually make a firm booking and deters frivolous abuse of the system.

3. The flight may be booked and payment details entered. Payment may be collected in one of two possible ways: (i) by entry of the consumer's card details, which are then used to pay for the ticket; or (ii) by selecting an option to pay for the ticket via a telephone call to the designated Expedia travel partner, which in the case of the UK is A. T. Mays. At present in the USA, over 90 per cent of customers who book travel products choose to enter their card details into Expedia rather than telephoning WTP.

In either case, following successful payment by the consumer, the ticket will be printed by A. T. Mays and despatched to the consumer's home address.

Tickets will only be despatched to the location that is registered as the cardholder's address. In the USA, an address verification system (AVS) allows a consumer's address as registered by Expedia to be checked against the cardholder's address as recorded by the card company's computer. However, this functionality is not presently available in the UK, or for that matter many other countries outside the USA.

To complete a booking, the consumer then specifies their personal details and preferences, such as the kind of seat they would like, the desired meal option and the frequent flyer number. However, virtually all of this information may be pre-stored in the consumer's personal travel profile held by Expedia. If this is the case then all the fields that are required to complete a booking will be populated automatically by Expedia from the profile. Once this has been done, the booking is complete and the consumer may elect to either quit the system or continue building their itinerary with other travel services, such as making bookings for hotels and car rental.

The Hotel Wizard

Microsoft has gone to great lengths to develop a comprehensive and up-to-date hotel information system and booking function, which is now an integral part of Expedia. The primary source of Expedia's hotel information is the Worldspan GDS (see Chapter 4). All of the information about hotels that is available in Worldspan is actually provided by the hotels themselves. A preliminary review by Microsoft, undertaken before Expedia's launch in the USA, highlighted a problem – much of the hotel information in Worldspan was out-of-date and required updating. So, before launching Expedia, over ten full-time Microsoft staff spent several months working with Worldspan's data management group and telephoning its participating hotels to clean up the data base. The team managed to review and update all hotel-related information prior to Expedia's launch – a considerable task. Procedural processes are now in place to ensure that Worldspan's hotel information is maintained and quality controlled as part of the day-to-day operation. The result is a powerful and user-friendly hotel booking capability that is an integral part of Expedia.

To add a hotel booking to an itinerary is very simple and straightforward. I would argue that it is far easier than trying to do the same thing via a high street travel agent. First, the consumer selects the Expedia Hotel Wizard. This can be done in relation to an existing air booking, in which case the system already knows much about the desired service, e.g. the city, the dates and the arrival time.

Once this information is available, either by direct entry using the familiar windows style GUI or from information previously entered, the Hotel Pinpointer may be selected. This is a very useful tool that helps the consumer locate a hotel in the area where their business trip or holiday is to be undertaken.

The first thing to be displayed by the Hotel Pinpointer is a Web page that on the right-hand side shows a map of the city in which the hotel is to be booked. Each hotel in the city is shown on the map by an unfilled small circle. A zoom feature allows the map to be expanded to show a wider area or focused down to show the locality of desired interest. On the left-hand side of the screen is a scrollable list of hotels in the city or area shown by the map. When a hotel is selected from the scrollable list by clicking on the hotel's name, a small red circle appears on the map showing the location of the selected hotel. This is a very powerful feature of Expedia's Hotel Wizard that enables an untrained consumer to make an effective decision on the best choice of location for their hotel in a given city. It also provides walking distances and times between the chosen hotel and any specified point in the city. This is accomplished very easily: having chosen a hotel, the consumer clicks on a point of interest on the map, say their office or a particular theatre. A heading box on the map then shows the walking distance and estimated walking time from the chosen hotel.

When a hotel has been chosen, another option within Hotel Pinpointer allows the consumer to view all relevant details that describe the hotel, such as: the address, the number of rooms, the facilities and amenities available to guests, the forms of payment accepted and the room rates. The choices now are either to book the hotel or to add it to the itinerary.

As with an air booking, if a reservation is required, Expedia will first ask the consumer to accept the terms and conditions that apply to the room and rate chosen. Then the required booking details are either automatically completed from the consumer's profile or entered field by field. Finally, the hotel room is booked via the Worldspan GDS Internet booking engine.

Car Wizard

This works in a similar way to the Hotel Wizard. The consumer chooses from a list of car rental companies or requests Expedia to show a list of car rental options in ascending order of price. Each option can be shown in more detail down to the level that includes information such as the type of car, its characteristics and rental rate. Again, the terms and conditions are presented in full for the consumer to review and accept prior to booking. A car rental service can be selected and either: (i) booked using the simple windows style GUI and the Worldspan booking engine, or (ii) simply added to the itinerary for booking at some future point in time. A related function that assists a car rental customer with their choice of route is Microsoft's Address Finder.

Address Finder

Microsoft owns the Autoroute software package and associated mapping data base. Expedia has packaged this with its data base of travel information to provide support for planning fly–drive holidays. This has been bundled up into a comprehensive mapping data base of over 500 destinations.

When consumers first log-on to the Address Finder, they select a destination and are presented with a 360 degree revolving image of a famous landmark or scene. This is an attractive way of introducing Address Finder's rich store of destination information, which includes country, region and city maps. In the USA, an address can be located by entering a zip code. The Autoroute function uses this to retrieve the appropriate local map and displays it as a Web page for the consumer with an indication of the desired location. This can be used to determine the best way to reach a destination by car.

Post-reservations support

Once a consumer has used Expedia to research and plan their trip and the booking process has been completed, Microsoft's travel partner comes into the picture to provide post-reservations support. This includes many servicing functions, the most obvious of which are payment processing and the delivery of travel documentation to Expedia's customers.

However, even before these events take place, there are some important customer servicing functions that need to be undertaken. One of the most important of these is the management of GDS queues.

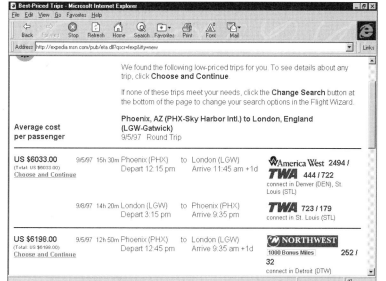

Figure 5.5 Flight Wizard – more flights (above)

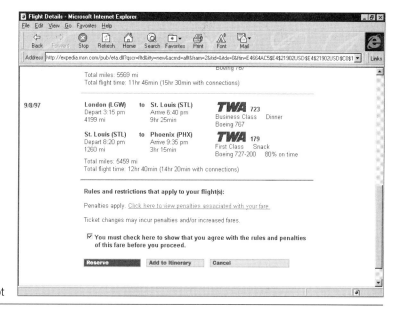

Figure 5.5 Flight Wizard – check to accept

When a reservation has been made by an Expedia customer, a PNR will have been created within the Worldspan GDS (see Chapter 4 for a more detailed explanation of Worldspan's booking system and PNR). When an airline needs to communicate with its customer it does so via the queue system. Queues are GDS tools that have been designed for use by travel agents (see Chapter 3 for more details). This aspect of customer servicing is little different with Expedia as compared with standard travel agency practices. Any changes to a customer's flight details are noted in the PNR by the servicing airline and a copy is placed on the travel agent's Worldspan message queue. This queue is 'worked' by Microsoft's travel partner, which, in the case of the UK, is A. T. Mays. Travel consultants in A. T. Mays review the Worldspan queues regularly and note any

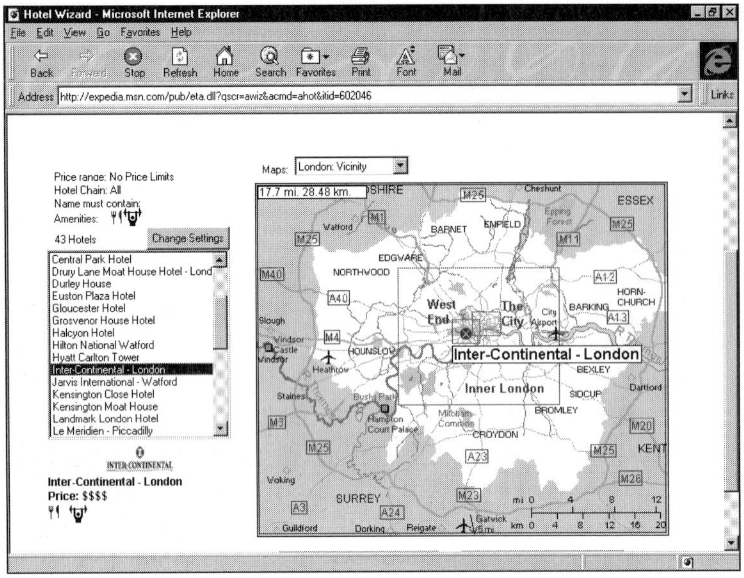

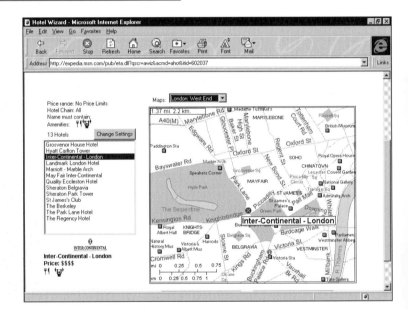

Figure 5.7 Hotel map – wide scale

Figure 5.8 Hotel map – zoom

significant changes. These are communicated to the customer either via e-mail or in the case of more urgent changes, by means of a telephone call.

Expedia and the future

The book is still open on how successful Expedia and similar Internet-based travel sites will be in the future. The initial indications are, however, encouraging for Expedia and other new intermediaries. But one of the issues that has only recently been identified is the ratio of 'look to book' transactions handled by GDSs like Worldspan. The price travel suppliers and GDSs have to pay for receiving more bookings directly from consumers is the increased overhead on computerized reservation systems.

By their very nature, consumers are less trained

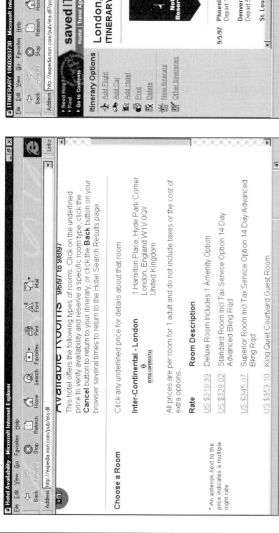

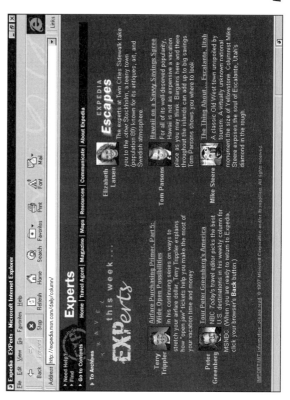

Figure 5.9 List of rooms and prices (above)

Figure 5.10 The itinerary (above right)

Figure 5.11 Experts magazine

THE INTERNET 205

in the complexities of travel than travel agents. Therefore they tend to do a lot more browsing and a lot less booking compared with a travel agent. But in so far as the GDS systems and networks are concerned, this manifests itself as an enormous increase in transaction volume that may well be out of alignment with historical booking ratios. At the end of the day this means higher costs for the GDSs and their airline participants due to the need for larger, more powerful computers and higher speed communications lines. Although this may be offset to some extent by the improvement in the price/performance ratio of IT, there remains the spectre of increased processing overheads and higher operating costs.

This issue will no doubt continue to be addressed over the next few years as electronic commerce grows and the new intermediaries develop enhanced capabilities for their interactive consumer networks.

An example of one such future booking facility being considered by Expedia is the provision of alternative options for those customers booking airline seats. At present, when a booking request is made, Expedia uses Worldspan to check availability on just the stated city pairs and dates. However, in the future a facility may be added that would ask the consumer a question of the form: 'Although an economic flight you have requested is not available on the date or between the city pairs you have specified, a good alternative is available on another day or between other cities close to your ideal choice. Would you like to consider these options?' This kind of functionality is rather complex to program and needs a lot of consumer research before it can become viable and/or practical. However, if it could be introduced, it would make the use of Expedia's Internet travel booking site that much more attractive to consumers.

Another enhancement that may be under consideration is the provision of contracted rates on hotels and car rental companies. This could be done in a similar way to the existing consolidated air fares data base facility. In the UK, Microsoft's partner A. T. Mays or even a specialized hotel company, could build a data base of contracted hotel rates. These would be special rates with a low price tag but with certain conditions only available to Expedia's customers. These special rates would be created and distributed in a similar way to contracted air fares with booking functions supported by the Worldspan GDS. Contracted car rental rates could work in a similar manner.

TRAVELOCITY

Travelocity (Fig. 5.12) is the name of Sabre's Internet site, which was established jointly by Sabre Interactive and Worldview Corporation in October 1995. These two key players combined forces to provide a powerful and popular Web site comprising over 200,000 pages, which was launched in March 1996 and that by November 1996 had already registered more than 450,000 members and received over 4.1 million visits. Travelocity is a 'do-it-yourself' travel site aimed at both individual leisure holiday-makers and business travellers. The two companies driving this new URL, known as **http://www.travelocity.com** are:

- **Sabre Interactive** This is a division of The Sabre Group and besides running the Travelocity product, it also markets EasySabre, which is described in more detail in Chapter 4 (see GDS – Sabre).

  Although Sabre Interactive is totally responsible for Travelocity, it buys specialist Web publishing services from Worldview Systems Corporation. This combination of expertise is one of the key success factors that contributes to Travelocity's broad appeal to consumers around the world.

- **Worldview Systems Corporation** This is a joint venture whose participants are Ameritech and Random House. It was founded in San Francisco in 1987 as an information publication and distribution company focusing on the travel industry. It provides up-to-date information on local events, attractions, dining, business services, night-life and shopping in thousands of destinations world-wide.

This new business comprises two main parts: (a) a consumer-facing world-wide Web site, and (b) a Web marketing business. Each of these two aspects

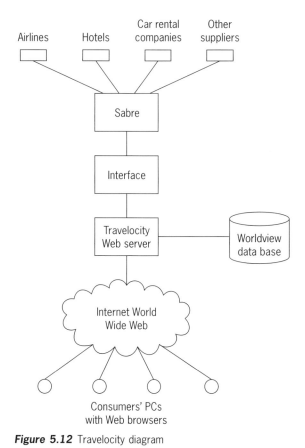

Figure 5.12 Travelocity diagram

of Travelocity are explored in more detail in the following paragraphs.

The consumer-facing side of Travelocity

Travelocity is a USA Web site that offers full access, on a controlled basis, to the Sabre system and is controlled by the Sabre Interactive subsidiary within The Sabre Group (see Chapter 4 for more details on Sabre's corporate structure). This site provides services that may be generally grouped under four main headings: travel reservations, destination information, chats and forums, and merchandise; each is explored below.

Travel reservations

To make a booking on Travelocity the user enters or selects the destination city involved in their itinerary. The user then enters other pertinent information, such as the number of travellers, company preferences and special rates that apply. Based on this information, Travelocity displays several choices. Additional details on rules and rates may be viewed before finalizing the booking. Travelocity then creates a detailed summary of each reservation, including a confirmation number received from each supplier. A complete itinerary may be printed at any time. The main products supported are:

- **Air**   The link between Travelocity and the Sabre GDS provides unbiased access to over 700 airlines of which direct bookings may be made on 400. Booking information (Figs 5.14–5.17) is automatically linked to a comprehensive data base of destination information maintained by Worldview. The Flightfinder function automatically searches for the lowest available fare between multiple cities and displays the three lowest cost itineraries. There are also diagrams of standard configurations for the most popular aircraft.
- **Hotels**   Travelocity provides real-time availability and rate information on over 32,000 hotel properties world-wide. The rates quoted by the participating hotels may include the following types: corporate, family plans, promotional, standard, senior citizen, convention and weekend specials. Certain hotels show street level location maps that are provided by Vicinity Corporation. Others, such as Hilton Hotels Corporation, Marriott International and the Ritz Carlton Company, show colour photographs of their properties.
- **Car**   The car rental booking feature of Travelocity provides real-time availability and searches for the lowest rates for more than 50 car rental companies world-wide. A variety of rates can be quoted including corporate and special rates, such as AAA and AARP, which are displayed from the lowest to the highest price. Users may book various types of vehicles, including sports cars and luxury cars.
- **Vacation packages**   This shows pre-negotiated accommodation packages together with photos of the destination, the accommodation and the facilities available. Alternative packages are searchable on destination and interest category.

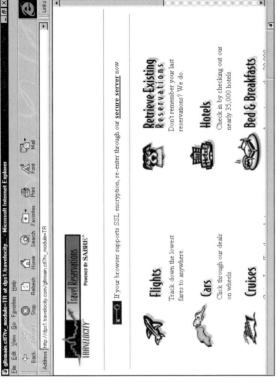

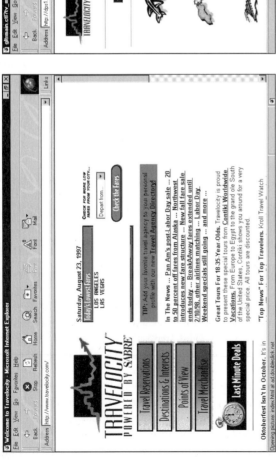

Figure 5.13 The Travelocity home page (above)
Figure 5.14 Travel reservations page (above right)

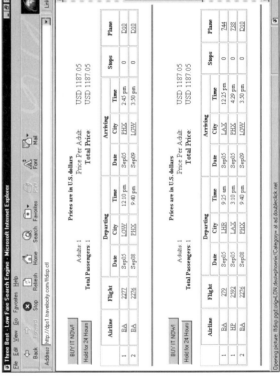

Figure 5.15 Air options page

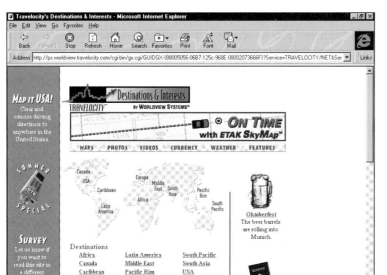

Figure 5.16 Mapping

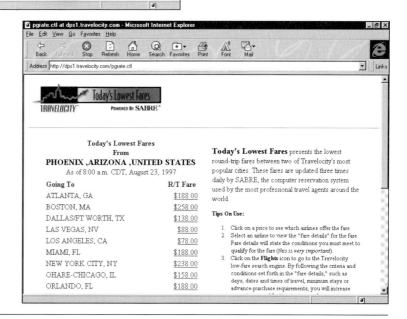

Figure 5.17 Lowest fares

Destination information

But Travelocity is a lot more than simply an interface to the Sabre GDS. Travelocity is truly a consumer-facing product that contains a great deal of searchable travel-related material. Examples include: (i) video and sound clips on over 22,000 destinations around the world, which provide in-depth facts and figures on a variety of subjects; (ii) merchandising services, which allow consumers to use Travelocity's Internet capabilities to purchase products and services with payment on major credit and charge cards; and (iii) articles written by experts on travel-related subjects.

The depth of Travelocity's data base may be illustrated by the following statistics that summarize the types of information available: 9,500 restaurants; 1,400 museums; 11,000 bed and breakfast properties; 3,000 theatre, dance and music performances; 4,500 condominiums; 13,000 golf courses; and thousands of exhibits, shows and festivals. Some of this information is provided by other sources with whom Travelocity has a commercial agreement. For example: Corel Professional Photos shows images of unique cultures and activities; Hotelogic shows amenities and contact information for over 30,000 hotels; IVN Communications provides more than 1,500 custom video clips, multimedia displays and still images; Magellan Geographix shows a library of city maps; and the Weather Service Corporation provides weather maps and forecasts for each continent.

Chats and forums

The site also hosts an interactive communication channel for users, which is best described as 'chats and forums'. This enables users to swap ideas and ask questions on a wide range of travel topics via Travelocity's bulletin boards. Travelive, a regularly scheduled live chat conference, allows users to discuss topics with leading travel experts. There are also features on places, people and travel trends around the world, with spotlight articles profiling a destination or topic of the month.

Merchandising

One of the most promising areas for the Internet, and certainly for Sabre, is merchandising opportunities. Travelocity features lists of merchants and products from around the world, such as: luggage, books, videos, travel products, accessories and other unique items. Relevant facts on packing and shipping are provided for each item.

At present, products and services purchased using credit and charge cards via Travelocity may be collected at an airline office, an airport or a travel agent. In the future it may even be possible to collect merchandise from a Travelocity ticket bureau. Such an operation would, however, need to be created by Sabre just for this purpose. However, like many new services on the Internet, Sabre is waiting until a clear pattern of demand is established before investing the substantial development and investment resources required to build this new infrastructure and distribution channel.

It is, however, recognized that at the present time there is somewhat of a consumer perceived barrier to paying for services on the Internet. Although barriers such as this are forecast to come down over the next few years as better encryption becomes more widespread and consumers become more confident of the Internet's security, in the short term Travelocity may well be used primarily for accessing information and planning travel. It is therefore quite possible that consumers will access Travelocity during the trip planning stages of tourism and then finally visit a local travel agent or some other retail outlet to purchase their tickets.

Another challenge for Travelocity is the degree to which it is customized for different areas of the world. Although, like most Web sites it is accessible globally, it is at present, i.e. mid-1997, customized purely for the USA consumer market. In this context, customization encompasses features such as the language in which the Web pages are displayed, the currency in which prices are quoted and the format of postal addresses. Customizing a Web site for true global use is a mammoth task that has associated with it a mammoth price tag. So, this development will undoubtedly follow an evolutionary path over a long period of time and will be driven by consumer demand.

Finally, Sabre has experimented with an interesting and innovative use of the Internet known as interactive auctioning. This is an electronic auction of airline seats, using the Internet as a communications medium. The way it works is as follows. An airline finds itself in the position of having a number of spare unsold seats on one or more of its flights, with only a short time to go before departure. It displays the details of these seats on the Internet, e.g. origin, destination, class, date, time, etc. Along with this, the Internet page invites consumers who are Internet users to make bids for the seats. At some point in time, the airline will review the bids received and sell the tickets to the highest bidders (whether or not there will be a reserve price is an open question, but I suspect somehow that there will be). Some people have received real bargains in this way and airlines have also benefited from the sale of seats that would otherwise have been empty.

Travelocity is also accessible from many other Web sites. These sites focus on providing specific types of users with targeted information on a variety of topics such as small business forums, links to other travel information providers and news services. Such sites have hypertext links to Travelocity. This means that visitors to these sites simply click on a particular sentence or key word and are then automatically connected to Travelocity.

Sabre Web Reservations

At present this is a service that is offered to travel agents in the USA, Canada, Bermuda and Europe. Australia, New Zealand and other countries will be able to subscribe to the service as part of a roll-out program that commenced in 1997. The service supports those travel agents who either have already set up their own Web sites or are considering one. The services offered by Sabre Web Reservations used to include full site development services, such as page design, navigation through multiple pages and links to other Web sites. However, more recently, Sabre has decided to focus on the primary customer demand, which is the need for links from the agent's site to Sabre's booking engine. There is a one-time set-up charge and then an ongoing maintenance fee for all Sabre Web Reservations services. The service was originally introduced in two phases, the first of which is now complete:

- **Phase I** This supported travel agents in their efforts to create personalized Web pages for display purposes only. It also enabled travel agents to receive e-mail from respondents who view their pages and wish to take some kind of follow-up action. An essential feature is the ability to monitor the hit rate on a travel agent's site. Also available to travel agents is Travelocity's own search engine called Travel Explorer. This searches the Travelocity pages for subjects and keywords specified by the user and returns a list of page references and available Web sites. As mentioned above, this was an early service offering that enabled travel agents to establish their own Internet sites.

    More recently, Sabre has recognized that many travel agents are now perfectly capable of independently creating their own Web sites and therefore the Phase I product offerings are now no longer available. Instead, Sabre has moved on to Phase II, which provides agents with a link to the Sabre booking engine.
- **Phase II** This enables travel agents with their own Internet sites to implement a link to Sabre's Travelocity booking engine product. Phase II products are marketed actively in the USA. The Travelocity dimension allows consumers who access a travel agent's own Web site, to link into Travelocity for reservations purposes. Customers can then search for and reserve the lowest air, car and hotel rates as well as special travel agency fares. All of this information is shown in their local currencies with local taxes. The resulting reservations are sent electronically to the travel agency for ticketing.

    This also allows consumers to pay for their products over the Internet, via Sabre. This whole process is controlled by core Sabre functions that communicate directly with the travel agent for payment and ticketing purposes. To use this service the travel agent must of course be a Sabre subscriber and possess an IATA licence that allows the agent to print airline tickets for their customers.

Travel agents are, however, exploring other ways of using the Internet in conjunction with Sabre. Because most of this development effort is undertaken by the travel agent, Sabre's role is now more of a supportive one, which really falls into the category of consultancy. However, it can be clearly seen that Sabre's underlying distribution strategy keeps the travel agent firmly in the loop, even though consumers may be able to book directly with them.

WORLDSPAN

Over the past ten years or so, Worldspan has developed and grown its own true global network in response to customer demand (see Fig. 4.17 which shows the Worldspan global network). This network can now support most of the common communications protocols, including those used by the Internet. It therefore provides Worldspan with an ideal springboard from which to exploit the Internet as a new distribution channel for its GDS services. This is a significant development because it expands Worldspan's travel agency world into a new dimension – that which is inhabited by that fickle of all users, the travel consumer.

There are really three avenues down which Worldspan drives its services on the Internet: (i) a straightforward subscriber service for travel agency users wishing to access the Internet; (ii) an alternative distribution channel for GDS services, which are provided via travel agents to consumers wishing to access the Worldspan system; and (iii) a third-party service that helps travel agents and other companies set up their own Web sites. Let's explore each of these three Worldspan Internet services in a little more detail.

Worldspan Internet for travel agents

Worldspan can provide full access to the Internet for its travel agency customers. This allows existing users to expand their booking PCs to become Internet browsers without the need for additional communication facilities. This is accomplished: (a) by using special software on the existing population of travel agency PCs, and (b) by using Worldspan's Internet servers with high capacity trunk connections into the Internet.

- **Gateway for travel agents** Travel agents use the Gateway Plus product (see Chapter 4 for more details), to establish a connection into Worldspan's global network by a variety of alternative methods. The two main methods are either by dedicated data lines rented from telecommunications suppliers or via dialled telephone connections on an as-needed basis. In either case, the travel agent may elect to use special Worldspan software on these PCs to access the Internet indirectly. The routing appears to be convoluted but is in fact extremely fast. Messages travel from the users' workstation PCs via their branch Gateway PC, into the Worldspan network and then via dedicated Internet Servers into the Internet itself. This allows travel agents who already have Worldspan PCs for information and booking purposes also to use those same PCs to access the Internet.

  But it is the branch Gateway server that provides some very special control functions. These functions have been designed by Worldspan to be of particular interest to the travel agent's head office management. The software running in the branch gateway server provides a high degree of management control over the services that are provided to end users in branches. For example, the branch gateway can limit the Web sites that are accessible by end users. This is especially relevant when a large multiple travel agent uses the Worldspan network to inter-connect its branches. In such cases, the multiple's headquarters management staff will almost certainly want to restrict the Web sites that staff in the remote branches are allowed to access.

  It could be, for example, that supplier and tourism information sites are perfectly allowable, whereas sports results and games sites would be out of bounds. The Worldspan branch gateway server is the means by which this level of access is controlled. In addition to this, the gateway server can also restrict the hours during which the net is accessible by certain travel agency end users. While controlled access during normal office hours could well be OK, access after 6 p.m. or before 8 a.m. could either be disallowed completely or totally open, depending upon the policy set by the travel agency management.

- **Internet servers** Once through the branch Gateway server, the end-user's Internet traffic is routed across the Worldspan network to an available Internet server with spare capacity. This type of server is dedicated to handling Internet traffic and is connected into the Internet by high speed telecommunication lines. Each server is itself a high speed, high capacity computer, dedicated to Internet processing. These powerful computers not only serve as an effective gateway into the Internet for all Worldspan travel agency users but they also provide an adequate level of security: and security is very important to ensure that, for example, payment transactions are secure, viruses are not downloaded and the travel agent's systems may not be accessed by unauthorized users.

The benefits of travel agents using the Worldspan Internet path are: (a) it eliminates users having to dial into their local Internet service provider, (b) it provides the agency's management with a high degree of control over how its staff use the Internet, (c) it provides a high level of security to the travel agent, and (d) it allows Worldspan customers to leverage their investment in GDS technology for Internet access.

Worldspan Internet for consumers

Worldspan's approach to consumer bookings over the Internet is inextricably linked to the travel agent community. While consumers may browse the Worldspan pages and peruse availability, when it comes to making an actual booking, a travel agent is always brought into play. Worldspan even goes as far as taking a consumer's card account details and then verifying them with the card company's own computer system. However, at this point, it offers the consumer a choice of travel agents from a list of pre-registered Worldspan subscribers. The consumer selects a travel agent that is, for example, either: (a) closest to the consumer's own home or office location, or (b) another agency with whom the consumer wishes to deal, perhaps on a mail-order basis. Once an agent has been chosen by the consumer, Worldspan automatically queues the booking to the agent for processing, ticketing and funds collection from the consumer.

Such an approach enables Worldspan to continue enjoying the support of a distribution channel that generates around 80 per cent of its bookings, while simultaneously marketing its services to consumers in new and innovative ways via the Internet. To a large extent, this strategy relies upon the travel agent for promoting the awareness of Worldspan's Internet service to consumers. Worldspan itself does not engage in the pro-active marketing of GDS services direct to consumers. This partnership approach works effectively and has so far proved to be mutually beneficial to both parties.

To do this, Worldspan has created its own infrastructure to handle consumer bookings over the Internet. This infrastructure is core to its Internet strategy and is based on an Internet booking engine (IBE). The IBE is a computer that is connected directly to the Worldspan host mainframe in Atlanta. It uses special interface software to front Worldspan's consumer-facing GDS service on the Internet. The IBE comprises two main components: (i) support for a user-friendly GUI browser for direct use by consumers, and (ii) a standard communications protocol called SMI, which indirectly links consumers to Worldspan via other Web site providers. Let's take each one in turn:

- **Direct – via browser interface** This type of IBE connection is aimed at supporting Worldspan's relationship with consumers, via the Internet. A key element of the software that runs on the IBE computer is the user-friendly browser interface. This supports an easy-to-use dialogue for communicating with the Worldspan host system via any of the commonly available Internet Web browsers, such as Microsoft Explorer or Netscape Navigator. It assumes that the end user will not be specially trained in how to use a GDS and makes extensive use of windows, drop-down lists, menus and check-boxes.

    Although the GUI is very user-friendly, it can be a trifle slow for an experienced user. It is for this reason that an alternative browser is planned by Worldspan, which will be offered as an optional product. This will incorporate native Worldspan GDS functions and will consequently be aimed at the more sophisticated user who may initially require some basic training before they can use it effectively. However, it will be significantly faster than the current Internet IBE browser.

- **Indirect – via SMI** This type of IBE interface is available to those companies wishing to connect their own Web site computers into Worldspan's GDS system. The communications protocol used to make this connection to the IBE computer is proprietary to Worldspan and is called SMI. This is a messaging standard that controls Internet-type messages flowing between computers. The two computers in this context are of course: (a) the Worldspan IBE computer; and (b) the Web provider's own computer, which, in turn, is connected to the Internet. In some respects SMI is similar to PADIS (see Chapter 1 – The TTI). It is an extremely successful protocol and is now widely used in the Internet industry. In fact, one of the reasons Microsoft chose Worldspan for Expedia's GDS booking engine was because of the flexibility and technical compatibility

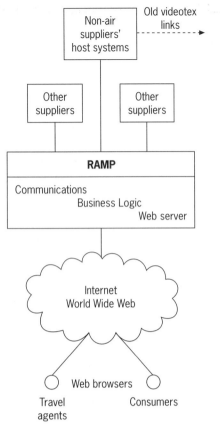

Figure 5.18 Worldspan's RAMP project

Worldspan's IBE handles the booking and ticketing of both airline and hotel products. Car rental functions are to be added soon (almost certainly by the time this book is published!). A great deal of effort is currently being directed towards enhancing Worldspan's Internet services and the project code-named RAMP (Fig. 5.18) will provide the supplier side for much of these developments. RAMP is a strategic system and is based on Internet technology. This, together with Worldspan's global network and GDS booking functions, should enable Worldspan to become a leader in Internet-based information and booking services to consumers.

Worldspan's third party Web service

When a company wishes to establish its own Web site, it faces some considerable challenges in the areas of skills and resources. There are the marketing issues to consider, the graphic design skills needed to create attractive and exciting Internet pages, the technical skills required to write programs in Java, the expertise needed to write hypertext with suitable links to other pages/sites and finally the operational resources needed to keep the site running effectively and the information up-to-date. To this list can often be added the technical complexities of inter-connecting a company's own product inventory system to the Internet. Worldspan is particularly active in two prime areas of this new market:

- **Travel agents** Some large and technically competent companies undertake this work all by themselves, often using in-house experts. Many large multiple travel agents therefore already have their own sites, several with links to Worldspan. However, for the smaller agency that wishes to focus on its core competencies, i.e. travel, Worldspan offers a new consultancy service. Using this service, the smallest of travel agents can set themselves up on the World Wide Web and compete directly, and on almost equal terms, with the largest multiple. The Worldspan service provides customers with specialist consultants in all the disciplines required to establish a successful Web site.
- **Non-air suppliers** There are many non-air suppliers using well established legacy systems to control their inventories of travel products,

of SMI (see page 196 for a fuller description of Expedia).

Worldspan's IBE can therefore be used by companies wishing to act as booking intermediaries. Examples of such companies include Microsoft's Expedia and travel agents themselves. Each customer of Worldspan's IBE service uses this interconnection to provide its own customized Internet booking facility. This allows companies to create proprietary Web sites with embedded links to Worldspan's IBE, just as though the whole site, including the booking service, was provided by the companies themselves. When a site is created especially for a customer like this, it is of course heavily branded for that customer. Once on the Internet, it then appears to a browser, i.e. a consumer, to be the customer's own site and is not branded as a Worldspan site in any way.

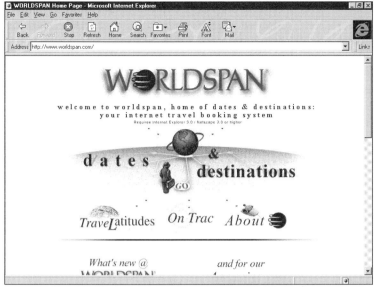

Figure 5.19 The Worldspan home page

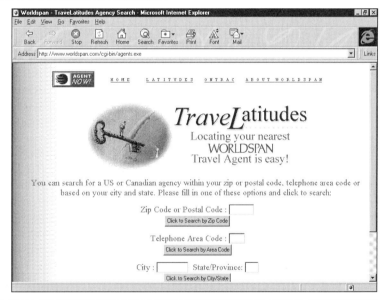

Figure 5.20 Finding a Worldspan travel agent

e.g. tour operators. In many cases these systems are distributed to travel agents via videotex. In such cases, Worldspan is able to offer these suppliers the ability to interface their systems to travel agents and consumers via the RAMP facility (which I first introduced in Chapter 4). A travel supplier wishing to expand its range of distribution options to other channels, such as the Internet, may thus contract-out the development of the required technical interfaces, to Worldspan. Because RAMP was designed to simplify this task, the supplier may concentrate on the commercial aspects of an expanded distribution channel without being burdened by the IT resource and skill availability issues so often characteristic of these projects.

As you will no doubt have gathered, RAMP is a key element in these aspects of Worldspan's Internet services to the global travel industry. Figure 5.18 shows an overview of RAMP and illustrates how the system works.

THE INTERNET 215

TRAVELWEB

The TravelWeb Internet site is one of the leading participants in a portfolio of new and alternative travel distribution channels marketed by Pegasus Systems. TravelWeb is a separate company wholly owned by Pegasus Systems Inc., the parent company of Thisco (see Chapter 4). Besides offering seamless connectivity to many leading hotel systems, it also has access to an airline booking engine provided by Internet Travel Network (ITN). The primary role of TravelWeb is to provide the technologically sophisticated traveller with a full-scale travel service via the Internet. Hotel bookings are serviced on a one-to-one basis with the consumer using Thisco's Ultraswitch technology to link him/her directly to the hotel system of his/her choice. Airline ticket sales are fulfilled with the participation of a USA based travel agent. But before we explore how TravelWeb is constructed, let's first take a brief look at TravelWeb's company history. A brief review of its background should help explain how it reached its position as one of the leading new Internet-based intermediaries.

TravelWeb first appeared on the Internet in October 1994 when it was positioned as an on-line catalogue of hotel products aimed at the travel industry. In December 1995 a pilot version of the hotel booking engine was Beta tested by a controlled group of Internet users. This was the first time that Thisco's Ultraswitch hotel booking system had been connected to the Internet. The test proved highly successful and so in March 1996 TravelWeb was officially launched with eight hotel chains available for on-line booking. The first live booking was soon received by TravelWeb and to the surprise of management, this originated from South Korea and was for a stay in San Francisco on 24 December at full-rack rate.

TravelWeb became an outstanding success over the first seven months of 1996 with over US$2.4 million in room sales being processed. By July 1996 a total of 16 hotel chains could be booked on-line via the World Wide Web. In August 1996 airline reservations and ticket purchase functions were added via the Amadeus System One GDS booking engine. This was replaced early in 1997 by a link to ITN, which is a private company operating links to most of the major GDSs. By October 1996 TravelWeb reached a year-to-date level of US$3.5 million in room sales and was averaging 15,000 individual visitors each day to its site. By the end of the year this had risen to US$6.5 million in booked room revenue. Since its launch in 1994, TravelWeb has experienced a 40 per cent average month-on-month growth rate for hotel bookings. Quite an impressive debut onto the World Wide Web.

Before we dive into the detail of TravelWeb, it is important to set it within the overall context of Pegasus' new distribution strategy. At present, there are broadly two classes (Fig. 5.21) of distribution channels that Pegasus' hotel customers can use to reach their consumers: (i) the classic GDS distribution system route; and (ii) a choice of several new alternate distribution systems, the prime one being the Internet. The first of these, GDSs, is covered in more detail in Chapter 4 – Distribution Systems (see Pegasus). I am going to concentrate here on the new alternate distribution systems, most of which are based on Internet technologies. Of these, TravelWeb is one of the leaders. But there are others. For example, besides TravelWeb, UltraDirect also supports the following alternative distribution system providers:

- **Preview Travel** San Francisco based Preview Travel has a customer base of 850,000 registered users, which is derived from two main sources: (i) AOL customers and (ii) the World Wide Web. These customers, most of whom are leisure travellers, are provided with hotel information and booking functions by Preview Travel via their link to Pegasus Systems' Ultraswitch.
- **Internet Travel Network** ITN is a company that provides Internet access to the GDSs via the World Wide Web. It has replaced TravelWeb's original connection to Amadeus System One and provides GDS access for other alternative distribution suppliers.
- **TravelNet** Pegasus provides TravelNet with a hotel booking system for its corporate travel management product. This allows business travellers to book a whole range of travel products themselves from their lap-top PCs, while retaining a travel agency in the loop to take care of ticketing, consultancy and account management.

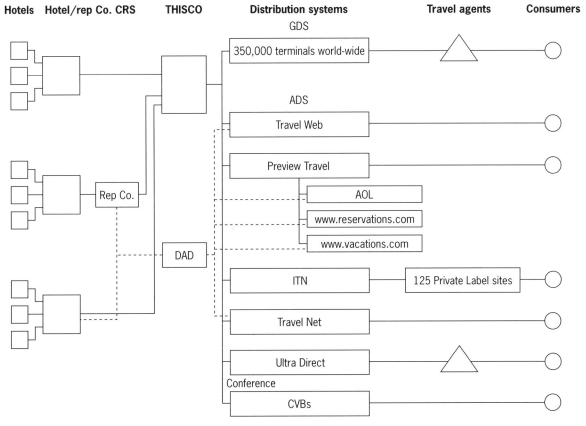

Figure 5.21 Pegasus TravelWeb transaction/processing flow

- **UltraDirect for travel agents** Although UltraDirect is the generic name for Pegasus' alternative distribution system product, this sub-product is distributed specifically to travel agents. It therefore provides Thisco's hotel participants with an alternative travel agency route to that offered by the GDSs.

All of the alternate distribution suppliers using UltraDirect have their own market of consumers that they address individually, and all are connected to the Ultraswitch for on-line seamless connectivity to hotel reservation systems. Besides reservations, however, a common requirement of all these alternate distribution systems is access to information on hotels. This common requirement, which is a key feature of UltraDirect, has been addressed by Pegasus through its new distribution access data base (DAD).

DAD is really a separate data base sub-system all of its own, which is connected to the Ultraswitch (see Fig. 5.21). In 1997 DAD stored information on over 15,500 hotels, each with text, photographic images and full graphics. The primary purposes of DAD are: (a) on the supply side, to enable hotels to update their non-dynamic information in a consistent and tightly controlled way with in-built quality control features; and (b) on the demand side, to enable alternative distribution channel end users to access both the non-dynamic information and the dynamic reservations functions supported by Ultraswitch. To provide this infrastructure, a network of four servers is connected to the Internet by a front-end communications router. Three of these servers are dedicated to information management and are connected via Netscape's LiveWire technology and a 100 Mb Ethernet LAN, to the DAD data base.

The fourth is connected directly to the Ultraswitch computer and provides a gateway to the seamless hotel reservations functions of Thisco. This subsystem provides some critical functions on both the supply and demand sides of DAD:

- **DAD supply** A key success factor is the remote authoring techniques supported by DAD. Remote authoring places responsibility for page changes firmly in the hands of the participants. Each hotel chain may use either: (a) a batch interface, which maps the hotel chain's own data base to DAD's; or (b) an HTML on-line editor connected to DAD for information maintenance. This approach minimizes the administrative overheads of TravelWeb and helps ensure that information is up-to-date and accurate. Hotel updates are first captured in DAD's Work In Progress data base and following quality control checks are then migrated to the live DAD environment.
- **DAD demand** Incoming messages from end users are routed to the appropriate DAD server which can then provide either: (a) hotel information services, which are supported by three servers, each with its own link to the DAD data base; or (b) seamless connectivity to 14,000 hotels via the fourth server with its connection to Ultraswitch (1,500 hotels are also bookable but only via e-mail). The information servers use Netscape's LiveWire to create pages on the fly by merging DAD data base accesses with standard HTML templates to form an Internet page that is transmitted to the end user.

The TravelWeb server is also linked to specialized booking engines, the most prominent of which allow consumers to book hotel rooms and airline flights themselves. TravelWeb uses Thisco's Ultraswitch for hotel bookings and ITN for airline bookings (see above). In addition to supporting bookings from straightforward inventory, there are certain special marketing opportunities that make it possible for hotels to sell distressed stock on the Internet. Distressed stock, in the context of the hotel business, comprises rooms that remain un-booked with only a few days to go. Such rooms can be heavily discounted and offered directly to consumers over the TravelWeb site.

One of the other main functions of the TravelWeb server is to act as a translator between: (a) classical text-based computer systems that support TravelWeb's host suppliers; and (b) the Internet's HTML to which all Internet users are connected. This translation function allows TravelWeb's host suppliers to communicate directly with the PCs of home and business consumers around the world. A more detailed description of TravelWeb's main components is as follows:

- **TravelWeb's information pages** TravelWeb stores static information on 60 chains and 15,500 hotels located in more than 125 countries, many of which are SMEs. The information stored about each hotel is rich in breadth and depth – a virtual electronic hotel brochure for each participant. Besides the kind of textual information expected of any computer system including, for example, name, address, room rates and facilities, there is also a rich set of multi-media enhancements. For example, there are colour photographic images of hotel rooms, restaurants, meeting facilities, local recreational activities, maps and much more. A customized search engine allows users to find a hotel by a wide variety of parameters including: geographic location, chain name, rate range, amenities and facilities. The TravelWeb site comprises approximately 65,000 World Wide Web pages of information on hotel- and travel-related subjects.

  Besides hotel-related information, TravelWeb also promotes a wide variety of advertisers and sponsors including AT&T, United Parcel Service (UPS), Access One, Aufhauser, Ceres Securities and The Sharper Image. All of this information is available via standard Internet browsers that allow consumers to navigate their way around the site easily.
- **The hotel booking engine** The TravelWeb Internet server is linked by high speed telecommunications lines to Thisco's Ultraswitch computer (see Pegasus in Chapter 4 for more details on this major hotel industry switch). It is through this link that consumers can book a hotel room from 14,000 properties that are part of 16 chains. The actual booking process is carried out between the consumer and the hotel chain's computer system, with no intermediate

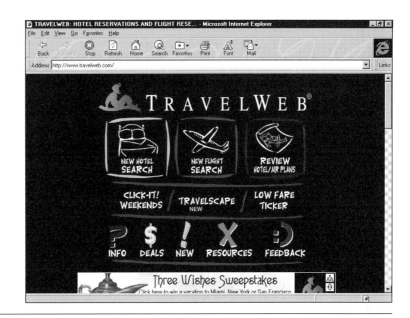

Figure 5.22 The TravelWeb home page

GDS involved at all. This *seamless connectivity* is about as close to a direct point-of-sale relationship with a prospective guest that a hotel could reasonably expect to achieve. Once a booking has been made, consumers may choose to guarantee their rooms by using TravelWeb's on-line plastic card authorization facility. TravelWeb therefore provides its participating hotel customers with a truly on-line confirmed booking service that is available to consumers all around the world.

The TravelWeb site is growing and developing all the time, usually in response to feedback from its site visitors. During Beta testing, for example, TravelWeb found that a great proportion of its site visits came from commercial Internet accounts, with most bookings occurring during business hours. Around 58 per cent originated from business travellers and 24 per cent from retail or leisure consumers. It was also found that 70 per cent of all TravelWeb bookers would have normally used an 0800 toll free telephone number to make their bookings – significantly, they would not normally have used a travel agent. About 15 per cent of users were located outside the USA, primarily in Japan and Canada. During the Beta test period, bookings were received from 29 different countries. Since the end of the Beta test the rate of cancellations has dropped from 51 to 19 per cent – a factor that reflects the increasingly serious level of use rather than the high level of experimental bookings made by people during the test period. In 1997, the TravelWeb site was averaging 33,000 visitors per day and generating over US$1 million in net reservations each week.

This feedback prompted TravelWeb to introduce more business travel oriented services. A prime example is The Business Traveller Resource Centre. This is a sub-set of TravelWeb's pages, which is aimed specifically at individual business travellers. It contains tailor-made pages of information on business travel topics and links to other sites on the Internet that offer products and services that may be of interest to business travellers. The 'special offers' category within The Business Traveller Resource Centre, for instance, provides some interesting promotional links. TravelWeb users can link to a merchandising site offering products at a special discount or alternatively to a sweepstake promotion organized by Preferred Hotels and Resorts Worldwide. There are also many other outbound links to services such as financial, computing/software, overnight package delivery, news, catalogue shopping, special fares and other promotions. At the last count, there were over 10,000 other Internet sites that incorporated dynamic inbound links to TravelWeb.

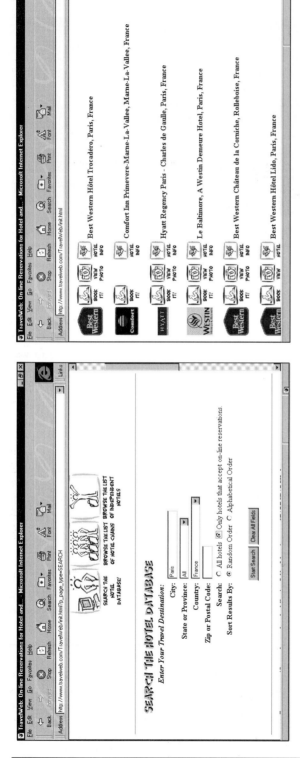

Figure 5.23 Hotel search parameters (above)
Figure 5.24 Hotel search results (above right)
Figure 5.25 Hotel photo

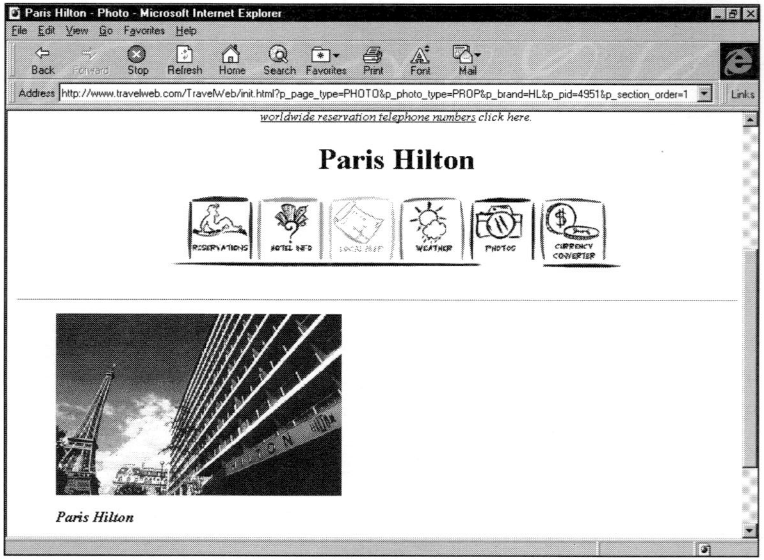

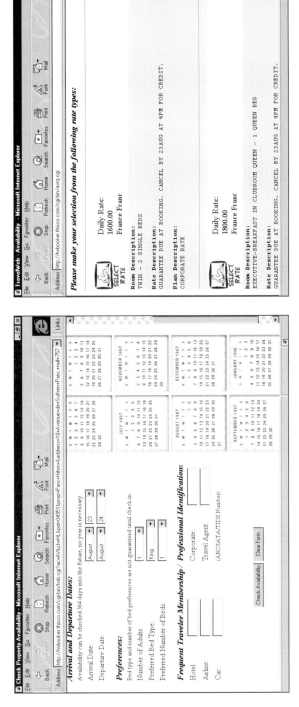

Figure 5.26 Check availability parameters (above)
Figure 5.27 Availability display (above right)
Figure 5.28 Booking screen 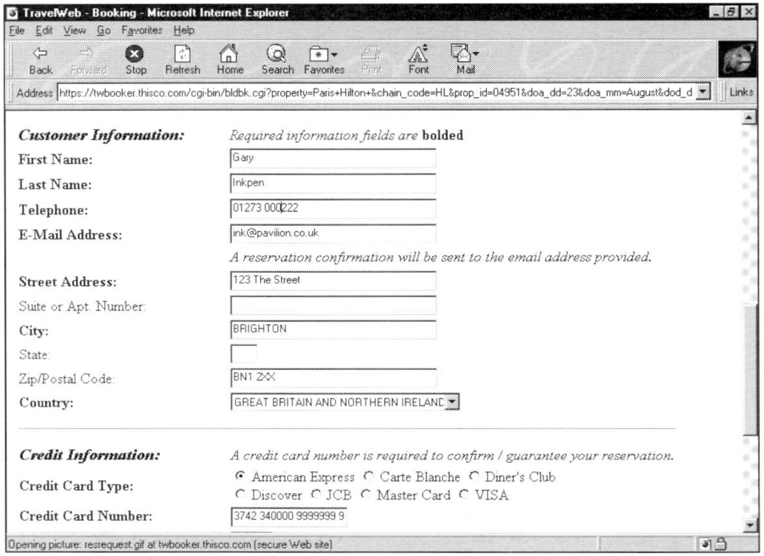

THE INTERNET 221

TravelWeb does not charge users for searches and reservations. The expenses of this site are funded by Thisco's hotel participants; and the reservation fees charged by TravelWeb to its participating hotels are less than those that are levied by GDSs for providing a broadly comparable service. TravelWeb booking fees are around US$2.50 plus an Ultraswitch fee of US$0.50 with no additional GDS fees payable by hotel participants. This also compares favourably with bookings received by hotels via telephone reservation calls, which average between US$10.00 and 15.00 (and up to US$30). TravelWeb's participating hotels also benefit from the following:

- **Potential market** Participating in TravelWeb opens up a potential market of 50 million Internet consumers in both the business and leisure markets, to participating hotels. This consumer base is truly world-wide and growing at a substantial rate.
- **Direct customer contact** The TravelWeb site provides participating hotels with the unique ability to hold one-to-one dialogues with their existing and potential customers. No other media provides this key-selling opportunity.
- **Reduced printing and distribution costs** Brochures can now be shown effectively on the TravelWeb Internet pages in full colour with pictures of rooms, locations and amenities. This reduces the need for high volumes of printed material currently used for promotional purposes.
- **Tactical marketing opportunities** Hotels can undertake their own innovative promotional activities on TravelWeb. This has a low overhead because it costs little to create and can be done within a very short period of time. The marketing of distressed inventory, as described above, is one good example.

Plastic cards accepted for TravelWeb bookings include American Express, Carte Blanche, Diners Card, Discover, Japan's JCB, MasterCard and Visa. Security is therefore a critical success factor. TravelWeb is controlled by Netscape's Commerce Server software, which has advanced Internet security features based on secure socket layer (SSL) encryption technology. Additional levels of security are provided by transaction authentication, data encryption, firewalls, a transaction history of activity between customers and hotels and, finally, trip-wires by hotels and TravelWeb to identify unusual activity. The hardware has changed several times in an attempt to keep up with the rate of growth of TravelWeb. The server is currently a Sun Enterprise Ultrasparc 3000 and this is the third upgrade since the site was first launched.

TravelWeb may be accessed by consumers using virtually any modern browser, although a secure browser is required to complete credit card guaranteed reservations. Browsers that enable users to take full advantage of TravelWeb's multi-media pages include Netscape Navigator 2.0, Microsoft Internet Explorer 2.0, as well as Macromedia's Shockwave for Director. All these browsers have an integrated e-mail facility for response and follow-up purposes. TravelWeb takes the e-mails it receives from consumers very seriously. In fact over 300 e-mails are received each day and each one is answered by TravelWeb within 24 hours.

Finally, a word or two on costs. Running a successful Web site is not cheap. Especially one that is dynamic, up-to-date and transactional. TravelWeb started life as an operation costing around US$110,000 to run in 1994. By 1995 this operating expense had grown to US$1.6 million and for 1996 the cost was over US$3.8 million. If TravelWeb continues to grow at the historic levels experienced to-date, we may not have yet seen the levelling of the operating cost curve. Future growth will always demand higher levels of investment in IT in order to keep pace with consumer demands as the Internet itself grows over the next few years.

Suppliers' Web sites

Suppliers are finding the World Wide Web an increasingly attractive directing marketing channel. While most suppliers would not consider it practical to distribute their entire product ranges directly to consumers, there are certain niche areas where direct selling is the ultimate route. The Internet offers suppliers an ideal opportunity to go one step further than advertising and sales promotion via the Web and use it for bookings.

This is, however, a significant extra step because it involves payment processing and an extra level of security. However, these functions are increasingly being provided by standard software like Microsoft Merchant Server. So, suppliers are experimenting with the Internet for the direct sales of niche products to both leisure consumers and business travellers. The following section contains several examples of suppliers' Web sites, some of which have been very successful in attracting and processing a significant number of on-line direct bookings.

BRITISH MIDLAND

British Midland launched an Internet Web site in December 1995 branded CyberSeat (Fig. 5.29), which is available at URL **http://www.iflybritishmidland.com**. It is interesting to explore the rationale that British Midland used to create this innovative new site, which incorporates full booking and payment functions. The starting point for our exploration is the business environment in which British Midland found itself during 1995. This was an environment in which the cost of sales was rising rapidly against an average of only £70 revenue generated from each ticket sale. When this was set against the company's associated internal processing the profitability of certain sectors of the business began to look marginal. British Midland also experienced a distancing of its sales and servicing staff from their customers. In fact, many pre-sales interactions with customers had virtually been lost in some cases. There was therefore a danger that British Midland would lose all opportunities to differentiate itself from its competitors.

Consequently, a review of British Midland's distribution strategy was undertaken. A fundamental objective, which was identified early in the project, was to reduce the cost of sales in order to improve yields and increase the underlying profitability of the business. One of the main distribution costs incurred by British Midland is GDS booking fees. At present these amount to a fixed fee of £4 per booking that, bearing in mind an average domestic ticket value, generates only a relatively low amount of revenue. But there are also some related concerns, the two main ones

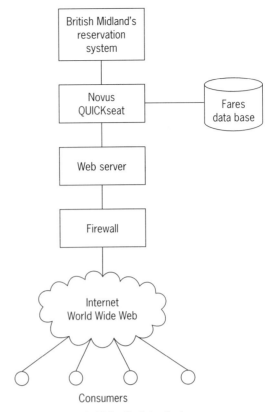

Figure 5.29 British Midland's CyberSeat

being: (a) the trend towards higher booking fees in the future, and (b) the fact that booking fees are fixed and not related to ticket value. While a fixed booking fee may be more acceptable to airlines with long haul routes involving higher ticket values and therefore higher revenues, British Midland exclusively fly short haul routes in the UK and Europe that generate low average ticket values. This makes a fixed GDS booking fee of £4 a very significant proportion of each ticket's overhead costs and was the primary reason why British Midland wanted to introduce an alternative to GDS distribution. On top of this, other distribution costs that are inherent in the GDS and travel agent channel, are also significant. These include communication network costs, travel agents' commissions and travel agents' override payments.

Along with this fundamental objective was a need to increase the effectiveness of the sales process and to increase the revenue generated per passenger carried. A closer relationship with the customer was also an important objective. The problem that British Midland faced in trying to pursue these objectives was that it was severely constrained by its participation in the major GDSs. Although GDSs are now virtually all neutral in terms of bias, they are nevertheless owned by airlines that are competitors of British Midland. It seemed to British Midland management that the GDSs offered limited opportunities for them to differentiate their airline from competitors. As a direct result of the large stake-holdings that major carriers have in the GDSs, their owner airlines are free to introduce GDS functions that suit their products and differentiate them from their competitors. These airlines can therefore use their GDSs as a means to steal a march on smaller niche competitors by introducing highly customized and specific new functionality; and not all of these new functions deliver differentiators that suit British Midland. So, in this environment, carriers like British Midland could only play a 'catch up' game and this was deemed unsatisfactory within the company. A new distribution channel like the Internet offered British Midland the opportunity to set standards for others to follow and thereby achieve a leadership position.

A number of associated business challenges were faced by British Midland management, not least of which was increased competition. Low cost start-up airlines, such as EBA and Easyjet, were getting established sooner than had been expected. These posed a threat to British Midland's core business – the domestic UK market. Also, in terms of competition, the Eurostar service to Paris and Brussels was beginning to threaten important parts of the company's European business. Other operating issues arising from the complexity of processing airline tickets and the ensuing congestion occurring in airport terminals with insufficient check-in counters, also needed to be addressed. Facilities such as this are expensive because of the high ground rents charged by airport authorities and the need for tight security. But as far as these two issues are concerned, there appears to be light at the end of the tunnel in the form of electronic ticketing (see Chapter 3). With electronic ticketing, the physical security and delivery problems associated with ticket issue could well disappear and airport check-in could be largely automated with self-service machines. So, with the dual pressures of increased competition and rising operating costs, the time seemed right for British Midland to consider a fundamental change in the way its product was distributed.

The main thrust of any new distribution method was not to take business away from travel agents but rather to relieve the pressure on in-house telesales units by giving customers an alternative to the telephone as a means of making bookings. In other words to go after independent travellers who would normally have telephoned British Midland and enable them to use a more efficient channel that could be serviced by electronic means. After all, travel agents cannot derive an enormous commission from low-value tickets without reducing their cost of sales. So, channelling these ticket sales via an automated route would not adversely affect the travel agency business. Nevertheless, it was recognized that CyberSeat could erode some of the value-added services offered by travel agents. The impact was, however, considered to be relatively small, especially in the early days of any new system.

In considering alternative distribution channels, the Internet appeared an attractive medium. Despite its relatively low numbers of users, its rate of growth was phenomenal and its potential for travel services was considerable. It embodied a comprehensive set of technological standards that reduced the risk of developing redundant applications. Also, there were a range of packaged software tools that could short-cut the development and implementation process. British Midland decided to experiment with the Internet route, but first of all set some important ground rules. The amount of investment in the preliminary system would be minimal and it would have to be up and running very quickly. The business to be targeted by this new channel was the high volume sales of straightforward airline tickets rather those with a higher price tag. This prompted management to start addressing the complexities associated with using the Internet as a commercial distribution channel. Complexities such as: (i) the geographical product distribution issue, and (ii) the potential security risks of taking payment over the Internet without being able to capture either a customer's signature or a card imprint. These two issues are worthy of some further analysis:

- **Geographical issue** Because the Internet is a global channel, it means that ticket prices must be set on a global basis. Instead of tickets for a flight being priced for the specific economic dynamics of each origin market, they needed to be set globally for all markets. This in turn means that foreign currency exchange rates need to be factored into the equation. It also means that, in the absence of electronic ticketing, physical ticket delivery to overseas customers must be available. For example, a customer travelling to the UK from a foreign destination, such as the USA, with a stop-over in France, would need to be able to collect their ticket to the UK from an airport in Paris. This resulted in a new set of procedures being developed by British Midland to support CyberSeat.
- **Security issue** The next issue was: 'How should payment for air tickets be processed over the public Internet infrastructure?' This issue was carefully considered and it was decided to: (a) only support secure Internet browsers, which incorporate SSL encryption technologies; and (b) to send critical payment fields, e.g. credit card numbers, expiry date and cardholder's name, across the Internet in separate encrypted messages. In this way, even if one of the messages were to be intercepted, not only would it be encrypted but it would only represent a part of the information needed to record a financial transaction. Finally, besides these Internet security devices, British Midland's CRS system is protected by a further three levels of security.

After much deliberation, investigation and research, all of the issues and obstacles were successfully overcome and the Internet route was finally decided upon. British Midland decided to use a multi-media reservations server using Netscape's Commerce Server as the back-bone for the new service. This would be the world's first airline booking system for the Internet with full on-line payment functions. Management decided that this would need to be compatible with future technologies such as interactive television and would of course have to support electronic ticketing and self-service check-in at airport terminals. An important requirement was that the system should be capable of building and maintaining a customer data base for marketing purposes. As previously mentioned in the above section on marketing on the Internet, this is a fundamental success factor in maximizing *pull marketing* opportunities. Before proceeding with the development, British Midland tested the market by undertaking a survey of their 'High Flyer' club members. This produced encouraging results. They found that around 21 per cent of their customers used the Internet regularly and that 72 per cent used electronic mail.

Following a review of the technical options available, British Midland management decided to develop an interface from its CRS to the World Wide Web. This would provide last seat availability and access to the latest fares, as they are introduced. Several years ago, British Midland decided to outsource its CRS operation to British Airways and use the RTB main-frame computer facility located at Heathrow. It is from the co-hosted RTB CRS computer that British Midland connects into GDSs like Galileo and Sabre (see Chapter 4). British Midland chose Novus, a Guildford-based international group of companies specializing in airline and other travel technologies, to help it develop its Internet channel. Novus developed CyberSeat to run on an IBM RS/6000 server that also uses DEC Alpha hardware, in a UNIX operating system environment. This is integrated with several Internet software products including:

- **QUICKseat** A seat booking application originally designed for the leisure market and developed by Novus. It is a tried and proven software product that has been used by several major carriers to distribute their air reservations products via videotex.
- **Novus Managed Internet Transaction Server** A software product that supports the development of commercial distribution products over the Internet. It makes the development process simpler and faster by the widespread use of proven sub-systems.
- **Netscape's Commerce Server** A software product that provides a secure environment for supporting commercial transactions with functionality to communicate with remote Internet

browsers. It also provides the core data base platform for new future Internet applications.
- **Novus Reservations Server** This is a key product that enables the user to interact with British Midland's CRS without having to understand the complexities of codes, transaction entries and travel industry jargon fully, all of which are an inherent part of airline mainframe booking systems. It provides: (i) easy to use format conversion routines for the translation of EDI and Internet-based protocols; (ii) rapid response times, using techniques such as the simultaneous processing of outbound and return flight segments; and (iii) provides resilient fall-back support in the event of RTB host line failure.

The server acts as a front-end processor between the British Midland CRS running on the RTB computer, and the Internet. Although the primary interface technology of the new server was based on TCP/IP, i.e. the Internet communications protocols, it would also be relatively easy for British Midland to also support emerging technologies like interactive television. The base application was kept as simple as possible and, for instance, supported full booking and payment but purposely excluded any booking changes because of the complexity of this function. Instead, customers were requested to cancel incorrect bookings and re-book. Despite obstacles such as a two to three week site registration process and following a two-months' development programme with only limited funds, CyberSeat was launched on the World Wide Web in late 1996.

CyberSeat contains a full range of booking and payment functions as well as a great deal of relevant information. Information that includes British Midland's domestic route network, its international route network, Diamond Service, Diamond Euroclass Service, Diamond Club, High Flyers, Timetables, frequent flyer information, phone reservations and customer feedback facilities. To use the CyberSeat Internet service, users proceed as follows:

1. Consumers first of all access the Internet using their PCs, modems and Internet service provider (ISPs) and open the site at **http://www.iflybritishmidland.com**.

2. Once the British Midland home page is displayed, the users click on the blue oval CyberSeat logo. This takes them to the CyberSeat front page (Fig. 5.30) via a hypertext link.

    At this stage the users need to ensure that they are using a suitable browser that must be, for example, Netscape Navigator Version 1.2 or higher. A help button allows users to access more information on what browsers and versions are supported by CyberSeat. This also allows users to download the latest version if desired.

3. From the CyberSeat front page, first time users or those unfamiliar with the site, may select the Easy Book button. More experienced users have the option of choosing the Quick Book button, which provides more functions.

    - *Easy Book* The user views the map displayed on the ensuing Web page which shows all British Midland's routes (Figs 5.31, 5.32). The users click on the origins and destinations of their intended journeys. Outward and return dates are keyed in.

        At this point the users may either choose to search all available fares or request the system to find the cheapest fare for the origins and destinations specified. The desired fares may be selected by entering the number of seats required followed by a simple click operation to confirm (Fig. 5.33).

    - *Quick Book* The users enter their places of origin and destination cities, travel dates and number of passengers. This results in CyberSeat displaying a table of flights.

        The users view the available flights from the table and may investigate each option in more detail. Eventually, a flight is selected for each booking (Figs 5.34 and 5.35).

At this point CyberSeat asks the users to enter their credit card details and to confirm that they wish to purchase the flight selected.

The users then select how they wish to receive their tickets. This can be: (a) by post to their home addresses (provided the booking is made at least seven days in advance); (b) by collection at the airport, i.e. ticket on departure; or (c) by collection from the customers' travel agents (in which case booking references are quoted).

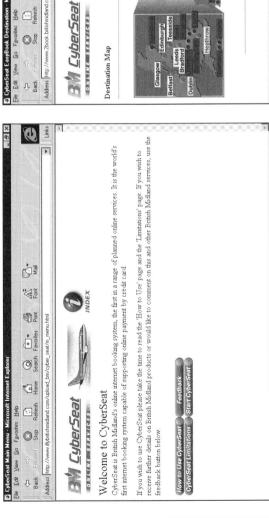

Figure 5.30 The CyberSeat home page (above)
Figure 5.31 Destination map (above right)

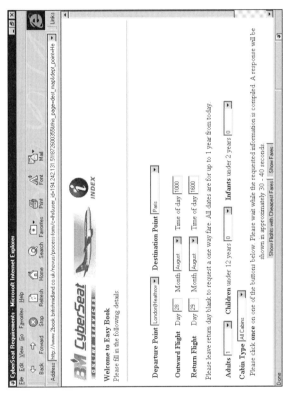

Figure 5.32 Reservations request details

THE INTERNET 227

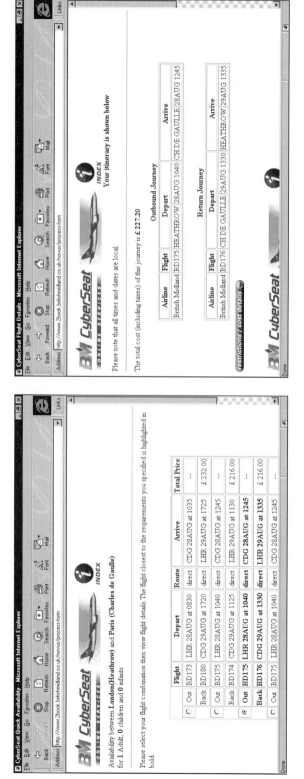
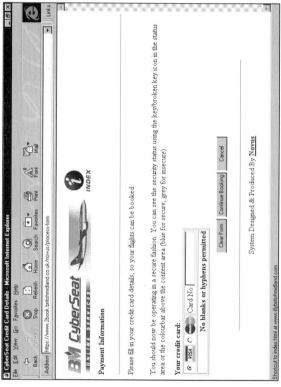

Figure 5.33 Fares (above)
Figure 5.34 Flight details (above right)

Figure 5.35 Booking screen

228 INFORMATION TECHNOLOGY FOR TRAVEL AND TOURISM

CyberSeat proved highly successful and became renowned as the first direct Internet booking system to be developed by an airline. It achieved 5,000 site visits per week and 400 bookings per month against an upper target of 750 bookings per month. This was regarded as particularly successful bearing in mind that electronic ticketing was not available at the time, which meant that ticketing had to be carried out mainly on a 'ticket on departure' basis. Interestingly, it was found that most of these bookings were made by regular frequent British Midland flyers who were making retail bookings. But being an early adopter of the new Internet airline booking technology, British Midland learnt some valuable lessons that could not have been derived from any other sources, such as textbooks, or the prior experience of others:

- First, British Midland underestimated the overwhelming response from customers in the form of e-mail. They received between 250 and 500 e-mail messages each day following the site launch. Most of these were as a direct result of people's interest in CyberSeat. The variety of these e-mail messages was found to be extremely varied. This posed a significant challenge to British Midland whose staff at the time had little experience of dealing with e-mail. It quickly became clear that these e-mail messages had to first of all be categorized and then dealt with by specialists in several areas.
- It was also found that most of the site hits were received over the weekend, mostly on Sunday. This was only natural bearing in mind the cheaper BT rates for local calls outside normal business hours. However, it meant that British Midland had to direct Novus to provide out-of-hours Web site operations coverage in order not to disappoint its customers with the risk of down-time.
- Another finding was the variety of browsers that site visitors were using. For booking and payment functions, British Midland had decided to standardize on browsers that incorporated a high level of security. These were invariably the latest versions of browsers available at the time. Many site visitors were using old versions of browsers that did not support secure encryption and these could not use the CyberSeat payment functions. In this context, users of Compuserve encountered particular browser incompatibility problems in the area of security. An option was therefore provided to allow users to download the latest version of Netscape's browser (a similar download service is also supported by Microsoft via its own Internet site).
- Originally, CyberSeat asked users to register before being able to browse the site. However, this proved cumbersome and instead, users were only asked to identify themselves as part of the booking and payment processes, using the details of their plastic cards.

Looking to the future, British Midland has been able to build on the success of its CyberSeat experience to plan ahead. It is considering a travel agent product that could be based on a new Intranet approach. In effect this would encourage travel agents to book British Midland via the Intranet instead of using their GDS connections. The Intranet would also have several other important spin-off benefits within the company, such as changing the culture to increase staff empowerment levels, increasing team working, fostering more open communications and sharing corporate information more widely and easily.

MARRIOTT

Marriott has been highly successful in using technology to market and sell hotel rooms and related services to customers around the world. The cornerstone of this distribution activity is Marriott's central reservation system, MARSHA (you can find more information on MARSHA and Marriott's interconnection to GDSs in Chapter 3). The latest version of MARSHA, known as MARSHA III, incorporates the functionality necessary to support Marriott's new Web site, which may be found at **http://www.marriott.com**. This is a popular Web site that was launched in 1996 and regularly receives millions of hits each month. These hits generate over US$1 million per month in hotel revenue and consequently rank Marriott within the world's top 5 per cent of all Web sites. The growth rate is also startlingly high at 100 per cent compound, month on month.

Marriott has been highly successful in using the GDS and HDS networks to distribute its accommodation services to travel agents around the world (see Chapter 4 for more information on GDSs and HDSs). This is illustrated by the fact that one in every five GDS bookings is for a Marriott lodging product. However, things do not stand still for very long and Marriott is faced with a rapidly changing distribution market, just as other travel industry suppliers are. The forces for change that are most relevant here are: (i) industry studies are forecasting that the number of travel agents will decline over the next few years, particularly in the USA; (ii) the Internet clearly has significant and proven potential as a distribution network for direct selling to consumers; and (iii) GDS booking fees cost several USA dollars per confirmed reservation, whereas the equivalent cost to make a booking on the Internet could be considerably less.

Alternative distribution channels other than GDSs therefore became a hot topic at Marriott a couple of years ago. The Internet was found to be especially attractive because it would enable Marriott to convey the details of its properties to the consumer in an interactive graphical way using pictures of properties and rooms, videos and virtual reality models, diagrams of floor layouts and maps of how to get there. Quick time virtual reality (QTVR), developed by Apple computers, makes use of 360° imaging technology, which enables a potential customer to actually look around the room they are considering booking by simply using a computer mouse. Also, the Internet's potential for direct relationship marketing was a powerful reason behind the company's decision to embark upon an Internet experiment. This resulted in Marriott's first World Wide Web site, which cost approximately US$1 million to develop and implement. By building on the success of this initial site, the development of subsequent versions has increased Marriott's Internet expenditure considerably.

Before its site could be created, Marriott had to overcome a significant technical architectural challenge. Its MARSHA system is based on operating software called transaction processing facility (TPF), which runs on an IBM main-frame. This is totally incompatible with the TCP/IP communications protocols used by the Internet. Although it was relatively straightforward for Marriott to connect an interface server to MARSHA for text, graphical images and information management, the reservations functions were another matter. To build its own Internet booking engine with an on-line interface to MARSHA could be done from a technical viewpoint, however, it would be quite costly. Marriott decided to postpone this major development until: (a) the demand for access to these functions increased, and (b) Marriott understood more about using the Internet as a marketing and booking channel. So, in the meantime, what was the answer to Marriott's Internet booking problem? Well, the answer was a very pragmatic decision taken by Marriott management, which was to use the Thisco hotel switch as the interface to MARSHA. The Thisco switch (which is explained in more detail in Chapter 4 in the section on distribution systems), was designed with an in-built capability to handle both TPF links to hotel systems and, via its TravelWeb booking engine, TCP/IP for Internet traffic.

So, Marriott's Web site is based on multiple Internet servers, located at its USA headquarters, that connect directly into the World Wide Web via the UUNET/Pipex ISP. (Fig. 5.36). These servers handle all incoming Internet traffic for **www.marriott.com** and respond directly to all information requests. They are fed with information from two sources: (i) a link to the MARSHA system that supplies information on such items as property descriptions, room rates, hotel addresses, facilities and so on; and (ii) other input supplied by picture scans, graphical images and mapping systems as well as some HTML text maintained by Marriott staff. However, when an Internet user wishes to view availability or make a booking, the server routes the enquiry via a third route – a direct connection to TravelWeb. Messages passing down the direct connection to the TravelWeb Internet booking engine are routed to the Thisco switch, which passes them on to MARSHA. The MARSHA system checks its room inventory data base and formulates a response, just as though it was a regular Thisco/TravelWeb reservation message. However, in this case the response is routed back to the Marriott Internet Web servers which route the message to the consumer. It may sound

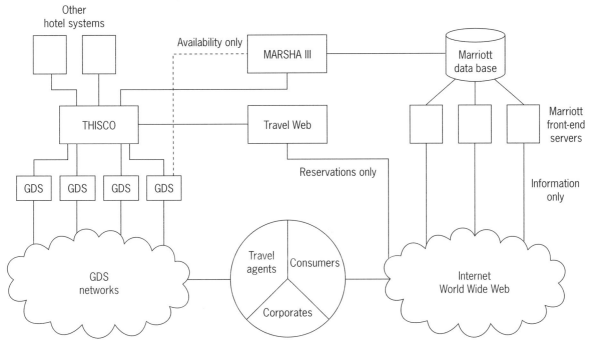

Figure 5.36 Marriott's internet connection

like a rather convoluted route but it still only takes between 2 and 3 seconds for MARSHA to respond to an Internet booking request with a confirmation number.

The beauty of this approach is that it maintains the stand-alone integrity of Thisco and TravelWeb. Neither of these systems need to hold a data base of rates or rooms. All data and inventory records continue to be held by MARSHA. This is an important point because it eliminates any problems that would undoubtedly arise from duplicating Marriott's hotel information on other servers. Another benefit is that it saves Marriott from having to develop a complex and costly booking interface to MARSHA. Having said this, if the volume of traffic handled on Marriott's Internet site grows substantially, then it may, at some point in the future, become attractive to develop a direct interface such as this. Only time will tell if this is economically feasible.

Besides being able to handle on-line consumer bookings automatically, there is one other important benefit of Marriott's Web site that I would like to explore in a little more detail. This is the production and distribution of printed brochures or what people in the industry call 'Collateral'. The kinds of brochures I am talking about here are not just restricted to a property flyers containing pictures and general descriptions. While these standard documents obviously exist, there are many more customized brochures that are printed specifically for corporate clients. These brochures include the usual pictures and descriptions, but they also contain a lot more. They invariably contain a full set of room rates that have been negotiated especially for the corporate company. Taken on a global scale, these brochures cost a small fortune to print and distribute. They also have a short shelf life. In other words because rates change and facilities are updated, the brochures quickly become out of date and must be scrapped. Not only is this a waste of the world's resources but it is also very costly. The Internet offers a solution to this problem.

Marriott views Internet sites as falling into one of three possible categories: (i) Shopping Malls, (ii) Supermarkets and (iii) Boutiques. The Shopping Malls are large sites that provide access to all kinds of suppliers; a particularly good example is Microsoft's Expedia. Supermarkets are sites that

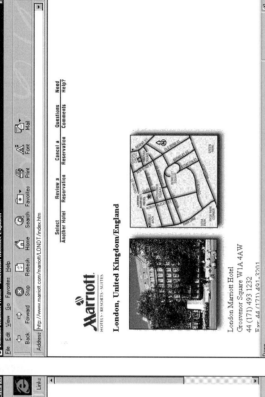
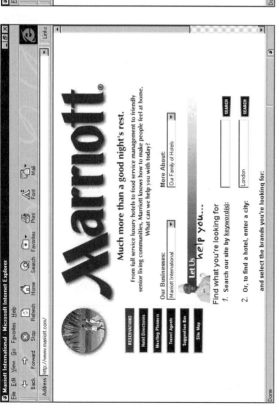
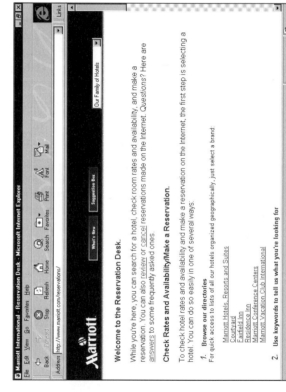

Figure 5.37 Marriott's home page (above)

Figure 5.38 London hotel rates (above right)

Figure 5.39 Reservations information

232 INFORMATION TECHNOLOGY FOR TRAVEL AND TOURISM

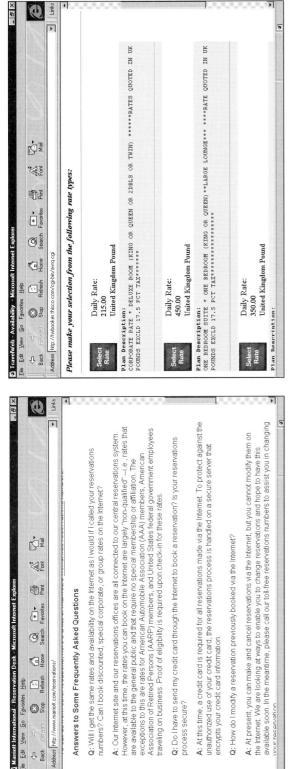

Figure 5.40 Q&A on booking procedure (above)

Figure 5.41 Rates in London hotels (above right)

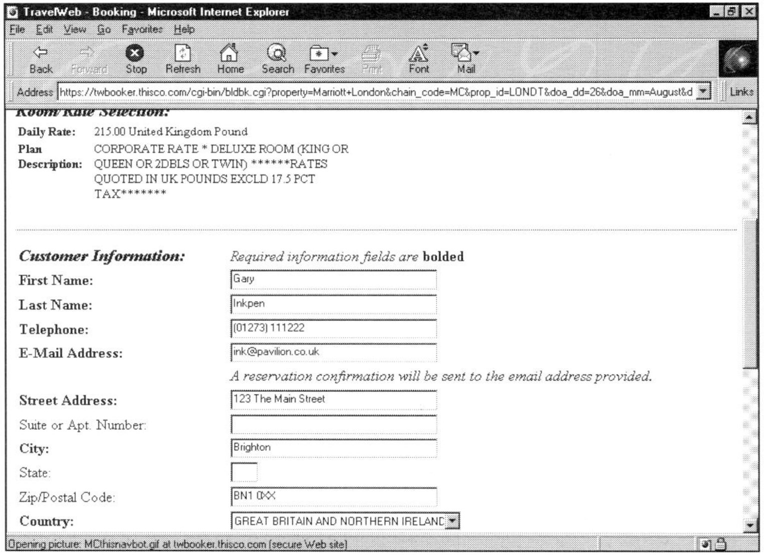

Figure 5.42 Booking screen

THE INTERNET 233

allow the consumer to purchase a wide variety of travel products; examples include TravelWeb and Travelocity. Boutiques are the smaller niche sites specializing in a single product only; good examples are Marriott and British Midland. This is a helpful analogy in today's retail Internet environment. Marriott belongs to several of these classifications. It participates within a Supermarket by way of its presence in TravelWeb but because it has its own site, it is also a Boutique. Boutiques can respond more quickly to environmental changes by introducing enhancements to meet the needs of the developing global hospitality market. Evidence of this can be found if we compare Marriott's original Web site with the latest version released in May 1997. The original site was highly customer focused and enabled visitors to carry out a wide range of functions including: (a) check availability, (b) view rates and conditions, and (c) book rooms. In May 1997 these basic functions were enhanced to include:

- **Interactive mapping** This is a USA-based mapping facility that is supported by a data base of 16 million points of interest and major business locations. The user simply enters his/her departure address and the site responds with a route map of how to find the nearest Marriott hotel. This map may be downloaded and printed by the user.
- **Enhanced search capabilities** A search engine has been introduced that enables the user to specify a number of search criteria including, for example; property features, meeting space attributes, nearby airports and geographic location.
- **Simplified reservation process** The number of clicks and keyboard entries required to book, confirm and cancel Marriott reservations has been reduced by enhancing the user/system dialogue.
- **Improved navigation** Some of the pages have been re-structured and re-indexed thus allowing users to find their way around the site more quickly and efficiently.
- **Meeting planning data base** A new section has been added to the site's data base that includes more detailed information for those who need to arrange meetings and conferences for their companies. The new information includes function room space, room dimensions, capabilities and floor plans.
- **Travel agent area and commissionable bookings** Marriott pay travel agents full commission on reservations made for all published transient rates that are booked via the Internet site.

Other services make use of the Web site infrastructure. For example, the secure payment processing functions have enabled Marriott to introduce the sale of Marriott Gift Certificates in denominations of US$25, 50 and 100, which may be paid for by credit card. Marriott is now increasing the use of e-mail for marketing purposes and plans to introduce some interesting new initiatives in the next phase of development. This will include a Concierge Service that will remind customers via e-mail of personal gift giving dates, anniversaries, birthdays and other events. No doubt Marriott will continue to develop and grow its site to meet the ongoing demand generated by Internet consumers. It will be interesting to observe how bookings shift between GDSs, travel agents and consumers as time marches on. No doubt Marriott, like many other travel vendors, would like to see a lot more business being done directly with its customers in both the leisure and business areas. If this does happen, the impact on GDSs and travel agents could be significant.

UTELL'S HOTELBOOK

Utell's Web site (Fig. 5.43) branded Hotelbook was launched in November 1996 and may be found at **http://www.hotelbook.com**. Utell intends this to become the world's premier hotel site on the World Wide Web. The number of locations featured will grow from 3,000 to over 6,500, thus embracing the entire portfolio of Utell's international hotel customers. The site is designed for use by all Internet consumers, be they individuals or travel agents. However, because only about 28 per cent of the world's hotel bookings come from travel agents, there is a significant opportunity to attract automated hotel bookings directly from the consumers, which represent the other 72 per cent. The following presents the major highlights of the Hotelbook site:

- **The basic Hotelbook service** Utell's participating hotel customers are allocated three Web pages within the Hotelbook site, free of charge. Each hotel is represented by at least these three free Web pages, which include:
  1. *Welcome* A page that shows a full colour 35 mm photographic image of the property together with a full textual description. A menu of further information is provided, along with the hotel's own e-mail address.
  2. *Features* Information that describes the hotel, its location, facilities and services using text and a graphical image (Fig. 5.44). Scrollable windows on this page show the hotel's features and services.
  3. *Rates* The rates for each hotel, which are shown within a series of pages automatically generated from the information stored within the core Utell system (see Chapter 4 for more information on Utell's systems).

  Consumers navigate their way around the site by means of a powerful hotel search engine specifically designed for Hotelbook.

- **Hotelbook's magazine** In addition to the product information, Utell's Hotelbook also includes travel news and information. This is sourced and edited by the Frequent Flyer magazine, which also provide sections on entertainment and current promotions. Hotelbook includes special awareness information on Utell International Summit Hotels, Insignia Resort and Golden Tulip Hotels, all of which are owned by Utell. Each of these pages allows each hotel to promote its own marketing partner, spread awareness of its special promotions, describe its products and distribute press information. The site also has a number of interesting features, two of which are: (i) a weather link that enables guests to review the weather reports for the time of their stay at their chosen hotel, and (ii) a rates conversion facility that enables customers to view rates in their own local currencies.

- **Electronic Brochure product** Participating hotels may elect to expand their coverage by purchasing five additional Web pages of their own. These can be used to promote information that is relevant to their own locations, such as:

  – *Meeting facilities* This can show images of meeting rooms, a description of the specialized meeting services available and the various meeting room hire rates.
  – *Location* This page can include a map of where the hotel is located, a description of how to get there and a list of nearby attractions.
  – *Room facilities* Pictures of the property's rooms can be shown as well as a description of the facilites available in each type of room.
  – *Dining facilities* Again, a full colour photographic image of each dining room can be shown along with links to other optional pages.
  – *Recreational facilities* Pictures and textual information enable the hotel's full range of recreational facilities for the use of its guests can be shown on this page.

  Extended hotel pages are particularly appealing to smaller independent properties that may not wish to invest in developing and running their own sites. Utell International is able to provide consultancy advice and guidance as well as Web page design services to hotels using these extended pages.

- **Group Display product** This is aimed primarily at larger hotel groups, i.e. those that are part of a group of ten or more properties. It enables them to promote their properties using a common corporate marketing message. This is supported by one of Hotelbook's optional features – the Group Display product, which is a sort of Web site within a Web site. This enables a hotel group to use several Customization features such as:

  – A branded home page of its own design (Fig. 5.45) – this is the first page that the consumers will see when they enter the URL of the hotel group (besides distinctive logos and product branding, this page can show special offers and promotions).
  – A customized colour scheme for all pages in the hotel group's site – this adds consistency and uniqueness of product from a marketing and product design perspective.
  – Supplementary pages to promote products – the hotel group may have special products,

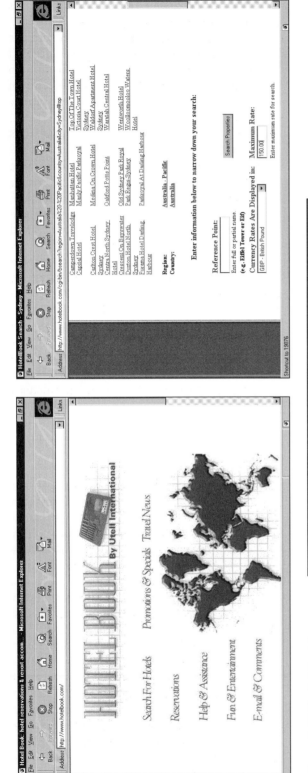

Figure 5.43 The Hotelbook home page (above)
Figure 5.44 Search criteria for hotel (above right)
Figure 5.45 Hotel page

236 INFORMATION TECHNOLOGY FOR TRAVEL AND TOURISM

unique to them. These may be explained and presented pictorially on special graphical Web pages.
- Partners and promotions – pages may incorporate hypertext links from the hotel group's own page or pages, to strategic partners such as frequent flyer sites.
- The ability to default Hotelbook's search engine to the hotel group's specific brand – this means that when consumers visit the site, their searches of the Utell hotel data base will always default to displays of the group's own properties.

The Group Display product is ideal for small- to medium-sized hotel groups because it allows them to enjoy the benefits of a full presence on the World Wide Web without the overheads of running their own sites.

- **Hotelbook reservations** To make a reservation via Hotelbook, a consumer has three options: (i) they may call any one of Utell's 52 telephone reservations offices around the world, (ii) they can send an encrypted e-mail message to the Utell Web server, or (iii) they can use Hotelbook's on-line booking system. Consumers who are nervous about entering their credit card details into the Internet will probably be attracted to the first option. However, there are many advantages to the second, more convenient method. One advantage is the return of a positive booking confirmation within 30 minutes of the original secure e-mail message being sent. However, because this is rather slow in today's instant 'here and now' business environment, Utell has developed a full on-line booking system. The on-line booking system produces a return confirmation within 7 seconds.

Hotelbook is marketed primarily through strategic business relationships. This means that Hotelbook can provide other Web site providers with a hotel information and booking system as an integral part of their site. This allows Utell International to benefit from the Web site's strong brand name and enables the Web site partner to offer a full hotel product, which may not be possible for them to do alone. For example, a national newspaper may have a site that enjoys a high hit rate on information that is not solely accommodation based. The newspaper may decide to add a 'places to stay' guide. This can be provided to their site visitors via a hypertext link to Hotelbook. The link would be almost transparent to the consumer who would see Hotelbook pages modified and customized to the newspaper's own particular 'look and feel'. Other examples may be drawn from airlines, car rental companies and tourist board sites.

The development of this site is an example of Utell's belief and commitment to the Internet. The reason I say this is because the site is not expected to generate significant revenues for some years. In fact, in its early years, Hotelbook will be very much a loss-leader product. Revenue streams are primarily expected to be derived by charging hotels a commission for reservations delivered via the Internet channel. However, a secondary source of revenue will come from selling the Group Display and Electronic Brochure products. Some revenue may also flow from offering the hotelier on-line advertising opportunities. All of these revenue streams will no doubt take some time to develop and will not become significant until the critical mass of the Internet is reached.

INTERFACING SUPPLIER SYSTEMS TO THE INTERNET

There are many countries where non-air products are distributed to travel agents and consumers by old technology, like videotex in the UK, or by proprietary national distribution systems, such as START in Germany and Esterel in France. These systems often limit their suppliers in terms of what can be offered to end users and how their services can be extended to other markets. End users frequently compare them to Windows-based systems and the Internet, against which they look decidedly dated. Take Videotex for instance. Many of the current videotex systems that are widely used by UK travel agents to book package holidays have been around for the past 20 years. These systems are cumbersome to use because they are character based, slow to respond to user's requests because they use old telecommunications technologies, subject to data corruption if accessed over dial-up lines and very limited in terms of their appearance. The new Internet technologies offer suppliers a solution to most of these problems, while at the same time opening up completely new distribution opportunities.

The supplier's problem to date, however, has always been; 'How can these new distribution technologies be used to boost bookings without incurring substantial development costs to replace in-house legacy systems?' One possible solution is to combine various new software technologies with standard Internet tools to produce an interface that supports both Intranets for private or limited access and the Internet for public access by consumers. This means that end users, whether they be travel agents or consumers are then able to access the supplier's core legacy system using standard Internet browser software that runs on virtually any PC. A new company that has recently entered this field is Gradient Solutions (a trading name of NewPage Systems Limited), based in London.

Gradient offer travel suppliers the opportunity to interface their legacy systems to the Internet while also improving the quality and usability of their booking screens. This has the dual benefits of: (a) enabling the supplier to continue using legacy booking systems without the need for any costly systems changes; and (b) allowing end users, whether they be consumers or travel agents, to enjoy the benefits of simple and dynamic Web-based pages of information for booking purposes. Gradient offers these services to suppliers in one of two possible ways, either:

- **Facilities management** The supplier contracts the development and operation entirely to Gradient who runs the interface software on its own computers. The Gradient computers are Sun Netra Web Servers, which use Cisco routers and fibre-optics to link both to the Internet and the supplier's legacy system, by high speed data lines. This computer has an uninterruptable power supply and incorporates firewall software to prevent unauthorized access to other parts of the system, virus detection routines, secure encryption algorithms and tape back-up systems. It runs 24 hours each day, seven days per week and reports on the number of end-user site visits and bookings made, for each supplier.

- **Supplier's Internet server** For suppliers that already have their own Internet or Intranet server computer facility, the Gradient interface software can be added. This software comprises several layers including: data communications, legacy system interconnection, legacy-to-Web middle-ware, added-value business logic and World Wide Web presentation. Once developed and loaded, however, the responsibility for running the network and handling the security issues lies firmly in the hands of the supplier's own IT department.

The interface software does more than simply convert a legacy screen format into an Internet page. It also enables new dialogues to be implemented by combining data from more than one legacy system screen into a brand new Internet page, complete with drop down lists, check boxes and radio buttons. An Internet-based approach also enables suppliers to distribute a great deal of descriptive information about their products to end users. This information can be created and stored using HTML techniques. This can be linked to booking response screens to create new items of information for users. Finally, the new pages can easily be 'e-mail enabled'. This means that when a user wants to receive more information, personalized to their own situation, they can request an e-mail response from the supplier. It is far easier and (perhaps more importantly), far more cost effective, for a supplier to respond to a prospective or current customer in this way, rather than by using the telephone.

But the overriding benefit of this approach is the ability of Internet-based technologies to broaden the reach of travel suppliers. A supplier may, for example, decide that the first step along the road towards a more widespread distribution strategy might be to open their system up to a specified group of travel agents, perhaps in a certain area of their home market. This is characterized as the Intranet approach. It allows the supplier to retain a tight level of control over who can access their system and what functions are provided. Later, when a sound base of experience has been accumulated, the supplier might decide to open the system up to all travel agents in their home domestic market as well as some overseas agents in other countries. Finally, the supplier has the option of allowing consumers to access the system on a global basis. This final step may involve some tailoring of the system to make the functionality less complex for the occasional, untrained users. The important point is, however, that the basic infrastructure can remain relatively unaltered. The supplier may continue to use their legacy system and is able to control the degree of system roll-out without being hampered by costly changes to their core system. In summary, a Web-based distribution system for travel suppliers offers the following advantages:

- The screens are easier to use than many legacy systems and other national distribution systems like videotex, START and Esterel, which means training is minimized.
- Screens appear more high-tech and can incorporate graphics and images that enhance the image of the supplier company.
- Several legacy system screens can be integrated into a simpler and more comprehensive end-user page with up-front editing that can speed up the booking process and reduce the transaction load on the supplier's central computer system.
- The booking process reflects the current business logic of the legacy system upon which the new Web-based distribution system is based.
- The Web page can be presented in the end user's own local language. Pages can be constructed as and when needed in most languages.
- Tariffs and fares can be displayed in the local currency of the country in which the travel agent or consumer is located.
- Core legacy system booking products can be integrated with fringe products such as travel insurance and foreign travel money to generate new revenue streams.
- The use of HTML techniques enables the supplier's Web site to incorporate an electronic brochure that describes the supplier's products in pictorial as well as textual terms.
- An on-line Web site enables suppliers to offer special promotions such as last minute bargains, late availability and the re-sale of cancelled bookings.

The key economic statistic that suppliers will no doubt use to determine whether or not to interface their systems to end users via the net, is the relative cost to receive a customer booking via the telephone versus the equivalent cost over the Internet. Because it is estimated that a typical telephone booking costs around US$10 and an Internet booking costs only US$0.50, you can see that there is a powerful argument for suppliers to consider this approach. The costs involved are really threefold: (i) there is the cost of developing the interface between the supplier's legacy system and the Web server, (ii) the facilities management charge for running the travel agent Web site, and (iii) a unit charge of around US$0.50 for each booking made over the network. With Internet-based solutions such as those offered by Gradient, it is possible that the long awaited migration from videotex to PC-based booking systems is about to commence.

Business travel on the Internet

Much of the above has focused on the way suppliers use the Internet to make direct contact with leisure travellers in their homes. But another significant opportunity is to use the Web to support business travellers and the companies for whom they work. Not only are suppliers entering this field but so are GDSs, travel management companies and new suppliers. Using the Internet for business travel functions is particularly attractive because: (a) business travellers are relatively sophisticated and are sufficiently confident to make their own travel arrangements, (b) business travellers often carry their own lap-top PCs with them when they travel, (c) many companies are seeking to use technology to increase the effectiveness of their travel policies, and (d) networking is an excellent way of integrating the complete business travel cycle from trip planning, through ticketing to expense reporting and administration. So, all in all, there are some very compelling factors that make the Internet an excellent platform from which to launch the next generation of business travel support systems.

As a result of advances in the field of technology, there are now a number of new travel oriented Internet sites and associated tools. While some of these are perceived as posing a threat to travel agents, some maintain the travel agent firmly in the loop between the customer and the suppliers. However, there is no doubt that the use of business travel Internet-related technologies will change the role of the travel agent considerably. One leading site is American Express Interactive (AXI), developed by American Express.

AMERICAN EXPRESS' AXI

American Express is uniquely placed to provide an integrated business travel management service because it operates two components that are critical to the success of a company's travel needs: (a) a global business travel service that is provided by a network of offices in most major countries of the world, and (b) a comprehensive range of card services, many of which are focused on controlling company expenditure. These two critical ingredients have now been combined with the technological capabilities of Microsoft to create the next generation of business travel services delivered over the Internet. It was in July 1996, that American Express and Microsoft announced a strategic alliance jointly to create an intuitive corporate solution for on-line air, hotel and car rental reservations (Fig. 5.46).

Over the course of the next year or so 'project Rome', as AXI was initially called, was developed by staff from both companies. In developing AXI, American Express has used the Microsoft Travel Technologies (MTT) platform, a suite of software products that specifically support Web-based travel applications. The result of the development programme is an Internet-based system designed initially for the USA corporate travel market called AXI. AXI was launched by American Express in July 1997 at a leading USA business travel conference held in St Louis in the USA. This initial product is designed for USA companies that want to provide their employees with the convenience of end-user travel management tools while at the same time retaining the control necessary to maximize their overall travel budget. American Express plan to launch an international version of AXI in 1998.

The AXI product is an integrated set of travel management services (Fig. 5.47) that uses the Internet as its distribution medium. It takes a holistic approach to business travel. By this I mean that it is built around the business travel life cycle, which comprises: (a) establishing and maintaining the company's travel policy, (b) supporting travellers with their trip planning activities, (c) making reservations and bookings either prior to the trip or modifying arrangements during a trip, (d) ticketing and boarding, (e) processing payment and expense reports (normally the paperwork bane of a traveller's life), and finally (f) providing management information that can be used by the company to negotiate better deals with suppliers and closely monitor internal expenses. Let's take these stages of the business travel life cycle in more detail and explore how AXI supports each one.

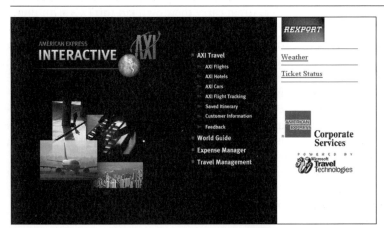

Figure 5.46 The AXI home page

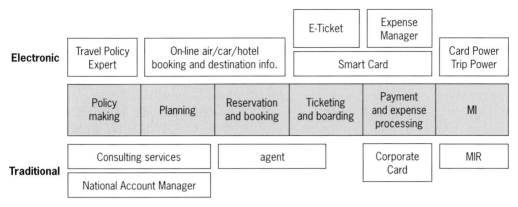

Figure 5.47 The American Express travel and entertainment management process

Travel policy

A travel policy is invariably established at main board level within a company. At this level, it usually comprises an overall set of guidelines outlining key directives such as when different classes of travel may be used by staff travelling on company business and the various entitlements of different grades of employees. These policy statements are expressed as unambiguous guidelines for travellers and define a set of rules governing how travel suppliers are chosen. Despite the fact that this process may appear simple, it is often quite challenging for companies actually to implement their travel policies consistently throughout their organizations. Although it is at the heart of virtually all business travel activities, it is surprising how many companies either do not have formal travel policies or which do not communicate the policies effectively to their employees. A cornerstone of AXI is therefore the Policy Editor, which automates this process.

AXI's Policy Editor supports the formalization, communication, execution and monitoring of a company's travel policy. Access to the travel policy maintenance functions are of course restricted to a senior level within the company's organization. This is often the designated travel manager, head of personnel or chief financial officer. AXI enables a data base to be maintained of travel policy parameters. This is sufficiently flexible to allow different groups within a company to each have their own slightly different travel policy. AXI recognizes different policy groups and other underlying environmental factors, such as the base currency, by means of codes assigned by American Express as part of the initial set-up process. Several other parameters and data elements combine to form a company's integrated travel policy, including for example:

- **Policy text** AXI's Policy Editor supports the inclusion of a company's full travel policy in textual form. The policy may be indexed and stored using HTML, which supports hot links to other related sections and relevant parameters within the Policy Editor. Eventually, AXI will be enhanced to include context sensitive help functions that support automatic back referral to specific sections within the travel policy, as appropriate to the user's query.
- **Preferred and excluded airlines** This is a powerful way for the company to keep a tight control on precisely which airlines its employees use for their business trips. Airlines can, under the complete control of the AXI user, either be included on the preferred list or specifically excluded. The old accusation frequently made by airlines during the negotiation process that the company has very little influence over which flights its employees choose, is groundless. With AXI, a company can instantly de-select a given airline or add a new carrier to its preferred list; any changes such as this take effect immediately. Similar functions also apply to hotels and car rental services (see the note on filtering below).
- **City airport selection** A travel manager may choose the precise airports that are included

in availability displays for any given city. For example, a regional airport may offer cheaper flights than a city centre hub. However, regional airports have the disadvantage that they are often not quite so convenient. The person setting the travel policy can choose which airports to include in the GDS displays shown to their travellers by AXI (see the note on filtering below).

- **Number of stops** The maximum number of flight stops may be specified within AXI by the company as part of its policy. This allows the company to decide the limit to which it is prepared to let its employees suffer multiple stops *en route* to their destinations, in order to achieve low cost fares. Generally speaking, the higher the number of stops, the lower the fares. However, flight stops increase travelling time and add to a traveller's discomfort. It is therefore important that their use is carefully controlled. AXI will not show alternative flights that feature more stops than the maximum specified in the Policy Editor (see the note on filtering below).
- **Filtering** Many of the travel policy functions supported by AXI employ filtering techniques such as those described above. Filtering allows a company to decide those suppliers, airports and travel arrangements that are both allowable and non-allowable, within the bounds of its travel policy. While the AXI technology supports filtering, the decision over precisely how the filtering parameters are used is totally under the control of the client company. If a company decides not to use filtering, then its AXI users will be presented with all options reported as available by GDSs and other information systems accessed by AXI. In many respects this is no different from the way companies enforce their travel policies at the moment in a manual environment. The difference with AXI is that the technology allows companies to be more successful in applying their policies in actual practice and this in turn allows them to control their travel expenditure more effectively.

The automatic application of an effective travel policy provides a company with a substantial bargaining lever in its negotiations with suppliers. Historically, suppliers have taken the position in rate negotiation meetings that companies have very little immediate control over their employee's travel decisions. It used to be very difficult to get employees to switch from using one travel supplier, to using another, e.g. to switch from using Airline A from City 1 to using Airline B from City 1 without impacting other airlines and cities. With widespread use of AXI, however, this is perfectly possible (Fig. 5.48). By making a number of simple adjustments to AXI's Policy Editor, the company can cause an immediate impact on the business delivered to specified travel suppliers.

Planning, reservations and booking

Trip planning is a vast area within business travel and it is closely integrated with the reservations and booking process. This is why I have merged these two stages of the business travel life cycle into this single section. Historically, these stages have arguably been a booker's prime time-waster because the tasks involved can mean long spells on the telephone explaining travel requirements to a secretary or a travel agent, which is then followed by frequent call backs and changes associated with fare selection. With AXI, the business traveller or their designated booker, e.g. a secretary or personal assistant, can cut out these time wasting intermediate steps by directly accessing travel information, fare data bases and availability information themselves. What's more, the AXI system enables the traveller's personal preferences always to be taken into account at each stage. Let's examine how AXI supports the provision of travel information and reservation services in a little more detail:

- **Travel information** For many travellers, the first stage in the trip planning process is to carry out some basic research on the destination areas included in their proposed itineraries. AXI provides access to Microsoft's global mapping and travel information data base, which I have explained above under the Expedia heading. For corporate travellers this can be extremely useful because it allows them to check information, such as whether or not visas are required for the countries they intend

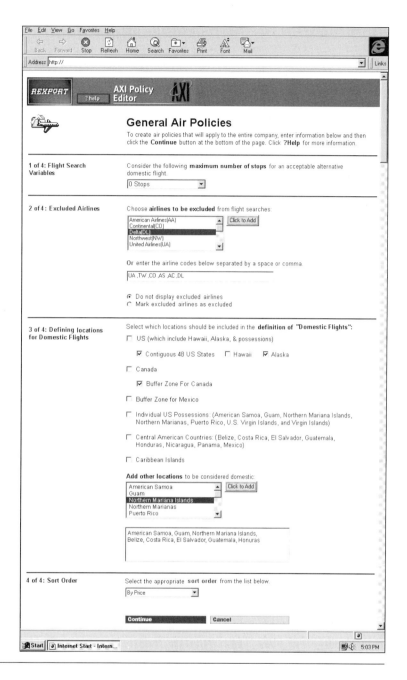

Figure 5.48 AXI general air policies page

to visit, what the weather forecast is for the region and what cultural events are taking place during their planned stays. The travel data base also contains a great deal of detailed information on restaurants and other attractions in the destination area.

- **Air** Travellers use the AXI GUI to define their requirements in terms of from/to city pairs,

date and time of travel, class of travel and many other parameters. When a traveller decides to request an availability display, AXI first consults its internal data base of specially negotiated fares for the itinerary specified. It uses these to construct an availability request that is sent to the GDS. In the USA, AXI uses the Sabre GDS; however, other major GDSs may

THE INTERNET 243

Figure 5.49 AXI seat pinpointer

well become available in the future as AXI is enhanced and extended over time.

The GDS responds with an availability display for each flight that meets the traveller's requested itinerary. Flights that are outside the company's travel policy may either be shown with a flag designating them as 'outside policy' or they may be excluded from the display altogether. The decision on which of these display options is implemented is made by the company and specified within the travel policy section of AXI. The availability display shows flights either: (a) in sequence on fare price; or (b) in sequence with those closest to the chosen itinerary first and those farthest away last. The prices shown on the display are those that are taken either from the negotiated fares data base or from the scheduled fare as stored within the GDS. Negotiated fares may be either those obtained by American Express and available specifically for its customers or those that have been obtained by the company itself.

The traveller simply selects their chosen flight and can then either: (a) store the flight details as part of an itinerary that they are building in AXI, or (b) proceed with making the reservation. To make the reservation, the traveller's preferences must be entered. These

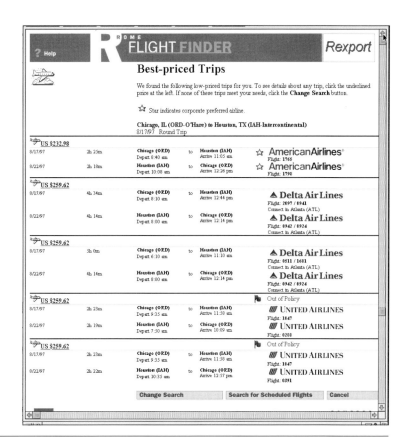

Figure 5.50 AXI best price tips page

can automatically populate many of the fields on the reservations page from a personal profile that each authorized traveller can maintain themself within AXI. This stores fields such as meal types, smoking or non-smoking preferences, the desired aircraft seating position, e.g. aisle or window, frequent flyer programme details and many more. Naturally, these pre-populated fields may be overridden by the traveller as necessary. When a reservations request has been successfully answered, i.e. the requested seat or seats are available on the flight, AXI's GDS response may in many instances offer the traveller a scrollable seat map of the aircraft (Fig. 5.49). This shows seats already reserved, those seats that are only available to members of the airline's frequent flyer programmes and other available seats on the flight. The AXI seat map therefore allows the traveller to choose their own seat: the availability of this function within AXI depends on whether or not it is supported by each airline's CRS.

- **Hotels and car rental** AXI uses a data base of hotels and car rental companies that is maintained globally by Microsoft as part of its MTT service. This data base is also updated by American Express with details of specially negotiated rates (Fig. 5.50). These special rates may in fact be of two main types: (i) rates that have been negotiated by American Express for general use by its corporate customers, or (ii) rates that have been negotiated by corporate customers themselves and are only available for their own use. Depending upon the authority of the end user, this data base may be searched and reviewed in many different ways. (From here onwards, I am going to be talking about hotels, as we explore how AXI works, but virtually the same remarks apply to car rental.)

When a corporate traveller requests AXI to perform a hotel search, the AXI server filters the data base to show the user only those hotels that their travel policy allows them to see and that meet their stated accommodation requirements. At its highest level, the selected data base listing shows summary information and possibly a picture of the hotel, in ascending sequence on room rate. (The choice of whether or not to include a picture on these pages is taken by the hotel or its parent chain company in conjunction with Microsoft who retains editing control over the hotel data base.) If further details are required, the user may either choose to view details of the hotel and its amenities or may choose to view a map using Hotel Pinpointer.

The mapping feature is similar in many ways to that already explained above for Expedia, because AXI uses the same MTT platform to support this function. The map shown on the AXI first response page pinpoints the selected hotel in a wide-area context that includes the chosen destination. The user can then choose to zoom in and view the hotel's location at closer quarters or use the mouse to determine how far the chosen hotel is from certain landmarks. The user also has the option of drawing a box on the map and then viewing all of the hotels that fall within this boxed area. This is a powerful yet extremely simple to use feature of AXI, which has the added benefit of allowing the user to print the map for inclusion as a part of their travel documentation (Fig. 5.51).

When a hotel is chosen by the corporate traveller, the first two things they will want to know are the availability of the required room in that property and the daily room rate. AXI firstly interrogates its hotel data base for property and rate information. Following this, it automatically links to the GDS for room availability information. The following situations may subsequently occur, depending upon the hotel and rate chosen by the user. Either: (a) if the hotel and the rate are present in the GDS, then the availability is shown with an option to book on-line, (b) if the rate selected by the user is not stored within the hotel's GDS record then the user is offered the option of sending an availability request directly to the hotel, or (c) if the hotel itself is not present in the GDS then a request can be sent to the appropriate American Express travel office for follow-up and booking.

All requests made via AXI are handled by GDS PNR queuing systems that may employ several communication channels including, for example, teletype, e-mail, fax or the telephone. It is interesting to note that in case (a) above, the hotel and the traveller receive confirmation of the booking on-line, but the hotel must then pay the GDS a booking fee; whereas in (b) and (c) the hotel must manually process the incoming request to make the reservation and the traveller must wait for a confirmation, but the hotel does not need to pay the GDS a booking fee. It will be interesting to see how these economic dynamics influence the future ways in which hotels choose to record their rates within the GDSs, in particular for customer bookings involving specially negotiated rates.

The entire planning, reservations and booking process is undertaken within the company's travel policy, as created by the travel manager. This means that the availability displays that are shown and the rates that are used are all filtered through the travel policy parameters (see the note on *filtering* in the travel policy section above). If, for example, the company has decided not to include a specific airline in its displays, then that airline will not show on a corporate traveller's availability display. Finally, AXI checks to ensure that the planned trip falls totally within travel policy with regard to fare, class and carrier.

The company's travel manager determines the appropriate action to be taken when an attempt to book an out-of-policy trip is detected by AXI. The action taken can vary in intensity from a simple warning to the traveller, right through to freezing the booking altogether. If the booking requires pre-trip authorization, AXI will ensure it is not completed or ticketed until the required level of authority has been granted. This is achieved by means of a message that is automatically sent by AXI to the person responsible for authorization. This person can view all the trip's details, including any explanations for the out-of-policy situation that the booker may have previously entered.

In certain situations, a company may allow an out-of-policy trip to go ahead. The company can use AXI to decide how it wishes to proceed in such cases. For example, AXI can either: (a) simply warn the travel manager that an out-of-policy trip is under way; (b) warn and document the out-of-policy booking and prevent it from proceeding; (c) document the situation but automatically authorize the trip, and so on. Each company can therefore use AXI to report out-of-policy situations as it sees fit for its travelling employees.

Ticketing and boarding

As electronic ticketing becomes more widespread, the issues associated with ticket delivery and boarding will recede into the background. Although AXI can handle e-ticket transactions, for the moment, the vast majority of airline tickets must be physically delivered to the traveller prior to their departure date. This may be accomplished by three methods: (i) ticket on departure – the ticket is collected by the traveller from an airline desk at the airport, (ii) collected from travel agent – AXI can queue the ticket for printing by an American Express travel agency location near to the traveller, or (iii) delivered to the traveller's home or office by secure express courier. For home or office delivery in the USA, American Express queues all tickets to the chosen carrier's central distribution hub, e.g. for Federal Express this is located in Memphis. The tickets are actually printed on-site in the hub, packaged and delivered overnight to the traveller's home or office. If at any time the traveller wishes to check on the status of their delivery, the AXI Web site home page contains a hot link to the carrier's own Web site. The Airbill Tracking number is used as the key to support enquiries from travellers.

Payment and expense processing

Payment for travel services can be supported by AXI in several alternative ways. Charges can, for example, be billed to the traveller's own American Express corporate card. This can be attractive from the company's viewpoint because it eliminates many of the accounts payable functions that are an inherent part of business travel. Or, certain expenses such as air and rail tickets can be billed to a lodge card. A lodge card is a single American Express card against which company travel expenditure is billed on a central basis, for all employees. The choice of payment method implemented by AXI is chosen by the client company in conjunction with American Express.

AXI also automates one of the banes of every traveller's life – the completion of an expense voucher following completion of the traveller's business trip. These functions are provided by AXI's Expense Manager application (Fig. 5.52). The primary source for the electronic expense voucher is the card charges that are submitted to American Express by service establishments, i.e. places where the traveller has used their card to purchase goods and services. When American Express receives these records of charge, either electronically or in paper form, they are input at regional operating centres around the world and eventually find their way into the corporate traveller's American Express card account. These charges may be viewed by the traveller and categorized for inclusion in their electronic travel expense voucher. Charges may be viewed in detail and the traveller may split them into the expense categories that their company uses.

In future, service establishment charge records may be available that are already split according to the services used. This is particularly relevant to hotel charges where the actual room rate may be only a small proportion of the total check-out bill. Not only will this facility make life easier for the corporate traveller but it will also enable the company's buyers to include other relevant expenditure in addition to room charges, when they negotiate future room rates with hotels and hotel chains. Finally, the traveller may create new entries for inclusion within their electronic expense voucher to record non-card expenditure, such as cash spent on taxis, tips and snacks. When the traveller is happy that their electronic expense voucher is complete, it is e-mailed to their designated authorizor (usually their line manager), for electronic approval. Approved expense vouchers are filed in a data warehouse that forms the basis for AXI's management information.

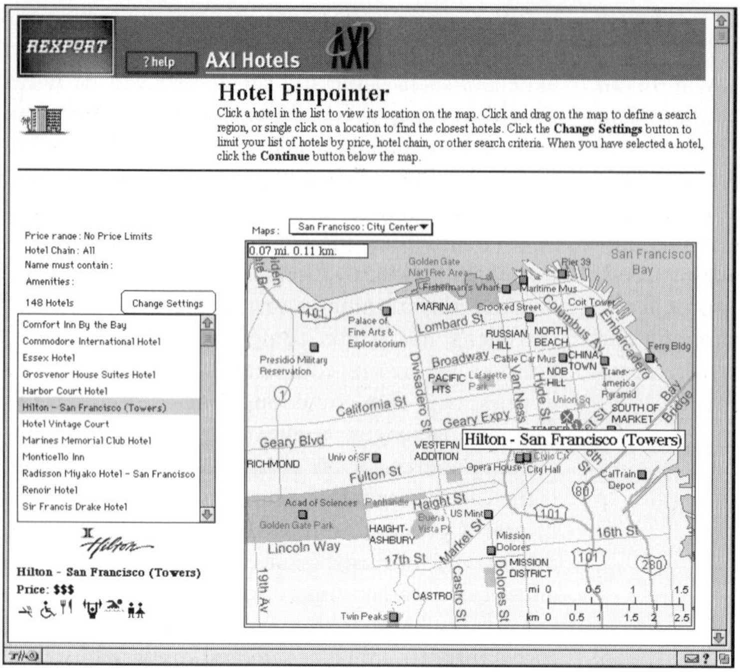

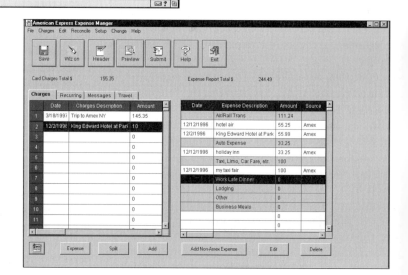

Figure 5.51 AXI hotel pinpointer

	Date	Charges Description	Amount
1	3/18/1997	Trip to Amex NY	145.35
2	12/2/1996	King Edward Hotel at Parl	10
3			0
4			0
5			0
6			0
7			0
8			0
9			0
10			0
11			0

Date	Expense Description	Amount	Source
	Air/Rail Trans	111.24	
12/12/1996	hotel air	55.25	Amex
12/2/1996	King Edward Hotel at Park	55.99	Amex
	Auto Expense	33.25	
12/12/1996	holiday inn	33.25	Amex
	Taxi, Limo, Car Fare, etc.	100	
12/12/1996	my taxi fair	100	Amex
	Work Late Dinner	0	
	Lodging	0	
	Other	0	
	Business Meals	0	
		0	
		0	

Figure 5.52 AXI expense manager

Management information

AXI has been integrated with several powerful management information applications that were originally designed to run in stand-alone mode on a company's own in-house PC. The functionality of these applications, known as Card Power and Trip Power, has now been migrated to AXI's Internet server, thus providing users with a choice of whether to run them locally or via the Web server. These software tools provide both internal and external management information functions. Authorized users can employ these management information support tools to select and process historical internal data in a variety of ways to measure travel expenditure and keep a check on the amount of business delivered to suppliers. Information on actual expenditure incurred can be extremely valuable to a company in negotiating the best possible deals with suppliers and in reviewing the effectiveness of a company's travel policy.

An important source of external travel industry data is also available to authorized AXI users, such as the company's financial officer or travel manager. These data are researched and published by American Express Consulting Services and profile the travel patterns of other companies, e.g. the Air Fare Survey Index. Such information can serve as an extremely useful yardstick for benchmarking exercises that establish the relative effectiveness of a company's travel policy *vis-à-vis* others in related fields. In particular, it allows the company to review its corporate rates for air travel, hotels and car rental services with industry averages by business sector and city pairs.

The technology

The AXI development uses the MTT platform for many of its functions. This is the same platform that was used to launch Expedia, Microsoft's own leisure travel oriented Web site. The web pages use frame technology throughout, which makes the system very easy to use and navigate. Frames allow users to drill down into the depths of data structures yet always provide a means to hot link into completely new areas. Under the terms of the joint venture contract, American Express has a two-year exclusive licence to the jointly developed corporate product. The AXI architecture comprises several different elements:

- **AXI Web server** This is an Internet server operated by American Express that runs software applications supporting: (a) a central data base of travel-related information, and (b) the core AXI processing functions. AXI uses Microsoft back-office server software to support the Internet, Intranet, or for that matter client/server technologies for networking and central data base access. These Microsoft products include Windows NT Server, Internet Information Server and SQL Server. In addition to this, Microsoft products are also used to connect into American Express' back-office systems, which feature a quality control application, electronic ticketing, a low fare search facility and support for a data base of special rates negotiated directly between clients and certain high volume travel suppliers. The core of the data base that stores hotel and car rental information is maintained centrally by Microsoft and distributed electronically to the AXI server.
- **Travel reservations server** The travel reservations server is located within Microsoft's computer facility in Redmond, Washington State. It actually comprises a number of server computers running software that interconnects AXI users with the GDSs, each of which uses different communications technology. Access to the GDSs provides AXI users with reservations functions for airlines, hotels, car rental services and, eventually, local supplier access in international areas.
- **Front-office** AXI's front-office client PC environment supports access to Internet technology and local processing using software that runs in the employee's desk-top or lap-top computer. The AXI PC client supports any browser that is HTML 3.0 compliant. The security and authentication standards used are SSL and Private Communications Technology (PCT). These enable credit and charge card transactions to be carried safely over the net. AXI also uses established American Express software products that provide company employees with a comprehensive business travel and expense management system.

All these products and remote data bases may be accessed via the Internet or a corporate Intranet from an employee's own desk-top or lap-top PC. So, as you will see from the above, AXI provides access to air, hotel and car rental reservations, a data base of company negotiated travel supplier rates, the company's travel policy, preferred supplier prompts, a wealth of destination information and the ability to track business travel transactions. The AXI service is a living product that will continually be modified and enhanced to provide applicability to other areas of the world. It will, for example, be adapted to show local language, value-added tax, foreign currencies and different postal code formats instead of USA zip codes. As the system is rolled out to other countries, it will connect into local supplier systems via its links with the world's GDSs. There are three connectivity

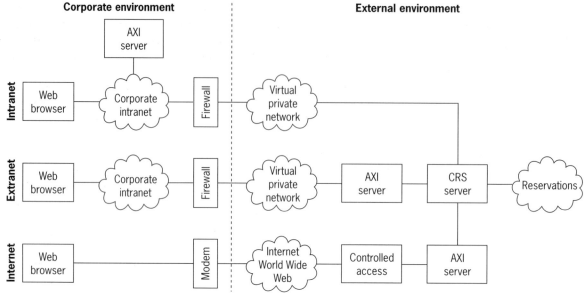

Figure 5.53 AXI network options

options that American Express customers may choose: Internet, Intranet or Extranet (Fig. 5.53). These are shown in Fig. 5.53 and are summarized as follows:

- **Internet** With the Internet option, both the corporate traveller and the company's head office staff access AXI via the public Internet. Dial-in access is provided by the company's chosen ISP. Using the Internet has the benefit of wide-scale geographical availability, including access services provided by ISPs with gateways in many countries around the world. Global access can be especially useful for business travellers who need to keep in touch with AXI via their lap-top PCs during trips.
- **Intranet** The Intranet option is available for companies that choose to run their own in-house communications networks with firewalls for protection against unauthorized access. Companies may therefore choose to implement AXI on their own in-house servers, which are kept up-to-date via information feeds channelled through head office connections to American Express.
- **Extranet** The Extranet option is very similar to the Intranet option except that the AXI Web server is actually run by American Express. This still allows the company's travellers to connect to AXI via its own private and secure Intranet, complete with firewall protection. In this environment, travellers access AXI via the company's private network, which is connected via the firewall with external computers such as the AXI Web server run by American Express.

To use AXI, a company must be a customer of American Express. In other words the client company must use the American Express business travel service and ideally should also be a corporate card-member. The company may then be granted a domain name by American Express that enables it to access the AXI Web server. This provides a client company with the automated core business travel services, but it still needs to be supported by a global network of human beings and servicing offices. After all, even in a fully automated electronic world, post-reservations support is critical. This is an integral part of the AXI service and is provided by the world-wide network of American Express offices and a 24-hour hot-line. It may be complimented further by a link between the company's corporate card and the travel information services of AXI.

The real issue for American Express is the degree to which companies will use AXI. Even for those companies that decide to implement the system, a significant internal selling job will need to be done on stalwarts. Despite the fact that AXI is so simple to use that it requires no formal training, there will no doubt be many employees who will resist change and continue with the old tried and proven travel services. However, there is a real impetus for companies to encourage the widespread use of systems like AXI because they have significant benefits, some of which may be stated as: (a) enabling a company's travel policy to become a powerful management tool and not just an administrative overhead; (b) substantially reducing the amount of time spent by travellers on completing trip requisition forms, travel expense vouchers and other forms of bureaucracy, thus increasing overall productivity; (c) helping a company obtain better deals from suppliers supported by accurate management information and the effective implementation of travel policy; and (d) delivering a more efficient and better informed travel service to business travellers. So, I think that companies will, over time, stipulate that an integral condition of being able to travel on company business is that the designated corporate travel system must be used for all aspects of travel, just like they stipulate that standard expense vouchers must be used to record expenses today.

RESASSIST

One such product is ResAssist '96, which was developed by the Travel Technologies Group (TTG), a USA company based in Dallas, Texas. This is a fairly new product, even in its home market of the USA. It is being marketed in the UK by ICC Travel Systems, the company that also sells the Concorde agency management system (see Chapter 7 for a more detailed presentation of the company and its main product – Concord).

ResAssist '96 is an end-user booking tool that is aimed at the corporate market. It is a product that travel agents can take ownership of, and market to, their corporate accounts. This enables business travellers to user their lap-top PCs to accomplish many travel booking functions themselves such as: viewing flight availability, changing existing bookings, making new reservations for flights, hotels and cars, and booking completely new trips. The product comprises a number of inter-related software products, many of which are industry standard, and access to TTG's Internet server on the World Wide Web. All of the following products may be offered by travel agents to their corporate customers:

- **Personal computer software** The ResAssist '96 software runs on a variety of PCs that may use Windows, UNIX or Macintosh operating systems. It is highly likely that most of the PCs running ResAssist '96 will be lap-tops because it is the frequent business traveller who stands to gain the most from this product. The user's PC will require one of the common Internet Web browsers. To access ResAssist '96 the traveller uses the Web browser to access the travel agent's Web site (see below) and, by entering their password, gain access to a main menu of options that includes: (a) Reservations – start a new trip, view or edit an existing trip; (b) Traveller – edit traveller's profile, select another traveller; (c) Destinations – view a destination, create a new destination, edit an existing destination; and (d) Company – edit company travel settings, create new contract, edit an existing contract. Some of these functions are restricted to certain personnel within the company, such as the company administrator. Others are fully accessible by all authorized travellers.
- **Travel agent's Web site** TTG offers travel agents two options for their Web sites. Either they may use TTG's service bureau Web site and effectively rent space from TTG or they may use their own Intranet Web servers which are owned and operated by the travel agency with an on-line link into the World Wide Web. In either case the travel agents will be using the TTG ResAssist '96 software to set-up and run the systems for their business travel customers. This software allows the company's travel policy, negotiated fares, preferred suppliers and other key parameters to be stored centrally and used to control all bookings undertaken by their business travellers.

- **Internet booking engine** Travel availability and reservations are provided by TTG's booking engine, which is linked to the major GDSs of the world. Wherever the travel agent's Web site may be, i.e. either on the TTG Web computer or on their own Intranet computer, booking requests are routed from the traveller's PC, over the Internet and into the TTG booking engine computer in Dallas. This computer system translates between the simple GUI format that is presented to users of ResAssist '96 and the more complex native commands used by the GDSs.

A traveller uses ResAssist '96 by first of all logging onto the Internet through their chosen point of presence, i.e. their Internet service provider. This may of course be accomplished from virtually anywhere in the world. They then enter their travel agent's URL, which will transparently link them either to the TTG Web site or to the travel agent's own Web server. In both cases the traveller is under the impression that it is the travel agent that is providing the on-line booking service, not TTG. Travellers are presented with simple user-friendly screens formatted in standard windows style, which they complete for booking purposes. First the date and time of travel is entered, followed by the destination city. This is either done by direct entry via the keyboard or by selecting entries from a list, e.g. destination cities. Behind the scenes, ResAssist '96 sends a message to its central booking engine computer in Dallas that is translated into GDS format and sent to the relevant GDS. After a few seconds, a response is received and translated into the simple format used by ResAssist '96, before being sent to the user's PC screen. This flight availability information is received by the user within a period of 10 to 30 seconds from the time of their original entry.

ResAssist '96 offers the traveller various choices based on the corporate travel policy, negotiated fares and company preferred suppliers. The contract fares, lowest applicable fare options and best time options are clearly identified. The traveller has the option of specifying a variety of sort options in order to help them decide the best travel alternatives. Once the traveller selects their choice of flights, a simple entry confirms the booking. A similar procedure is then followed for hotel and car rental services. Again, the ResAssist '96 responses are policy compliant showing negotiated vendors and pricing. Finally, live seat maps retrieved dynamically from Sabre, Apollo and Worldspan allow seat selection to be made. By using the ResAssist '96 user-friendly interactive dialogues over the World Wide Web, the traveller can build up a complete itinerary of their planned trip. When a booking is required, ResAssist '96 brings the travel agent into the loop.

The travel agent is brought into the booking loop by means of standard GDS queuing features. Each PNR created by the traveller using ResAssist '96 is automatically queued to the travel agency, whether it be located in the traveller's home town, another city or even another country. This is just one of the powerful GDS tools that TTG has used to enhance the level of control over the booking process. ResAssist '96 therefore enhances control over the business travel life cycle by means of two key features: (i) it allows the company to maintain close control over travel policy, and (ii) it allows the travel agent to keep a quality control check over all bookings.

The corporate travel policy is built in to ResAssist '96 by the company administrator, often the travel manager. This is done via the travel agent's Web site, as described above. The administrator logs onto the site and then uses a Web page editor to enter key policy parameters including, for example, allowable fare classes per employee grade and length of flight, negotiated air fares to be used on certain journeys, negotiated hotel room rates, preferential car rental rates and preferred suppliers. All of these fields are stored within the travel agent's Web site and subsequently referenced during the booking process undertaken by the traveller.

The quality control checks that ResAssist '96 supports, are possible because all bookings and changes to booking files are channelled via the GDS queue management system, to the company's travel agent. This means that when a traveller uses a lap-top computer to make a booking, it is queued to the travel agent for quality control checks prior to ticketing. These quality control checks may include, for example, ensuring that the traveller's department code is present in the

PNR, checking that the traveller has the required travel documents for the trip and ensuring that all MIS data are correctly recorded as per company policy. Ticketing need not be restricted simply to the travel agency location. It can for instance be queued to the nearest airport to the traveller for processing as a TOD. This is a good illustration of how productive it can be for a travel agent to be kept in the loop between the traveller and the supplier and consequently protect their source of income, i.e. commission from suppliers on the sales of product.

This type of technology brings several important advantages to all participants. It provides companies with an effective way to implement their corporate travel policy. From the business traveller's viewpoint it provides direct access to on-line supplier information without having to spend a long time on the telephone, possibly to an agent back in the company's home town. Finally, from the travel agent's perspective, it takes away a great deal of the routine administrative work associated with making and maintaining a booking.

TRAVELNET

TravelNet is a product of Reed Travel Group (for more information on Reed Travel Group please see Chapter 3 – Suppliers). Based in Santa Clara, California, TravelNet developed an Intranet-based product for the business travel market. In January 1997, TravelNet was acquired by Reed Travel Group and became an integrated brand within the enlarged group's portfolio of travel-related products. TravelNet is a corporate booking travel management system that delivers benefits to business travellers by allowing them automatically to book air, hotel and car reservations directly from their desk-top or lap-top computers via a company's corporate Intranet. Using an Intranet provides companies with a higher degree of security and control over access to their travel information and booking mechanisms. However, companies have the choice of making their corporate Intranet – along with its TravelNet system – available to travelling employees via the World Wide Web. Users must, however, have an authorized user log-in, password and corporate identification to access TravelNet. From a corporate perspective, TravelNet allows travel managers to access up-to-the-minute reports to help better manage policy compliance, supplier utilization and travel costs.

The TravelNet software, which runs on the travellers' PCs, is compatible with Microsoft Windows, Apple Macintosh and UNIX-based operating systems. Using their PCs, users can access their personal profiles stored on the TravelNet server, which also holds their trip and expense histories. The workstation software allows users to specify their itineraries using a windows GUI, which is very easy to use. The basic booking screen has five vertical action buttons down the left side of the screen (New Trip, Change Trip, Get Calendar, Help and Exit) and six horizontal menu items across the top (Air, Car, Hotel, Trip Notes, Itinerary and Reserve). The body of the screen contains a set of fields that depends upon the specific combination of action button and menu item selected by the user. Finally, the desired itineraries are checked against the company's travel policies, which are stored as part of the TravelNet data base residing on the corporate Intranet.

Once a booking request has been formulated, integrated with policy and checked by the user, it is transmitted to a GDS for availability checking in real-time. Both the Sabre and Apollo GDSs are supported by TravelNet. A response from the GDS is received within a matter of seconds. It shows several options, each of which is ranked by the degree to which it conforms to the company's travel policy. This display also shows availability and other important information as presented by the GDS and TravelNet. For example, contract fares – which may have been specially negotiated by the company – are included in the availability display. The traveller can then use TravelNet to explore many 'what if?' scenarios, such as: 'Will it change the cost if I stay over Saturday night?' Finally, once an itinerary has been built, TravelNet automatically processes travel authorizations.

TravelNet uses the Intranet server to store the company's travel policy and collect management information on actual trips undertaken. Several reports are produced that may be either viewed on-screen or printed for use by the company's travel manager. Many alternative report formats are available, including tables and pie-charts. These reports measure policy compliance, show airline market share, vendor utilization, travel patterns and trip expenditure – all of which are vital if a company is to negotiate the best possible deal with its travel suppliers. Other reports show negotiated rate utilization, total trip expenditure and travel policy exceptions. These are used by divisional line management to control costs and keep travel expenditure within operating budgets. The data reported on corporate travel reports may be taken either from bookings made via TravelNet or from bookings made through the company's travel agent. TravelNet may also be integrated with many third-party expense reporting systems.

Tourism on the Internet

The Internet is a natural medium for tourist organizations. It enables a country or area to create an encyclopaedia of information and even booking functions that can be distributed to every part of the globe accessible via a PC and modem. There are many Web sites devoted to tourism and it has been impractical for me to begin to address more than two of them here. However, I have included two relatively new and important sites that illustrate the power and reach of the Internet: the British Tourist Authority's **www.visitbritain.com** site and the Ireland national tourist board's award winning site **http://www.ireland.travel.ie**.

In reading how these sites operate I would suggest you look out for an interesting and recurrent theme that is one of the main issues facing tourism on the Web. I am talking about the ability of tourism Internet sites to facilitate bookings for their visitors. Most tourism Internet sites contain a fair amount of information on accommodation services, usually at the low end of the price scale. The kinds of establishments that fall into this category are the bed and breakfast houses and small independent hotels. Most of these do not have any kind of automation simply because they do not need it. At present, there are few third-party booking services with automated systems that could be connected to the new Internet sites. Naturally, if the visitor wishes to stay at one of the larger chain hotels then on-line booking functions via the Internet may well be available, e.g.

see Marriott in Chapter 5. However, this is not often the case for the SMEs, which comprise one of the most popular sectors of the market for tourists to Great Britain. So, the nub of the issue I am addressing here is: 'How can a tourism Web site visitor book accommodation with SMEs in the destination country of their choice?'

One possible way is for the tourism Web site organization to obtain a computerized inventory control system that would enable it to process on-line reservations for rooms. There are several software packages that could support these booking functions and that could be connected to the main Web site server. The problem is that if the tourism organization were to embark on this course of action then it would need to engage in all the usual commercial activities that are a part of running a business. It would, for example, have to be contracted to card companies, contracted to the accommodation establishments, operate substantial computer and telecommunications resources, charge a commission, accept some form of liability or at least responsibility for the quality of service provided to customers and last but not least, it would need to generate a profit. These commercial activities are in many cases incompatible with the role of national or regional tourism organizations. Their constitution usually contains some form of not-for-profit business objective.

Accommodation booking services are, however, a fundamental requirement for most visitors. While some sites try to support bookings by means of electronic mail and faxes, these are nowhere near as satisfactory as an on-line booking system that immediately guarantees the visitors the accommodation they need. There is therefore an opportunity for a third-party company to provide an automated Internet booking service for SMEs. If this could be done then the tourism site could 'point' to the booking site whenever the visitors reached the stage of wishing to make firm bookings. This would seem to offer some important benefits to all the parties involved: (a) it would leave the tourism organization free to focus on its core role, which is the promotion of tourism in domestic and overseas markets; (b) it would enable the booking service company to enjoy a new revenue stream, which would not be bundled up with the tourism organization's finances; and

(c) it would provide visitors with an on-line booking tool. Only time will tell if such a service is justifiable.

THE BRITISH TOURIST AUTHORITY

The Internet could be the solution to the information distribution problem that UK tourism organizations have been seeking for the past ten years or so. It is for this principal reason that the BTA, has focused significant attention and resources on Internet technologies. These technologies open up a new distribution channel enabling information to be supplied direct to consumers and other companies such as inbound travel trade suppliers, incentive travel organizations, conference organizers, overseas tour operators, suppliers and the travel trade in general both at home in the UK and also overseas. Opening up such a direct channel offers the promise of significant increases in tourism and enhanced promotional opportunities for Great Britain 'limited'.

For more general information on the remit and functions of the BTA you really need to read Chapter 2. This should serve as the required foundation for this section, which explores the opportunities that the Internet poses for the BTA and how the organization is actively following these up.

The ETB's Intranet experiment

The first step towards realizing its goal of distributing UK tourism information via the Internet, was the ETB's *Intranet* experiment. The domain chosen was the area of nation-wide tourist information on Great Britain, so desperately needed by England's or TICs (see Chapter 2 for a description of TICs). Although larger TICs usually have their own richly populated PC data bases of local information, there is precious little on the UK at large. So, while TICs can provide assistance with local services, they find it difficult to advise tourists who call into their offices requesting guidance on other parts of the UK to which they may be travelling as part of a touring holiday. The idea behind the Intranet experiment was to fill this information gap by providing TICs with access to a sub-set of TRIPS. Briefly, TRIPS is a national UK data base of information concerning accommodation, events, attractions, English language schools and other tourist-related data that is collected by the ETB. For a more in-depth analysis of TRIPS, see Chapter 2.

For the initial experiment, four TICs were chosen. These are situated in Canterbury, Weston-Super-Mare, Greenwich and Islington. Two further TICs are to be added soon, in Manchester and Hexham. The objective of the experiment was to distribute a sub-set of the TRIPS data base to these TICs using Internet technology. Each participating TIC was equipped with a PC, an ISDN connection, a modem for back-up purposes and access to the Internet via Demon, a UK Internet service provider. The end-user browsers supported by the Intranet experiment were Netscape Navigator Version 3 and Microsoft Explorer Version 3.

The ETB set up its own Intel-based Intranet server that was hosted on a facilities management basis by an external software company. This company used a server with a high capacity telecommunications link to the Internet that supported a transmission speed of 2 Mb/s. Web servers such as this have the capability to store vast amounts of information within a searchable data base environment.

For the purposes of the Intranet experiment, the ETB decided that enquiries would be answered by trained TIC staff who would be the sole users of the system (direct use by consumers was deemed to be outside the project's scope). However, even with a user-friendly Intranet front-end, TIC tourism officers would still have needed special training in order to use TRIPS effectively. This is because TRIPS is a sophisticated system with powerful search capabilities, some of which require many parameters and are too complex to be mastered by occasional users. So, to make productive use of the full TRIPS data base, a considerable amount of training and regular use would have been necessary for all users. Because this was considered impractical for a number of reasons, another solution had to be found.

The solution adopted was simply to use only a selected sub-set of the full TRIPS data base. The ETB therefore developed an extract program that selected only those items from the TRIPS data base that were considered germane to the majority of tourist enquiries handled in TICs. An extract such as this could be presented to TIC officers in a simple format and the system could then be used intuitively by a wide range of users with varying degrees of knowledge. Likewise, a simplified approach was taken to Web page formatting. Photographs and other multi-media techniques were specifically excluded. Only limited graphics and symbols were used and information was presented simply as text entries, just like existing information handbooks. Once this extracted and re-formatted information has been derived from TRIPS, it is periodically downloaded from the BTA computer to the Intranet Web server.

The Intranet experiment has also opened up the opportunity for TICs to enhance the ETB site with their own specialized local services. As already mentioned, TICs use a variety of PC technologies to store details of local services. In the future, one of the objectives of TICs could be to migrate this information onto the ETB's Intranet. The Microsoft product branded Front Page is being actively considered as the means of realizing this objective. Besides helping rationalize the information stored on a multitude of local TIC systems around the country, this development has the added benefit of also making even more local information available to other TICs and, in fact, to all users who the ETB decides can access its site.

The current technologies used to access the Internet are often criticized for being too slow. If all local information (currently stored on in-house PCs within TICs), were to be placed on the Intranet then it might be thought that frequently needed information would take longer to retrieve. However, storing all local information in this way need not slow down the speed in which information is accessed by TICs. It is possible to download and store that section of the Intranet information that is pertinent to a particular TIC. This downloaded information is stored on the local PC hard disk and as such is accessible within a fraction of a second.

The Intranet experiment has proved highly successful and has provided the ETB with proof that this technology can deliver real benefits to TICs. For the first time, TICs now have access to information on all areas of England; a significant improvement for any tourism-related organization. Of course, this needs to be viewed in the context of the wide variety of different types of TICs. There are some that are too small to justify an Intranet link. However, for most TICs the benefits of the ETB's Intranet link are significant. One of the spin-off benefits is that it allows them to network. This means that they are able to communicate with each other electronically and exchange information and files across the network. There are several areas for the future development of the Intranet including, for example, multi-media technologies.

The VisitBritain site

The BTA recognizes that the Internet is central to its information strategy. There are currently 50 million Internet users around the world, of which 2.5 million are in the UK, and the numbers are growing rapidly all the time with over 90,000 joining every day (source: IHBRP, Inteco Corp, 1997). With this in mind, it established an Internet site in 1996, populated by static information on Great Britain. Since its formation, the site has been progressively enhanced and developed. Although very rudimentary, it proved to be a worthwhile experiment that enabled the BTA to assess the potential implications of this new technological opportunity and decide upon its strategy for a longer term presence on the World Wide Web. Research completed in March 1997 revealed some interesting statistics on site visits, which are summarized in Table 5.3. Visits to the site increased from 70,000 per month in 1996 to an average of nearly 400,000 in 1997. A new site was therefore developed with the support of Interactive@Brann, an independent company who was also responsible for the highly successful EURO 96 Web site.

The VisitBritain site was formally launched on 16 July 1997 by Chris Smith, Secretary of State for Culture, Media and Sport. The site's address is www.visitbritain.com – a new world-wide Internet site that replaced the old experimental pages at this URL. This is a consumer-driven site it was developed using the experience gained from: (a) the ETB's Intranet experiment described above, and (b) the initial rudimentary site established in 1996. Its 40,000 pages of content are based largely on the information sources provided by the TRIPS data base, as described fully in Chapter 2. This

Table 5.3 Nationality of visitors to the 'old' BTA Internet site

Nationality of BTA's site visitor	%
USA	61.00
UK	13.00
Canada	5.00
Australia	5.00
Japan	2.50
Netherlands	2.50
Germany	1.50
Sweden	1.50
Brazil	1.00
Finland	1.00
Singapore	0.75

makes VisitBritain possibly the largest and most content-rich Web site originated in Britain. The site therefore contains information on the principle headings of accommodation, events, attractions, English language schools and other activities (see Chapter 2 for more information). The important point to remember here is that the TRIPS information upon which this site is based is kept up-to-date as part of the embedded life cycle of the BTA's and ETB's ongoing tourism operations.

The VisitBritain site is entirely consumer driven and is based on Internet frame page architecture. Its welcome page shows a map of Great Britain, which is the initial consumer interface page. This supports a variety of drill-down features, keywords and search engines. Special promotions are displayed to users at all times by means of Java applets that run moving image sequences across the screen. Information is generally divided into the following main categories: (i) Great Britain as a destination, (ii) regions of Great Britain, and (iii) special promotions:

- **Destination Britain** Britain as a destination presents the consumer with general information about Britain and with pertinent facts needed when planning a visit. Hot links to other sites are also embedded within many VisitBritain pages. Internet versions of successful BTA campaigns, such as the Movie Map, i.e. locations that have been the subject of films and TV movies, British Arts Cities, and Style and Design. There are also special interest sections for those keen on cycling and walking.
- **Regions** The areas of Britain begin with a page that shows a map of the UK, sub-divided into the major geographical areas. When a consumer clicks on an area of interest, a more detailed map is displayed. This process continues until a choice of specific tourist information is presented.
- **Special promotions** Special promotions show the consumer information relating to bargain breaks, special deals offered by suppliers and other incentives that are designed to encourage tourism to the UK. There will be up to ten new special offers every month from the travel industry and a media room will give journalists access to all the latest BTA press releases, media briefings, travel stories and even video footage.

The BTA site also incorporates a powerful and easy to use search engine. Consumers select areas of interest simply by clicking on check boxes and radio buttons. Specific place names may also be entered. The site then responds with all of the desired services that meet the search criteria. If necessary, a complete list of suppliers that also meet the criteria may be displayed. Many of these also provide graphical images and maps of their products or sites. Consumers may elect to store the results of their searches in private itineraries known as 'virtual brochures', which may be constructed as they browse the site. At the end of their sessions, the accumulated set of information comprising their virtual brochures may be sorted into itinerary sequences and then either: (a) stored on the site for future reference, or (b) downloaded to their PCs for local storage and printing. A virtual 'shop' is also provided that enables consumers to purchase books, guides and gifts with secure on-line payment using their credit cards. The following is a quick summary of the main action button that site visitors may click on to see what is available on the VisitBritain site:

- **Introducing Britain** A whistle stop tour to give the uninitiated a taste of all that Britain has to offer.

- **The Shop** Visitors can choose from 39 books, guides and gifts. Secure on-line payment can be made by credit card, phone, fax or e-mail.
- **Facts and Figures** For students, the travel industry and the more serious minded tourist, this part of the site features detailed information on Great Britain, its constitution and the economy.
- **Special Offers** Up to ten new offers every month from the travel industry – tour operators, hotel groups, destinations.
- **Search** The VisitBritain search engine can select items of interest from the 40,000 pages that make up the site.
- **Destinations** England, Scotland and Wales and all their regions, areas and towns, including many links to other sites.
- **Home** This is a tab that is always displayed on every page and enables visitors to return to the home page from wherever they may be within the site.
- **Activities and Attractions** Places to visit, food and drink, culture, sport and other things that potential visitors would be interested in.
- **E-mail** Site visitors may easily send the BTA an e-mail simply by clicking on an envelope icon at the top of the page.
- **Full Index** This contains an alphabetic index of all the pages of information that are held on the site.
- **Virtual Brochure** Visitors can accumulate information within their own virtual brochures but they must first register with the site in order to use this facility.
- **Map** Click on the map image to zoom in on regions and areas. From icons on the map visitors can access any of the 40,000 records with information describing England, Scotland and Wales. This includes accommodation, events, places to visit, TICs and language schools.

One of the most important features of the VisitBritain site is the accommodation section. This can be reached by several routes, the main ones being either the search engine or the map. Once a site visitor has specified the area and type of accommodation required, they are presented with a list of several possible places to stay. Each shows a full set of details about location and amenities, as well as the price. Because the VisitBritain site is not currently connected to any on-line booking system, this information is static. So, it is impossible at present to know whether or not the room in a particular establishment is available for certain specific dates. The site does, however, put the visitor in contact with the selected property by one of several means. First of all, if the property has an e-mail address then a request for reservation is e-mailed to them. Otherwise, if the property has a fax machine, a pre-formatted fax message is automatically sent. Finally, if the property has neither e-mail or fax then the VisitBritain site will generate a printed reservation request that is delivered by regular mail. While this is adequate for the present, an on-line booking facility would be preferable. This is something that the BTA may therefore well have under consideration for the future.

A particularly important feature of the site is data base marketing. This is made possible by the customer registration process. While casual site visitors may browse the site and obtain a great deal of useful information, the more serious site visitors are encouraged to register. Site registration must be completed, for example, before a visitor can start building their virtual brochure. Registration entails the site visitor entering some data (including their e-mail address), which enables a personal travel profile to be assembled by the BTA. When coupled with information on how this person actually used the site and what pages they viewed, this should enable a very powerful and accurate customer data base to be accumulated. As I have already mentioned in the section on marketing of this chapter, this is a critical feature of any Internet site. It should enable the BTA to target customers from all around the world with e-mail containing information that is particularly relevant to its particular interests in Great Britain. The marketing and promotion opportunities that could be generated as a result of this data base are enormous: in the not too distant future, this will enable the BTA to program the site so that each time a registered individual visits, it will show them the special interests they asked to see.

Companies can advertise on the VisitBritain site in a number of ways. First, there are special offers and promotions. A company can have its special offer mentioned on the home page and featured in the section reached by the Special Offers button. Then there are banner advertisements. These appear on the home page and other strategic points throughout the site. They allow the user to be routed automatically to an advertiser's own site by simply clicking on the banner itself. The BTA will also allow whole sections of its site to be sponsored and interactive partners with their own Web sites can participate in collaborative promotions with the BTA. Finally, there is the search engine. Companies can have keywords of their choice included within the VisitBritain search engine parameters and also in hot links throughout the site.

In designing the VisitBritain site, the BTA recognizes that consumers in different parts of the world will need to view it through a local gateway. By this I mean that they will need to be able to see certain items of information that are particularly relevant to them and ideally communicated in their local language. The BTA therefore intends to develop customized market gateways in North America, Japan, Australia and Singapore. These will allow overseas BTA offices to work with local partners to provide travel information and special offers translated into the home language and tailored towards its specific customers in these important inbound market areas. Opportunities also exist for British companies with a strong presence overseas to work with the BTA in all major markets. There will also be hot links into parts of the main site that most appeal to the market segments already identified by BTA as priorities. The BTA will promote the VisitBritain site by including the URL, i.e. site address, on all 128 titles of its literature, which are published in 25 languages producing 18 million copies world-wide.

Eventually, once the consumer aspect of the site is up and running, the intention is to develop a market facing aspect of the site. This will be a virtual Intranet because it will embody pages that can only be accessible by registered trade bodies approved for access by means of password control. This site will, for example, provide special rates on accommodation and attractions that may be of interest to a tour company considering creating a UK inbound package holiday for sale in its local overseas market. There will also be general information that tour companies need to create tours and facts on specific market segments. In short, the site will provide tour operators and other companies with all the information they need to bring visitors into the UK.

The technology

The 1997 VisitBritain site is totally new in terms of its presentation and user interface. Although designed using the experiences gained from the 1996 Intranet experiment, the new site offers some significant improvements and enhancements, many of which are based on Microsoft SeQueL Server technology with Internet Adapter. This handles user registration and holds 40,000 records of hotels, places to visit, events, TICs and English language schools. Behind the scenes, Microsoft Usage Analyst allows the BTA to monitor who is using the site and how they are using it. The BTA has applied to one of the industry's leading Web site auditors, ABC Electronic, for certification of its traffic audit, which is generated internally.

The VisitBritain Internet server is a Compaq Proliant 200 MHz Pentium Pro with 120 Mb RAM and 20 Gb of RAID disk array that is currently run on a facilities management basis, just like the ETB's Intranet experiment. However, the BTA is considering moving its new server so that it is co-located at an ISP's premises. The current Web site uses Microsoft's Internet Information Server (IIS) Version 3, which is used to deliver pages to site visitors. Active server pages (ASPs), customize the data base content and allow pages to be built on the fly.

A software product called MapObjects ActiveX and its components provide the interactive maps for the VisitBritain site. Internet Map server ActiveX's components talk to the Internet server applications program interface (ISAPI) filter of IIS to integrate MapObjects with IIS. This system processes and creates the maps in parallel with the Web server allowing more users to be processed simultaneously. Black Diamond's Surround Video is used to bring images of Britain (including the Silverstone racing circuit and Avebury's stone circles), to life.

Like the experiment, end users may use Netscape Navigator Version 3 and Microsoft Explorer Version 3 to browse the new BTA site. These browsers support the frame-based construction that has been used to build the pages for this site. Less sophisticated browsers are also supported but these will not allow the user to enjoy the frames, Java applets and Giff images that really bring Internet pages to life.

Implications for Great Britain's tourism

In my view, these new developments in IT that the tourism organizations of Great Britain are pushing forward, offer significant potential for radical change in the way tourism is promoted and supported around the world. In this section I will give my own opinions on some of these implications.

Looking to the future, I think it could well be that the VisitBritain Internet site will eventually replace the PIMMS marketing and brochure distribution system (see Chapter 2 for more details of PIMMS). Instead of overseas consumers contacting their local BTA offices either by telephone or by visit, they could instead access BTA's site on the Internet from their home or office PCs. Access to the site could in many instances provide the answer to the consumer's or the travel company's query. In cases where a brochure is still required, the request could be logged by the Internet site and fulfilled centrally. This will depend upon the growth in the number of consumers who are able to access the Internet. This in turn will undoubtedly depend to a large extent upon the wide-scale availability of enabling consumer technologies, such as net-PCs and digital interactive television.

In my view, this could have far reaching consequences for the BTA's *modus operandi*. Depending upon how the service is received and on the take-up of the Internet overseas, this development could result in substantial structural implications for the BTA's organization and deployment of resources. I think there are at least two significant opportunities that the Internet offers the BTA: (a) electronic publishing, and (b) central distribution. I'll consider each opportunity in turn:

- **Electronic publishing** BTA offices often stock up to as many as 300 different brochures. Now, brochures are costly to produce and incur an overhead in distribution, stocking and logistics. They also need to be regularly updated and consequently there are occasions when substantial amounts of out-dated stock may be destroyed. So, publishing this kind of information electronically offers the BTA a chance to reduce these costs and at the same time provide more up-to-date information for consumers and travel companies.

  It is perfectly possible for much of the information included in brochures to be made available on the Internet, including photographs, pictures, maps and graphics. In fact by comparison with paper publications, the Internet offers a wider range of media for bringing electronic brochures to life, including sound bites and interactive dialogues. Besides on-line use, many travel companies and an increasing number of consumers can already download information for local viewing and printing. As the Internet gains in popularity and usage, electronic publications of this type will become ever more popular.

  Many of the BTA's publications are printed in several languages, in addition to English. But because the generally accepted language of the Internet is English (or rather American!), it is quite possible for the electronic brochures to be published on the Internet in English. In the future, it may even be possible to use special software to translate English language text automatically into a foreign language. This is yet another factor that should allow the variety of different brochures to be reduced.

  When all of these things happen, I think the BTA could be attracted to undertaking a redesign of its range of paper-based publications and streamline them significantly. No doubt certain items of information will always need to be printed on paper. However, a significant proportion could well be produced and distributed electronically, direct to consumers and travel companies. Because printing and distribution costs are a major item of operating expense, the BTA could use the funds thus liberated for further promotional programmes that encourage tourism to Great Britain.

- **Central distribution** If the VisitBritain Internet site is indeed successful and is heavily used, then I can foresee a situation where overseas BTA offices could well require fewer local resources.

Instead of visiting or telephoning the local BTA office, consumers and travel companies would access the VisitBritain site on the Internet. Either the site would provide the information required or it would support automated brochure requests. In my view this could create the demand for a new central distribution facility whose role would be to fulfil end-user requests for brochures and other paper-based products.

This central facility would receive requests for brochures and other items of information from consumers and travel companies in other countries. Such requests could be effectively dealt with by a high volume fulfilment service. I think this new fulfilment service could be provided in one of two ways: (a) it could be set up by the BTA itself using in-house resources, or (b) it could be out-sourced to a private company and provided to the BTA on a facilities management basis. It is quite possible for this distribution facility to be created in regional areas or even possibly on a global basis. However, the overriding factor will be the need to supply the customer with the information requested in the shortest possible time.

These are just two examples that I have constructed to illustrate how the Internet could pose a significant opportunity for tourism companies and organizations. An opportunity for them to improve their interactions with customers while re-structuring their organizations for increased productivity. Such services on the Internet have the advantage of being centrally controlled yet can be accessed by a variety of means. They can be accessed from a consumer's home PC or a company's office workstation. They can also be used by BTA staff in overseas offices to service those customers who do not themselves have access to the Internet. Finally, they can be piped into a customer service kiosk that may be located either within the BTA office or in a public area such as the local high street or an airport.

Besides the provision of tourist information on Great Britain, a key requirement from a consumer's viewpoint is the ability to book services via the BTA's Internet site. However, although the site may well incorporate so called 'hot links' to the reservation systems of major companies, there is no such facility for SMEs. So, for example, even though it would be extremely useful for consumers to be able to view bed and breakfast establishments, make a selection and then book a room for a particular range of dates – this function is not supported directly by the BTA.

The reason for this is principally because the BTA is not a commercial organization with a charter to compete with the private sector. It is primarily funded by the UK Government and is not driven solely by the need to make a profit. Having said this, it needs to generate sufficient revenue in relation to the grants it receives. This enables as much burden to be taken off the UK taxpayers as is feasibly possible within the bounds of the organization's charter. As such it would not therefore be within the BTA's goals to provide a commercial revenue generating booking service on behalf of suppliers, which could end up competing with the private sector.

But even though the BTA's Internet development programme has no specific plans for the provision of an on-line supplier booking function for SMEs, there is nevertheless a market need for such a service. This offers a real opportunity for a third party service provider. Such a company could create the infrastructure necessary to enable bookings for SMEs to be taken over the Internet. Consumers could use the BTA site to view areas of the country, browse alternative suppliers in the SME category and then be linked automatically to the third party company's booking site on the Internet. However, at present, I do not know of any plans to develop such a capability.

Access to a supplier's product inventory is a natural extension of the information services provided by the VisitBritain site. It would enable consumers to go one step further than basic information gathering and allow them to use a booking engine to buy products and services directly from suppliers. In all probability, a third party booking site would be a separate Web site owned and operated by the service provider. It would take a service already selected by a consumer and show the up-to-date availability. The consumer would then be offered the ability to confirm the booking and guarantee the reservation by paying a deposit using a credit or debit card. The

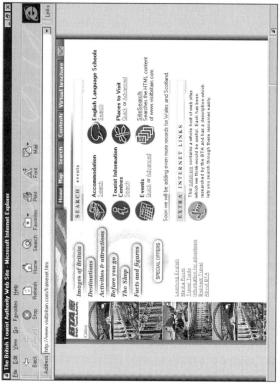

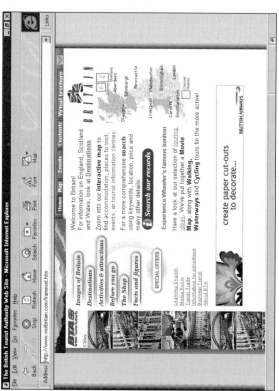

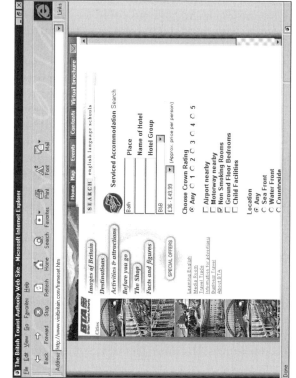

Figure 5.54 VisitBritain home page (above)

Figure 5.55 Search page (above right)

Figure 5.56 Search parameters (bed and breakfast)

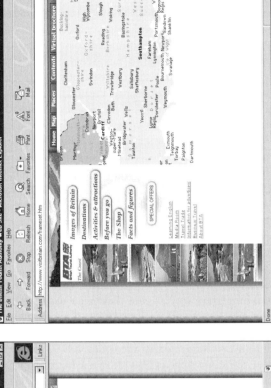

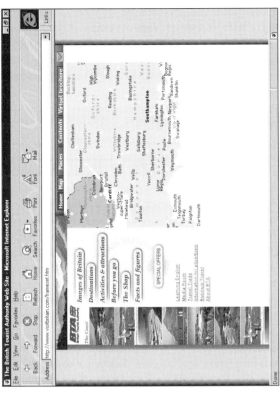

Figure 5.57 Bed and breakfast in Bath (above)
Figure 5.58 Area map (above right)

Figure 5.59 Events in Bath

THE INTERNET 263

company could then use some form of EDI message for booking purposes, like the TTI RESCON approach (see Chapter 1) and thereby derive a commission from the SME supplier. The BTA would no doubt encourage the development of a service like this because it promotes the UK as an inbound destination for overseas visitors, is an excellent extension of the BTA Internet site and yet does not directly involve the BTA in a commercial activity.

Finally, the feedback mechanism is an important feature of Internet technology that should not be overlooked. This is the ability of the Internet software products running on the server to keep a log of how many times the site has been visited and how users navigate their way around the pages of information. This feedback allows the BTA to keep track of how consumers are using its site, what the most popular pages are and how the data base is being searched.

GULLIVER

Gulliver is the name of Ireland's national tourist information system, which is described in detail in Chapter 2. The re-engineered Gulliver system has provided a springboard from which to launch a very effective Web site of information on the island of Ireland. In December 1996, Gulliver was launched on the World Wide Web. It is accessible via the Tourism Brand Ireland Web site and its URL is http://www.Ireland.travel.ie. It has already won an International Gold award as the world's best tourism Internet site at the ENTER '97 Tourism and Technology conference in Edinburgh and was selected by Yahoo as its 'world-wide pick of the week'. The information contained in the Gulliver data base provides the foundation for this Web site. However, at present it is only the items of static data that are replicated for display on the Web. Topics include sections on how to get to Ireland from different countries, places to stay, things to do and general tourist information.

The Internet has opened up Gulliver to users around the world who are offered a wide variety of information about Ireland that is portrayed using modern technology-based media. Even from an internal perspective, it has delivered many benefits to Bord Failte (the Irish tourist board). For example, it has enabled the central Gulliver data base to be accessed from a variety of local tourist offices, each using different technologies, e.g. Apple and various PC systems. Browser technology, such as Microsoft Explorer, has provided a platform from which these end users can benefit from full inter-operability at minimal cost. But in terms of international usage, it has been found that 75 per cent of the site's visitors to Gulliver are USA-based, a market of prime importance to Bord Failte. Besides text-based information extracted from the Gulliver central data base and formatted for Internet browsers, the Web pages also show pictures of properties, famous sights and accurate maps. In only a short time, the Gulliver Web site has become a comprehensive, interactive multimedia brochure on the subjects of Irish tourism, goods and services.

Users may select hypertext links to browse through the site and find the information they require or they can use a powerful form-based search engine to find more specific data. This search engine enables users to navigate their way easily around the massive Gulliver data base. The front page of the search engine shows a map of the island of Ireland, which clearly identifies all 32 counties. On the same page is also shown a list of topics of interest and activities such as, for example, horse-riding, golf, sailing, pubs and so on. To use the search engine effectively, a topic 'box' is checked and a county is selected from a drop-down list box. The instruction to search the data base may then be given and after only a few seconds, the users are presented with a set of customized information pages and references relating to their enquiries.

An additional innovative feature of the site is the ability for users to build up electronic personalized brochures and itineraries covering their planned trips. As pages of interest are found, they may be selected for storage and accumulation. These pages may, for example, include maps, pictures and accommodation details. Once a user's search is completed all accumulated pictorial itinerary information pages may be downloaded and printed by the user. There are even plans to develop the site further so as to support electronic bookings (which are in any event already a feature of the Gulliver core system – see Chapter 2).

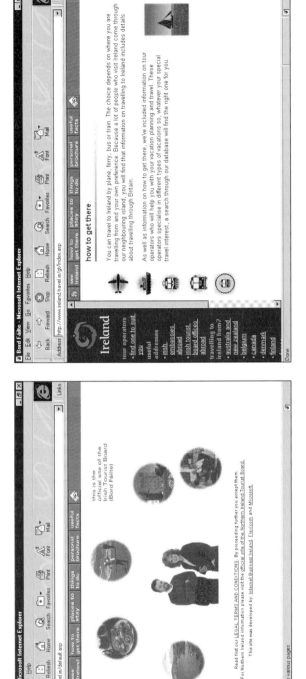
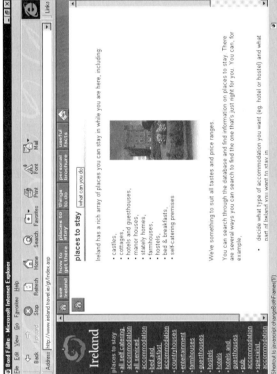

Figure 5.60 The Gulliver home page (above)

Figure 5.61 Gulliver – How to get to Ireland (above right)

Figure 5.62 Gulliver – Places to stay

THE INTERNET 265

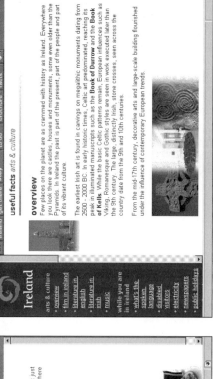
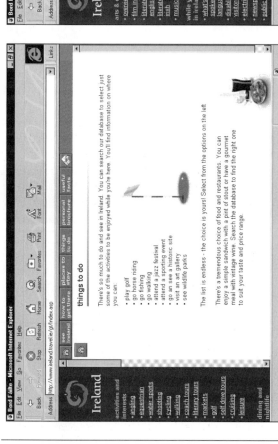
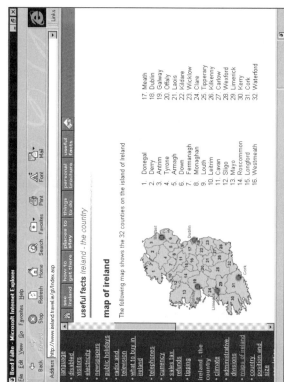

Figure 5.63 Gulliver – Things to do (above)

Figure 5.64 Gulliver – Ireland Arts and Culture (above right)

Figure 5.65 Gulliver – Mapping

The Gulliver site was developed over a period of just six weeks, starting in mid 1996, by a team formed from Bord Failte, the Gulliver Tourism information and reservations network, Microsoft Consulting Services, Internet Business Ireland and Flexicom. The heart of the site's functionality is made possible by the Gulliver data base. This was jointly developed by Bord Failte and the Northern Ireland Tourism Board as part of the revamped Gulliver project. It is maintained by Microsoft's SQL Server and is a distributed data base with a central version that is used to drive the Web pages. The central data base serves as a publisher of information, which is replicated and stored at remote sites where it is used by local tourist offices. Locally applied updates are consolidated at the central site, which is updated overnight. The updating of the central data base and the functions that keep track of remote changes are handled automatically by Microsoft's SQL Server.

The underlying technology used to construct the Gulliver Internet site is based upon the Microsoft BackOffice set of products. A key part of the site's platform is the IIS. This runs under Windows NT Server Version 4.0, a leading Microsoft operating system. IIS supports the maintenance of static information content and its presentation as Internet pages. It also manages the exchange of information between the Web server and the main data base controlled by Microsoft's SQL Server as described above, which is so critical for 'on the fly' Web page construction. This is a technique that allows the core data to be maintained separately from the Web page formats. Separation of these functions allows end users to maintain and update the information content without having to worry about its effect upon the Web page layout.

Another key IIS function is the ability to recognize the end user's browser and optimize the Gulliver pages for that particular site visitor. This allows site visitors who may be using an old version of an Internet browser to access fully the information content of Gulliver. The main drawback experienced by these users will be the possible lack of some images, especially those animated graphics driven by Java applets and their exclusion from any possible on-line payment or booking functions that may be introduced by Gulliver in the future. Finally, ActiveX technology is used to provide 360° Surround video. This enables site visitors to control their view of a wide-angle picture by using a PC's mouse. This is an impressive feature that can help convey a sense of actually 'being there' to the end user by allowing them to pan around a colour photograph of Ireland.

One of the next steps in developing Gulliver is to add a route-planning engine. This will allow visitors to specify the places they wish to visit and request Gulliver to work out the optimum routes, based on several parameters specified by the site's visitors. It will be of special assistance to anyone who does not have an in-depth knowledge of Ireland. Also under consideration is an on-line booking service for accommodation. However, these and other potential enhancements require a substantial amount of funding, and it is the source of the required investment that is the main issue for the future development of Gulliver. This growth and development is, to a large extent, dependent upon Ireland's plans to privatize the system and its related infrastructure. If a suitable commercial owner/operator can be found then it is highly likely that the Gulliver site will be enhanced and broadened even further.

Travel information on the Internet

There are a number of sites on the Internet that provide travel-related information. These sites are extremely useful to travellers during the planning stages of a trip. It would be virtually impossible for me to review every single information site on the Internet. However, here are just a selected few of them.

WORLD TRAVEL GUIDE ON-LINE

This is an interesting and very informative site with vast amounts of information available via an excellent indexing system. It is provided by AT&T and the Columbus Group and may be found at **http://www.wtg-online.com** (Fig. 5.66). The basis of the site is the information gathered by Columbus Press and used to publish its travel book entitled *The World Travel Guide*. Like the book, the site contains maps, pictures, climate charts, health/visa requirements, tables and data for every country

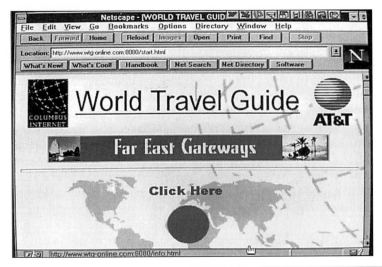

Figure 5.66 The World Travel Guide home page

of the world (including Antarctica). There are two search engines. The first locates a country and a topic and the second is branded HotelFinder:

- **Country/topic** Countries are selected by either clicking on a map of the world or by selecting the first letter of the country name. In either case, an index of countries is displayed from which any one may be chosen. At the country level, a variety of topics may be explored.
- **HotelFinder** As the name implies, HotelFinder navigates a large inventory of the major hotel chains. The display shows the hotel name, the city and the telephone numbers (both voice and fax). A useful facility is the option to specify a hotel chain and to include key words for searching purposes.

In addition to comprehensive information on each topic, the user is presented with a choice of products and suppliers related to the topic. Categories include transport, accommodation, business, essentials, social, addresses, travel and resorts. These categories present valuable information and in many cases also direct the user to either: (a) the latest relevant product information held within the site, or (b) to external sites on the World Wide Web. Two sections on travel news are presented: (i) World Travel Guide News – travel news and stories from around the world, updated daily with

Figure 5.67 World Travel Guide – example screen 1

Figure 5.68 World Travel Guide – example screen 2

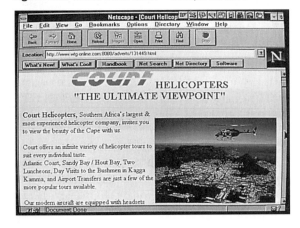

hotlinks to the core site and contact addresses; and (ii) World Travel Guide Features – a section of monthly editorials on a range of travel subjects linked to the Information Provider pages of the site. Both sections attempt to encourage users to visit the site and add value to information providers.

WEISSMANN TRAVEL REPORTS

Weissmann Travel Reports is a part of the Reed Travel Group. It is a leading provider of destination information to the travel industry and is accessed by millions of consumers via on-line computer services and Web sites, software applications, print publications and privately branded licensing agreements. The data are available by country profile (every country in the world), state/province profile (USA and Canada), city profiles of the most visited cities world-wide and cruise port-of-call profiles (all major ports in North America and the Caribbean). In all, the data bases include comprehensive unbiased and frequently updated information on more than 10,000 cities. There are several printed publications available from Weissmann Travel Reports. For the purposes of this book, however, I have chosen to focus on its electronic products, which may be summarized as follows:

- **Weissmann for Windows** The Weissmann Travel Reports, which are distributed by print subscription services, also are available via a Windows program that allows users to draw from all data bases to produce highly customized reports for travellers. This electronic publication is available on a subscription basis and the data is updated monthly.
- **Weissmann Travel Reports on-line with System One/Amadeus** The information in the publication entitled *North America Profiles and International Profiles* can be accessed on-line with System One/Amadeus, via a main-frame computer. This information can be customized by travel agents for their clients.
- **Travel Corner, America On-line** Weissmann Travel Reports has hosted *Travel Corner*, a core travel service on the 'travel channel' of America On-line (a leading ISP), since 1993. Features include portions of the destination databases, a Late Breaking News service, Virtual World promotional opportunities for travel suppliers, an Exotic Destination Message Centre, contests, Web links, Ask Arnie travel advice column, Top 10 Picks, Destination of the Month, an electronic travel photo album, a mechanism for ordering individual destination reports and advertisements with click-through Web links.
- **Travel Corner on the web** A Web site that includes some of Weissmann's America On-line features as well as original travel feature articles written by Weissmann editors, an interactive Travel Personality quiz, reviews of selected travel Web links, slide shows and an interactive directory of subscribing agencies.
- **Weissmann Travel Reports' Internet licences** Weissmann Travel Reports has licensed portions of its information to several travel-related Web sites.
- **Weissmann Travel Reports on CD-ROM** A read-only CD-ROM version of Weissmann for Windows. Sold exclusively to the library and education markets.

The subscription services have a total circulation of more than 4,000 travel agencies, mostly in the USA. More than three million verifiable impressions of Weissmann Travel Reports are made annually via consumer on-line services. Although Weissmann does not accept advertising in print publications, it does accept inserts into its update packages. It also supports promotion and advertising for on-line computer-based electronic publications as well as for its own Web site. Weissmann Travel Reports offers a service for Web buttons and banners that supports advertisements for both its America On-line areas and its Web sites. Virtual World is a service offered in conjunction with an established multi-media advertising agency to create promotional destination areas for Weissmann's consumer on-line services.

6
Networks

Introduction

Telecommunications technologies hold the key to the effective use of IT in travel and tourism. These technologies provide the connection path between suppliers, intermediaries and consumers. Logically, one could argue that the Internet should have been included here. However, my reason for not doing so is primarily because in so far as travel and tourism is concerned, the Internet is a subject in itself. A subject that has as much to do with marketing as it does with the technology used to connect the suppliers with the consumers of travel and tourism products. Not that the other communications technologies do not have anything to do with marketing – they certainly do. However, generally speaking, the marketing opportunities outside the Internet are much more oriented towards interactions which take place between the network suppliers and the travel and tourism companies, rather than consumers. I don't mean to imply that there are no consumer marketing issues with other networks, there probably are; however, they are not as fundamental and as far reaching as those posed by the Internet. In any event I hope this goes some way to explaining why this chapter concentrates on other non-Internet communications technologies.

Even without the Internet, the range of communications technologies that fall into the area covered by this chapter are enormous. Far too many for me to cover adequately in this book. However, I have tried to select those areas where telecommunications services are fundamental to any company or individual in the field of travel and tourism. The first of these, video-conferencing, is one which has long threatened the very business of travel itself. I therefore thought that this would be a good place to start this chapter.

Video-conferencing

Video-conferencing is often seen as a technological application that poses some real threats to business travel. It is a technology which allows people to communicate so effectively that they do not need to travel in order to meet each other face-to-face. Instead, they can use television screens linked by high speed telecommunication lines to hold discussions and assess the impact of the conversation on participants. This ability to view the reaction of the other person is important, since it is thought that 70 per cent or more of our interpersonal communication is perceived non-verbally. This is why video-conferencing is such a richer communications media than the plain old telephone.

Although video-conferencing technology has been around for a number of years, it is just about set to become a commonplace next generation business tool, just like spreadsheets and word-processors. A tool which holds the key to future corporate survival and competitiveness. Why? Because firms which effectively exploit video-conferencing will be quicker to react to business changes, quicker to bring products to market, more effective in controlling projects and more efficient in deploying their management's time. What most firms do not realize is that they already have much of the basic infrastructure in place which allows video-conferencing to be implemented very cost effectively.

Many large multi-national companies have set up their own in-house video-conferencing studios just for this purpose. They are becoming increasingly used in the current environment of cost cutting and expense account cut-backs. There are also independent studios that can be hired for smaller companies which cannot afford to set up their own video-conferencing facilities. These may be rented out on a time used basis. On the face of it these rates seem fairly expensive. However, when compared against the cost of international travel and the time lost by highly paid staff, they begin to look a lot more attractive. So, video-conferencing is a definite long term opportunity for the business travel market of the future.

Video-conferencing is all about using technology to enhance the ability of people to meet with one another, regardless of geographic location. Video-conferencing enhances this ability by: (a) the use of technologies which have been around for some time but which have only recently attained a viable cost/performance ratio, (b) the agreement by major industrialised countries of certain key telecommunication standards, and (c) the definition of video-conferencing standards between major suppliers. The following are some of the triggers for change:

- Computer processor chips such as Intel's Pentium are now extremely fast and capable of handling the thousands of complex compression calculations necessary to send video images down communications links cost effectively.
- Camera technology has progressed to a level where very small devices can be obtained for around $200 which capture images in digital form for processing directly on Personal Computers (PCs).
- The ISDN communications standard (more on this later), is a telecommunications protocol which has been accepted and agreed by almost every industrialised country in the world. It provides a common 'language' for computers to talk to one another at very high speeds.
- The video-conferencing industry has agreed standards which govern how systems from different suppliers can 'talk' to one another using standard protocols.

There are now approximately 90,000 video-conferencing installations in daily use throughout the world, i.e. 1996). By the end of 1997 this is expected to reach 180,000 and by 1998 the total is forecast to reach 350,000. This exponential growth is borne out by the commercial exploitation of the technology in different parts of the world. For instance, in the US, AT&T Global Business Video services provides 500 publicly available video-conferencing rooms right across the USA.

Like most technologies, video-conferencing will not deliver any meaningful benefits unless it is used effectively. And to be used effectively we have to ask, what is the underlying objective of the technology. Well, it is all about helping groups of people communicate with each other. Another term used for this activity is 'meetings'. So, lets take a brief look at meetings and examine how video-conferencing can help.

Meetings management

Meetings Management is so critical to the success of video-conferencing but is so often overlooked that I consider it worthwhile spending just a few paragraphs discussing it in a little more depth. First of all there are the different types of meetings which video-conferencing supports especially well and then there is the crucial aspect of meeting planning.

First of all, let's take a look at the different types of meetings and how video-conferencing supports each one. There are really three main types; (i) group meetings, (ii) one-to-one meetings including one or more desktops and (iii) multi-point meetings. Each of these is described as follows:

- **Group meetings** This type of meeting occurs with several people (usually between six and eight), at one end of a communications link, usually sitting around a table in a studio. At the other end of the link is a similar group, again probably in a video-conferencing studio. Historically, this has been the way in which companies have used video-conferencing to date.
- **One-to-one meetings** Meetings of this type occur between two or more individuals in remote locations, using the PCs situated on each person's desk. It is this type of meeting that will almost certainly grow explosively in the next few years.

- **Multi-point meetings** These are meetings of people in various geographic locations in a conference, usually with a designated chairperson co-ordinating the meeting. Again, it is highly likely that these types of meetings will become a commonly accepted way of doing business over the next few years.

Now, let's consider how meetings should be planned. This is important because if video-conferencing is to be effective, the organization of the meeting must be carefully planned and well controlled. This may sound like 'motherhood and apple pie', but it is very important and is surprisingly often overlooked by many companies. For a meeting to be effective some fairly simple steps need to be taken:

1. A co-ordinator needs to be appointed to plan the meeting and undertake the following actions. This is often the chairperson, but is sometimes an attendee with a different role, sometimes known as a facilitator.
2. A decision needs to be taken on whether this is a one-off meeting, a series of several meetings or regular periodic meetings.
3. The meeting needs to be carefully planned. This means that the goal, objectives, scope and end products of the meeting should all be clearly defined. This will often involve one-to-one sessions with the key players or members of the management team involved.
4. The co-ordinator must decide upon the most appropriate venue for the meeting, i.e. date, time and place. It is at this point that a decision on the use of video-conferencing facilities versus physical travel should be taken.
5. The required participants need to be identified and their superiors consulted to schedule their attendance.
6. A meeting package should be circulated in advance containing the following:
   - An invitation letter that identifies the meeting attendees and clearly states the venue, e.g. date, time, place, resources, as well as the targeted end time.
   - An agenda stating the subjects to be covered, who is leading each agenda item and the amount of time allotted.
   - The goal, objectives, scope and expected end products of the meeting.
   - Any relevant briefing papers pertinent to the meeting.
7. The co-ordinator needs to book the required resources. This may mean booking a conference room or a video-conferencing slot.

Much of this may sound obvious to the enlightened readers of this book. However, let me assure you that many companies do not follow these steps and do not devote anything like enough time, effort and attention to the management of meetings. Unless this is done, however, video-conferencing will often be a waste of time and money; and the pity of this is that the blame for an ineffective meeting is often attributed to the technology.

TECHNIQUES

Video-conferencing is not simply just about people seeing and hearing each other from one side of the world or country to the other, it is about sharing information. Many video-conferencing systems provide a 'whiteboard' facility that enables what one person 'draws' on their PC to be viewed by the other people attending the virtual meeting, on their PCs. Application sharing enables all people in the session to use a common software application, e.g. word processor or spreadsheet, as well as shared data.

A good video-conferencing system, together with these tools, enables groups of people in different geographic locations to work jointly on a single document, spreadsheet or other product. It is possible for the video image to be minimized, i.e. shown in a small window on the screen, so that other shared information, e.g. the document being jointly worked upon, can be seen by all 'attendees'. Each user has control of these features on their own workstation.

These features enhance the scope for collaborative development of products and services by people in diverse geographic locations. Such techniques are rapidly becoming a fact of every day business life. A company that uses video-conferencing to achieve these ends will often gain a significant time advantage while also keeping its travel and subsistence costs to a minimum. Two crucial factors in a competitive world.

THE STANDARDS

Historically, companies have used in-house video-conferencing studios to communicate only within their own organizations. However, the emerging deployment of video-conferencing is pushing these boundaries outward. Many companies, for instance, now use video-conferencing to communicate with their major suppliers and increasingly, their key customers. This is made possible by the use of standards, not just in telecommunications, but within the video-conferencing world itself. These are some of the standards that make video-conferencing a common language for the global business community:

- **ISDN** The UK's own standard has now matured and a version has been agreed by most industrialized countries in Europe, North America and the Far East. It provides a standard and very fast dial-up communications technology that is used to support video-conferencing sessions around the world.
- **Video-conferencing standards** H.320 is the commonly accepted standard for interpreting video-conferencing sessions on all types of hardware and software. A future standard, T.120, expands this into data sharing and 'continuous presence'. (Continuous presence is a feature that allows all participants in a multi-point call to be visible on each screen in the session, in their own windows.) Other standards include H.323 (LAN video-conferencing) and H.324 (the plain old telephone service, POTS – a protocol that may be used to carry video-conferencing for residential users).

THE EQUIPMENT

Video-conferencing is implemented on PCs, which in today's technical environment, usually communicate via dial-up ISDN. In the future, more use will almost certainly be made of interconnections via company LANs, which even today are bridged across company sites via WANs. It is, however, worth bearing in mind the demands placed upon a LAN by video-conferencing applications. Video-conferencing, like other image technologies, is very resource intensive and the impact on a LAN's load must be carefully considered. If this is not done then there is a real danger that the LAN will quickly become overloaded and all users will suffer slow response times. Hence the use of H.323 video-conferencing over LANs and WANs is really for the future.

The minimum IBM-compatible PC specification is usually a 386SX processor with at least 20 Mb of hard disk storage and 8 Mb RAM (Apple has its own Mac specifications for video-conferencing systems). In other words, only a basic PC is needed although this does assume that an additional hardware card is installed (see below). A PC requires the following enhancements to enable it to become a video-conferencing work-station:

- **Camera** This is a compact device that either sits on top of the PC in the case of desk-top systems or is strategically placed in a studio for group or multi-point systems. Some camera systems are very sophisticated and have a 'follow me' system for roving speakers inside a studio.
- **Microphone** There are many types of microphones. The best ones used for group or multi-point meetings automatically adjust for those speakers who roam around the room, and produce a constant level of sound. The audio channel runs at 7 kHz and is full duplex, i.e. concurrent two-way conversations are possible.
- **Software** This comprises special purpose computer programs that: (i) handle the audio and video images, (ii) control the compression of digital images, and (iii) oversee transmission protocols. Intel and Microsoft are seeking to make software image compression widely available over the next few years, thus making PC hardware boards, used by some video-conferencing suppliers, redundant (see below).
- **PC board** Many present-day video-conferencing systems use hardware, in the form of PC add-on boards that perform various functions. These include interfacing with the camera and microphone as well as executing the video compression/de-compression calculations. These calculations require the repetitive execution of some very complex algorithms and at present the most effective way of doing this is by using a dedicated processor chip on a PC plug-in board.

- **ISDN connectivity** The PC running the video-conferencing application needs to 'talk' to the PC at the remote location. Once the video-conferencing software has formed an image for transmission, the PC needs to send this to the ISDN port. It does this either via a directly connected ISDN line and terminal adapter on the PC itself or, more likely, it routes the transmission via the LAN to an ISDN gateway on the WAN operated by the company.

Typical video-conferencing systems range from under £1,000 for a desk-top system (excluding the PC itself), to around £40,000 for a studio system. There are several suppliers, including: PictureTel, Intel, Vtel, Compression Labs and GPT. In addition to this cost must be added the ISDN telecommunications expenses, which are explained below.

VIDEO-CONFERENCING COMMUNICATIONS

Speed is a key factor in video-conferencing. As expected, the faster the system runs, the better it appears *but* the more it costs. With most video-conferencing solutions there is a variety of speeds that can be chosen for different uses. For example, with a group or multi-point virtual meeting, a relatively fast speed of 30 frames per second is required to make all movement continuous and smooth. However, this requires a communications capacity of 384 Kb/s, which requires the use of a service such as BT's ISDN-30 (see below).

However, for desk-top PC to PC sessions, a slower speed of 15 frames per second is often perfectly acceptable. The only drawback of this somewhat slower speed is a rather jerky movement of the image and lip movements, which can become out of synchronization with their associated voice. The benefit is, however, that only a 128 Kb/s communication capacity is required, such as that provided by ISDN-2.

Both the above ISDN product offerings are therefore fundamental components of video-conferencing systems. ISDN is a digital switched communications technology, which is now widely available in the UK and most other countries. The basic usage charge is time-based and is exactly the same as for normal telephone calls. It is provided in the UK by BT, Mercury and some cable television companies. BT's offering comprises two optional services: ISDN-2 and ISDN-30.

- **ISDN-2** This uses simple twisted pairs of copper wires that run from the user's ISDN termination point to the nearest BT exchange. (Most BT exchanges are now digital.) Each line supports up to eight extensions and two sockets are provided from each line box installed on a customer's premises. This means that two PCs may operate at the same time, each at a speed of 64 Kb/s, or alternatively a single PC can run at 128 Kb/s. ISDN-2 involves a one-time charge of around £400 for a connection with £84 per quarter for ongoing line rental.

  Using video-conferencing with ISDN-2 at 128 Kb/s will work satisfactorily, albeit with the drawbacks mentioned above, e.g. jerky image movements. However, this may be perfectly adequate for one-to-one desk-top sessions. But for really smooth video approaching television quality, ISDN-30, which offers speeds of at least 384 Kb/s is needed.

- **ISDN-30** This uses various communications technologies to link a 2 Mb line from the BT exchange to the customer's premises. The link provides a minimum of six channels each of 64 Kb/s or 384 Kb in total. Up to 30 independent 64 Kb channels can be combined to form a fewer number of faster speed links. Two versions are available: ISDN-30 (I.421), which is a global standard, or ISDN-30 (DASS2), which was the original UK standard, now superseded by I.421.

THE BENEFITS

Put simply, the benefits of video-conferencing are those that an enterprise would realize if all its activities and those of its customers and suppliers were under one roof. It is the plain old ability of people to meet and talk about business problems that they share in common; and yes, there are the benefits of reduced physical travel as well, although these are not necessarily the predominate ones. The following summarizes some of the main benefits of video-conferencing:

- **Builds commitment and trust** It has been proven that when a person has eye contact with another, then any promise to carry out a task is usually fulfilled. This is not necessarily the case with a telephone call, which is less personal. Video-conferencing builds trust and reinforces commitment among participants.
- **Controls disparate projects more effectively** Many companies undertake projects that are formed from team members located in geographically dispersed offices. Video-conferencing allows project team members to 'attend' more frequent project co-ordination meetings at less cost.
- **Improved knowledge distribution** Video-conferencing is particularly effective in a distance learning environment. This enables, for example, product knowledge to be demonstrated far more effectively than by written instructions or costly travel. It also allows staff throughout an organization to learn about a new system or procedure that is being implemented.
- **Better support for teleworking** Many companies are beginning to use staff who do not regularly attend the office. There are many examples, including computer programmers who work from home, sales representatives who are located in remote areas and so on. These people can keep in regular contact and close touch with headquarters using video-conferencing.
- **Access to remote experts** With new technologies and emerging financial instruments, specialists are becoming ever more important to companies. However, if these specialists are remote from possible users then their knowledge is wasted. Video-conferencing allows knowledge and skills that are possessed by a few experts to be readily available throughout a company.
- **Accelerated decision making** Video conferencing supports meetings that need to be organized very quickly. When an important decision needs to be made in a short period of time, many key staff in various locations need to be involved. With video-conferencing people can be in touch almost instantly to discuss and analyse information required to make an important decision.
- **Less money spent on travelling** A company's travel budget represents a significant amount of directly controllable expense. Although video-conferencing will never completely replace travelling and face-to-face meetings, it can substantially reduce this controllable expense.
- **Less non-productive time spent travelling** Probably more important than travelling expenses is the time wasted by staff sitting in airport lounges, aeroplanes or trains. Video-conferencing enables highly paid and valuable staff to be more productive by reducing wasted time.

CONCLUSION

To be effective, video-conferencing must be embraced by senior management. It is simply no good just investing in the hardware, software and telecommunications technologies alone. Management must be proactive in directing the staff of its company to change the old ways of doing things. Some real directives need to be communicated to the workforce in order to capitalize on the video-conferencing opportunity. For example:

1. Regular management committee meetings need to be re-appraised to see if video-conferencing could eliminate the need for all attendees to be physically present in one location.
2. Line managers with travel budget responsibility should be tasked with reducing travel expenses by making more effective use of video-conferencing in their day-to-day project review meetings and other group sessions.
3. Innovative ways need to be found of making the services of experts more widely available throughout a company that is comprised of widespread places of business.

Companies need to recognize the importance of video-conferencing to their businesses. We have only to look at other related technologies, such as PCs and office productivity tools that provide word-processors and spreadsheets. Such tools are widely used and regarded as 'business as usual'. So in the same way, video-conferencing will shortly become a standard business tool.

However, if video-conferencing is to be implemented successfully by companies and used to its best effect to deliver real benefits to business, then it must be carefully planned. A strategy needs to be formed for using video-conferencing, the appropriate system suppliers and associated standards need to be chosen carefully, the impact on existing systems needs to be weighed-up and, finally, the company culture must be adapted to embrace this new paradigm in order to derive the maximum benefits and increase the enterprise's overall success rate.

THE FUTURE

In the future, it is highly likely that faster ISDN communication services will be available at cost levels far lower than today, in relative terms. This will make video-conferencing capable of producing even higher quality images with more sophisticated techniques based on the T.120 protocol. Virtual meetings involving people from several locations, will become commonplace.

This, along with the relentless reduction in the unit cost of computing, will generate an ever increasing growth in the use of video-conferencing around the world. The knock-on effect of this is likely to be twofold: (i) it will increase the amount of personal communication between people in diverse organizations while supporting a higher degree of collaborative working, and (ii) it will reduce the level of business travel.

The lesson is clear: travel agents need to watch carefully how their customers use video-conferencing. They need to do this in order to understand the technology's strengths and weaknesses and to adapt their service strategies accordingly. However, while video-conferencing will no doubt reduce the incidence of business travel, there will almost certainly continue to be a need for face-to-face meetings between people in different geographical locations. But before travel agents can breathe a sigh of relief, there is the little matter of the Internet to deal with!

Electronic mail

Electronic mail (Fig. 6.1) is a technology that is in its infancy in the travel business at present. The closest thing to it that most travel agents would recognize, is the old telex facility. However,

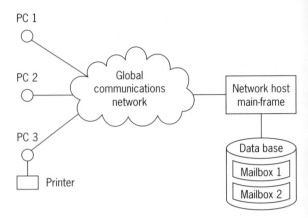

Figure 6.1 Electronic mail

electronic mail is far more sophisticated and is an ideal communications media when a dialogue with another time zone of the world is involved or when the person being contacted is travelling extensively. It can be more efficient than the telephone because it is a more concise method of communication and is therefore cheaper. Further, it doesn't require the addressee to be present at the other end of the line when a message is sent. Finally, it also has the benefit of producing an electronic copy of the dialogue or indeed a paper copy if required. There is an ever-growing band of executives who wouldn't leave home without their portable PCs with electronic mail access. This is an important fact that needs to be recognized by travel agents who have some large corporate customers in the business travel field. In other words, electronic mail is a powerful vehicle for keeping in touch with important business travel clients as they travel around the world.

To use electronic mail you need a terminal that is invariably a PC (in fact in the case of electronic mail, a portable PC offers many advantages), e-mail software that is usually bundled up with most Internet registration kits, a modem, a telephone, i.e. a telephone handset or a dial-up plug point, and registration with one of the many electronic mail services, many of which are ISPs. Each person registering with the electronic mail service provider is given a 'mail box'. This is a file into which all messages, both incoming and outgoing, are placed. In actual fact a mail box is a part of a network company's data base that is reserved for your incoming and outgoing messages.

So, when you wish to send a message to someone, you use your terminal to dial into the nearest node of the ISP. This is the nearest computer to where you happen to be located, which is given in a little quick reference booklet when you register with the ISP. Having signed on, you then specify the mail box or e-mail address of the person to whom you wish to send a message. The system can help you do this by displaying a list of all registered users and offering you the option of selecting the correct one. You then key the message into the PC and it is stored by the ISP computer and placed in the mail box of the person you specified. Then you sign off. That's it! When you wish to receive your messages you simply dial into the nearest node, sign on and request all your messages. These will be displayed one at a time as you request them. You may, of course, print them out if you have a printer attached to your terminal.

There are a number of electronic mail service providers in the UK and most of the popular ones are also ISPs. Each one offers national coverage so that for most locations only a local telephone call will be required to access a company's communications computer. Many also offer international access. The ISP is connected to the global network, known as the Internet, using links to computers in other countries. If your ISP has an international presence, then if, for example, you are travelling in another country, you can keep in contact by carrying a portable PC with you and dialling into the local network node in the country in which you happen to be. This can be done outside normal office hours. You can then access your mail box and retrieve your messages as well as send messages. If your ISP does not have an international presence then you will need to dial long distance back to the ISP's home base in the UK.

In some countries your PC may not be able to plug straight into the telephone socket. In order to get around this problem all you need is an acoustic coupler. This is a device onto which you place the actual telephone handset. It transmits by 'shouting' the modem voice sounds generated by the PC into the voice piece of the handset and receives by 'listening' to the sounds coming out of the ear piece of the telephone handset and sending these to the PC's modem. It is possible in fact to obtain a special-purpose portable PC with an acoustic coupler built into it as an integral feature of the device.

Electronic data interchange (EDI)

Did you know that around 70 per cent of the information keyed *into* computer systems was originally printed out by another computer! What a waste of time! Also, 40 per cent of invoices sent or received by UK companies contain errors. EDI is a technology that aims to solve these problems by allowing computers in different companies to pass information between them without the intermediate steps of printing and data capture. This can only be done by using internationally accepted standards for the formatting and transmission of transactions. In other words a common language must be used. The best way of illustrating this is to take an example of two companies, call them A and B. Assume that Company B purchases a product from Company A. Now if both companies and their respective banks keep to the EDI standards then the following are possible:

- **Invoicing** Company A sends an electronic invoice to Company B. This is accomplished by the computer of Company A transmitting an invoice message to the computer of Company B.
- **Payment authorization** When Company B receives this electronic invoice it first needs to display this upon a computer screen where it is authorized. The authorizer views the invoice on the screen and if it is OK, issues an authority to pay. Payment is made by sending an EDI message to the bank instructing payment to be issued to Company A. Company B also generates a remittance advice that it sends to Company A, again by computer-to-computer data transmission.
- **Settlement** Bank B, when it receives the payment instruction, issues an inter-bank credit transfer to Bank A. In time, it will notify Company B of this transaction by means of an electronic bank statement.

- **Receipt** Bank A receives the credit transfer and credits Company A's bank account. Again, in time this will be reported to Company A by means of an electronic statement of account.
- **Reconciliation** Both companies use their computers to reconcile their bank accounts automatically using the electronic statements that they receive from their banks via data transmission.

All that is required to do this is agreement by both companies that they will use EDI principles to settle invoices and for both companies to have the necessary EDI software. In the UK travel industry, the large tour operators are seriously considering batch EDI between themselves and travel agents. Also, both tour operators and airline GDSs are considering interactive EDI for reservation message exchange. In both cases a network will be needed to exchange all these standard EDI messages. Let me explain these terms in a little more detail:

- **Batch EDI** The example I gave above involving invoicing between Company A and Company B, is known as batch EDI (Fig. 6.2). It enables the above exchange of regular financial transaction to take place between organizations without a single piece of paper being printed or processed manually! But if this is to happen successfully then it is critical that the companies involved use a standard layout for an electronic invoice and remittance advice and that the banks use agreed inter-bank standards. These types of standards are called batch EDI standards.

    Batch EDI would be an ideal way for tour companies to collect funds from travel agents for holidays booked by their customers. Gone would be the familiar invoice from the tour company for the deposit followed by another for the balance of the holiday. This would be replaced by a conversation between the tour company's computer and the travel agent's back-office system. All the agent would see would be a list of all the payments that had been authorized and paid via EDI.
- **Interactive EDI** Another type of electronic data interchange is interactive EDI (I-EDI). This is a standard that sets out how computers pass messages to each other as events actually

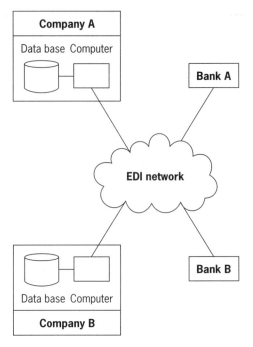

Figure 6.2 Diagram of batch EDI

happen. The EDI standards bodies have defined the precise layout of these electronic documents or messages, on a global basis. There are just one or two world-wide organizations that set these standards. In the UK, it was the ferry companies that first set some standards for conversational EDI, which were known as the UNICORN standards. Fully interactive EDI, however, was developed separately by an offshoot of the United Nations EDIFACT standards committee under the umbrella of the ISO 9735 standard (International Standards Organization). The I-EDI standards have been developed in close conjunction with the relevant industry specific standards group. In Western Europe, the travel, tourism and leisure message development group is known as MD8.

The GTI initiative (an early attempt to migrate the UK travel industry from viewdata to PC technology), intended to use the I-EDI standard as the basis for communications between the travel agent, the tour operator and the airline GDSs. The advantage of this approach would have been that it would offer an open set of standards that any company would be free to use. A closed set of standards, by comparison, would, for example, be the Thomson TOP viewdata standard. This is a *de facto* standard that has obviously been set by a single company and although highly successful is therefore more difficult for others to implement and may be less acceptable to other companies.

- **EDI networks** Now, if a company is to use EDI to communicate with most of its suppliers then it will need a communications facility. It probably will not want to set up a complicated and costly series of links with each of its trading partners and will try to avoid the complications of dial-up if possible. So, how will it communicate with other EDI companies? One answer is via a value-added network (VAN – see previous section for a description of VAN suppliers). The VAN acts as a postman for a company's EDI data. It receives a series of files of computer information that is addressed by the originating company to each destination company. The VANS' company stores the files just in case any get lost and then sends them to their respective destination companies. While connected to the VAN, EDI files sent to it by other companies may also be received.

There are major cost savings and operational benefits that can be gained from using EDI. These benefits become quite substantial if a high volume of transactions is involved. So it is the large trading companies that stand to gain the most from EDI. It is for this reason that some large companies have used their muscle to force their smaller suppliers to implement EDI. In fact it has become one of the terms of business for some companies, especially in the motor manufacturing industry where there are a relatively few large buyers and a lot of small suppliers. Just lately, however, even small companies have come to appreciate the benefits of EDI and are proceeding apace with implementation on their PC systems.

EDI is already widely used, especially in the manufacturing industry. It is expected that the distribution and transport sectors will lag behind slightly, although as we shall see later there have been some interesting new developments for travel agents in this area. One such initiative that unfortunately fell at the first fence was being spearheaded by a consortium of leading suppliers and was known as GTI (an early attempt to move the UK travel industry from videotex to PC technology). In general, however, EDI is set to grow very rapidly over the next few years. The reasons for this are: (a) the effect of the single European market, (b) the connection of small computer users to large corporate hubs, (c) changes in the way EDI is sold, and (d) advances in international standards.

In summary, all you need to implement EDI is a computer (which may be a PC, a mini- or a main-frame), some special purpose software that is now readily available on the IT market as a package and a communications facility, i.e. a modem and a communications line. You may also need to register with one of the UK VANs although this is not strictly necessary because EDI between only a few companies may be achieved by direct contact. Of course you will also need the commitment within your company to make it work, which in most cases means allocating sufficient time and money to allow staff to be trained adequately.

Teletext

Teletext is a relatively simple yet effective technology that has made a dramatic impact on the sale of leisure travel products in the UK, principally packaged holidays and flight-only air travel; although many specialized activities, such as UK holidays, cruising and skiing are also sold via the service. It is a one-way communications channel used for a rather specialized form of television advertising. Teletext Limited is an independent company owned by Associated Newpapers, Philips Electronics and MVI Limited. This company provides the teletext service on ITV (Independent Television) and Channel 4, the two commercial broadcasters in the UK. The teletext service reaches over 60 per cent of all UK households and provides its services to consumers free of charge. Over 18 million people each week and 9.4 million people each day watch teletext pages and of these, seven million use teletext to help them choose their holidays (Fig. 6.3). Independent surveys indicate a high level of satisfaction with the media and 98 per cent of bookers saying they would use teletext again.

Figure 6.3 Holiday teletext pages

In the UK, many tour operators and travel agents use teletext to advertise their late availability holidays. Teletext is a particularly suitable medium for this because advertising copy can be created very quickly and broadcast to consumers within a matter of minutes; and when the offers have expired or there are no more places left on specific holiday departures, then the relevant teletext pages can be modified easily and the outdated holiday departures quickly removed. Air travel seat-only operators work in a very similar way. Again it is the rapidity with which teletext advertising pages can be updated that is so attractive. One of the reasons teletext has been so successful can be attributed to its simplicity. This arises from the way the technology makes clever use of the television signals that are broadcast to 18 regions of the country (Fig. 6.4). In general, this is how it works:

Figure 6.4 The UK teletext regions

- **Page creation**   All teletext pages originate from the Teletext Limited headquarters in London. Some are keyed in from copy that is provided by contributors although many are sent in real-time from outside editorial or advertising agencies. The teletext holiday pages start at the main index page, which may be found on ITV page 200. From here there are many different categories of holiday information, e.g. holidays in the UK, holidays abroad, flight only, tourist exchange rates, etc. There are a number of different types of teletext advertising pages:

- *Source page* The source page provides the advertiser with a dedicated page number. Extra frames can be added that flip over automatically in sequence if the advertiser needs to promote additional information. Potential customers are directed to the advertiser's source page through signposts on teletext's editorial pages and/or cross-referrals from other media. Or, as in the holiday classified section, it provides a page number for a particular type of holiday or destination and the frames on the section carry advertising from clients who all offer products within that market.
- *Signpost* A signpost is the strip along the bottom of an editorial page, used to direct viewers to an advertiser's source page. Advertisers select the position of the signpost to meet their marketing objectives by utilizing the mass audience of the most popular pages, e.g. TV Guide, or targeting a specific audience, e.g. City News, Sport pages, etc.
- *Mini-page* A mini-page is an advert that makes up one-third of an editorial page and is available on a limited number of high profile sites. A mini-page can be used either to direct viewers to an advertiser's source page (like a signpost) or as a stand-alone advert to generate direct response.
- *Interleaf* An interleaf is a full page advert that flips over automatically between editorial frames. Again, positioning will reflect the advertiser's targeting objectives and could appear, for example, within Channel 4's 'Personal Finance' or 'It's Your Life' feature pages.
- *Sponsorship* An advertiser can sponsor editorial pages or sections by including its name or logo on the screen. Through accreditation, the advertiser can effectively build brand awareness and identity by associating its brand with a sporting event, weather forecast or any other teletext feature.
- **Page distribution** Pages are then sent to 27 local television transmitter sites around the country, via the BT network. The information is sent independently to each transmitter site, allowing for a truly local service. Each region has around 4,000 pages available at any one time.
- **Page transmission** Pages arrive at the transmitter site and are broadcast alongside the television signal. They are transmitted to the home television receiver as a hidden part of the standard picture signal, in a continuous stream of pages that is constantly available.
- **Page reception** Teletext pages are received by the consumer's aerial and transmitted to their television set (which must of course have a teletext feature), usually in the home. While in normal mode, the teletext pages are not visible to the viewer. It is only when a special hand-held remote control device is used to switch the television into teletext mode, that the pages may be seen.
- **Page navigation** A viewer can select the pages they wish to see by entering the page number via their hand-held remote control device. These pages may be viewed free of charge at any time the viewer wishes. Pages are navigated by means of menus of page numbers and topics, as follows:
  - The viewer first of all accesses the front page menu for the service by pressing the text button on their remote control.
  - They are then guided through the service by a series of index pages that can be called up by entering the relevant three digit number.
  - An index shows a series of different pages, which are called source pages.
  - Source pages have a number of extra pages behind them that flip over automatically providing additional information.
  - Most teletext televisions include a Fastext facility. Four coloured prompts at the bottom of the teletext screen correspond to the buttons on the hand-set, allowing the viewer to move around the pages more conveniently.

A teletext screen is made up of 22 rows, each of 39 characters. Each of these characters is formed from six pixels arranged in a grid comprising three pixels high by two pixels wide. A pixel may be regarded as simply a pin-point of colour in a specific position. Characters are just standard representations of pixels within this grid. So, other shapes, besides characters can be constructed by arranging the pixels in each grid in a different way. Each line is defined in terms of how it is to be displayed on the screen, by five special character positions that occur at the beginning of each line. These are known as Design Codes and there may be up to five of them that are used to define the colour of the background, the colour of the text, the colour of the graphic, whether it is to be displayed as single or double height and, finally, whether it is to be flashing or not. Eight colours are supported by teletext: blue, black, yellow, green, cyan, white, red and magenta. The rules governing teletext formats were established as a standard some time ago and the problem is that the pages look somewhat outdated, especially in comparison with GUIs and the Internet. However, a solution is in sight with the advent of digital broadcasting – a new technology that holds the promise of higher quality page graphics.

There is little doubt that the advent of digital broadcasting will open up new opportunities for teletext. Digital broadcasting supports a higher bandwidth for transmission purposes. This allows more flexibility, an increased number of channels and a higher definition in the pages that can be assembled and broadcast to viewers' televisions. The new service will have almost limitless character fonts and colours. It will support photographic images and potentially, in the future, also sound and video. With the advent of intelligent set-top boxes for digital broadcasting, teletext could well be integrated with the Internet. This would be an excellent way to build upon the success of the current teletext service. It would allow consumers to view high quality information pages and then follow-up with an enquiry that would link them automatically with the Internet. A link such as this could provide the interactive on-line dimension that a broadcast medium like teletext currently lacks. It would enable consumers to book as well as look. (Teletext Limited currently operates its own Web site at **http://www.teletext.co.uk**.) The current Web site includes a comprehensive guide to holiday destinations and other general interest pages. This could be enhanced as and when the demands of digital broadcasting materialize. Some of the other existing and future uses of teletext are:

- **Cross-referral advertising** Advertisers can also benefit from teletext by integrating it with their corporate advertising campaigns which may use other media. They can, for example, refer to a teletext page from their television, press and/or radio campaigns.
- **Regional advertising** Advertisers can target their products to consumers within certain regions of the country. Tour operators and travel agents can, for example, advertise late sale bargains that depart from the closest regional airport to a viewer's home.
- **Interactive teletext** This is a relatively new development by Teletext Limited and makes simultaneous use of the telephone. Viewers access teletext then use a touch tone telephone to call a number shown on the teletext screen. This number accesses a computerized voice response system that prompts the viewers to select certain pages using their remote controls. Viewers react to the aural or on-screen commands via their touch tone keypads, to interact with the service. A related service branded Talk Back was launched in 1995. This enables users to enter coded responses via their telephones and a premium rate number. Teletext Live offers interactive games with daily and weekly prizes. Interactive transactional services such as Take Off, which is offered by Teletext Limited and airline networks, enables viewers to see pages of flight details and make bookings via a touch tone telephone using their credit cards.

Teletext has been a highly successful medium for advertising holidays within the UK market. So successful that many travel agents and operators have set up special telephone service centres to handle incoming calls generated by viewers who view travel-related teletext pages. These centres sometimes use special purpose software applications that support teletext sales. Applications which, for example, enable the supplier to maintain a data base of teletext pages, by region, so that they can answer customer enquiries easier and more efficiently. Callers often begin their enquiries to these centres with: 'I know I saw *xyz* product on teletext but I can't remember the details, so can you tell me more about it'. The operator may access this data base in a variety of ways by, for instance, teletext page number, product, page display date, departure date, departure airport and so on. The fact that such systems have been developed at all illustrates the success of teletext within the UK travel industry.

Videotex

The term videotex is the international word assigned to the technology that is often called viewdata in the UK (Fig. 6.5). Its use in UK travel and tourism started with BT's Prestel service (Fig. 6.6). However, videotex systems are also used within the travel industries of other countries, e.g. Germany's Bildschirmtext and France's Minitel. The following explanation of this technology is taken very much from a UK perspective.

Viewdata is really a special kind of communications protocol that requires a television, a telephone line, a keyboard and a special modem. Originally in the days of Prestel these were all separate items, with the keyboard and modem being incorporated into a single unit. Then with the rapid growth of Prestel in the business world, a number of major television manufacturers and specialist suppliers began selling dedicated videotex terminals. Some of these recognized the shortage of desk space in a travel agency and designed compact models such as the Sony KTX9000, which proved very popular in the 1980s. More recently, with the advent of PCs we have seen an increase in the use of PCs as viewdata terminals by means of a special purpose 'card', i.e. printed circuit card, in the back providing the necessary communications capability.

Unfortunately, there is neither the time nor the space to consider the history of how viewdata came to be a dominant technology in the UK leisure travel business. What we are more concerned

Figure 6.5 Videotex

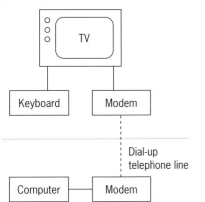

Figure 6.6 The early Prestel service

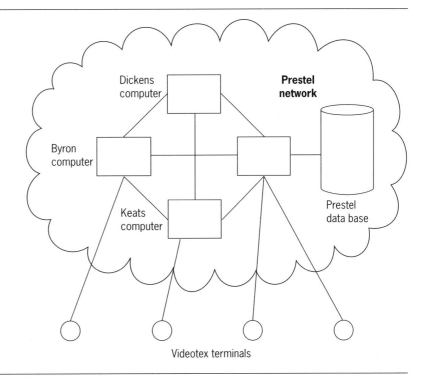

about here is how this technology is used by the leading leisure travel supplier companies and what its strengths and weaknesses are. This will allow us to discuss possible replacement technologies, such as the Internet, in later chapters.

WHY VIDEOTEX?

To understand the commercial aspects of tour operations, put yourself in the position of a supplier of leisure travel products, whether they be packaged holidays, independent tours, ferries or airline seats. Probably your prime objective is how to reach your distribution channel, which in this case is the travel agency network, in the most effective way so that the people who sell your products can do so easily and cheaply.

The leisure travel supplier's viewpoint

As a fully automated operator, you almost certainly have a computer system that is used by your reservations staff to sell your product. This computer system has a data base of the products available, full descriptions of the products and the current status as regards availability. This is called an inventory of products. The inventory is added to whenever your marketing staff create a new product. Similarly, the inventory is reduced whenever you sell a product from it. In between times, the status of the inventory items change as customers place options to buy certain products. Everything feeds off this data base. When an item is being priced before being sold, it is on the data base in skeleton form only. Then when it is sold it will trigger the production of an invoice to the travel agent and eventually the production of tickets and the appropriate accounting entries. So, you might ask, what else can be done with the system?

The sales and reservations process

Well, having computerized, the travel supplier will no doubt look at where the operating costs are heaviest and try to reduce them. Besides the cost of the actual product itself, one of the highest areas of cost will be in product sales otherwise known as the reservations area. This needs to be staffed up and available to take calls throughout the working day. The reservations area must have the capacity to meet peaks and troughs and consequently has a high operating cost structure associated with it due to training and staff turnover factors. It also needs to have a large number of telephone lines available because the worst thing during a peak booking period is for a travel agent to get a busy or engaged tone when the reservations centre is called. This may be enough to make the travel agent try a competitor. Finally, there is the problem of trying to persuade the travel agents located in some of the more remote areas in relation to your reservations centre, to pay for the phone call to enquire about a booking.

So, the travel operator thus establishes that the reservations area is the next target for automation. But wait, why not take a leaf out of the airline's book and let travel agents do the reservation themselves by giving them a computer terminal. Good idea, until you add up the cost of giving 7,000 travel agents one or more terminals each. However, most of those travel agents already had a terminal that they were using for Prestel. This was thanks to a deal between Sealink and Radio Rentals that provided agents with a very cheap rental arrangement. The videotex technology used by Prestel was simple and the number of terminals installed in agency locations was sufficiently great to give a wide enough coverage, so the operator decides to follow this route. The next problem is how to get the operator's computer system to 'talk' to the viewdata terminals in the travel agency offices. The answer is a special communications computer, WAN telecommunications and some special software (Fig. 6.7). Let's see what this involves in a little more detail.

Accessing videotex systems

First, let's consider the communications computer, which in this context is often known as a viewdata front-end computer. (I use viewdata and videotex terminology interchangeably throughout this section although strictly speaking 'videotex' is the internationally accepted term for this technology.) This is usually a separate computer but it is sometimes a program running inside the computer that controls the operator's data base. The prime purpose of this computer is to act as a translator. It takes

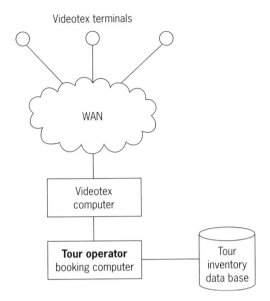

Figure 6.7 Videotex distribution systems

the operator's computer system dialogue, which is used by the sales staff in their reservations centres, and converts it into a simple set of options easily understandable by a travel agent. Furthermore it edits entries as they are made by agents and restricts the functionality of the system to those entries that the tour operator wishes to allow travel agents access to. There is another dimension to this translation process that also converts the technology from the operator's in-house computer system to videotex. Right, we are now in the situation where our in-house reservations computer has an interface that supports access by viewdata terminals in travel agents. But, how will these agents access our computer?

There are really two options: (a) travel agents can dial directly into the operator's computer using the public telephone network, or (b) the operator can connect the computer into a VANS and have travel agents dial into the VAN and access the operator's computer via the VAN. Consider the first option for a moment, which I will call direct dial. Although direct dial may seem straightforward, there are some significant drawbacks.

With direct dial, the travel agent must use the viewdata terminal to dial the operator's computer. Because operators are spread around the country, there is a good chance that a number of agents will have to pay for a national call as opposed to a local call. Travel agents being what they are, i.e. sound business people, they will avoid having to pay a long distance national call if another option is available. Like, for instance, a competitor's system that is available on a local call basis – a situation the operator will avoid at all costs. Then there is the problem of how to cope with the peak booking times. If the operator's system is to be able to answer all incoming calls during these peaks then a maximum number of call answering exchange lines will be needed to take the maximum viewdata call volume. Also, a large capacity front-end computer will be needed to receive the calls, translate them and feed them to the operator's main-frame computer. The combination of an exchange line and the input 'plug' on the front-end computer is usually called a 'port'. If the operator is to provide a system that is always there when the agent needs it then a large number of ports is going to be needed. Outside of the peak times, this capacity is going to lie largely unused. Because the price of providing a port is substantial, then this is a high price for an operator to pay.

Alternatively, using a VAN eliminates most of these worries. Most of the large VANS like AT&T and Imminus provide local call coverage throughout the UK. This means that virtually all the agents who will dial-in to the operator's system will do so at local call rates. This is accomplished by the VANS placing strategically located communication node computers all over the country. They are placed such that travel agents only need a local call to access them. Once into the VAN, the travel agents' messages and dialogues with the operator's computer are routed through the VANS' network and into the computer of the operator. Now these node computers provided by the VANS have a great deal of capacity with which to deal with incoming calls. So, the likelihood of a busy or engaged tone being received by an agent is reduced to a minimum. Nevertheless, the operator's main-frame computer still needs the capacity to deal with the peak call volume, but this workload is somewhat reduced in comparison to a direct dial approach because the VANS' wide area network does some of the routine processing for the operator.

Finally, from an agent's viewpoint, it is more convenient to access operators via a VANS. The reason for this is that once an agent is into a VAN, it is a lot easier to switch between the various systems without having to re-dial. The agent only needs to select the operator's system to access and the system becomes available. For operator systems on direct dial the agent must ring off from the VAN or the operator currently being accessed and make another call to the direct dial operator. So unless the operator is very large, e.g. Thomson Holidays, the most cost effective access method is via a VANS.

SUPPLIERS' VIDEOTEX SYSTEMS

Having given you a potted history and rationale for the use of videotex, or viewdata as it is more often referred to, now should be a good time to explore some of the popular systems currently in use. All of these systems are used by suppliers in the UK where viewdata is a dominant technology for the leisure travel industry, for the reasons given in the preceding section. Now although people have been saying for years that viewdata is about to die, reports of its death are greatly exaggerated. It is still very much alive and kicking (at least it was in 1997 when this edition of the book was written). Nevertheless, the writing is on the wall for viewdata and it is simply a matter of time before it is replaced with a newer, faster and more effective technology. But before this can happen, there are some substantial barriers to change that must first be overcome. I shall discuss this issue throughout the chapter but first, let's start with that sector of the travel industry that is the mainstay of viewdata – tour operators.

Tour operators

There are more tour operator reservation systems accessible via videotex than any other type of travel supplier. The background to this situation and the history of how the technology evolved is described in Chapter 1. Some of the larger operators run their own videotex networks that are accessible to travel agents on a direct dial basis with the added advantage that only a local telephone call is involved. However, most operators connect their systems to one of the leading videotex VANS such as AT&T or Imminus.

There are many different tour operator systems that can be accessed by travel agents. Most of these systems are based on videotex or viewdata technology. I first introduced the history of viewdata in the travel industry in earlier chapters in very general terms. Now it is time to be more specific. With so many tour operator videotex systems, however, it is impossible for me to cover them all in a book like this. I think the important thing is for you to understand the background and operation of at least one frequently used system because most of them are very similar indeed. Only the screen formats and some of the entries differ from system to system. The example I have chosen is the Cosmos system.

Cosmos

Cosmos started distributing its package tours to travel agents via viewdata in 1988 (see Chapter 3 for more details about the formation, background history of the company and a review of the IT that is used by Cosmos). The way this is done is via the company's ICL VME main-frame, which is linked via multiple front-end communications processors to both of the main videotex distribution networks in the UK – AT&T and Imminus (see details later in this chapter). Each front-end processor has a link to each network (Fig. 6.8), thus ensuring maximum availability to travel agent customers in the event of a failure that could occur in any one of the link's individual components.

The Cosmos viewdata booking system is similar in many ways to other tour operator reservation systems that are based on videotex technology and also, for example, to Thomson's TOP as described above. It allows agents to view availability, obtain information on holidays, make bookings and view the status of previous booking. However, one significant feature of the Cosmos system is that it also supports the RESCON standard that I described in Chapter 1 under the heading of TTI. This enables travel agents who book Cosmos tours to download transaction details automatically into their back-office systems. RESCON thus allows travel agents to print customer documentation and capture accounting information without any additional re-keying effort.

Cosmos, like many other viewdata booking systems, is plagued by a relatively new phenomenon

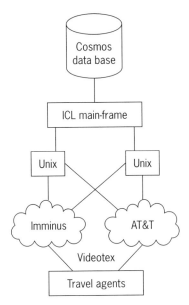

Figure 6.8 Cosmos network diagram

sale that can be booked via its viewdata system. Cosmos is no exception to this and, again like many of its competitors, has a late availability display that may be accessed by travel agents who are users of its viewdata system.

Because the proportion of late availability sales made by Cosmos is so significant, it is not surprising that a relatively sophisticated search system has been developed to support this sector of the market. A special screen has been developed, which is almost a standard used throughout the package tour industry. This is the late sale matrix. It provides travel agency users with a structured search of all late availability holidays available from Cosmos. These are presented as a series of hierarchical viewdata screens that show different types of late availability. The agent enters the search criteria, such as the preferred destination (if any), the departure date and the departure airport. This search request is sent to the Cosmos main-frame, which accesses its data base of holidays and selects those that meet the stated requirements.

At the highest level, a screen is displayed that shows a matrix with: *rows* containing the lowest-priced packaged holidays for various destinations and lengths of stay; and *columns* containing different combinations of accommodation, e.g. operator ratings such as SC for self catering, 1/2 Star, 3Star, 4Star, etc. The travel agent may select a particular row and column for further exploration. When this occurs, a refined search request is again formulated and sent to the host system that repeats the data base search but without the extraneous holiday information, i.e. the packages that are of no interest to the agent. However, the actual search function is almost identical in terms of processing power and IT resource requirements as the original. Finally, the search process ties up a front-end computer port and a route through the third-party network, all the time that the travel agent is in session with the host computer. All in all, a very resource hungry process.

– the emergence of late availability scanning systems. It is therefore worthwhile examining this subject in more detail because it raises a number of important issues that will influence the possible future development of videotex-based tour operator booking systems. An analysis of one of these late availability systems, AT&T's travel late availability search (ATLAS), is presented in more detail in the next section. However, for the moment, let's consider the problem from the tour operator's viewpoint.

The market for late availability holidays has grown steadily over the past few years. Between 25 and 35 per cent of Cosmos' holiday sales are now concentrated within the late booking sector. Late availability holidays are those that would be called 'distressed stock' in many other industries. They are holidays that have a short shelf life and unless they are sold within the ensuing few days, will expire, i.e. the flight will have departed with empty seats and the hotel rooms will possibly remain empty. Most late availability holidays are those that were originally published in holiday brochures but remain unsold with only a few days to go before departure. There are even some so-called late availability holidays or packages that are specially constructed for this new market. In any event, a tour operator will invariably have a number of late availability packages available for

While the use of the late sale matrix if just about acceptable for travel agent use, the bane of a tour operator's life is the increasing popularity among travel agents of computer-controlled scanning bureaus. These bureaus have sprung up to support the late availability market and provide a scanning service to travel agents. The bureau will retrieve the late sale matrix from several tour operators and construct a generic late availability matrix that includes input from several tour operators. Now, despite their popularity with travel agents, tour operators do not like these scanning systems for two main reasons: (a) they compare an operator's product with its competitor's on price alone and exclude other intangible factors such as service quality, and (b) they consume vast amounts of computer resources that are costly and can impact the distribution of full price holidays. However, despite these drawbacks, the scanning bureaus nevertheless represent an incremental selling opportunity for tour operators such as Cosmos. Inclusion in bureau late sale displays increases the likelihood of a travel agency customer choosing to book Cosmos and increases the size of the distribution channel. But the price is a high one in terms of IT resources.

Cosmos has estimated, for example, that scanning systems cost it around £80,000 per year in third-party network charges alone. That's just the charges that AT&T and Imminus levy for carrying scanning messages across their networks, i.e. it excludes the cost of processing these transactions on the host main-frame and front-end computers. Because the requirement for late availability searches is so popular, Cosmos has devoted considerable systems efforts to responding to search requests as efficiently as possible. As a result, the average time to create, format and present a complete late sale matrix, ready for transmission to the end user is just 0.14 seconds. The bottom line problem can be summarized as follows. Substantial resources in terms of IT skills, computer processing power and network usage have been dedicated to the late sale market in order for Cosmos to remain competitive. The problem is that on average, Cosmos receives only a single booking as a result of every 30 late sale display requests. In other words, the sales conversion ratio is not particularly high in relation to the resources expended. However, late sales are now a firmly established market sector and it seems that search bureaus are here to stay for some time.

Nevertheless, for the reasons outlined above, Cosmos would ideally like to provide late sale information in some other way. This is because videotex technology does not lend itself to the kind of efficient data base searches that the late sale market really needs. I explained the reasons for this above. Each late sale matrix, whether it be for the top level display or a subsequent specific display, requires almost the same amount of computer processing. So, each request involves a full search and selection from the Cosmos data base. A far more efficient approach would be for Cosmos to provide a full late sale matrix once only for each request. The recipients would then be able to store the data and extract only those parts that they need for their own particular customer display purposes. There are two main ways in which this could be done: (a) an extract file could be formed and transmitted to the late sale search bureau for their processing, or (b) the information could be formatted as an Intranet page. Taking each in turn:

- **Extract file** Cosmos could extract a file containing all data on late sale package holidays. This could be sent to the late sale search bureaus. A related issue here is the file format. As late sale search bureaus increase in number, so the different file formats also increase. From the tour operators' viewpoint, a standard file format is preferable. So, tour operators are directing a fair amount of effort within standards organizations like the UK's TTI, towards the establishment of a common file format based on EDI messaging standards (see Chapter 1 for more details of the TTI, and for more information on a specific late sale search bureau please read the section on AT&T's ATLAS product that appears below).
- **Intranet** The second alternative, which is based on Intranet technology, is very interesting and is worthy of a little more analysis. Intranets offer tour operators some significant opportunities for the future distribution of their products to travel agents. Videotex technology has some significant disadvantages that are explored in more detail in the ensuing sections. On the other hand, the Internet as it stands right now, is somewhat unreliable and is perceived to be insecure for commercial transactions. But Intranets overcome many of these problems. They are run as secure and reliable communications networks, controlled by a supplier's own

systems and available only to selected users. The only problem that remains to be overcome is, unfortunately, a pretty fundamental one. This is the size of the travel agent population that can access Intranets from their points-of-sale. This topic is addressed more fully in the section on Imminus that is given later in this chapter. But if we imagine for the moment that travel agency groups might be willing to subscribe to the kind of network services that support Intranet access, then tour operators such as Cosmos could provide a far higher level of service to their travel agent sellers using new technology that is more efficient in terms of resource utilization.

Take for instance the late sale matrix I discussed above. In an Intranet environment, the travel agent would be presented with an Internet style page with several push button options. One of these options would be late sale details. When the travel agent selected this page and completed the basic search criteria, the tour operator's system would respond with a page containing an extract of its data base containing late sale packages. The travel agent would be able to search this page for the best deal for their customer without tying-up the tour operator's computer system. The local search would be done by a combination of the browser, the locally stored data and locally executed programs, i.e. Applets. The end result would be a better looking page that would be fully scrollable thus providing a faster service to travel agents and less usage of computer and network resources for the tour operator.

Avro, Cosmos' sister company, has tried to address this problem using different technology. Its problem is almost identical to Cosmos' except that its product is seat-only air travel. This is sold on a similar basis to package tours and has its own late sale market. Avro has provided ISDN access to the late sale data base for its high volume specialist seat-sale outlets. These outlets simply initiate an ISDN call to the Avro computer that connects them to the data base. They then use terminal emulation technology to download a file of late seat-only bargains in a standard format into the subscriber's PC. The plan is to use EDI standards for the transmission of this information, which should make it even more appealing to subscribers who may face having to download similar files from other seat-only suppliers. This overlaps in some ways with the UNICORN standard that I covered in the TTI section in Chapter 1. The basic rationale for this approach is that it is more efficient for both the supplier and the bureau. From the suppliers' viewpoint it allows their computers to support multiple sessions with several users without tying up communications resources, and enables just a single file format to be used. From the subscribers' viewpoint it enables them to receive the file quickly and assuming standards are taken up widely, allows them to process information in a single format.

Overall, Cosmos recognizes the likely emergence of Intranet technology as the way forward for travel agency booking systems. There is little doubt that an Intranet approach could provide the next generation of booking and information systems that travel suppliers could use to replace viewdata. However, this is very much a 'chicken and egg' problem. Without sufficient numbers of travel agents able to access an Intranet booking system why should Cosmos, for example, invest significant amounts in developing an Intranet front-end to its booking system. After all, the existing viewdata approach seems to be working well for the majority of agents and many of these are still running old PCs that are not sufficiently powerful to make effective use of Intranets. On the other hand, an increasingly significant number of agents now have latest generation PCs at the point-of-sale and some are already experimenting with the Internet, which uses technologies identical to Intranets. The question is: 'Who is going to make the first significant investment in Intranet technologies – the travel agent or the travel suppliers?' To safeguard against being caught out by a rapid shift towards Intranets within the travel industry, Cosmos is experimenting with the technology and thereby hopes to ensure that it establishes the necessary in-house capabilities, resources and skills to exploit these new developments, as and when they may occur.

It could well be, for example, that Cosmos may use an Intranet solution for its own in-house telephone sales operation. This would allow

Cosmos to provide telephone sales operators with a new and superior booking and information service while at the same time enabling the company to gain valuable experience of Intranet technologies. If the in-house Intranet proved successful, then Cosmos could choose to broaden its Intranet access to encompass certain important travel agency business partners on a selective basis. Such an approach could conceivably also be followed by other tour operators and this could build the islands of Intranets that I discuss further in the section on Imminus below.

Late availability search systems

Over the past few years, a new sector of the leisure market has emerged that goes under the name 'late availability' holidays. This is the term used to describe the sale of a tour company's distressed inventory – in other works, packaged tours that remain unsold with only a few days or hours to go before departure. Late availability products include each of the two main services that make up a tour, i.e. charter flights and hotels. The market for late availability is now firmly established although its popularity varies with customer demand and the number of packaged holidays that tour companies create for their inventories. I introduced late sale search systems in the Cosmos section above. This examined the technologies that support the late sale market from the suppliers' viewpoint. In this section, I'll be looking at it more from the viewpoint of the travel agent. It used to be the case that travel agents would go to great lengths to sell a last minute bargain to a customer wishing to go anywhere at anytime within the next few days (or hours). They would desperately call several tour operators and make quite a few viewdata calls to tour operator systems in their search for a bargain. Nowadays it is far simpler thanks to a few specialized late availability service companies. One good example is AT&T's ATLAS.

ATLAS

ATLAS is a viewdata system that may be used by travel agents to find last minute bargain holidays for their customers. The service is provided by AT&T in conjunction with an independent software company. This type of system is especially

Figure 6.9 ATLAS title screen

Figure 6.10 ATLAS system index

useful during the peak season when travel agency branches are busy and when a rapid customer turnaround time is essential. Instead of the travel sales consultants having to either telephone or use viewdata to access each tour operator's system, they simply use ATLAS (Fig. 6.9). The system automatically polls participating tour operators and stores the latest situation on last minute holiday bargains in a central data base.

So, at the heart of the ATLAS system is a data base of packaged holidays and flight-only special deals, all of which have been reduced in price due to the fact that their departure is imminent (Fig. 6.10). This data base is stored on a computer that is an integral part of AT&T's national network. Travel agents may access this network using their viewdata terminals or PCs with videotex emulator cards. They use these terminals to either dial into the AT&T network or they use their dedicated Direct Service link to connect to

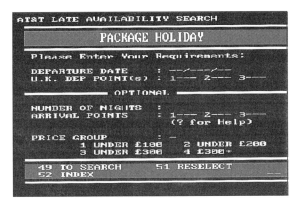

Figure 6.11 ATLAS search criteria

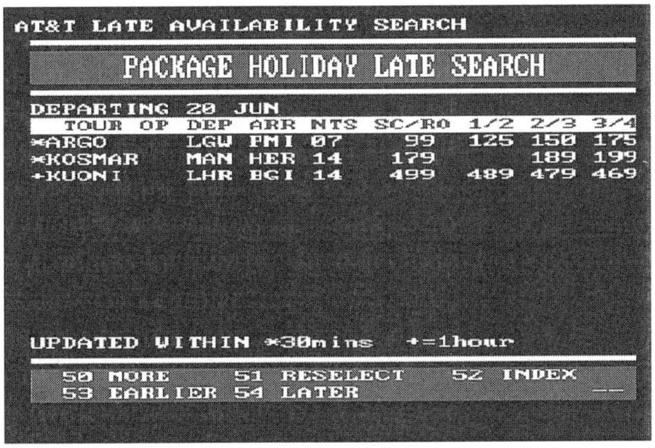

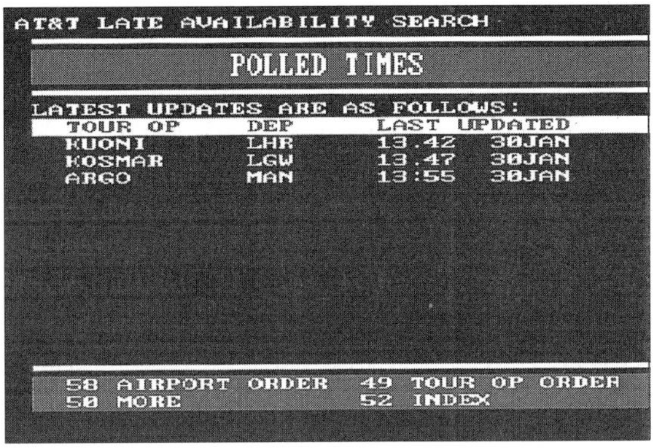

Figure 6.13 ATLAS polled times

Figure 6.12 ATLAS search results

ATLAS (Fig. 6.11). The details of the holidays are downloaded from the participating tour operating companies' systems by AT&T's special polling software, directly into the ATLAS data base (Fig. 6.12). In fact, the latest information on last minute bargains is kept up-to-date by polling the tour operators' computer systems every 30 minutes (Fig. 6.13).

Because the data base is so vast, it is vital that a good search method is provided to travel agency users of ATLAS. This is provided in two ways: (a) a pre-set filter that shows data base entries in certain sequences in an agency's branches, and (b) a search engine with criteria entered by individual sales consultants. Travel agents can use the ATLAS customization parameters to tailor the information displays to meet the particular requirements of their own branch. This allows travel agency branch management to decide which tour operators they wish their travel consultants to see and the order in which holiday options are presented to their customers. For example, holidays may be presented in either ascending price sequence or alternatively in preferred tour operator sequence. Additionally, each sales consultant may select entries from the data base using a number of key fields: (i) departure date, (ii) the airport from which the customer wishes to depart on their holiday (up to three departure airports may be specified), (iii) up to three alternative arrival destinations, (iv) the number of nights, and (v) the customer's budget in terms of certain pre-set price bands.

The results of the data base search are displayed, one holiday to a line, on a standard viewdata screen. This display includes the price and departure details of each holiday that matches the customer's stated requirements. It then provides the travel agent with several further options, which are to: (i) search for more holidays, (ii) search earlier, (iii) search later, or (iv) specify another set of search parameters and repeat the process. If necessary, the sales consultant may choose to display a screen that shows when ATLAS was last updated with a particular tour operator's late availability information.

Ferry companies

The ferry companies were in fact the instigators of videotex in the travel business, way back in the 1980s. Ever since those early days, they have continued to play an important part in the development of the technology and the associated standards. Although many ferry bookings are still taken over the telephone by reservations staff, the level of videotex usage is now sufficiently high for the ferry companies to continue the growth and development of videotex. In fact there is a fair amount of co-operation among the ferry companies on important issues such as standards.

The AT&T FERRY# system

Six ferry companies and the Travellog systems company collaborated in the development of a standard booking system based on videotex and accessible on the AT&T network. This common reservation system called FERRY# was launched in November 1992. It is available to any travel agent with a videotex terminal or a PC with a videotex capability, which has been registered to use the service. There are now eight ferry companies accessible via FERRY# and they are:

- Brittany Ferries.
- Hoverspeed.
- North Sea Ferries.
- P&O Ferries.
- Sally Ferries.
- Scandinavian Seaways.
- Sea France.
- Stena.

A cornerstone of the common system is the UNICORN standard. This is a standard that has been agreed among the major ferry companies and that sets out how the data are formatted during the booking process. The UNICORN standards, which were developed by the ferry companies themselves, are based largely on EDI principles, which are a set of standards governing the interfacing of computer systems owned by different companies (see Chapter 1 – Standards and the TTI group).

The FERRY system is a very good example of how standards can benefit the user of reservation services in the travel industry. A travel agent using the FERRY service on AT&T need only enter the customer's booking details at the start of the reservations process. Information such as the ferry operator required, dates of travel, passenger details, size and type of vehicle and accommodation needed, are therefore entered just once. When the first ferry company system is accessed, the AT&T system sends the data keyed to this company's system. Then, when the agent selects another ferry company the same basic booking details are automatically sent by AT&T in standard EDI format to this ferry company's computer system. The customer's booking details are already present in the AT&T system because they are automatically stored in the standard UNICORN format as a by-product of the initial transaction. The travel agent sees only the different route information and tariffs offered by the system being currently accessed. The only way that this has been achieved is by the adoption of standards by the ferry operators concerned. The benefit to the travel agent is that the whole booking process is accelerated and productivity is increased.

Seat-only air sales

The seat-only air travel market is serviced by operators who are known as consolidators. Let me explain what this business is all about before I even mention the technology that is used to support it. In summary, consolidators are entrepreneurs who buy unfilled airline seats from the airlines and tour operators and sell them to the general public. These seats are sold to consumers via travel agents using viewdata technology.

The consolidator has the job of selling the seats to travel agents. The question we are interested in is: 'What kind of IT system is used by the seat-only operator to support these business functions?' And the answer is – the same type of system that tour operators use. Well, if you think about it the whole process is not unlike the sale of a package holiday, except that only a single travel component is involved instead of several, as is the case with an all-inclusive holiday. The type of system used by a tour operator shares a common need with the seat-only operators to maintain an inventory of travel products and then to sell the products to travel agents. There are many tour operator systems available on the travel automation market, but I shall not be evaluating them in detail in this book because I propose to focus on those systems that are of prime interest to travel agents.

However, I should like to mention briefly that one popular system is Autofile. Autofile has a well established tour operator systems product that maintains a file of inventory (in this case airline seats) and supports the sale of these seats as well as the ticket printing, documentation and associated processing that goes with it. It has enhanced its core system with something that it has called a 'transparent link'. This is an interface to certain CRSs that allows the seat-only operator to place bookings on scheduled flights. The seat-only operator will usually have many reservations operators whose job it is to receive incoming telephone calls from travel agents and will use a computer terminal to access the tour operator system that holds their inventory of air seats.

This is accomplished by bolting on a videotex front-end to the tour operator system being used by the seat-only operator. This front-end needs to have a sufficient capacity to handle the expected number of travel agency enquiries that are generated using videotex terminals on the leisure travel counter. The whole process is, from a strictly IT viewpoint, very similar to the tour operator's scenario. The principal difference being the format of the videotex screen, which is often fundamentally different from that used by a tour operator.

Scheduled airline reservations via viewdata

The leisure travel market is changing quite rapidly in many ways. One of these changes is the increase in people making up their own independent holidays. In such cases a person will decide upon the itinerary themself and then ask a travel agent to book the various products and services for them. Independent tours, as this type of business is called, is a valuable source of revenue for the travel agent, but it does involve a great deal of extra work and expertise.

This is where IT comes into play, because using travel technology is an important way to lessen the work involved and gain access to sources of information to provide the necessary expertise. Now, because one of the base products in many such independent tours is an air ticket, access to airline systems is required. But, in many cases leisure travel agents do not have a GDS because it is too costly and does not get used frequently enough to justify itself. So, what to do? The answer is that there are now some very sophisticated viewdata airline booking systems for both scheduled, charter and consolidated airline seats that are available to travel agents at very little cost. The purpose of this section is to explain these services and systems in a little more detail.

Viewdata scheduled airline seat reservation systems

The airlines and other distribution companies have recognized the potential of the growing seat-only market that has been generated to a large extent by the independent tour product. In fact as far as scheduled airline ticket sales are concerned, there are other sources of business besides independent tours. There is also the small self-employed person who needs to travel on business and the customer who needs to fly out to a foreign destination to visit their villa, friend or associate. There is therefore a good business case for an investment in systems and distribution channels to make scheduled airline seat sales easily available to travel agents. Below are given some of the companies and systems that are now available.

Scheduled airline seat reservations are not the sole province of the CRSs of the world. Airline seats on scheduled flights can now be made by travel agents using viewdata technology. One prominent UK system that supports this capability is EasyRes.

EasyRes

EasyRes is a product of Reed Travel Group (see Chapter 3 for more information on this major travel information company). Reed Travel Group's EasyRes is an innovative product, which was launched in the UK in 1988 and is aimed at UK ABTA travel agents in the leisure travel sector of the market. EasyRes is free to all ABTA and IATA travel agents via viewdata networks, i.e. Imminus and AT&T. It is a videotex-based, on-line scheduled airline seat reservation product. EasyRes provides guaranteed fares and instant confirmation of reservations at time of booking, for more than 45 scheduled airlines for short- and long-haul flights. It also includes access to charter and consolidated flights. One of its fundamental design principles is that it has to be easy to use. In fact, it is so easy that there is no need for a travel agent to attend a special training course of any kind. A simple booklet is enough to train the average agent in all the system's functions. Although EasyRes provides access to hotel and car booking facilities, in this section I am concerned primarily with the EasyRes air system.

The success of the EasyRes product is demonstrated by the fact that 90 per cent of UK travel agents, including many of the large multiples, use EasyRes regularly. 'How can it be free?', I hear you ask. The answer is that like GDSs, Reed Travel Group derives its revenue from the participating airlines and not from the travel agents. For every segment, i.e. for every city pair, sector or leg, booked via EasyRes, the airline pays Reed Travel Group a booking fee. This is what goes to cover the operating costs of the system and contributes towards generating a profit for the Reed Travel Group business.

EasyRes started out as a simple single class system with just three airlines connected to its central switch. Now, it is a multi-class system with more than 45 scheduled airlines bookable via videotex. The way this growth occurred is interesting and is I think worth a closer look. In 1989, a major European airline, whose telephone sales had become 'overloaded' as a result of its fierce competition with charter carriers then operating on routes between the UK and its major home cities, took the decision to endorse EasyRes as the preferred way for leisure travel agents to make bookings on these services. It positively discouraged these agents from using the telephone. So, while business travel agents could still use GDSs and members of the general public could still telephone their reservations' centres, all other bookings rapidly started to come through EasyRes. This allowed the airline to cut down the size – and, therefore, the cost – of its manual reservations operation in the UK. Following its lead, several other major European airlines have adopted this policy. The result is that agents are encouraged to utilize technology whenever possible, thus reducing low-yield reservations telephone calls.

In February 1993, EasyRes Plus was launched. This offered several significant enhancements which included: (a) last seat availability, (b) up to seven seats bookable in one transaction, (c) the inclusion of transfer connections, and (d) display of the airline's own record locator on completion of a PNR (see Chapter 4, Distribution Systems, for more information on these terms). In other words, EasyRes Plus offered agents true 'last seat' availability on the airlines connected to it. This is effected by a direct computer link between EasyRes and the airlines' host reservation systems enabling EasyRes to see exactly the same availability as the airlines' own reservations staff.

For smaller airlines that do not have the direct link, availability is maintained on EasyRes through what is known as Availability Status (AVS) messaging, which enables airlines to control what flights and booking classes they wish to sell via EasyRes in a manner identical to the way they do on a CRS. An AVS message is a message originated and sent by an airline when there are only four seats remaining on a flight. When this happens, the flight obviously is becoming full and each reservation is checked on-line before being booked. So, EasyRes was beginning to look much more like a true CRS in terms of booking features but with the advantage that it was much simpler to use. Although the list of participating airlines is impressive, there is one airline in particular that is missing. British Airways has its own viewdata booking system for its own product, which is called BALink. This is a system very similar to EasyRes.

One of the key features of EasyRes, which makes it so attractive to leisure travel agents, is the fare-driven display. The agent always sees the cheapest fare that meets the client's requirements, matched with real availability, which can then be booked if required. This is no trivial task to provide by means of an automated system. To give you an idea of what is involved in providing a fare-driven display, let me give a quick explanation of the processing steps that support this function.

First, Reed Travel Group needs to store all the latest fares on all routes for all the airlines participating in EasyRes. This consists of several thousand fares, which are stored on Reed's large main-frame data base. This data base is one of the most up-to-date sources of fares information in the world and is current to within a matter of hours. Then, the system has to look at what the agent has keyed in on the 'availability request' screen and build a table of the fares on that route for all airlines that fly that route. The airline reservation systems, which are connected to EasyRes by high speed data lines, are contacted. Their availability for the route in question is retrieved and stored in the EasyRes main-frame computer. The system then associates the appropriate fare with each flight that is available. The resulting information is sorted into sequence, with the lowest fare first and the most expensive one last. The resulting information is sent to the travel agent's videotex screen and this is what is called a 'fare-driven' display. In summary then, the major steps are:

1. Agent requests a fare-led display.
2. The system displays the route.
3. The agent selects the airline.
4. The system checks outward and return availability and does a fares check.
5. The system matches the information and constructs a composite display.

EasyRes includes access to airline, airport and destination information and, besides air and hotel, also provides access to car rental systems. It supports every ticket type from domestic shuttle to transatlantic flights, full fare to consolidations and also special offers. Agents are notified automatically through the system of any airline schedule changes and, through a link to Sabre, can take advantage of automated ticketing. The EasyRes system has the word 'easy' in its title because that is just what it is like to use – easy. An airline booking, for example, may be made using just six simple screens. This is how you would use EasyRes to make a flight booking:

- **Access the network** Use a videotex terminal or PC with viewdata emulation to access one of the travel networks, e.g. AT&T or Imminus. Sign-on to the EasyRes system using your agency identification and password.
- **Choose, from main menu** From the EasyRes main menu screen, select Option No. 1 (flight booking).
- **Specify requirements** The system will present you with a screen that needs completing as with a form. You will need to enter the departure and destination airports/resorts and the dates of travel. Choose 'Availability' or 'All Fares' options.
- **Select airline** Next, EasyRes will display a list of airlines that you may check for availability on the route you have already specified. It is interesting to note that this display is shown in random sequence by airlines in order to comply with guidelines on non-biased displays (see Chapter 1). Every time you request a display such as this, the sequence of the airlines displayed on the screen will probably be different. Thus, there is no bias with EasyRes.
- **Select flight** The system responds with a list of all available flights and asks you to choose 'All Fares' or, expand the details of selected flights. If 'All Fares' is selected, a list of all available fares on a particular flight combination is displayed, along with the key restrictions of each fare. Agents, on selection of the fare of their choice, will then be taken automatically into 'display flight' details, as described below.
- **Display flight details** The expanded flight details are now displayed. This shows all the details of the flight, including the currency of the fare, restrictions, other charges, minimum check-in times and baggage allowances. At this stage, the flight can be booked or the 'All Fares' option can be selected as described above.
- **Enter passenger details** Having selected the booking option, a screen is presented that asks for the passenger's details to be entered. Once this has been done the flight is booked and the airline's booking locator or PNR reference code is displayed on the screen (Fig. 6.14).

It is as simple as that. There are just one or two things to point out, though. First, although you can retrieve a previous booking for display purposes, you cannot change a booking. The only way to accomplish this is to cancel the booking and start all over again; and, it must be noted, that under half of the participating airlines support cancellation as a function on EasyRes. So, if the airline does not have a cancellation facility, then you would need to telephone the airline to cancel the booking before starting to re-input the new one. Then, there is the need, eventually, to produce a ticket. EasyRes does not have its own automated airline ticketing facility but provides one through its link to Sabre. Ticketing options are: (a) write the ticket manually, (b) queue the booking to a ticketing agent using Sabre via EasyRes' link to Sabre, or (c) if your agency has another GDS – perhaps in the business travel department – then you could create a 'ghost PNR' and use that to print the ticket. It should be noted that unless the ghost PNR is cancelled after ticketing, it will

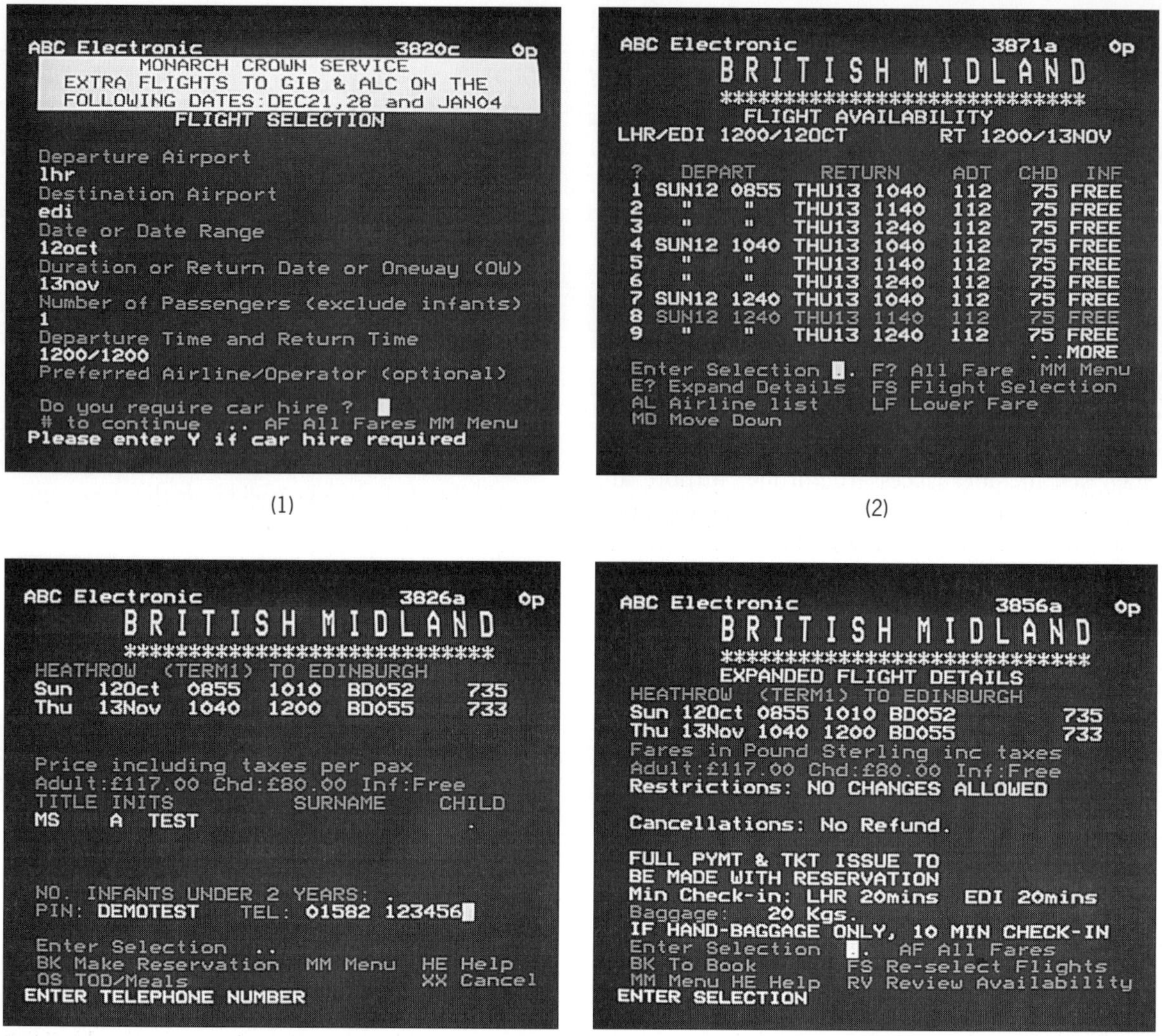

Figure 6.14 The final EasyRes booking screens

cause the airline a problem by causing it to pay an unnecessary double-booking fee to the reservation system provider.

EasyRes was further enhanced in time for the UK World Travel Market in 1992. This saw the launch of HotelSpace on EasyRes. Developed to make hotel bookings as quick and easy as airline reservations, this service provides a simple link for agents to access Utell International's hotel data base of more than 6,500 properties using EasyRes' well established screen formats and system logic. Access to this hotel information is through the code UTL. As well as providing a booking facility, HotelSpace also offers agents instant commission through the Utell Paytel system.

Looking at the product from an airline's viewpoint, EasyRes offers some significant benefits. For example, the EasyRes system also provides management information to airline hosts that paints a picture of where its bookings are coming from. This information is shown by county, city and agency. Sales figures for the month and year to date are shown. MIS reports available to participating airline hosts are:

- **Agent booking analysis by county** This shows bookings by county for every ABTA travel agency that has accessed EasyRes. Any timespan may be requested, along with year-to-date figures.
- **Booking analysis** A detailed month-by-month analysis of bookings by route and class.
- **Summary booking analysis** A one-page month-by-month summary of the booking analysis, highlighting both 'through' and 'connecting' flights.

The EasyRes system is available free of charge to any UK ABTA travel agent with a videotex terminal. The system is distributed via the main UK VANS, i.e. AT&T and Imminus. The EasyRes system is connected to each of these networks via high speed data lines. At the core of the EasyRes system is a powerful Amdahl main-frame computer capable of performing 56 MIPS. This mainframe is itself connected to each of the participating airlines' reservation systems. There are ambitious plans to develop EasyRes further. Under consideration, for example, are a new fare and availability search that will make it even easier to find the cheapest fare available on a route, and an Internet interface.

Worldspan

Worldspan supports access to its GDS via viewdata technology, using a product called Worldspan View. This allows travel agents who cannot justify the expense and training overhead required of a dedicated GDS system to nevertheless gain access to many of the productivity and customer service advantages of GDS technology, using their existing equipment. This approach is perfectly suitable for those agencies that typically generate a relatively low volume of scheduled air bookings.

With Worldspan View, either a viewdata terminal or a PC may be used as the primary workstation for accessing the Worldspan network. However, it is not the native GDS system, along with all its rather complex keyboard entries and associated screen displays, that needs to be learned. Instead, Worldspan has developed a special interface for viewdata users. This guides them through the enquiry, booking and ticketing processes using a specially developed and simplified user–system interface. This interface, although somewhat slower and less functionally rich than the native Worldspan system, allows the travel agents to use the GDS in a similar way to other viewdata host systems.

Access to Worldspan View may be either directly into the Worldspan network or via the AT&T viewdata network. Access via AT&T is achieved in two alternative ways: (a) dial-up – the users dial into the AT&T network and log-on in the normal way using their pre-assigned user-IDs and passwords, or (b) direct connect – the users do not have to dial into AT&T and simply log-on as normal. Once logged-on, Worldspan View may be accessed by selecting the WSP# host service from the AT&T network main menu.

Railtrak

Railtrak is a simple viewdata-based train reservation and information system provided by British Rail. The technology is very similar to that used throughout the tour operators as described above. Travel agents access the system using a simple viewdata terminal. The viewdata terminal in the travel agent's office dials into the agent's chosen videotex network. By selecting BRL for British Rail, the network connects the agent's viewdata terminal to a front-end British Rail videotex computer. This computer acts as an interface between the viewdata technology and the British Rail mainframe computer in Nottingham. This interface computer translates and converts from the mainframe screens into viewdata screens and vice versa.

Figure 6.15 The GTI concept

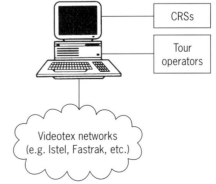

Once a connection is made, the Railtrak system offers the agent some useful facilities, which I have described in a little more detail as follows:

- Seat reservation.
- Sleeper reservation.
- Boat train reservations.
- International reservations.
- Motorail availability.
- Help.
- Mailbox.
- Information.
- Reservation sales.
- Training system.
- Password control.

The Railtrak system is a useful facility that is aimed at travel agents who have a fairly low volume of British Rail business. For the higher volume agents, there are several other systems that are probably more relevant. The main one being the main-frame rail system distributed via the GDSs (see Chapter 3 for more details).

Hotels

Utell may be accessed via AT&T or Imminus. A videotex front-end computer system acts as an interface between Utell's system and each of these videotex VANS. The end-product that the travel agent sees is branded as a Utell International service within the EasyRes system offered by Reed Travel (see previous section on Reed Travel). This is called Hotel Space. Both these networks access the same core Utell system. Making a reservation via a videotex system couldn't be easier. The whole process simply consists of the following steps:

1. Sign on and select the HotelSpace main menu.
2. Enter client's name, arrival date and either the number of nights required or the departure date. Enter the city only.
3. Select the required location of the hotel from a list presented by the system.
4. Select the hotel required from a list presented by the system.
5. Enter the number of rooms required and indicate via the BK entry that you wish to make a booking.
6. Enter any special requests or messages. Select your preferred payment method.
7. If you select full payment enter your reference, telephone number and BK to continue. Otherwise enter credit card details in addition to reference and telephone number.
8. The booking is now confirmed and the screen can be printed to form a hard-copy record for your files. An entry of SS will provide further information on commission collection.

THE PROBLEM WITH VIEWDATA

The basic problem with viewdata is that it is a redundant technology. However, because most tour operators have viewdata reservations capabilities and there are several thousand viewdata sets in travel agencies around the country, there is an enormous force of inertia that will resist any change of technologies in the area of leisure travel. Having said this, there are some pretty substantial pressures building up from a variety of interested parties for a change. Let's consider some of these from the viewpoints of the players involved:

- **The travel agent's view** Although viewdata is easy to use and cheap, it is rather cumbersome and slow. It also clearly looks outdated compared with the new graphics-based PCs that we all see around us nearly every day. Another factor is the spread of PCs among travel agents for a variety of applications such as GDSs; general office automation, i.e. word processing, electronic mail and spreadsheets; and back-office systems. Finally, the travel agent is faced with having to become familiar with a wide range of different viewdata systems as well as other technologies. Each system is to a certain extent different and this is what causes the problems.
- **The tour operators' view** Tour operators would dearly like to get out of viewdata technology but they have been hoist by their own petard, so to speak. Although they would like to print some form of charter ticket or holiday voucher for their products at the point-of-sale, the problem is that you can't easily do this with viewdata unless they use a product like RESCON (see TTI in Chapter 1). Then there is the processing burden that the tour operators carry on their own systems in order to support

interactive viewdata for travel agents. If only all travel agents had PCs then a great deal of this processing could be off-loaded onto them with the consequent reduction in technology overhead costs at the tour operators' end.

- **The airline CRS's view** One of the lessons learnt by the airlines from the recession was that it is not a good idea to be over reliant on the business travel sector of the market. It was precisely this sector that went belly up during the recession when companies cut back ruthlessly on travel and entertainment. The GDSs need a low cost point-of-sale workstation that will be acceptable to leisure travel agents in terms of cost and functionality.

Now if only the travel agent were to have a PC already on the premises . . . I hope that by now you are detecting a common wish running through the minds of the management of the tour companies and the airline GDSs. In the case of the GDSs, a PC in the agency would allow the leisure travel agent to use the sophisticated functionality enjoyed by the business travel agent. The GDS would only have one system to support, thus minimizing ongoing support costs and improving customer service. One of the key technologies that will be instrumental in accelerating the shift from videotex to PC-based systems throughout the travel industry, is to be found in the services offered by today's communications networks.

Communication networks

Telecommunications is a vast subject and quite a complex one too. Besides that, it is developing and changing as fast, if not faster, than the computer technologies that we hear about so much. Not only is the technology changing rapidly but the services are evolving at an ever increasing pace. It seems that the telecommunications marketing people are becoming extremely innovative. So much so that there is a real abundance of different ways to communicate with another party across the country or indeed the world. These new services offer a variety of communications methods and more importantly from a business persons viewpoint a variety of tariff structures.

Nowhere is the subject of telecommunications more important than in the travel services industry. This industry is seen by telecommunications companies as one of the most significant areas for future growth. So, it is no wonder that in the travel agency market, there are a number of suppliers offering some labour saving and sophisticated methods of communicating with travel principals around the world. Not just around the world either. Even across the UK there are a variety of methods that may be used to contact airlines, tour companies, hotels or even other travel agents. There are now several VANs that offer users, small or large, direct access to travel booking and information systems.

This chapter is devoted to these VANs and telecommunications suppliers. Naturally I have focused on those VANs that provide specialized services to the travel trade. But once again, as for the preceding chapters, the following should not be taken as any kind of a survey or a recommended list of telecommunications suppliers. My objective in presenting these services is to give you an idea of what is available on the market at present and to help you understand the kinds of things you can do with a good communications network.

As I mentioned before, the UK telecommunications infrastructure was created by BT. But in the new competitive environment in which we find ourselves, BT is required to supply telephone lines to other telecommunications companies. Suffice it to say at this point that as a result of the opening up of the telecommunications business in the UK there are now several VANs offering specialized services for travel agents. Besides BT itself, there is AT&T and Midland Network Services (MNS). Each of the travel agency service offerings of these companies is presented in the following sections.

CONCERT

On 4 November 1996 BT and MCI announced plans to form Concert, the world's first global communications company. Combining the global assets of BT and MCI, as well as the companies' 25 global ventures and 44 international alliances,

Table 6.1 Concert – the combined companies

Summary statistics	MCI	BT	Concert
Annual revenues (US $ billion)	18.5	24.5	43
Customers (million)	21	22	43
Employees	55,000	129,000	184,000
Countries with:			
International offices	70	30	72
International ventures	4	16	19

Note: Concert
Since this book was written, the proposed merger between BT and MCI has not proceeded as originally planned. However, the text describing the general market aims and objectives of international telecommunications companies are nevertheless relevant to information technology within travel and tourism.

Concert will begin operations positioned to rapidly grow its 6 per cent share of the US $670 billion world-wide communications market. Fuelled by increasing privatization, widespread deregulation and technology innovations, this market is expected to grow to US $1 trillion by the year 2000.

Concert will have its headquarters in London and Washington DC and its stock will be traded in London, New York and Tokyo. The company will be co-chaired by Bert C. Roberts Jr and Sir Iain Vallance and led by Sir Peter Bonfield as Chief Executive Officer and Gerald H. Taylor as President and Chief Operating Officer. BT and MCI, both of which will continue to operate in their respective countries under their existing names, make up two of Concerts five operating units. The other units are International, Operating Alliances and Systems Integration. Table 6.1 gives some idea of the overall scale of the combined companies.

Concert will offer an integrated set of products and services including local calling, long distance, wireless, Internet/Intranet, global communications, conferencing, systems integration, call centre services, multimedia and trading systems. Together with its ventures and alliance partners, it will reach 80 per cent of the global communications market on its first day of formation. It will be the second largest carrier of international phone traffic in the world and its Internet network, the largest and fastest in the world, will make Internet access available from all regions of the globe. The wireless market is expected to double in four years from 165 to 334 million subscribers. Concert will be in a position to capture a large share of this market via its operations and ventures in Asia, Europe and North America. The Concert Communications Company will begin operations with one of the most advanced portfolios of global networking services for multinational businesses. The top 5,000 multinational companies account for US $100 billion of annual global telecommunications revenue.

Finally, Concert's systems integration unit will be a US$2 billion enterprise, ranking it among the top five global IT service providers. The unit will employ 10,000 professionals in 120 locations world-wide, dedicated to providing a full spectrum of global IT solutions. This is one of the most significant outcomes of the Concert merger. It is expected to result in the creation of a computer services company called MCI Systemhouse. This new company will be created from MCI's existing Systemhouse division (which was formed from the takeover in 1995 of SHL Systemhouse) and BT's Syntegra division. Much of the success of MCI Systemhouse has been attributable to the handling of its outsourcing business. However, in the future, this new Concert service provider could well be the genesis of some exciting new travel and tourism products. After all, it is ideally placed to exploit these markets with its significant internal development resources and a truly world-wide customer base. Only time will tell.

Line One

As you would expect, BT is an ISP. I thought it would be worthwhile describing at least one ISP in the book and as Concert will be handling the majority of the world's Internet traffic, I thought that it would serve as an excellent example of one

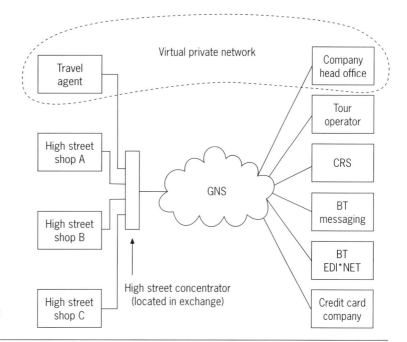

Figure 6.16 British Telecom's high street network

of the many ISPs available to a consumer. The BT ISP package enables a consumer to get connected to the World Wide Web and enjoy a range of information services that are provided by BT's Line One Internet site. To use Line One, a consumer needs: (i) their own PC, which should be a 486 or higher running windows with a minimum of 8 Mb RAM; (ii) a modem, preferably capable of a speed of 28.8 Kb/s; and (c) a dial-up telephone line. The Line One package consists of browser software, which is provided on either a CD-ROM or diskette with full instructions. Once the software has been loaded, the user may then access the Line One site on the Internet, which offers the user the following pages of information:

- **News and sport** These pages have feeds from many popular news services and newspapers. They also incorporate weather forecasts, but only for the domestic area (international areas may be specifically selected).
- **What's on** A guide for entertainment services, e.g., cinema, dance, theatre, music and clubs. Also provided is a television guide and reviews of shows and other events.
- **Family** A range of games, bulletin boards and other general areas of interest, such as horoscopes, and areas of interest to younger consumers.
- **Reference** Information that includes an encyclopaedia showing a great deal of useful travel facts and figures as well as financial services and many stock, share and foreign money prices.
- **Home shopping** This includes a ticket buying service, a telephone shop, a wine club and other systems.
- **E-mail** Line One provides Internet e-mail and other communications services, such as bulletin boards and chat forums.
- **Internet** Full access to the World Wide Web and all its sites.

AT&T

AT&T is a large global telecommunications company that in the UK provides some important services to the travel industry. In fact AT&T has a turnover of around US $52 billion per year from all its world-wide businesses. AT&T was founded back in 1925 as part of what became the giant Bell corporation in the USA. In fact 'Ma Bell', as it was known, became so big that the US Government decided that it was in danger of acting as a monopoly and passed legislation that broke it up into separate regional telecommunications companies. AT&T is one of them – and one of the biggest in fact.

The current UK company has its origins in Istel, an IT service bureau originally owned by the British Leyland (BL) group. Istel provided computer services and telecommunications for the BL group and later, around 1979 when the company was devolved, also provided services outside BL on a commercial basis. The name Istel was adopted in 1984. In 1987 the company was the subject of a management buy-out from the Rover Group, which took the company into the private sector where they became a public limited company. The final move was made in November 1989 when AT&T purchased the company, which then became AT&T Istel. Subsequently, the Istel part of the name was dropped and the company is now known simply as AT&T.

In the UK, AT&T offers its services to companies in the manufacturing, healthcare, retail, finance and insurance industries as well as travel. AT&T first started providing services specifically for the travel business around 1978, when it were competing with Prestel. In comparison with Prestel, however, AT&T seemed to offer more on-line tour operator reservation systems and as a result it became increasingly popular with travel agents. The AT&T network now carries more than 70 per cent of all holidays booked electronically in the UK. Virtually every major tour operator is connected to the AT&T network, as are 90 travel principals and over 130 other service providers. Approximately 3,000 of the UK's top travel agents are hardwired into the AT&T network (see the following section on Direct Service below). It now operates one of the largest private digital wide area networks in Europe and is of course a fully fledged VANS provider.

AT&T's Direct Service

This is one of AT&T's key services. It connects 3,000 of the UK's top travel agents to a multitude of travel supplier systems. The AT&T Direct Connect Service enables travel agency branches to be 'hard wired' directly into the AT&T network by leased data lines. Using leased or rented lines means that the agents no longer have to use their terminals to dial into the AT&T network each time a supplier's system needs to be accessed. The Direct Service guarantees instant connection to the AT&T network without the engaged or busy tones so frequently experienced at peak times. It also provides the agent with a good quality connection without line noise, which tends to corrupt the characters appearing on the screen. So, provided the travel agents make more than a certain number of dialled calls each year to AT&T then direct connection will actually save the agencies money in telephone call charges. Most of the major multiples are 'hard wired' into AT&T via the Direct Service.

The Direct Connect Service operates as follows. AT&T installs a Direct Service multiplexor in the agency. This is a special type of communications controller that supports up to eight devices, each of which may be either a viewdata terminal or a PC with a videotex emulation card. It allows each and every one to use the service at the same time. A modem is also installed in the agency and this is connected on the one side to the multiplexor and to the data line on the other. The data line is leased from BT and runs from the agency to a so called 'donor site'. It is called a donor site because it is not actually an AT&T owned location and is often a travel agency that happens to be located in a conveniently central position within a region. The donor site is reimbursed any extra operating costs by AT&T. This link is so transparent that most locally connected agents are not aware that their data lines run through another agency. Indeed, there is no reason why they need to know because the donor site has no access to the data at all. The donor site acts as a kind of hub and is itself connected to the AT&T network by a high speed data line.

Summary of AT&T

AT&T also provides several travel-related telecommunications services that are covered in other chapters of this book. For example, the FERRY# reservations system is supported by eight ferry companies who have all agreed to use a standard booking format and method (see Chapter 1). The Internet-based World Travel Guide On-line system provides information on up-to-date airline fares, a country gazetteer, the world's weather, tourist exchange rates for foreign currencies and car hire details (see Chapter 5).

IMMINUS

Imminus is a separate business activity within a company called General Telecom, which is itself a wholly owned subsidiary of the General Cable Corporation. This parent company is 40 per cent owned by a major French utility company. Besides General Telecom, the General Cable Corporation also owns Yorkshire Cable and the Cable Corporation, and has a 40 per cent share of Birmingham Cable. So, Imminus is very much a key part of a large international telecommunications business. The company's origins are, however, worthy of consideration, particularly because it now services a significant proportion of the UK's travel network market.

Originally, the company was known as MNS and until 1993 was a wholly owned subsidiary of Midland Bank. MNS was a telecommunications company or more specifically, a VANS that had its origins in the Thomas Cook multiple travel agency chain. MNS was formed in 1984 when Thomas Cook was a part of the Midland Bank Group. The rationale for the formation of MNS was based initially on the core communications network that Midland Bank had built up over several years. This network connected not only every Thomas Cook agency but also every high street branch of the bank. This formed an excellent nation-wide infrastructure to use as the basis for selling communication services to other companies. A Midland Bank internal audit in the early 1980s identified the high level of investment and expenditure in this area and recommended the setting up of a new company to sell spare capacity to others, as a separate and new business.

This was allowable under the prevailing rules of the UK Government's telecommunications rules and regulations. Put simply, these rules stated that a communications company would be allowed to compete with BT in the area of telecommunications services provided that certain guidelines were followed. One of these was that the communications resources rented from BT for resale to others at a profit, were only allowable provided that a value was added to the base resource. The addition of computer processing to communications was one such example of an added value. This bundled package of communications and computer processing therefore formed the basic services offered by MNS, and for that matter several other new VANS that started trading at roughly the same time.

The next significant event in the company's history occurred in 1995 when the management of Imminus decided to buy the business from the Midland Bank. Following the management buy-out, the company continued to grow in many sectors of the communications market, particularly in the travel industry. Then in March 1997, the General Cable Corporation acquired all of the outstanding shares of Imminus. It integrated Imminus within the General Telecom subsidiary that has responsibility for voice communications and, through Imminus, telecommunications services for the UK travel industry.

Imminus has developed a strategic programme for the future growth and development of its telecommunications services for the UK travel industry. I am going to present an overview of the company's strategy by describing its travel products in the context of an evolving telecommunications infrastructure. The starting place for this review is the portfolio of tried and proven network services that Imminus has operated successfully within the UK for many years.

The established products

Imminus continues to provide the UK travel industry with the well established videotex network services that have now been in use for some 15 years or so. These comprise several products, all of which are currently supported by a national X25 network. This back-bone network is now well established throughout the UK and is the platform that supports most of Imminus' current product line. So, before we explore the products in more detail, it is first necessary to understand a little more about the communications network on which they are currently based.

The X25 network

These services are based on an X25 network that has been around ever since it was originally developed by Midland Bank. This network, which is now called Fastrak, used to be called Midnet and was one of Europe's largest and most advanced private data communications networks. The network is managed 24 hours a day, 365 days per year. Besides supporting the UK travel industry, it supplies communication services to customers within the following industries: banking, retail, financial services, motor trade and distribution. The core data network is based on X25 packet switching technology supporting both videotex and asynchronous terminal access.

Probably one of the most significant events in the history of Fastrak was the interconnection of Thomson Holidays. This occurred in April 1993 and was more than just the interconnection of the Thomson Holidays reservations computer. Imminus was successful in obtaining the contract to run all of Thomson Holidays' UK communications network. This is known as a facilities management (FM) or outsourcing arrangement, which is becoming increasingly popular with many companies of late. The reason FM arrangements are becoming so popular is that they allow a company to hive off administrative operational functions and allow it to concentrate on its core business. In the case of Thomson Holidays, this is of course the sales and marketing of packaged holidays. So, as a consequence of this telecommunications FM deal, Imminus assumed responsibility for all of Thomson Holidays' nation-wide network of data lines and related equipment. The Imminus X25 network now has around 7,500 UK travel agents who regularly use the service as well as between 80 and 90 other tour operators besides Thomson Holidays.

The Fastrak network is a national UK network of high speed telephone lines and other communications equipment. Control of the network is critical and a network management centre located in South Yorkshire is the hub of these activities. From this centre, Midnet management can monitor the performance and availability of the entire network. With its links to other networks overseas, Midnet provides a growing capability for international operations. The centre employs over 200 specialist managers with technicians working in shifts to provide a high level of network availability and comprehensive management information reporting for measurement purposes. From a security viewpoint, Midnet uses the most sophisticated security controls available, including passwords with reversible coding systems to provide additional safeguards. The Imminus market share of the UK's travel industry network traffic currently stands at around the 45 per cent level.

This network allows tour operators' computer systems to be connected on the one side and travel agents' on the other. Although both parties connect into the same network, the way they do this is different in each case. Take tour operators. Tour operators connect into the X25 network using high speed dedicated leased data lines, each of which is capable of supporting multiple simultaneous videotex conversations with a number of travel agents. However, the number of such conversations (more often known as sessions), is limited by the viewdata technology involved. This requires the tour operator to hold one of its communications ports open all the time that a travel agent is holding a session with the tour operator's booking system. Although this works satisfactorily enough, it is somewhat wasteful in terms of resources, particularly those involving computer time and network usage. Travel agents are connected into the network in a number of ways, depending upon which Imminus product they use. So, now is a good time to consider these products in more detail.

Fastrak
Fastrak provides travel agents with dial-up videotex access to all the major tour operators, ferry companies and scheduled airline services. Fastrak is based on a strategically located network of 90 viewdata communication nodes throughout the UK, which provide local call access to around 98 per cent of all UK travel agents. Most towns in the UK have a network node, i.e. access point, and many of the major towns have two nodes. So, for an agent to gain access to Imminus' X25 network, a local call is all that is needed. This local call is made via the BT network and is charged separately by BT on a time used basis, as any other telephone call. To use Fastrak, a user needs either a viewdata terminal with integrated modem or a PC with a viewdata card and modem. Generally speaking, Fastrak operates at an up-link transmission speed of between 1200 and 2400 b/s.

Fastdial

Imminus also offer asynchronous dial-up access to the X25 network via the Fastdial product. This is designed to send high volumes of data over dial-up connections to remote host computers within the UK. With speeds up to 14.4 kb/s, users benefit from faster data transfer, thus reducing PSTN network access charges and the need for direct connectivity. This product is not widely used within the UK travel industry at present.

Fastlink

This is Imminus' product name for its Direct Connect Service and the product was originally launched in January 1993. It is an important part of Imminus' travel networking products and services and therefore it is worth considering in some detail. If a travel agent is using Fastrak extensively and is therefore running up a substantial dialled telephone bill, then Fastlink could actually save them money. The reason for this is that it does away with the variable cost of a dial-up telephone line and replaces it instead with a leased telephone line that is permanently connected to the nearest X25 communications node. To use Fastlink, a travel agent needs the following items of equipment:

- **Chameleon Linkmaster** This is a purpose built PC that manages all of the networking services on the travel agency site. The Chameleon Linkmaster allows (Fig. 6.17) the data line to Imminus to be shared by a LAN or several standard videotex terminals in the agency. Chameleon Linkmaster is a PC product that is designed to evolve over time. The hardware is based on multiple processor cards. This is like having several computers within a computer, with each one mounted on its own printed circuit card. The hardware is extremely reliable because it is estimated by Imminus that the mean time between failures (MTBF) is around 20,000 hours. In other words, if the system is used during normal office hours, there will on average be one hardware failure every nine years. The software is written in the industry standard 'C' programming language, which enables Imminus to enhance and refine the system continually over time using widely available expertise and development productivity aids. Communication speeds are variable up to 64 Kb/s.
- **Leased line** The leased line is normally rented from BT but could in theory be supplied by any of the national telephone service providers such as Mercury. The leased line is permanently connected to the Imminus network via the nearest node.
- **Cabling** The travel agent's equipment is interconnected with Linkmaster via special purpose cables installed within the office.

Chameleon will support videotex technology for as long as the industry requires it and will

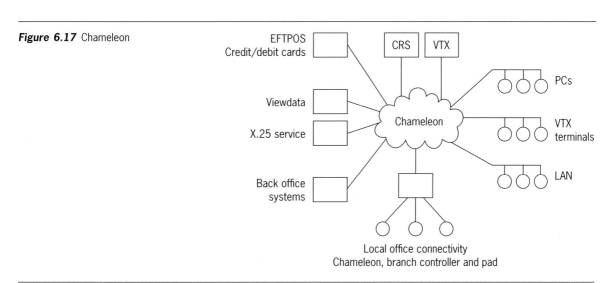

Figure 6.17 Chameleon

also provide travel agents with the means to migrate to more sophisticated travel technology as it becomes available. This means that videotex systems can be accessed using the PC, and the old videotex screens will appear and be usable just as they have always been on the old 'dumb' viewdata terminals. The difference with Chameleon is that as suppliers convert their systems to PC-based technology, the agent using Chameleon will not need to change any equipment in the agency. Using the Chameleon PC hardware, a set of customized operating system software and communications resources, the Chameleon Linkmaster product can provide its users with a range of functions, some of which are optional. These functions are being continually updated and enhanced. While it would be impossible to describe each function in detail, the following is an overview of the most important ones:

- **Support of videotex terminals** Up to 64 videotex terminals may be supported from each Chameleon PC. Each videotex terminal can carry out a 'conversation' with a host system, independently of the others in the agency. Any terminal that conforms to the Prestel Terminal Technical Guide may be connected to the Chameleon Linkmaster PC.
- **EDI** An optional extra is ferry ticketing. If this is required, an EDI printed circuit board needs to be installed in the cabinet. This supports the secure transmission and printing of tickets from certain companies that have implemented the UNICORN standards promoted by the TTI group. This enables the system to print the ferry tickets available via certain ferry operators and other evolving remote ticketing and printing systems planned by leading suppliers.
- **Support of other PCs belonging to the agency** The Chameleon PC can be interconnected to other IBM-compatible PCs in the agency via either a direct connection or a LAN. These PCs should have a good colour monitor and support for videotex emulation software. For PCs running Windows, Imminus suggest that a good package is 'SoftKlones Talking Windows' (see below for more details). Direct connection is achieved by connecting a cable from the serial port of the Chameleon PC, i.e. the RS232 port, to the serial port of the agency's PC. Connection via a LAN is by installing a LAN printed circuit card in the Chameleon PC and cabling this to the LAN server PC.
- **GDS link** The Chameleon PC supports access to the major GDSs, via the Imminus network.
- **Automatic password insertion** The Chameleon PC can be instructed to insert the agency's password automatically into the sign-on dialogue with host systems. This can save time and improve operator productivity.
- **Local screen dump** The system can dump an image of a screen to a printer in the agency. This is accomplished without the need to communicate with the host system and can be a useful facility for producing a hard copy of a booking for a paper-based customer file.

The Chameleon PC also supports remote network management, which eliminates the need for the equipment to be physically changed with each major network enhancement. A few years ago, this used to require a site visit from an Imminus engineer. However, the Chameleon PC can now receive new software updates automatically via the Imminus network. New editions of software are simply transmitted from the Imminus network management centre's host main-frame computer, to Chameleon (this is known as downloading). The new software is then immediately available for use by the customer.

An important optional feature of Chameleon that I mentioned above, is a software product called Imminus Talking Windows (a SoftKlone UK Limited product). It can run in any of the PCs that are connected to Chameleon. Talking Windows is a Microsoft Windows-based PC emulation package that allows simultaneous access to multiple tour operators' systems and supports the RESCON standard (see TTI in Chapter 1). The Talking Windows product allows users to quickly compare holiday details offered by different tour operators and to cut and paste information into other applications. It also supports data transfer from a PC connected to an agent's LAN, across the Imminus network to a remote PC host. This can be used for applications such as data collection and management information collection.

A key advantage of Chameleon is that it makes inter-branch communications easier and more cost-effective for multi-branch travel agencies. Multiple travel agents can use Linkmaster to make use of private branch communications to maximize host connectivity and minimize branch costs. In the past, independent agencies have been considered too small to posses the in-house expertise needed to set up and run private communications networks. But with managed networks such as Imminus, this is entirely possible and is in fact encouraged by the provision of consultancy services to such agents. Independent multi-branch agents can interconnect their agencies via Chameleon, which may be connected directly to their head offices if they are nearby or via the Imminus network if the agencies are distant from the head offices. Either way, the head offices are able to communicate with all branches to, for example, receive consolidated sales figures or to allow branches access to the agencies central back-office computers.

The Fastlink Direct Connect Service was first launched to travel agents in 1989. In 1997 there were around 1,300 agencies directly connected to the Imminus X25 network using Chameleon's branch controller. Some of the major multiples and most of the independent regional multiple branch agencies use the service for inter-branch communication and for gaining access to suppliers' systems. Other agencies simply use Linkmaster to minimize their telephone bills and provide their sales staff with a fast, reliable and secure method to access the major industry reservation systems.

- **Chameleon's videotex support** Because videotex is expected to be around for a few years yet, it is important to understand a little more about how the Chameleon product supports videotex. After all, it is important from the leisure travel agent's viewpoint that the dialogue with host reservation systems is fast, accurate, secure and reliable. Chameleon Linkmaster controls the communication between the videotex terminal user and the videotex host connected to the network. When a user enters the mnemonic of a host system from the menu page, a call is established to the videotex host system. The welcome page of the host is transmitted to Chameleon and displayed on the user's terminal.

Communication between the user and the host is carried out by a series of requests and responses and by control and display commands. Chameleon sends and accepts these display and control characters from both the host and the user and displays the information on the user's terminal. The user can enter data into information fields displayed on the screen (for example the departure date), or can respond to requests for further information. In addition, the user may enter commands to correct or change the information displayed or to end the videotex call. During the call, Chameleon displays status messages at the foot of the user's screen. These messages can take the form of information received from the host or transmitted from the Chameleon PC in response to an error.

Chameleon maintains a screen image of every videotex display in use. This enables information to be validated locally and an error message displayed if the information is incorrect. It also allows immediate responses to be made to the user's commands without the need to send the information to the host. This is a feature that can significantly enhance the speed and therefore the productivity of videotex users throughout a leisure travel agency.

- **Chameleon's ticket printing functionality** Chameleon Linkmaster conforms to the specification defined by the TTI for EDI ticket printing (see Chapter 1). A ticket message received from a host, e.g. a ferry operator, is accepted by Chameleon and printed via the connected ticket printer.

The user communicates with the ticketing host via viewdata and requests a ticket to be printed. On receipt of this request, the host sets up a call to Chameleon and transmits the information using the defined EDI format. EDI contains pairs of messages that determine the start and end of the transaction (in this case the ticket), the start and end of the data and finally the data. Chameleon responds to the instructions contained within the EDI message, translates the data into a format recognized by the printer and sends the information for printing. All of this exchange of data between the agency's Chameleon and the host system is secure and accurate because it is controlled by the standards embodied in the TTI procedures.

Chameleon will support multiple incoming ticket requests, queue print requests when printers are busy, search for available printers (if more than one is being used) and reject calls if printers are off-line or faulty. In such cases of printer unavailability, the ticket information remains on the host to be requested by the user when the printer next becomes available.

The new Imminus products

Imminus, like most companies in a fast moving market such as leisure travel, keeps a close watch on its customers. The critical factor for a service provider such as Imminus, is the way in which its customers' business requirements evolve over time. The only effective way to keep in touch with customers and to be able to develop products and services to meet their needs is by talking with them. Several years ago Imminus therefore undertook a major quantitative survey of 800 travel agents. This survey was undertaken by an independent company called Travel and Tourism Research Limited and involved the completion of a detailed questionnaire. In addition to this, Imminus also did some qualitative data gathering and interviewed 30 leading travel agents to solicit their views on the true business needs of technology. The results of the survey were carefully analysed and Imminus found that:

1. Agents wanted a system that was as future proof as possible. They didn't want to invest in a system that quickly became obsolete with the emergence of some currently unknown technology.
2. They wanted to be protected from conflicting standards like the old Betamax versus VHS issue. This is especially relevant in the travel industry at present where different groups such as TTI and EDI are all developing standards and systems.
3. More agents were investing in PCs as the replacement technology for videotex. They therefore wanted something that could access videotex host systems but that would also run on their in-house PCs. An important conclusion was that the majority of agents would be PC-based within three years.
4. Agents wanted a system that would enable them to link their front-office systems with their back-office systems as well as their LANs.
5. A system was needed that could meet the challenges of a new breed of customer that was becoming more frequent; this was the independent traveller who needed an agency with access to a wide range of separate travel products, not just packaged holidays alone.
6. Many smaller agents wanted the functionality of GDSs but without the need to invest in costly GDS equipment and a time consuming staff training program.

In addition to this kind of customer feedback, Imminus is as aware as many others within the field of travel and tourism in the UK, that viewdata is yesterday's technology. The trouble is, there is a large and well established user base that is going to be difficult to migrate towards the newer technologies. Particularly if a new investment in end-user IT is involved. The attitude of travel agents and tour operators may be summarized by the question: 'If it ain't broke, why fix it?' The only real incentives that would cause travel agents and tour operators to change to a new replacement technology are reduced cost, increased speed of service and improved servicing quality. Despite this barrier to change, Imminus has recognized that change will nevertheless happen and if it is to remain competitive then it must develop an infrastructure that will support the next generation of communication network services. Further evidence for the need to develop new products and services can be gathered from a simple market analysis. The UK market for travel agent and tour operator communications is estimated to be £26 million per year. However, Imminus expects this to be eroded by 20 per cent due to new technologies. So, for all of the above reasons, Imminus has invested in new communications technology to help customers enhance their service levels and begin to use new applications. The new Frame Relay Network is at the heart of these new services;

- **The Frame Relay network** Imminus has created a new network in the UK, which coexists with the X25 network. The long term view is that this network will eventually replace the X25 network completely. However, for the next few years Imminus can only expect a gradual migration by travel agents and tour operators. The new Frame Relay network uses technology that supports a number of new telecommunications protocols including TCP/IP (the Internet protocol), ISDN and other more efficient transmission methods. User locations access the network via a communications device called a router. The router enables a user's LAN to connect into the Imminus Frame Relay network via a dedicated leased line. Once connected to the network, the user can access other devices via their router, which is also connected to the Imminus network. The principal advantages of the new Frame Relay network are that: (a) at a transmission speed of 64 kb/s, it is faster than the X25 network, and (b) it handles multiple concurrent sessions more efficiently than today's networks.

For example, it allows a communications session between an end-user terminal and a host computer to take place without tying up dedicated resources. With viewdata, a session between a viewdata terminal and a host tour operator's computer ties up a host communications port for the duration of the session and keeps a network path open until the session terminates. With a Frame Relay session, the user's terminal, i.e. PC, sends a request via the network to a host system. The host responds to the request but then instead of keeping a communications channel open, waiting for a reply, it moves on to service the next incoming request from another user. This reduces network traffic and consequently end-user communications charges.

To take another example, a router could be installed in a multiple travel agency's head office. This would act as a communications gateway between: (a) the head office servers and main-frame computers, and (b) the multiple's population of travel agency branches. Each branch would have its own router that connects into its local branch LAN, which in turn supports PCs at the point-of-sale and other devices. With this kind of network, head office can either broadcast information to all branches or send data to specific point-of-sale PCs in selected branches. Conversely, branches can send their end-of-day results to head office whenever they wish to, for central consolidation: it also enables branches to connect into suppliers' systems via gateways established at head office level. Imminus' customers can obtain help in accomplishing this by using the Fastroute service (see below).

However, in order to support the gradual migration of travel agents and travel suppliers away from the X25 network and onto the new Frame Relay network, Imminus has developed some transitionary products. I will be describing these in more detail below. However, one of the basic features that these new products require is a bridge from the old to the new. This is accomplished by means of a high speed communications link between the X25 network and the Frame Relay network, which runs under yet another communications protocol called asynchronous transfer mode (ATM). There is no need for us to go into ATM in more detail here. Suffice it to say that ATM allows high speed networks of various kinds to be interconnected.

- **Fastroute** This is Imminus' managed router service for customers with large networks of remote LANs, which need to be inter-networked by a wide area network. Inter-networking allows computers and terminal devices to communicate regardless of the technical architecture that supports them. Fastroute uses the local routers installed in customers' sites to control access to the core Frame Relay network that carries traffic at high speed between sites. Fastroute has been designed as an all embracing service that takes away the day-to-day operation network responsibility from customers. It is implemented in four steps: (i) a requirements audit, which determines the routers required to support the network and enables an initial cost estimate to be quoted; (ii) technical consultancy, which designs the customized router network based on traffic flows and results in a detailed cost quotation; (iii) project implementation, which installs the hardware and configures the operating software; and, finally, (iv) total service management, which runs the service for the customer 24 hours per day, 365 days per year. A good example of a travel customer currently using Fastroute is Airtours.

Imminus has recognized that not every travel agent will invest in the routers and higher specification PCs needed to get the best out of the new Frame Relay network. It has therefore developed an alternative method for travel agency users to access both the existing X25 network and the new Frame Relay network from their currently installed PCs. They are able to provide these new access services because Imminus is a licensed public telephone operator with its own links into the Public Telephone Network operated mainly by BT in the UK. This allows users with Chameleon Linkmaster to opt for direct access to both Imminus networks. All that is needed is a software parameter change. This causes all communications traffic to be routed via a point-to-point Imminus network path. There is even a migratory product for the dialled Fastrak product that operates as follows:

- **Fastrak Direct** Travel agents using dial-up viewdata for enquiries and bookings can save time, money and effort by using Fastrak Direct. This service allows dial-up viewdata users to select Imminus instead of BT as the local exchange line carrier into the Imminus network. The net end-user benefits of this are that: (a) local access calls to Imminus are 30 per cent cheaper than normal; (b) Imminus is totally responsible for the end-to-end service for its customers with virtually no BT involvement at all; (c) depending upon the end-user's PC specification, the service supports transmission speeds of between 28.8 and 33.6 kb/s; and (d) the end-user pays on a per minute basis with no minimum number of minutes.

  Imminus has implemented Fastrak Direct using an interconnect processor that has a direct route to both the X25 network and the Frame Relay network. All the user has to do from their terminal is dial a prefix to the Imminus network node number. This prefix causes the local call to be routed via the Imminus interconnect processor and from there into the appropriate network. This is similar in concept to the way in which Mercury customers dial a prefix to by-pass BT's local access service.

The creation of this new networking infrastructure allows Imminus to migrate travel agents and travel suppliers to newer and more efficient communications technologies. This in itself has spawned several other new services that can be delivered over the new network structure (Fig. 6.18). The size of the new service opportunities is reflected in some of the terminology used. For instance, the area within Imminus that provides the computer power for these new services is called the Server Farm. It comprises a room that houses a number of servers, each of which is dedicated to its own particular function. The Server Farm is connected to both the X25 network and the newer Frame Relay network. This dual connection allows any of Imminus' customers to gain access to the new services. In actual fact, there are many such new services and it would be difficult to present each one in detail in this section; but here are a few:

- **InTouch** This new service, designed specifically for the travel industry and launched in the early part of 1997, already had over 1,000 travel agency and tour operator users registered by the middle of its first year. It uses Microsoft Exchange to provide users with a full range of e-mail services, but with some significant special features. For example, it allows viewdata terminal users to send and receive e-mail to any other user who has an e-mail address, whether it be on a corporate X400 network or on the Internet e-mail network. InTouch supports most popular e-mail software packages, such as Microsoft Mail and Lotus CCMail. This allows a number of applications to be implemented such as: (a) allowing tour operators to broadcast sales and administration messages directly to the viewdata screens of their travel agency partners, (b) enabling multiple travel agents to send information from their head office to outlying branches, and (c) collecting information from branches that needs to be delivered to head office.
- **Sky** This is a server that links into Sky Television's teletext service. It allows an authorized user of Imminus' Frame Relay network to receive enquiries which have been keyed by Sky television's audience. This works in a very similar way to teletext and can be used to support response advertising campaigns undertaken by Sky's corporate customers. One

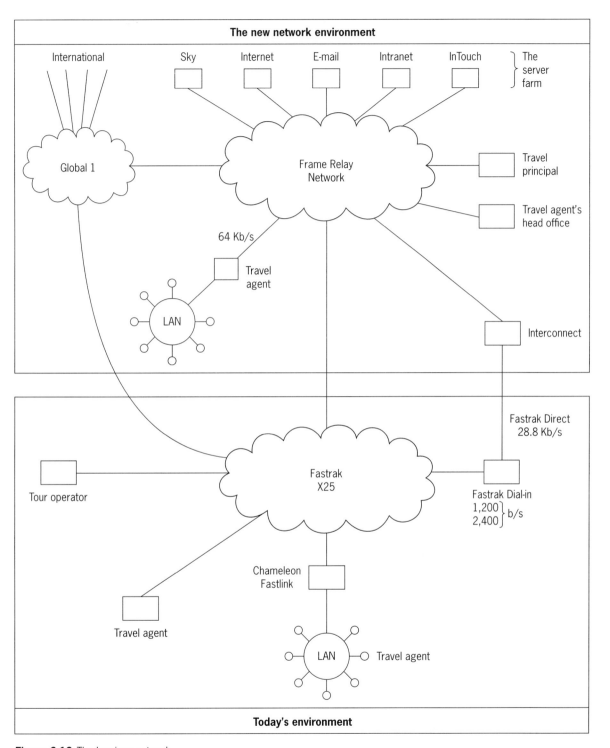

Figure 6.18 The Imminus network

example of an application that makes use of this channel is American Express' foreign currency ordering service. Sky television viewers can use their television's keypad to request foreign currency and travellers cheques. The request transactions are captured by the Imminus server and may then be picked up by American Express for fulfilment processing.
- **Internet** Users of both the X25 and Frame Relay networks may gain access to the Internet via a gateway into BT's Internet server. This Imminus network gateway is available to all subscribers on both networks and is protected by firewall products that limit unauthorized access in both directions, i.e. from the Imminus network to disallowed external sites and from the Internet to internal Imminus addresses.
- **Intranet** Travel agents that decide to set up their own Intranet-based on Imminus technology can use a specially designed server to communicate with tour operators via EDI messaging standards. The server receives inbound messages destined for a tour operator and converts these from HTML format into EDI format. The messages are passed from the server via the Frame Relay network and to the tour operator's system. Return message flows work in a similar way. This describes how a so called 'Intranet island' can be created by a user – in this case a travel agent. An Intranet island is a closed community of users who all deploy common Internet standards and widely available Internet software products in order to share information and messages between them.

The concept of Intranet islands is an interesting one and has given rise to discussions about the Travel Industry Intranet. Imminus can see that there are compelling reasons for groups of its travel industry customers to set up secure and reliable Intranets based on Imminus technologies and networks. For example, a vertically integrated tour operator could decide to distribute its package tours to its own in-house travel agency chain using an in-house Intranet. With this approach, each travel agency PC terminal at the point-of-sale would use a standard software browser to interact with the tour operator's system run by head office. This would completely replace viewdata and deliver some important operational advantages, such as more efficient use of network resources, improved information displays with graphical images, local scrolling of information pages with no central system overheads and many more. The system would look just like an Internet application but it would not necessarily be connected to the World Wide Web at all. It would be restricted to accessing head office data and would have firewalls protecting it against both: (a) access to the Intranet from outside the company, and (b) access to other external systems from inside the company.

Once our theoretical company had set up this Intranet, initially for its own internal distribution purposes, further extensions could be considered. For example, a separate and independent travel agency chain that sells a high number of the company's tour products could be invited to join the Intranet. The travel agency chain would have to commit to Intranet standards for its point-of-sale PCs, but this is not as onerous as it sounds because they may in any event already be using browsers for Internet purposes. Then the company owning the Intranet may decide that it wishes to standardize its GDS access on the same distribution technology. As outlined in Chapter 5, most major GDSs already have an Internet booking engine product and so it would be quite possible for them to provide their information and reservations services in this way via a dedicated Intranet link at headquarter-level. So, I hope you can see that it is possible for a company's Intranet to grow and expand under its own control, using secure and reliable technologies, to form an Intranet island of travel services distributed throughout a network of owned and affiliated travel agencies, close trading partners and suppliers.

Imminus intends to not only provide the necessary communications network resources to enable its customers to create their Intranet islands, but it will be able to go one step further. It will be in a position to interconnect consenting Intranet islands into so called Extranets, which collectively are referred to as the Travel Industry Intranet. In the medium term, Imminus intends to provide whatever support is necessary to allow customers to create their Intranet islands using the Imminus network and even to interconnect them for cross-company communications. Longer term, the interconnected Intranet islands may wish to be connected to other islands supported by other network companies. When this happens we will have seen the birth of the Travel Industry Intranet. It is interesting to compare and contrast this with the aims and objectives of the abortive GTI initiative. While the underlying objectives of GTI and the Travel Industry Intranet are very similar, the Intranet approach that is only just emerging in the travel industry, is far more pragmatic, has a wider appeal throughout the industry and is less of a 'big-bang' approach. It therefore may well succeed where GTI failed. Only time will tell.

International Imminus services

So far, I've talked only about Imminus' domestic UK customers and suppliers. However, Imminus has a very substantial international portfolio of services. Its Global-1 network is a joint venture of major telecommunications companies, which offers international connectivity to 800 different locations in 44 countries and an additional 120 countries through agreements. The members of Global-1 include Sprint of the USA, Deutsche Telecom of Germany and France Telecom. This inter-networking alliance is based on the worldwide standardization of Sprint's leading edge communications technology. In the UK, Imminus is a reseller of Global-1's services. Access to Sprint's global network (SprintNet), is from the Imminus 14.4 kb/s asynchronous dial-up network (Fastdial) and dedicated X25 lines. Companies' networking requirements are addressed by the three main services offered: Custom Link Service, Data Call Plus (DCP) and Global Data Connect (GDC).

- **Custom Link Service** This service allows customers to configure their own global managed data network using SprintNet. Three main networking configurations are supported: (i) point-to-point connectivity, (ii) centralized or 'Star' networking and (iii) meshed inter-networking. Charging is based on a fixed fee for unlimited traffic across the network.
- **DCP** This is an asynchronous dial-up service to Sprint's Global Data Network. Users connect their devices by dialling the nearest Imminus/Sprint access point via the PSTN. Usage is billed at a fixed hourly rate, regardless of the amount of data transmitted. This allows customers to predict communications costs easily regardless of the data application.
- **GDC** This is a dedicated point-to-point data communications service designed for customers with low volumes of data to be transmitted. It is billed on a per kilosegment basis for the traffic sent over the connection (a kilosegment is 1,024 segments of data).

These services enable customers to create their own private networks without the management and ownership overheads normally associated with such endeavours. This can be particularly attractive to tour operators with offices in destination areas that need to communicate with their home headquarters. Access to the Global-1 network is of course two-way. In other words, subscribers who use Global-1 overseas can access the Imminus network in the UK. All of these services are full MNSs that enable customers to focus on how their telecommunications application are used rather than worrying about network performance, fixing network faults and monitoring individual network component billing.

Finally, Imminus has demonstrated its commitment to the UK travel industry by several partnership ventures. It has, for example, worked with ABTA's Travel Training Company to develop an on-line training facility that can be accessed either from viewdata terminals, a company's Intranet or the World Wide Web. This is a sophisticated real-time training mechanism that, given the Internet or Intranet access methods, can provide pictures and other graphical images to enhance the learning experience for the student. The training facility incorporates several management features that assist course leaders to administer the progress of their students. Imminus is currently working on voice-based coaching features and video-conferencing, which will both enhance this innovative approach to travel training.

Conclusions

As you will see from reading this chapter I have not been very brave about foretelling the future of network automation beyond stating the obvious move to PCs from videotex. The reason is that it is very easy to be proved wrong by actual events when trying to do any form of crystal ball gazing in the field of technology in travel and tourism. However, I hope I have alerted you to some of the major issues that the industry is facing at the present time and some of the factors affecting the trends that may well have a significant impact on the future. One thing is certain and that is that you must always be looking to the future whenever you consider anything in the area of IT for travel and tourism. The only effective way of doing this is to keep up with developments as they happen or as they are debated. A good way to do this is to get yourself on the circulation lists of several of the leading computer and telecommunications magazines, as well as the travel trade press. Reading these publications will at least bring to your attention the level of debates on the emerging PC software technologies, the new peripheral devices, evolving telecommunications technologies and the price of PC hardware.

7
Travel agents

In this Chapter I am going to discuss and explore the various technologies that are used by travel agents as part of their day-to-day business operations. We have already seen how the distribution systems allow agents to access and sell products from a variety of suppliers. Now it is time to see how travel agents actually record these sales and process the resulting financial transactions. The technology used for this function is called the 'agency management system'. Choosing a good agency management system is one of the most important and strategic decisions a travel agent will have to make: with over 700 general purpose accounting system software packages on the market and a growing number of front-office support tools, it is by no means a trivial task. In this chapter I hope you will gain a basic understanding of the components of such systems, which should enable you to understand how travel agencies identify their requirements and evaluate the various products on the market.

Before we embark upon our review of the type of agency management systems available to travel agents, it is important that you first understand that there are different types of travel agents for different types of travel businesses. These distinctions are important because there are different types of agency management systems for each of these types of businesses. So, what are the different types of travel agents and how do their business needs for IT differ?

Types of travel agents

There are almost 5,000 ABTA registered travel agency branches in the UK. This network of agents is dominated by only a few of the larger chains usually referred to as the multiples. A multiple in this sense is taken to mean a travel company with more than 20 branches. Table 7.1 shows the top few multiples (by number of branches), and the rest, which is made up of a number of other smaller multiples (sometimes also known as miniples), plus other independent travel agents.

Table 7.1 ABTA travel agents

ABTA member	Total agent branches
Lunn Poly	796
Going Places	712
The Thomas Cook Group Ltd	386
MTG (UK) Ltd	291
Cooperative Wholesales Society Ltd	236
American Express Europe Ltd	190
Travelworld (Northern) Ltd	94
Midlands Cooperative Society Ltd	83
Carlson Wagonlits Travel UK Ltd	78
IT Travel Ltd	74
United Northwest Cooperatives Ltd	60
Hogg Robinson (Travel) Ltd	53
Bakers World Travel Ltd	49
Woodcock Travel Ltd	40
Portman Travel Ltd	40
Sub-total of multiple travel agencies	3,182
All other agents	1,590
UK grand total	4,772

(*Source:* ABTA, 1997)

Within this overall community of almost 5,000 agency branches there are several different types of travel agency outlets and each type has its own way of using IT. One important and very relevant fact here is that the vast majority of all UK agency outlets have an IATA licence that enables them to issue airline tickets on their own premises. It is therefore important that I explain the type of location and the kind of business undertaken by each.

THE BUSINESS TRAVEL CENTRE (BTC)

Well, as the name implies, this type of travel agent is one that is dedicated to business travel. Because it usually deals with its customers over the telephone, the actual choice of location is not critical. In fact a BTC is often situated in standard office premises off the high street where the rent is as low as possible. In the USA, for example, some large agencies have set their BTCs in towns in the middle of the country where communication costs, office space and wage rates are low.

BTCs often bear a passing resemblance to a stock exchange dealing room in the City of London. They tend to be noisy, frantic places littered with the most advanced technology in the industry. In general though, the average BTC looks much more like your average office than the kind of travel agent we are used to. The staff will probably each have their own airline reservation terminal or at least will share one between two. In many cases they will use telephone headsets, which leaves both hands free to work the CRS visual display unit (VDU) and write while talking on the telephone with their customers.

There will often be a separate equipment room where the computer and telecommunications kit is housed. This room will usually house a fast set of fast printers for automatically producing airline tickets, invoices and itineraries on continuous computer stationery. A dedicated computer operator or operations person will be found in the larger BTCs.

THE HIGH STREET TRAVEL AGENCY

The average high street travel agency is probably well known to most readers. It tends to be quite large and often has three distinctly separate departments. On the ground floor is the leisure travel shop, which sells holidays and other travel services to customers who walk in off the street. Also on the ground floor will be the bureau-de-change if the agency operates one of its own. Upstairs or perhaps in the back of the premises will be the business travel department. Most travel agency locations of this type are operated by the major multiples.

It is the ground floor travel shop that stands out to passers by. This will invariably be ABTA registered and will usually have its own IATA licence, which is of course shared by the business travel department. The principal technology used to service the leisure travel business is viewdata, although at least one CRS airline reservation terminal is available for shared use by the sales staff. In many cases the viewdata screen is on a swivel so that it can be used by the sales person and then shown to the customer.

Also on the ground floor is the bureau-de-change. This usually comprises only two or three teller positions and is located to one side of the leisure sales area. If a bureau-de-change service is provided, then it is law that the exchange rates and commission charges must be on display to customers. So a rate board is an integral part of the layout in this area. The operators are often called tellers or cashiers. In some travel agents the tellers are responsible for collecting payments from customers for their holiday bookings. Tellers will use a dedicated terminal to transact their business. This is very compact because bureau-de-change 'cages' are often quite small and space is a premium.

Upstairs or at the rear of the premises will be the business travel department. Its customers nearly always contact the agents over the telephone and so the agents do not need the shop front. However, it is usually convenient to share some of the facilities of the leisure operation and make use of economies of scale especially in terms of rent. Members of this department will use CRS airline reservation terminals for the bulk of their work.

INPLANT

An inplant (or implant as it is sometimes known) is a feature of business travel. It is a sub-branch of a travel agency and it is located within the premises of one of the agent's large corporate customers. Because this involves considerable extra overheads it is only justified for the really big business travel customers. It is staffed by the travel agent's employees, and airline reservation terminals are often installed within the inplant.

The large multiples operate the major inplants and some of these have an IATA licence in their own right. This means that airline tickets can be printed on the spot for their customers. If no IATA licence is held, then the inplant will make the reservation on its CRS terminal and the nearest travel agency branch office will access the CRS from its own terminal and issue tickets for delivery to the inplant. This is known as the main and satellite office type of operation.

BUREAU

This is similar in many ways to an inplant but may have nothing to do with a bureau-de-change. It is a leisure travel agency located within a major shopping store or some other high street retail shop. It functions just like a high street travel agency except that it is located inside another retailer. In other words, a bureau will usually concentrate on selling leisure travel to passing trade within the store. Sometimes, however, the bureau will also provide a business travel service to the staff of the store in which it is located, e.g. in large stores there are important staff called 'buyers' who travel extensively around the world buying merchandise for sale in the store and are therefore prime business travel customers.

BUREAU-DE-CHANGE

Bureau-de-changes are usually small kiosk-type premises located on the high street, which provide currency services to the general public. They are predominantly found in destination cities because they profit to a large extent from the purchase of foreign money and the encashment of Travellers' Cheques for inbound visitors to the UK. They also offer to sell Travellers' Cheques and foreign currency.

These days even the smallest bureau-de-changes use computerized equipment to process transactions.

As previously mentioned, they must by law display the exchange rates and commission charges that form part of their terms of business.

THE INDEPENDENT HIGH STREET TRAVEL AGENT

These are the small travel agents that make up the majority of outlets in the UK. They are known affectionately in the USA by the term 'Ma and Pa businesses'. These travel agents may be members of trade associations like ABTA and IATA or they may not. In most cases they do not possess an IATA licence. However, many are registered with ABTA and some may be members of NAITA.

In the UK some people call unlicensed agents who are not members of any recognized trade association, 'bucket shops'. These types of travel agencies used to have a negative image in the trade several years ago, although in most cases they are now recognized as respectable independent agencies. The term 'bucket shop' arises from the type of business they undertake, which tends to be soley the cheap and cheerful package holidays to the sun or seat-only packages (see Chapter 3 for a fuller description of the seat-only business).

Although one can never generalize too much, the small independent travel agent does not usually handle business travel to any great extent. This is primarily due to the lack of an IATA licence, which is costly to obtain because it requires a good, safe and fully trained staff in the area of airline ticket issuance.

Like most other types of outlet in the leisure travel business, these agents use viewdata as the principal technology. The reasons for this are described in a subsequent chapter of this book. There is therefore a considerable stock of viewdata equipment tied up in these agencies throughout the country. Most use BT's dialled telephone service to reach the viewdata systems but some have hardwired links into third-party networks (see Chapter 6).

Automation of agencies

Travel agents use a variety of technologies to help them access the information they need to service customers and process transactions for accounting purposes. The kinds of systems they use fall into two main areas: front-office systems and back-office systems. This division is becoming increasingly blurred but it nevertheless remains a pretty good way of distinguishing the two main categories of technology used by most travel agents. There are certain functions that clearly fall into the front-office category, such as airline reservations and tour bookings. There are also some that are undisputedly back-office systems, such as the general ledger and management information. However, there are those that float in between the

Figure 7.1 Front-, middle- and back-office systems

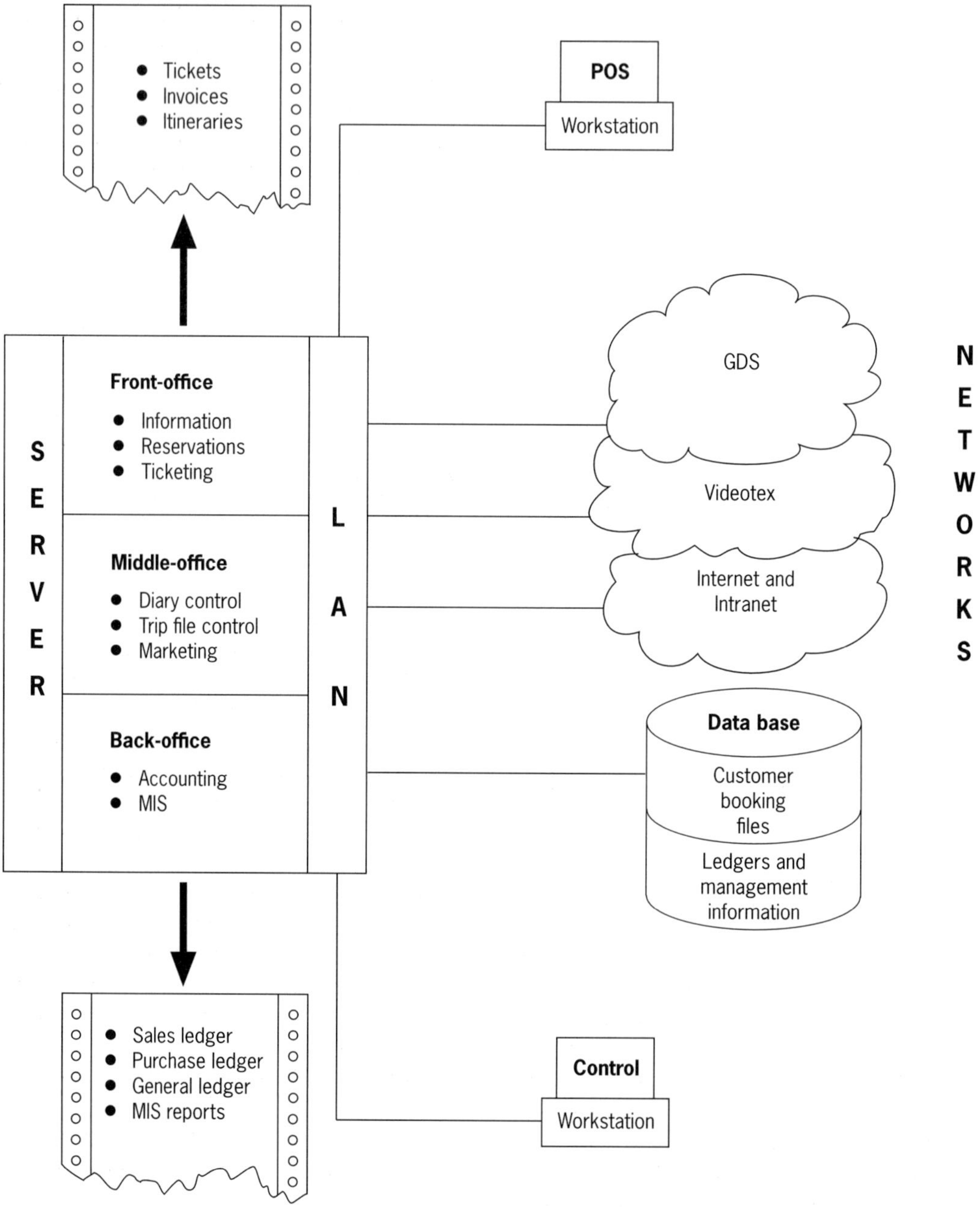

two categories, such as client files and automated diaries. Some people might describe these as mid-office systems. Figure 7.1 shows how the front-, middle- and back-office system functions coexist within a travel agency. However, for the sake of simplicity I am going to talk just about front- and back-office systems, but please bear this blurred distinction in mind while we discuss the functions and products available in each area.

FRONT-OFFICE SYSTEMS

A front-office system is one that has little to do with accounting and much more to do with servicing customers. The ultimate front-office system is of course the GDS PC or the videotex terminal. These devices are the travel agent's virtual window onto the market-place of all commonly required travel products and services. But reservation terminals are now so complex to use effectively that specialized system tools are being introduced to help travel agents become more productive. These system tools are particularly prevalent in the USA where, due to the dominance of air travel, the travel agent's use of the GDSs is relatively sophisticated compared with other areas of the world. First, let me identify the main types of system tools that fall into the 'front-office' category. I think these are generally as follows:

- Reservation terminals, e.g. GDS PCs and videotex terminals.
- Point-of-sale assistants, i.e. software products.
- Software robots.
- Automated quality control software products.
- Customer documentation, e.g. air tickets, itineraries and quotations.
- Client name, address and booking files.
- Automated diary functions.

If you consider the above types of systems, you will no doubt observe that I have covered many of them in previous sections of the book, e.g. GDSs, client files and many other related functions in Chapter 4, Videotex in Chapter 6 and so on. But there are one or two particularly interesting areas that I have not yet covered. These are the front-office functions that fall into the category of point-of-sale assistants and software robots. Such tools are commonly used by travel agents in the USA but are only beginning to make an appearance in Europe and other parts of the world. Here is a presentation of just two of them.

Point-of-Sale assistants

A point-of-sale assistant is a software product that helps to make travel sales consultants more productive by automatically alerting them to certain pre-set conditions. The product that I am going to describe is one good example of this; it is called CRS Screen Highlighter (Fig. 7.2). It is a product that was developed by the Travel Technologies Group based in Dallas, Texas, and marketed in the UK by ICC Concorde.

The underlying objective of CRS Screen Highlighter is to guarantee the reservations accuracy of travel sales consultants and ensure that customers receive the highest possible level of service. It does this by: (a) carrying out a series of pre-set checks on the booking records received from GDS systems, and (b) popping-up appropriate messages on the travel sales consultant's PC screen. The following are just a few examples of the kinds of automatic checks that can be performed by CRS Screen Highlighter. It can for instance:

- Call the travel consultant's attention to the fact that today is the last day on which certain special fares can be purchased.
- Highlight penalties and restrictions on routes and fare categories that relate to the PNR received from the GDS for the current booking.
- Keep track of all the unused and non-refundable tickets that a traveller accumulates thus enabling them to be exchanged for a valid ticket at the appropriate time.
- Summarize the agency's special fares and negotiated rates on routes and carriers that are directly relevant to the current PNR.
- Alert a consultant to switch sell a preferred airline when an override threshold is about to be reached.
- Automatically detect bookings for very important persons (VIPs) by recognizing titles such as CEO and vice-president (VP) as well as picking out certain travellers by their frequent flyer numbers.

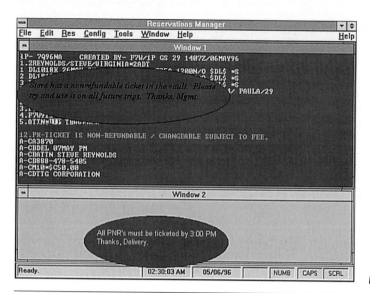

Figure 7.2 CRS Screen Highlighter – screen 1

- Instruct the sales consultant to follow a particular sales process for certain pre-defined customers, i.e. by customer name.
- Notify the consultant if a particular corporate customer has reached a serious credit position that threatens to jeopardize the agent's cash flow and financial risk levels.
- Inform travel consultants about visa requirements on journeys overseas for the current booking.
- Recognize certain categories of bookings and offer pre-set customer service advice as determined by the agency's management.
- Remind sales consultants of tasks that must be completed by pre-set times of the business day.
- Display a daily broadcast message to the sales consultants within an agency that eliminates the need for paper circulars and shouted messages.

The CRS Screen Highlighter communicates with the travel sales consultant by means of pop-up messages that are displayed on the PC screen. It goes without saying I suppose, that the travel agency needs to be using one of the major GDS systems with a PC as the terminal device. The message is popped-up as an overlay to the windows reservations screen. Each message is tailored to the particular check that has just been carried out. The consultant just has to read the message and then hit a pre-set key, e.g. the 'escape' key, to remove it. The CRS Screen Highlighter program comprises two main parts: (i) the control program that runs on the agency's server, and (ii) the operational program that runs in each GDS PC:

- **The control program** This stores all of the point-of-sale checks that are to be performed for certain categories of PNRs. Each point-of-sale check is defined on a single window. The window contains two search criteria that may be linked by the logical operators 'only', 'and' as well as 'and not'. Searches may start at the beginning of a PNR line only, anywhere in the PNR or only at the end of a PNR line. The second part of the window defines the pop-up message that is to appear on the GDS PC screen if the search criteria are successfully detected. The message may be formatted in a variety of ways and in a number of different colour combinations. Several action buttons are available that support the quick creation of a library of point-of-sale checks. The choice of 'escape' key may also be defined by the control program.
- **The Terminate and Stay Resident (TSR) program** The CRS Screen Highlighter operational software that runs in each GDS PC is a TSR program (Fig. 7.3). This is a special kind of computer program that is loaded when the PC is powered on and remains active, even while other applications are running. This program

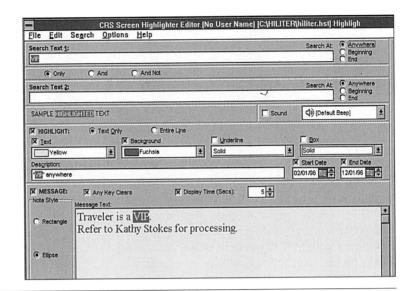

Figure 7.3 CRS Screen Highlighter – screen 2

'wakes up' whenever a PNR is received by the GDS PC. The PNR is scanned for the search criteria as originally created by the control program, as described above. If a search is successful, the TSR displays the appropriate message for the condition detected. Once the operator presses the designated 'escape' key, the TSR terminates but stays resident within the PC awaiting the next PNR to be received. From a technical perspective, it accomplishes this functionality by intercepting the Windows screen buffer, altering it to show the desired message and then repainting the screen after depression of the 'escape' key.

These point-of-sale software assistants are written in a programming language that is especially efficient. This allows the programs to execute in the shortest possible times. So, provided the number of searches is reasonable, the GDS PC user will not notice any degradation in its response times. The software runs within Microsoft Windows Versions 3.1 and 3.11.

Software robots

These are even more sophisticated products than the reservation assistants that I have described above. Software robots grew from automated quality control systems. These were systems that automatically intercepted booking records received on a GDS PC and performed checks on the PNR to ensure that all fields were compliant with an agency's quality control programme. However, software robots go substantially further and undertake operations normally carried out by travel consultants. Being more complex, they require a far higher level of technical competence on the part of the travel agency. Nevertheless, given that the agency has at least one 'super user' who also possesses a fair degree of IT skill, software robots can increase the productivity of a travel agency by an enormous amount.

The example I am going to present here is another of TTG's products, i.e. the Centrally Oriented ResReview Edition (CoRRe – see previous section for a description of TTG and its UK distributor). CoRRe was developed in C++ and runs within a Microsoft Windows environment. It requires its own dedicated workstation PC, which itself is connected to the travel agent's GDS, i.e. Sabre, Amadeus Apollo, Galileo or Worldspan, although at present it has only been ported to work with Galileo outside the USA. This workstation, which has its own unique GDS terminal address, needs only exist in a single location within an agency. Even if the agency has several branches, the CoRRe PC need only be installed at headquarters. This is because the underlying operating philosophy behind CoRRe is the automated working of GDS queues. Queues can be accessed from any location and shared within a single agency (or indeed by any group of agencies that are affiliated). The GDS queue system is therefore an integral part of the entire CoRRe operation (see Chapters 3 and 4 for a description of GDS queues).

The key to effective use of CoRRe is a queue structure that optimizes the way in which the product works. However, whereas quality control software products are based on working directly with GDS queues, CoRRe works them in a far more sophisticated way. It takes active queue items and builds its own internal data base that mirrors the active PNRs on all queues at certain times. This allows the program to process more than just a single function on a PNR, which may in fact be present on more than one queue. All modules within CoRRe therefore process each active PNR, thus minimizing activity on the GDS and maximizing the system's internal efficiencies.

Once the queue structure has been set up, CoRRe is 'programmed' with the functions required by the agency. I use the term 'programmed' because the way in which these checks are specified is so detailed that it closely resembles a computer programming language; and the skills needed to create these functions are very similar to those used by programmers, e.g. how to structure a program, how to organize and name the data, and so on. This is why the travel agency needs a so-called 'super user' who is also very IT-literate. The checks that CoRRe can perform automatically are virtually any of the checks that a human operator can perform on a GDS queue. These checks can be tailor-made for each customer serviced by the agency. For example, whereas one company may wish all its departments to comply with an overall pre-set corporate travel policy, others may have a different set of travel guidelines and rules for each department and even for certain individuals. CoRRe can be set up to support separate checks for each customer. Also, the frequency and timing of checks can be pre-set. A checking function can be set to be activated every few hours or within a certain time of the trip departure schedule. The functions provided by CoRRe fall into two general categories: (i) PNR checks; and (ii) PNR finishing routines. Some of the standard library functions available are:

- **QualityCheck** This allows the agency to check every booking automatically for certain pre-defined quality control checks. For example, the presence of certain fields and the automatic completion of certain PNR field entries, depending upon pre-defined rules.
- **SeatFinder** CoRRe will automatically search the GDS for the kind of seat required by the customer. This is an example of a PNR finishing function that produces as its end-result, a seat assignment for the customer selected according to their pre-defined preferences.
- **FareFinder** The system automatically searches the airlines and the fares on the GDS for the lowest possible fare for the journey specified within the PNR. It can also do this for a specified alternate itinerary. All low fare options are obtained and stored for later review by a sales consultant.
- **Clearance** The system works on a wait-list queue and repeatedly attempts to clear wait-listed flights and fares. This involves the controlled initiation of repeated availability request messages to the GDS until either the time limit expires or the flight can be booked.
- **UpGrade** CoRRe automatically upgrades frequent flyers into first or business class in compliance with airline and GDS rules. It identifies these customers by comparing their name with their client profile details.
- **ForeCast** This is a pre-trip report generator. It allows the agency to print a detailed itinerary in a customized format. Many other reports can be constructed and produced by CoRRe either for display on the screen or printing in the agency.

CoRRe works each of the travel agency's queues in turn. At the appropriate pre-set time, it reads each item on the queue, which is of course a PNR, and processes it according to the queue type. Let's take, for example, a PNR that does not have a seat assignment. This is a PNR finishing function. CoRRe will read the queue and select the first PNR. It determines the customer's seating preferences from the appropriate client profile record and formulates a customized seat request message. It then sends this request message to the GDS. When a reply is received, it will check to see if the requested seat has been reserved and act accordingly. If the seat has been reserved it will place the updated PNR on an 'actioned' queue. If no seat of the required type is available, then CoRRe may either hold the PNR on a queue for further processing, i.e. try again later, or place it on a 'failed seat assignment' queue for subsequent manual follow-up.

Another example – ensuring that a customer's travel policy is enforced; this is a CoRRe PNR check. Enforcing a company's travel policy is an important servicing function that is expected of most travel agents in the business travel sector. CoRRe can automatically perform travel policy checks without the need to involve travel sales consultants. It can, for example, work a queue of pro forma bookings created from two possible sources: (i) skeleton bookings made by travel consultants, or (ii) bookings made by the company's travellers using their lap-top PCs and special end-user software products (see the Chapter on GDSs for details of these). In either case the PNR is retrieved from a 'pro forma' bookings queue. CoRRe automatically checks the PNR against the company's stated travel policy. For example, the flight time and class of travel is checked. If first class is specified and the flying time is under eight hours then CoRRe will re-book the flight in Club or economy and cancel the original booking. Changes such as this can be identified for later reporting if necessary.

Finally, one more example – getting a seat on a busy route. Again, the travel sales consultant will have created a booking record specifying the customer's itinerary and preferences. However, if there was no availability at the time the booking was attempted, then the PNR will get placed on the wait-list queue. CoRRe will process each item on this queue in turn and try to find an available seat on an acceptable flight. Each PNR on the queue is retrieved and used by CoRRe to construct an availability message. This is automatically sent to the GDS and the resulting response is carefully analysed by the program. If availability is shown, then the booking is placed on the 'actioned' queue. However, if no seat was booked, then CoRRe can either make several more attempts to obtain a seat using different flight times and/or routings. The actual number of attempts and the degree to which the customer's itinerary is to be modified, can be pre-programmed into CoRRe. The system can also be programmed to carry on trying to find a seat and to alert the travel sales consultant only within a certain number of hours before departure.

There are many other examples that I could give on how these kinds of software robots can be used within a travel agency. However, I hope the few I have outlined above will give you some idea of the potential power of these products. But like many good things, there is unfortunately a down side. The GDSs don't like their systems to be hit by a high volume of messages. From their viewpoint, every message should ideally be a one-hit booking on a flight – any additional hits are just an overhead. But from the agents' perspective, they don't get charged by the number of entries they make on their GDS PC terminals and so why should they worry about re-trying a wait-listed flight every five seconds (to take an extreme case). So, there is apparently nothing to discourage agents from flooding their GDSs with thousands of 'hits' to get just one difficult booking for an important customer. However, since the introduction of software robots in the USA, GDSs have indeed introduced penalties on agents that have high hit rates. In fact, generally speaking, if a USA agent generates more than an average of 250 hits per segment then the GDS will levy a penalty charge to cover the extra cost of processing. It is for this reason that CoRRe and other software robot products carefully count the number of hits that they have automatically generated and advise the travel agent when the threshold of 250 hits per segment is about to be reached.

Software robots are, as I have said before, sophisticated products. They hold the promise of several significant benefits for travel agents. Benefits such as: (a) very high productivity rates because only the most simple of PNRs need be created manually when the booking is first made, with other more detailed time-consuming entries completed automatically by the software; (b) the ability to de-skill consultants from a technical aspect, thus allowing them to focus on developing their customer servicing skills and enhancing their knowledge of the industry; and (c) closer adherence to quality control checks and travel policy compliance, which would be impractical or not economically feasible to undertake manually. However, in order to realize these benefits, a holistic approach needs to be taken to agency operations. To use these software robots effectively, a travel agent's whole business really should be re-engineered. A new approach to work practices should be taken and a new breed of travel agency staff must be developed. This new breed will have to include at least one specialist IT-skilled 'super user' who will become an increasingly important member of the travel agent's workforce in the future.

BACK-OFFICE SYSTEMS

First question: 'What is a back-office system?' Well, it is really a system that processes the sales generated by a front-office system, such as an airline terminal or a videotex system. The terminology is American and stems from any type of sales operation where the front-office is the part of the business that the customer sees when goods are being bought. In other words the shop front. As soon as the sale has been made the paperwork is passed to the back for processing. The back-office is always the hidden office at the back of the store where all the boring old accountants sit doing the books. They take the sales receipts generated by the front-office sales staff and enter them into the company's books of account.

The term back-office is more relevant in the USA, where cheap office space is easily available and separate back-offices are a reality. This is not so in the UK travel industry where a travel agency has just one location, which is the front-, middle- and back-office. But anyway, in spite of the inaccuracies of the term, most people now know what a back-office is: a back-office system is a computer and special purpose software that automatically does the accounts and controls the books. In this book I refer to these type of systems as agency management systems because I feel that this is a more appropriate and up-to-date term.

In the travel agency business, agency management systems have not been around all that long. Certainly not as long as airline reservation systems and not much longer than videotex. The large travel agency multiples were really the first to have back-office systems, which they usually developed themselves. These were usually main-frame computers that processed sales forms mailed in to the company's headquarters each day. These forms were captured onto computer readable media and input to the computer. The computer used a set of programs to process the data, store it on data bases and print books of accounts.

One of the first commercially available back-office systems available to travel agents was the Document Printing Agency System (DPAS). It was originally marketed by a company called CCL and was used by many agents, including some of the multiples like Thomas Cook and Hogg Robinson. DPAS wasn't too bad at just printing airline tickets, invoices and itineraries but it left a lot to be desired in the area of travel agency accounting and management information.

The problems of DPAS were symptomatic of the problems experienced with many travel agency systems that were marketed at that time. They all suffered from several problems inherent within the very core of the business. For example: (a) most travel agents operated their businesses in different ways from each other; (b) the way a multiple did its accounting and processing was quite different from the way independent agents did theirs; (c) the travel business is fairly complex with a wide range of products and services, each with different settlement methods and accounting rules; and (d) technology was not as advanced as it is now and it was difficult for system suppliers to provide what was needed at an acceptable cost with the required flexibility to change quickly to evolving business needs.

With the advent of the PCs, however, things are beginning to look up a bit. These are becoming more affordable, there are now many more suppliers available for software packages and the technology has enabled some complex systems applications to be developed and maintained in a current state so as to keep pace with developments in the industry. I would, however, conclude that the quality of travel agency back-office systems is not as high as it could be. They are nothing like as sophisticated as the airline systems and compare poorly with other industry systems. I put this down to the complex nature of the travel agency business, which itself is evolving rapidly, and the widely differing requirements of agents in different sectors of the business. However, this situation is in the process of being rectified by some exciting new products that I shall be discussing later.

Well, I hope that this chapter gives you a broad understanding of the way the travel agency business is structured and a brief insight into the history of IT in travel. I shall wherever possible in the subsequent sections of the book give you as much background information on the history of the companies that supply some of the leading travel systems. This will I think help you to understand how the current situation has arisen and may help you project future developments and directions. Now, in the next section I will present you with a quick *tour de force* of back-office terminology. It is important that you read this section carefully because I will be using some of the more technical terms throughout the remainder of the book.

We discussed agency management systems and what they were in general in Chapter 1. Well, now its time to take a closer look at them. An agency management system is usually viewed as the unglamorous part of travel agency automation. Despite this misnomer, agency management systems play a crucial role in the successful running of a travel agency business. Automating the back-office allows an agency to concentrate on the personal travel services that are the key to profitability and growth. By contrast, an old and cumbersome back-office can really drag an agency down. I know of one agency in the business travel field that took six weeks and a lot of manual effort to produce a special report for one of its corporate customers. This involved one of the staff going through hundreds of old and dusty manilla client folders to extract certain information that was tabulated, cross-cast and then typed-up into a special report. This kind of activity is just not feasible in today's fast moving and competitive travel services world.

Unfortunately, travel agency management systems have had a bit of a chequered history in the UK. There have been some really bad ones that have been oversold by overzealous computer system salesmen: and then to be fair, there have been many travel agency managers who have been too taken with the pizazz of front-office systems and who have consequently neglected the back-office aspects. These are some of the main reasons why the image of back-office systems in travel has to some extent been tarnished. In this chapter we are going to look at some of the major agency management systems on the market at present. But first, let's consider the primary functions of agency management systems and define some of the terminology. The main functions may be summarized as:

- **Accounting** Probably the thing that most people associate with an agency management system. The accounting system is an electronic set of books that records the financial state of the travel agency and controls its business operations. Accounting functions can be categorized simply as comprising the sales ledger, the purchase ledger and the general ledger. Incidentally, a ledger is no more than a 'book' containing debit and credit book-keeping entries. In the case of an agency management system the ledger is an electronic book stored as a data base on a computer.
- **Management information** Sometimes known as a Management Information System (MIS), this comprises the information that is needed by: (a) the management staff of a travel agency in order to run the business effectively, and (b) the business travel customers who need information on their employees' travel patterns in order to optimize the discounts that they can negotiate on travel products and services. This is particularly important in the field of business travel where corporate customers now expect very sophisticated management information on their businesses and also in supplier tracking where complex commission structures are involved with different revenue levels depending upon business volume. In many cases a computer system that includes a large data base is the only way that MIS can be accumulated and presented to its key users in an easily readable form.
- **Marketing** Agency management systems are ideal vehicles for marketing programmes. This is because a great deal of the information needed to run an effective marketing campaign is available as a by-product of front-office systems, accounting and MIS sub-systems. At present it is true to say that the average travel agent

does not use the marketing potential of its travel agency management systems to anything like their full extent. I hope that reading and assimilating the information in this section will open up a few innovative approaches to marketing, using the power of IT and a good agency management system.

The first area that needs to be covered is BSP. The reason for covering this subject up-front is because it represents a significant part of the average travel agent's business, i.e. the sale of airline tickets.

However, there is one thing that virtually all types of travel agents have in common. This is the need to settle-up with the airlines for the tickets that they, the travel agents, sell to their customers. The airlines have clubbed together to address the administrative problems associated with this settlement process and in most countries of the world now operate a standard system, i.e. BSP. It is vital to understand this process before we start reviewing agency management systems in detail.

THE BANK SETTLEMENT PLAN (BSP)

BSP is a crucial part of most travel agents' operations and is a major influencing factor on agency management systems in particular. It is for these reasons that I am going to explore BSP UK in a fair amount of detail before I launch into agency management system functions. Because it is often said that in order to understand the present, one has to study the past, I will start this section on BSP with a brief history of how it developed. If we go back in time to the 1940s when the travel business was in its infancy, travel agents did not hold stocks of tickets. Whenever a ticket was needed it was requested from the airline concerned who wrote the ticket and returned it to the travel agent, usually by post. It was not long before airlines realized that this was an onerous task that would best be handled by travel agents. This was considered reasonable because it was one of the tasks the airlines could reasonably expect in return for paying commission fees on airline tickets sold by travel agents.

So, travel agents were issued with blank airline ticket stock. A travel agent that was licensed by IATA, kept a stock of tickets at the discretion of each individual airline for whom it acted. Each airline's ticket stock was different because it contained the airline's logo and branding design. The stock control procedures applied equally to the agent and to the airline itself. Such procedures alone were therefore a considerable task to manage, especially for agencies that supported a wide range of airlines. This was a task that could have typically involved an agent controlling 50 different ticket stocks, and to control each ticket stock the agent had to keep a record per airline showing the stock received, transfers of stock to other departments for ticketing, items of stock destroyed for various reasons and of course stock used to issue tickets. The administrative overhead was enormous. In addition to the stock control procedures there were the settlement tasks that needed to be performed by the agent. The travel agent was responsible for collecting payment from the customer for the airline tickets sold by the agency and for onward remitting payment to the airline concerned. Airlines with a far flung travel agency distribution policy suffered in a similar way.

The settlement procedures were perhaps the most time-consuming and labour-intensive tasks of the whole airline sales business. Each month the travel agent had to retrieve the audit coupons of each ticket sold in the period. These had to be batched up by airline, sorted into ticket number sequence and add-listed on a calculator. Then each airline's settlement form had to be completed for all sales issued on that carrier's ticket stock for the month. This settlement form had to balance to the batch of ticket stubs attached. Finally, the travel agent had to issue a cheque to each airline for the amount of the tickets settled that month. The cheque, settlement report and ticket stubs were then mailed to each of the airlines. In some of the larger travel agencies there were people whose sole job it was to settle air ticket sales and administer the resulting queries and problems.

If you think that was bad, you can imagine the problems faced by the airlines. They were receiving settlement reports and ticket stubs from thousands of travel agency locations around the country. All these batches had to be balanced and checked before being keyed into a primitive computer accounting system. Then there was the workload associated with dealing with individual payments from all the travel agents throughout the UK. The bank reconciliation task alone was a mammoth undertaking. Also, a great deal of the procedures used by the airline accounting departments in those days were based on manual methods and used only the most primitive of computer systems by today's standards. In summary the whole system was a burden that was not sustainable from both the airlines' and the travel agents' perspectives.

There had to be a better way. Several IATA members agreed with this sentiment in the early 1960s. A new approach became a talking point in IATA committee meetings. Eventually one such meeting of marketing and revenue people produced the concept for a proposal of what was to become known as the bank settlement plan (BSP). There were several rather bureaucratic steps that needed to be undertaken before the first BSP could be implemented and the procedure is similar even nowadays. In order for a country to adopt a BSP the following need to take place:

1. The national carrier in a country calls all other airlines operating in that country to a meeting to discuss the possibility of BSP.
2. Assuming that the required level of support for a BSP scheme is agreeable, a feasibility study panel and several working groups are formed.
3. A study of the costs is carried out and items such as postage and bank processing capabilities are evaluated.
4. A study report is produced containing detailed estimates of operating costs and a general statement on the feasibility of introducing a BSP in the country concerned.
5. Finally, the airlines vote on the motion and if approved, BSP is introduced. It is not mandatory for airlines to participate in a BSP scheme.

Since that time in the early 1960s, the BSP approach to airline ticketing and accounting has been established successfully in 53 countries around the world. The first country to adopt a settlement plan like BSP, was the USA. However, due to complications surrounding the IATA lead body, mainly involving anti-trust laws, the scheme is called the Agents Reporting Corporation. It is the biggest scheme in the world and involves some 30,000 travel agents and around 200 airlines. Outside the USA, the UK is the next biggest BSP scheme in terms of the number of transactions; although Japan is close with the largest scheme by value. There are 11 BSPs in Europe, six in the Far East and a further 32 in other areas around the world. There is a clear trend for the number of BSPs to shrink due to consolidation and mergers. But although the number of schemes may decrease, the volume of BSP transactions is expected to grow over the next few years. There may, for example, eventually be a single European BSP of which the UK would be a participant; time will tell.

The BSP was introduced into the UK in 1984. With the benefit of hindsight, it would have been better to have done so several years previous to this date. The UK was a mature market when BSP was originally introduced. CRSs and technology were widespread and it was difficult to fit the BSP needs into an existing sophisticated structure such as this. To some extent the industry is paying the price for this now with several changes to the way the scheme operates. It would have been far easier to introduce BSP into the UK before CRSs and automation but there we are, you can't turn the clock back. After all, BSP itself is a part of IATA, which is in turn owned and controlled by the world's airlines (see Chapter 1 for a description of IATA).

The UK's BSP is run as a commercial operation within IATA. As such it is measured on its financial performance to a large extent. Although it is not charged with making a profit it must balance its funding resources with the costs of operating the scheme. An individual is appointed by IATA to manage the scheme in the UK and this person is known as the BSP Plan Manager. The main function of the UK's BSP Plan Manager together with a relatively small group of staff is really to co-ordinate the service providers that make BSP work. The BSP Plan Manager is responsible to IATA and the local airlines for the smooth running of the whole operation. Besides co-ordinating with the service providers, the BSP Plan Manager also monitors and controls the ticket stock issued to travel agents.

In most cases, BSP is primarily interested in the settlement of the fare shown on the ticket less

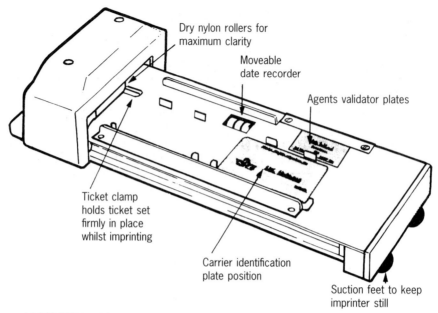

Figure 7.4 The model 760 BSP imprinter

the commission as entered by the travel agent. BSP does not police the fares used nor the commissions deducted. Once BSP has captured and billed the ticket for settlement purposes, any further disputes are sorted out between the travel agent and the airline concerned. The Plan Manager has substantial disciplinary powers to ensure that travel agents and airlines comply with BSP's rules. The service providers to BSP in the UK are GSi, Barclays Bank plc and Aeroprint. Their respective roles within the scheme are:

- **GSi** A data processing and systems bureau. GSi is a French company contracted to provide a service to BSP UK. It has considerable experience in running other BSPs in Europe. The software used by GSi to process the UK's BSP was originally developed to support the Belgian BSP and as such is tried and proven.
- **Barclays Bank plc** One of the leading UK clearing banks that processes the direct debits and funds transfer services between the travel agents, the airlines and BSP itself. Barclays is of course a member of the UK Bankers' Automated Clearing Scheme (BACS), which enables inter-bank transactions to be processed using modern technology.
- **Aeroprint** A document printing and distribution company. It distributes stocks of standard tickets to travel agents and advises BSP of all such stock movements.

One of the cornerstones of BSP is a common ticket. This is a ticket that is not branded for each airline, but is issued in a standard format and is plated to denote the airline that is providing the service. Plating is a term that needs explaining. An airline may issue travel agents with a carrier identification plate (CIP). This is used for manually issued airline tickets. The CIP is an aluminum plate about the size of a credit card that has the name of an airline embossed on it. It is used in a 'validator' or 'imprinter', which is rather like a credit card embossing machine (see Fig. 7.4). When a blank BSP ticket is put into the validator along with the appropriate CIP, it can be used to imprint the name and other details of the airline onto the ticket, thus validating it. The ticket then becomes a ticket issued on behalf of the validating carrier or airline. For automated tickets, a table of data containing the plate information is built into a computer and this is used to print the name and other CIP details onto the ticket. For automated tickets the ticket stock is of a continuous type so that it can be

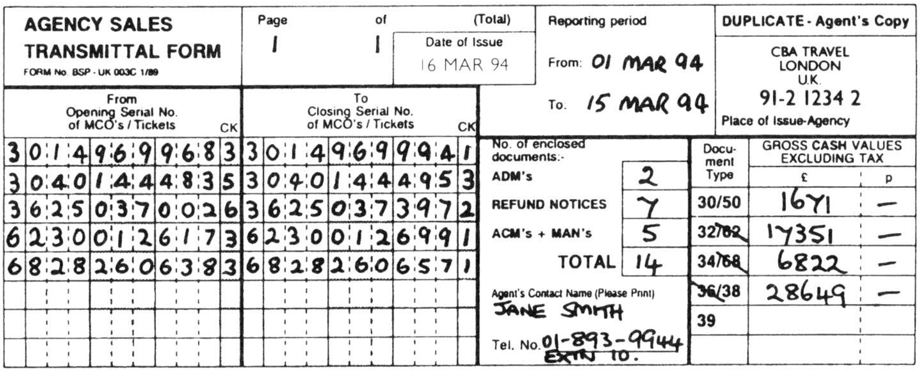

AGENCY SALES TRANSMITTAL FORM
FORM No. BSP - UK 003C 1/89

Page 1 of 1 (Total)

Date of Issue: 16 MAR 94

Reporting period
From: 01 MAR 94
To: 15 MAR 94

DUPLICATE - Agent's Copy

CBA TRAVEL
LONDON
U.K.
91-2 1234 2
Place of Issue-Agency

From Opening Serial No. of MCO's / Tickets	CK	To Closing Serial No. of MCO's / Tickets	CK
3014969968	3	3014969994	1
3040144483	5	3040144495	3
3625037002	6	3625037397	2
6230012617	3	6230012699	1
6828260638	3	6828260657	1

No. of enclosed documents:-	
ADM's	2
REFUND NOTICES	7
ACM's + MAN's	5
TOTAL	14

Agent's Contact Name (Please Print) JANE SMITH

Tel. No. 01-893-9944 EXTN 10.

Document Type	GROSS CASH VALUES EXCLUDING TAX £	p
30/50	1671	—
32/82	17351	—
34/68	6822	—
36/38	28649	—
39		

Figure 7.5 An example of a completed sales transmittal form

printed by a dot matrix printer connected to a computer. But BSP handles other forms of stock besides just airline tickets. It also handles the following:

- **Multiple purpose document (MPD)** This is really like a voucher that has a form of value. It is issued by a travel agent or an airline and may be exchanged either for a ticket-on-departure arrangement or a non-ticket type of service. For example, in cases where the travel agent has under-collected the correct fare from the passenger or for hotel accommodation provided by the airline in the event of an unscheduled stop-over.
- **Agency debit memo (ADM)** This is almost identical to an invoice. It is raised by an airline, for example, when money is owed to it by a travel agent. In many cases an ADM is used to correct mistakes that have resulted in the airline being underpaid by a travel agent.
- **Agency credit memo (ACM)** This is almost identical to a credit note. It is raised by an airline, for example, when a mistake has been made and the travel agent has for some reason been charged too much for a particular service.
- **Refund** This should be fairly self-explanatory. It is a credit for a ticket that has been either wholly or partly unused and is submitted to the airline for reimbursement. Refunds are passed through BSP as a separate batch containing all returned or unused flight coupons.
- **Universal credit card charge form (UCCCF)** This is a form that is used whenever the airline ticket is paid using a credit or charge card. In some BSPs the UCCCF is embedded within the airline ticket stock. There is always a separate form for manually issued UCCCFs in all BSP schemes.

Manual agents report their airline sales to BSP twice each month on pre-determined dates. Period 1 is from day 1 to 15 of the month and Period 2 is from day 16 to the end of the month. The travel agents report sales to BSP for each of these two periods by batching up the audit copy of the ticket coupons, add-listing the coupons to get a total value and completing a sales transmittal form (STF), as shown in Fig. 7.5. The STF contains the 'from' and 'to' common ticket numbers and the total value, for each ticket type. It is of course important to point out here that these batches are for a mixture of all the airlines on whose behalf tickets were issued during the period. There is no need to separate batches by individual airline within the BSP methodology. The batches are sent to the BSP processing centre, which is operated on behalf of BSP by GSi.

GSi receives the batches of work and keys data from each ticket into its front-end capture computer system. Each batch is balanced against the STF before being allowed into the main billing system. GSi staff then sort the ticket stubs physically into airline sequence, which allows them eventually to deliver the stubs to the airline whose CIP was used by the travel agent who issued the ticket. The GSi main-frame computer system sorts and processes the ticket images and stores them on a large data base ready for the month-end processing run. Each month the GSi system produces three main outputs:

- **The travel agent's billing analysis** This report is sent to the travel agency and contains the details that comprise the amount due for the current month's airline business. It contains a total that is used to debit the travel agent's bank account directly within a period of three working days.
- **The airline billing analysis** This is sent to each of the airlines participating in the BSP scheme. It contains the total amounts that they will receive when the travel agents have been debited for the month's airline sales.
- **The credit card billing report** This is sent to the credit and charge card companies. It contains the amounts that travel agency customers have charged to their plastic cards for the purchase of airline tickets. The card companies use this report to pay the airlines and to bill their card members.

GSi produces the three types of reports, as well as many others, and sends them to the interested parties. A file is created by the GSi system which contains a debit for each travel agent that has submitted transactions. After a period of three days, this file is transmitted to Barclays Bank plc who in turn enters it into BACS. This allows the transfer of funds between the travel agents and BSP to take place. The clearing scheme automatically passes a debit to the travel agents' bank accounts and a credit to BSP's bank account. When this has been accomplished successfully, the final stage in the process consists of a funds transfer operation between BSP and the airlines. The BSP's bank account is debited and each airline's bank account is credited.

The travel agent receives a copy of the billing report from GSi each month. This report should mirror a report produced by the travel agent (usually an output from an agency management system), showing the total airline ticket sales for the month. The travel agent reconciles the internally produced report with the GSi billing report. Reconciliation is really no more than checking off each matching item on both the GSi report and the travel agent's report; which, as mentioned above, should ideally be produced by an agency management system. If any problems are detected then the travel agent has time to liaise directly with BSP to resolve any discrepancies before its bank account is automatically debited.

Automated travel agencies have an even easier time of it. Incidentally, what I mean by an automated travel agent in this context, is one that uses a GDS to generate an airline ticket. The GDS may either actually print the ticket itself or it may generate a PNR ticket image that is subsequently printed by an agency management system. So, for example, a computer system that prints an airline ticket from entries keyed solely by the travel agent would not be considered an automated travel agent for BSP purposes. The GDSs supported by BSP are Galileo, Sabre, Amadeus and Worldspan. Automated travel agents are encouraged to use automated reporting. This means that the GDSs feed ticketing data to GSi directly on a daily basis using telecommunications, i.e. via data lines from the GDS computer to the GSi computer. In this environment, the travel agent does not even have to submit any ticket stubs or batch control forms.

Automated reporting is at present being rolled out to as many travel agents as possible. Of the 4,300 IATA-approved travel agency ticketing locations in the UK, approximately 2,500 now use GDSs to produce airline tickets. However, the old 80/20 rule applies here (or almost anyway), i.e. these 2,500 locations account for around 80 per cent of the ticket volume.

The travel agency multiples and some agents using their own in-house agency management systems, used to follow a different method of reporting, called 'Method 1'. This is an old and decaying approach that has virtually been replaced by automated reporting. Method 1 BSP reporting entailed submitting a magnetic tape of ticket images that had already been captured, usually by an agency management system of some kind. Although this saved BSP a great deal of keying effort it also involved some other undesirable administrative operational problems such as reconciliation. Consequently, BSP is working hard with all such agents to migrate them over to the more streamlined automated reporting method described above.

More recently, BSP has introduced support for European STP (see Chapter 3 for more details on STP). This capability, sometimes known as

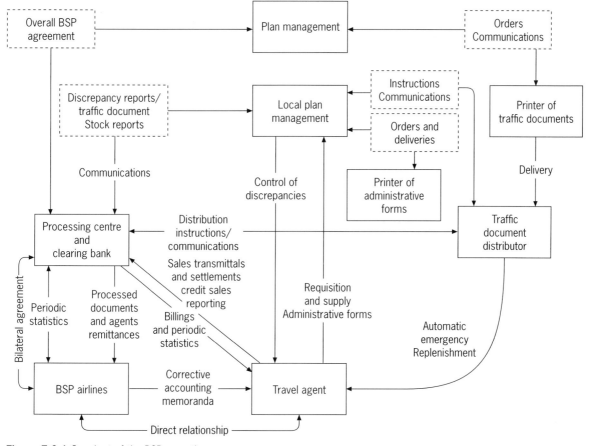

Figure 7.6 A flowchart of the BSP operation

cross-border STP, is best illustrated by an example. Let us say that a travel agent in London operates an in-plant in one of its customer's offices in Paris. Now, say an employee of the Paris company wishes to travel by air from Paris to London, returning to Paris. The travel agent in London may make the reservation and instruct its GDS to print the ticket in the Paris in-plant. This ticket is actually printed with a fare field that shows French Francs. However, it is automatically reported as a Sterling ticket sale by the GDS, as part of the London travel agent's BSP return. The exchange rate used for the conversion is that which is set by IATA on a daily basis. The London-based travel agent then settles for the ticket in the normal way, in Sterling. European STP provides travel agents and their customers with an effective ticketing service that minimizes administration. This is currently made available by IATA to any country within the European Community although consideration is also being given to other non-EC countries, depending upon demand.

BSP's enhancements

BSP's overriding strategic objective is to convert as many travel agents as possible to automated Reporting. This is extremely important because the current manual operation undertaken by GSi is severely strained by the ever increasing mountain of paper in the form of air ticket stubs that must be keyed, physically sorted and then filed. BSP was therefore disappointed to witness the demise of the GTI initiative (see Chapter 6 for a more in-depth description of GTI). GTI would have helped accelerate the take up of PCs by agents and with it, more widespread access to GDSs by leisure travel agents. Because access to GDSs allows more agents to migrate onto automated reporting as described above, you can see where BSP UK is coming from.

It is this underlying desire to get as many travel agents onto automated reporting that influences some of the future developments that are being considered by BSP at present. After all, if administrative costs can be reduced then this is an advantage, not just to the airlines, but also to the industry. In a true competitive environment this should lead to lower product pricing and more discretionary funds for incentive programmes that boost sales. There is therefore a set of service enhancements that BSP is considering, which are designed to be of interest to travel agents and should encourage them to use GDSs. Here are some of them:

Table 7.2 UK BSP travel documents processed in 1996

Type of document	Millions of documents	%
Automated tickets		
OPTAT	11.4	63
ATB2	2.5	13
Manual tickets	1.8	10
MPDs	1.7	9
Refunds	0.7	4
ADM/ACMs	0.2	1
Total documents	18.3	100

(Source: IATA: UK BSP)

- **Automated reconciliation** In the future there will be further enhancements to the BSP service. One of these enhancements is concerned with the reconciliation process that I mentioned previously. As you will probably recall, this is the reconciliation between the GSi-produced billing report and the report produced by the travel agent's own agency management system. Already GSi can produce a diskette containing all the details of the transactions shown on the travel agent's billing report. This diskette is in a format acceptable to Galileo's PAMS system.

    A travel agent using PAMS who has the optional Report Writer facility can automate its reconciliation process using this diskette. The PAMS system itself will already have a record of all air tickets sold by the agency during the month. This will have been created as a by-product of the ticketing and accounting process. By inserting the GSi diskette into the PAMS PC, the PAMS system can check off each of the GSi air sales billing records against the internal data base of air tickets issued. Any discrepancies where a ticket image does not match, are highlighted for user follow-up. This is a useful feature that can save a great deal of time at month-end when there is a fair amount of pressure to reconcile the GSi billing report within the three-day window before the direct debit hits the travel agent's bank account. In the future, BSP intends to support other formats of diskette besides PAMS so as to make this feature more widely available on more agency management systems from different suppliers.

- **ATB2 tickets and printers** In the UK, BSP officially endorse the ATB2 as the preferred ticket for use by travel agents (there are many reasons for this, which I explain in Chapter 3). Therefore, when they were first introduced in the early 1990s, the airlines subsidized the cost of ATB2 printers that were purchased by UK travel agents. Although this subsidy has now ended, it resulted in there being a total of 688 locations using ATB2 ticket printers in the UK, i.e. as of the second quarter of 1996. It can therefore clearly be seen that the ATB2 format for airline tickets is in the early stages of a national roll-out programme. This is best illustrated by Table 7.2, which shows the volume of BSP documents processed in 1996.

    This table shows that of all the 14 million automated tickets printed under the BSP scheme, only around 18 per cent were produced using ATB2s. The remaining bulk of airline ticket production from travel agents in the UK was 82 per cent. So, there is still a long way to go for the ATB2 format ticket. It is, however, expected that OPTAT tickets will steadily be replaced by ATB2 tickets over the next four years. However, this migration will depend to a large extent upon how quickly electronic ticketing takes off. If e-ticketing grows quickly then the spread of ATB2 technology will be correspondingly slower (see Chapter 3 for a description of e-ticketing).

Some travel industry pundits consider ATB2 tickets to be rather user-unfriendly because the information printed on them is less easy to read for the average traveller. They also have the disadvantage of requiring a sophisticated ticket printer, which costs around £3,500 versus £800 for the old dot-matrix type printer. However, the widespread use of ATB2s is inevitable because the airlines have invested substantial funds in ground operations equipment to process them. ATB2s have a magnetic stripe on the reverse side that is encoded by the ATB2 ticket printer at the time the ticket is issued. This enables airlines to use ATB2 tickets at check-in and also at the departure gate. In fact, British Airways is one of several airlines that has already installed reading or scanning devices at airports, in preparation for ATB2 tickets.

But despite their user-unfriendly appearance, ATB2 tickets will no doubt be demanded by most frequent travellers before very long. The reason I say this is because when ATB2 tickets start being used, passengers will notice their advantages and demand that their travel agents issue ATB2 tickets for them. A business traveller will, for example, only have to see their fellow travellers using ATB2 tickets being whisked through check-in while they are left in a long and slow-moving queue, before they demand that their travel agent provide them with new style tickets. The ultimate threat being that the traveller will move their account to another agency that can produce ATB2 tickets. So, unless e-ticketing takes off soon, it will be just a matter of time before ATB2s become widespread.

The advantage for both BSP and the travel agent of the ATB2 ticket over the old standard BSP ticket is that there can be no mistake about what ticket number is recorded on both: (a) the actual physical ticket given to the customer, and (b) the ticket as recorded by the back-office system (whether this be a GDS or an agency management system). The problem that sometimes occurs at present is that continuous ticket stock in the agency can sometimes get out of alignment with the system. The system registers that it is printing ticket Number 1, whereas in fact there was a printer jam in the agency and ticket Number 1 was wrecked – the travel agent operator advanced the ticket stock in the printer to start at the next available ticket which was, for example, ticket Number 2. From this point onwards, all airline tickets issued by the agency will have an incorrect ticket number assigned to them by the GDS and/or agency management system. This can cause havoc if it is not spotted before BSP reporting is carried out. Even if it is trapped before reporting, it is a real nuisance for the travel agent to correct. However, this problem cannot occur with ATB2 tickets because the special purpose intelligent printers encode the same ticket number onto the ticket's magnetic stripe as it transmits to the GDS.

BSP would like as standard, an enhancement that would exploit the technology of ATB2s even more fully. If the pre-printed stock number shown on all ATB2s could also be encoded on the magnetic stripe on the reverse side, then BSP's stock control procedures would be greatly simplified. As a ticket was being printed, the stock number could be read by the ATB2 ticket printer. As part of the ticket issue process, the stock number thus derived from the ticket could be inserted into the ticket image recorded in the GDS system. In this way, BSP could be provided with the electronic information which it needed to control ticket stocks, with the minimum of manual effort.

- **Airline MIS** The airlines would like to be able to get their hands on more management information regarding the flights actually ticketed and paid for by their customers. This kind of information is key to the marketing efforts of the airlines that, as I have explained before, is in turn so key to profitability and competitiveness in the open skies deregulated environment of today (and tomorrow). At present, GSi can provide the airlines with only limited data on what is known in the industry as a 'short record'. These records refer to data that have been generated from the printing and capture of airline tickets and there are really three main types:

- *Short record* This is basically the ticket number and the value of the ticket. It is really no more than the basic accounting data needed to report and settle an airline ticket. As such it has only limited interest to an airline's marketing department.
- *Long record* This comprises the short record plus some routing information e.g. the from and to city pairs of each flight sector. This is more interesting to airlines because this kind of data can be used to produce some interesting marketing analyses on, for example, the volume and value of airline ticket sales on certain routes flown during special promotional campaigns.
- *Super long record* This comprises the entire ticket image and contains all of the information shown on the ticket. The super long record will only be available where sales are reported via a GDS. This is the 'gold dust' that is sought by the sales and marketing departments within the airlines. With this class of information, all kinds of business analyses are possible.

Already some super long records are being stored on large data bases by airlines. These data are captured and stored for up to one month by GSi, after which it is then up to the airline concerned to store the data for longer periods for analysis purposes, as they see fit. The intention is to enable airlines to build up a historical analysis of their sales for MIS purposes. This is because one of the key dimensions of any MIS reporting system is time. Data in a time frame can be used to study trends and performance in relation to other external factors such as the state of the economy.

- **Support of other documents** At present only a limited set of documents can be processed automatically by BSP. This is due mainly to the limited capture facilities of the various GDSs for anything other than airline tickets. In the future, BSP intends to encourage the GDSs to provide the wherewithal to support the capture of all other BSP documents at the point-of-sale. This will help reduce agents' administrative time and will reduce yet further the paperwork overhead experienced by GSi.

- **Non-airline suppliers** BSP has been an undoubted success. The scheme has reduced costs for participating airlines and has made life easier for the travel agent. It has therefore attracted the attention of other travel industry suppliers who would dearly love to discover an easier way to settle the sale of their products and services to travel agents. From BSP's view, any new participant represents an opportunity for it to defray its costs among a wider user base. BSP would welcome a new non-airline supplier provided it can be accommodated without incurring any significant implementation expenses for the airlines. In Canada, for example, rail and coach companies participate in the BSP. In South Africa, rail, coach and ferry companies settle with travel agents via BSP.

So, around 1994, BSP UK invited a number of non-airline travel-related companies to consider participating in BSP. Although there was a lot of interest, there were far fewer companies who were prepared actually to commit corporate funds to a feasibility study. But there were some who did, and in fact such a study has already been completed. This looked primarily at the ferry, rail and hotel industries in the contexts of travel agency settlement and accounting. The main issues addressed by the study team were:

- *Complexity* The existing BSP processing systems were designed for the airline business and work well in this context. To add a new non-airline company introduces a whole new set of requirements, such as different data elements, different commission structures and different ticket rules. Because the GSi processing system is built around main-frame technology that was developed several years ago, it suffers from the disadvantages of old technology such as being expensive to modify and requiring long lead times to accommodate changes. Some very innovative thinking will be needed to overcome these obstacles at a cost that is acceptable to the member airlines of the UK's BSP.
- *Coding systems* The airline industry is not alone in experiencing a fundamental problem in the area of code assignments. IATA is running out of codes that can be used to

Table 7.3 ATB2 printer hoppers

ATB2 printer hopper	Type of document printed
1	IATA airline tickets on ATB2 stock
2	SO-ATB stock for non-air carriers
3	Blank card for use as an itinerary (or any other non-value item)

describe new cities, companies, services and so on. For example, what code should be used to describe the new channel tunnel terminal in France? Any logical choice might well clash with an existing city code somewhere else in the world. Then there is the clearing house code, which is part of the ticket number printed on airline tickets. This is used to identify the issuing airline and is 125 for British Airways, for example. There are very few spare codes left to be allocated to new airlines, let alone new non-airline companies. So, the issue of a new coding system to support an expanded BSP is a real challenge.

- *Printers* Travel agents really want a single ticket printer in their agencies that can print all kinds of tickets: the potential non-airline BSP members would not want to have to ask travel agents to purchase and install a separate ticket printer just for their stock. This would not be economically feasible for the travel agent. So, ideally, an ATB2 printer loaded with two kinds of stock would be the ultimate solution. There would be two hoppers containing: (i) a common ticket stock, and (ii) a stock to be used for printing itineraries. However, in order to achieve this, a common ticket design would need to be agreed within the industry.
- *Common ticket stock* Allied to the above discussion of printers there is the issue of ticketing documentation. This sounds quite straightforward but is in truth so complex and difficult to achieve that one could be forgiven for saying it is totally impractical for the foreseeable future. The difficulties lie in getting all the various industries and competitors within each industry to agree on a common design for their tickets: there is substantial evidence within the travel industry to prove that such standardization efforts usually come to naught. However, there are moves afoot to accomplish this seemingly impossible task. A universal multiple purpose document is being considered by one of the industry's working committees. This would support airlines, hotels, car rental companies and other carriage and service companies. The functions of an MCO would also be incorporated. There is also a reasonable chance that intelligent ticketing will become a reality (see Chapter 3). Only time will tell whether these standardization efforts will eventually bear fruit.

However, despite all the above-mentioned issues, difficulties and challenges, BSP has nevertheless succeeded in signing up Eurostar as the first non-air participant in the UK's BSP. Initially BSP is supplying only a ticket management service on behalf of Eurostar. This first phase of Eurostar's implementation involves BSP distributing a special kind of ticket for use in travel agents' ATB2 printers called a surface operator ATB (SO-ATB). These tickets may be used by automated agencies that use a three hopper ATB2 printer, configured as illustrated in Table 7.3.

The SO-ATB ticket is printed under the control of the Galileo GDS. The travel agent must therefore be a Galileo subscriber to have the capability of automatically printing Eurostar SO-ATB tickets. The actual ticketing process is controlled by Galileo's NVP capability (see separate section on Galileo in Chapter 4). The travel agent selects the Eurostar host system on their office Galileo PC. This connects them via the NVP directly into Eurostar's TSG system. Once connected, a reservation can be made and Tribute can direct a ticket to be printed on the agent's ATB2 printer. It is expected that between three and four million tickets will be printed in this way each year.

Eurostar is currently in Phase I of its UK BSP implementation programme, i.e. as of the second quarter of 1997. Currently, for example, the agent/carrier funds clearance functions normally provided by BSP, continue to be performed directly by Eurostar. So, although a travel agent may book a Eurostar seat and produce a ticket, the agent cannot settle that ticket via BSP. Settlement processes like this will be introduced as the next phase of Eurostar's participation in the UK's BSP.

This next phase is only likely to proceed when Eurostar establishes a true CRS connection from Tribute directly into the Galileo GDS computer. When this happens, Eurostar will be allocated codes that are consistent with the airlines. This will enable Eurostar to be ticketed on normal airline ticket stock, which means that its tickets may be produced from Hopper 1, just like an airline ticket, thus freeing up Hopper 2. The major feature of this change to the ticketing process is that it allows Eurostar to participate fully in the funds clearance services of BSP. When this happens, Eurostar tickets will be processed exactly like any of the airlines that participate in BSP, from a ticketing and administration viewpoint. This will benefit Eurostar and agents alike in reducing administration, streamlining cash flows and simplifying ticket production. Having said this, some smaller travel agents with a mixed business/leisure profile may well object to parting with their funds earlier than at present. Larger business travel agencies with a higher volume of rail business will be the real winners.

Other non-air carriers are considering joining BSP, particularly the ferry companies, such as P&O and Stena. UK rail companies may also join, depending upon how progress is made towards connecting the 'old' British Rail main-frame computer in Northampton into Eurostar's Tribute system. This will allow the five major train operating companies (TOCs) that are members of ATOC (a kind of IATA for the UK railways), to participate in BSP and enjoy the benefits of: (a) travel agency ticket stock management, (b) ticket management functions for rail stations, (c) automated ticketing, and (d) funds clearance for UK sales outlets. If this happens, it may well eventually replace the Rail Settlement Plan operated by the TOCs.

Other productivity enhancements

BSP has identified several enhancements and new developments that are being gradually implemented over the forthcoming five to ten years. These are all part of a cost control programme designed to minimize the distribution costs incurred by IATA's members and other non-air carriers that participate in a BSP scheme. A summary of the main actions is as follows:

- **Agent reporting** BSP would like to move agents to a weekly reporting cycle. This would spread the workload and reduce BSP's costs. After all, BSP's costs are funded by IATA, which in turn is funded by the airlines. With airlines putting their distribution costs under the microscope, the cost of ticket management and reporting is a significant cost element for them.
- **Automated document production** BSP would like agents to be able to enjoy increased levels of automated document production. For example, MPDs remain high volume documents that could theoretically be printed automatically in a travel agency. This would also have the spin-off benefit of including MPDs in the automated reporting function.
- **Automated refunds** It is quite possible that refunds could be automated using the GDSs. At present handwritten forms are raised and processed by airlines and travel agents. This is a laborious process that could be handled by new GDS computer applications and intelligent ATB2 printers.
- **Increased agency automation** Although BSP has no direct control over the level of automation used by UK travel agents, it is in its interest to see the level of automation maximized. Only by doing this will the proportion of tickets that are automatically captured and reported be increased. It is, for example, conceivable that BSP could in the future make a charge for manually issued tickets. This could be justified on the grounds that it costs BSP more to process a manual ticket than it does an automated one generated by a GDS. After all, BSP already levy charges for exceptional items such as an unreported ticket enquiry, late ticket sales reporting and missing traffic documents.

- **Support for the Euro currency** BSP is currently considering how its systems should be modified to support a common European currency unit. In this new environment, customers will be able to pay for airline tickets in either their local currencies or in Euros. BSP's systems will therefore need to be modified substantially to support both different currencies and the Euro and may in the future, for example, need to produce two sets of billing reports.

It is interesting to speculate on the possibilities for an EC-wide BSP, although at present, i.e. mid-1997, I know of no plans for this at all. It would, however, seem to make sound practical sense because with a common EC currency and banking system, the basic infrastructure to support an EC-BSP will effectively be in-place: it is an attractive proposition for airlines to consider because the economies of scale could be substantial. This could drive down the unit processing costs associated with selling airline tickets and thereby ease the pressure on distribution costs incurred by IATA's member airlines. Nevertheless, despite the fact that I think this remains a distinct possibility for the very long term, there are many major issues to overcome before an EC-BSP could become a realistic possibility.

This has been a very brief overview of the UK's BSP. However, it should provide you with a basic understanding of how this important element of the travel industry's main product, i.e. an airline ticket, is handled. Because BSPs in other countries operate in a broadly similar way to the UK, you should by now understand the fundamentals of the world's airline ticketing and administration functions. BSP itself is significant enough to make it one of the main functions that a travel agent would want to automate. This is where agency management systems come into the picture. So, it's now time to take a more detailed look at what kind of functions are provided by these systems.

FINANCIAL SYSTEMS

Once a booking has been made by a travel agent, the more mundane side of the travel business must be dealt with (some would say this is the more exciting part because it involves the collection of funds and therefore the generation of profits). Financial services support the process of accepting customers' payments and remitting funds to suppliers as well as providing added-value services, such as bureau-de-change. Many of the larger multiple travel agents have their own in-house systems and even their own communications networks to help accomplish this. But the smaller high street agents are able to enjoy similar services from several service companies; take, for example, AT&T. It offers travel agents two financially oriented services: Transtrac and The ABTA Single Payment Scheme, both of which are described below:

AT&T's Streamline Transtrac

This is a service available on AT&T, which is in fact provided by National Westminster (NatWest) Bank plc and American Express. It is a viewdata-based credit and debit card authorization system that accepts all major credit and debit cards, including Mastercard, Access, Visa, Delta, Eurocard, American Express and the UK's 13 million Switch card holders. Transtrac provides sales and refund transaction processing at the point-of-sale. It is unique in that instead of requiring a dedicated point-of-sale (POS) terminal, it only requires a viewdata terminal capable of displaying the Pound (£) symbol; not all of them do and it is recommended that a printer is available. The way it works is:

- **Log on** The travel agent accesses the AT&T network in viewdata mode by either telephone dial-up or by means of the Direct Service. Once logged on to AT&T, the Streamline Transtrac service is selected and the travel agent's user number is entered. A password must then be entered in order to authenticate the agent.
- **Transaction selection** The user is provided with a screen that allows up to five main types of transaction and some administrative functions to be selected. The transaction options are: (a) transaction with authorization, (b) transaction with voice authorization, (c) authorization only, (d) transaction pending completion, and (e) IATA BSP UK authorization.
- **Transaction screen** The Streamline Transtrac system then displays a screen into which the user keys the required transaction, which may

Figure 7.7 ABTA's Single Payment Scheme

be for either a sale or refund as appropriate. Information such as the card number, card issue date, expiry date, amount and transaction reference are entered and checked by the system.

- **Authorization screen** The transaction is routed via AT&T to the NatWest computer, which in turn contacts the card issuer's computer for authorization of the card. If this is received without any problems then an 'authorization approval' message is displayed on the screen. Otherwise the transaction is referred and will need to be followed up by the agent manually.
- **Print screen** The travel agent is then required to print the viewdata screen on two-part paper for the client to sign. One copy may be given to the customer and the other retained by the travel agent for storing in the customer's booking file.
- **End of day** At the end of the business day the travel agent will need to make the day's transactions available for processing by the card scheme group, i.e. by the card issuer. These amounts will be credited to the agent's account by the card issuer who in turn will bill the customer, i.e. the cardholder.

The above is of course only an overview of what the Streamline Transtrac service provides on AT&T's viewdata network. It is an innovative use of the technology and since the service was launched in mid-1992, several hundred agents have started using Streamline Transtrac. The main benefits delivered by NatWest and Amex to the agent are: (a) a low and guaranteed price for card transactions, (b) guaranteed payment for all authorized card payments without the problem of unpaid cheques, (c) easier and quicker administration of card transactions, (d) no daily visits to the bank for card transactions alone, and simpler/quicker reconciliation at the end of each day.

ABTA's Single Payment Scheme

This is a viewdata service operated by AT&T that allows travel agents to consolidate their payments and support direct debit schemes controlled by tour operators (Fig. 7.7). The service was developed at the instigation of ABTA and with the close involvement of AT&T. ABTA wanted to see agents using a smooth settlement system that would be similar in concept to IATA's BSP.

AT&T has 132 tour operator customers who are able to access AT&T's national network, to supply billing information to travel agents. This service is not restricted to tour operators who interconnect their systems into AT&T's network for videotex reservation purposes. The service is open to virtually any UK tour operator. This allows them to enjoy the benefits of the direct debit scheme, which itself simplifies administration and minimizes bank charges. The way this works is:

- Travel agents make bookings in the normal way, either by telephone or via viewdata booking systems. Besides bookings, travel agents

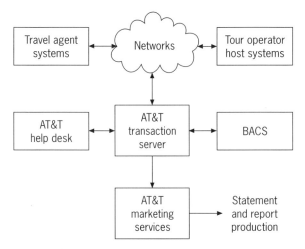

Figure 7.8 ABTA's Single Payment Scheme – diagram

may record cancellations and adjustments to their customers' holiday plans.
- Each week, tour operators create a set of files on an AT&T computer system for each travel agency with whom they deal. The files are created automatically by a tour operator's application program and are transmitted to the AT&T computer that supports the Single Payment Scheme. This file contains:
  - Debits for advance booking deposits, final balances and other regular financial transactions that have arisen since the last file was created.
  - Balances outstanding for departures up to and including 42 days as on Friday of the current week.
  - Credits due to agents. These may be due to cancellations or overpayments.
- Travel agents (and operators) can view the files and peruse the contents between 07.00 hours on the Monday of each week until 16.00 hours on the Wednesday. Travel agents may adjust the pro forma entries on-line using screen editing facilities that are part of the system. An example of an adjustment would be the deletion of an item that has already been paid or is not now due.
- All transactions remain on the AT&T system until a pre-determined cut-off time, which is currently 16.00 hours on the Wednesday of each week. Prior to the cut-off, the transactions may be accessed by logging-on to the AT&T network and then viewing the transaction file entries using an on-line terminal.
- When the cut-off time is reached, the AT&T system collects all non-zero transactions and prepares a BACS file. This file comprises a number of single consolidated debit instructions, each of which involves both the relevant tour operator and travel agent in a single payment.
- Once this file has been created, AT&T transmits it to the BACS computer by 21.00 hours on the Wednesday of each week.
- As part of this processing, the AT&T system produces several operational control reports. For instance, it assembles a weekly exception report for tour operators detailing amended transactions. These are usually used by the tour operators to generate entries into their in-house ledger systems. Also, a confirmation of transaction report is generated for use by both agents and operators.
- Funds transfer occurs by 09.30 hours on the Friday of each week. The travel agents' holding accounts with ABTA are debited and the tour operators' accounts are credited by 09.30 hours on the following Monday.

The Single Payment Scheme has several important benefits to travel agents and tour operators. It improves a tour operator's cash flow, it allows several transactions to be consolidated into a single debit/credit, which minimizes bank charges and simplifies paper flows. It allows on-line credit control to be effected by both parties and enables valuable management information to be produced as a by-product of the process. It also allows tickets to be released faster because the tour operator does not have to await receipt and clearance of final payment cheques. This is especially important for late bookings where time is usually very short. Finally, it allows administrative costs to be reduced to an absolute minimum for both travel agents and tour operators.

ADMINISTRATION SERVICES FOR HOTEL COMMISSION

In relation to the revenues derived from the sale of airline tickets, hotel commissions appear to be almost insignificant. However, now that the airlines are introducing commission capping for certain types of tickets and other margins are increasingly being squeezed, travel agents are becoming much more focused on all non-air commission sources in order to protect their businesses. However, in the past there has been little incentive for travel agents to make hotel bookings because of the difficulties in collecting their commissions. There are many reasons for this but they all add up to the same thing – hotel commissions are costly and time-consuming for travel agents to collect. It is worth examining some of the principal causes for this:

- **Low value transactions** Individual commission payments are often relatively small because many bookings are for just one or two nights' accommodation: from the travel agent's perspective, it is not therefore economically feasible to spend too much time chasing commission payments for such small amounts.
- **No record of bookings** In many cases no physical records are made by the travel agent for hotel bookings. The travel agent either telephones the hotel or makes an entry via a GDS as part of booking a flight. Consequently when it comes to taking stock of which hotels owe commissions to the agent, there is no supporting paper mechanism that even allows the agent to see at a glance what is owed to them by hotels.
- **No shows** Because most hotel bookings do not require an advance payment, there is little incentive for the travel agent's customer actually to turn up. So, agents often spend considerable effort chasing payments from hotels only to be advised that the client did not show up and therefore no commission is due.
- **Inefficient processing** Where commission payments are made by hotels in other countries, they are usually issued in a currency that is foreign to the travel agent. This can be costly for the agency to convert via its local bank. Also, the number of cheques raised by hotels and sent to travel agents is high, with a single cheque being needed for each customer's stay.

These problems caused many travel agents to regard hotel bookings as low priority. This could therefore be one of the reasons that travel agents are responsible for only 28 per cent of all hotel bookings. A solution to many of these problems is, however, now available from companies that have recognized the potential benefits to both hotels and travel agents of an effective commission collection system. I have included two examples of systems that support hotels and travel agents in tracking commission payments: one of these is provided by the Hotel Clearing Corporation (HCC) and the other by Utell.

The HCC

The HCC provides travel agents with a hotel commission collection, reconciliation and payment service. It was formed in April 1992 by many of the same hotel companies that founded Thisco (see Chapter 4 for more information on Thisco and its parent company, Pegasus Systems Inc.). HCC's mission statement is: 'To provide the most effective hotel commission management reports and consolidated commission payment system in the travel industry so that travel agencies and hoteliers can realize improved efficiencies and profits in managing the commission process.' In other words to collect commission payments from hotels and pay them to travel agents following reservations that they make on behalf of their customers.

HCC is a growing business as evidenced by its compound annual growth rate of 74 per cent over the last three years. Today, more than 65,000 travel agencies and 42 major hotel organizations world-wide rely on HCC to collect over US$111 million in hotel commission payments.

HCC works in partnership with Citibank in the USA and also with participating hotels. Travel agents may either be registered for the service with HCC or they may be non-HCC agencies to which the hotel chains wish to pay their commissions directly (Fig. 7.9). In summary:

1. Participating hotels capture commission booking transactions as a by-product of the check-out process on the first business day of the following month. They send these transactions to their head office for consolidation.
2. The hotel chain's operations centre consolidates transactions on behalf of all its properties and passes them to HCC on the fifth business day. The data identify the travel agency that originated the booking.

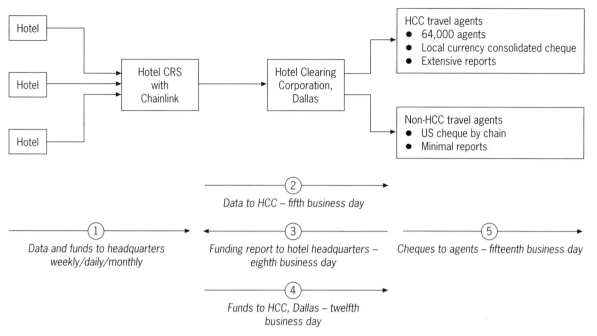

Figure 7.9 HCC data/funding flowchart

3. HCC receives the transactions from all hotel chains and accumulates the information. A funding report is produced each month and sent back to the hotel chain by the eighth business day. This is, in effect, a request for payment.
4. The hotel chain's processing centre sends payment to HCC's bank for all travel agency commission payments for the processing period, as detailed on HCC's funding report. These funds are received by HCC by the twelfth business day.
5. HCC produces consolidated payments of hotel commissions, less HCC's processing fee, which it either sends to travel agents by cheque or by electronic funds transfer (EFT). These payments are for all commissions due during the processing period and are denominated in the travel agent's own local currency. Travel agents receive commission payments by the fifteenth business day.
6. HCC also provides a set of management information reports for both its hotel and travel agent customers. These reports are generated as a by-product of the whole transaction processing cycle and can be very informative. Travel agents may also elect to receive a diskette that can be used to reconcile HCC commission payments against their own records.

The primary advantage for the travel agent is that a single local currency payment is received from all participating hotels promptly, i.e. by the fifteenth working day following the customer's check-out from the property, without the need for any follow-up action by the agent. Agents may also simplify their commission reconciliation procedures either by: (a) receiving an HCC diskette that they input to the reconciliation function within their agency management system, or (b) using special reconciliation software provided by HCC that uses the monthly HCC reconciliation diskette to produce control and tracking reports. Other advantages for agencies arise from the wealth of management information on the bookings they generate. Information such as the source of reservations, e.g. by telephone to property, by toll free telephone to a reservations centre or by a GDS, and the status of each booking, i.e. whether it was cancelled or the customer was a no-show.

From the hotel's viewpoint there are also many advantages. Again, the primary advantage is the elimination of labour intensive clerical tasks that have historically been needed to track and pay agencies for their commissions. Hotel chains are keen to reward agencies for delivering customers to their properties because rewards encourage future sales and therefore increased revenue. Hotels also receive some valuable management information that enables them to identify their most productive travel agents, monitor average room rates and thereby support yield management activities.

Utell

Utell also provides a hotel commission collection system. As with HCC, the Utell service aims to overcome the hotel commission collection problem that has historically been experienced by travel agents. In the past this has been a major deterrent that has suppressed agents from making hotel bookings due to the administrative overhead involved in collecting individual commissions from hotels. Utell International has a solution to this problem in the form of two products:

- **Paytell** Paytell is Utell's commission payment system for travel agents. It enables travel agents to collect advance deposits from their customers. This is often equivalent to a customer's first night's accommodation fee, although it may be any amount the travel agent considers appropriate (full pre-payment in the agent's local currency is also possible). This deposit guarantees a traveller's hotel reservation and also allows the travel agent to settle for the booking in their local currency. The key benefit to the travel agent is that the hotel commission is deducted from this advance payment (or fully pre-paid booking). Currency exchange losses and lengthy payment delays are therefore avoided.
- **PayCom** If a customer prefers to settle their bill on departure and not pay a first night's deposit, then the PayCom service is available to the travel agent. It enables Utell International's hotels to pay the travel agents commission through Utell's local office in the currency of the travel agent's country. As a result exchange loss and bank charges are avoided making reservations through PayCom hotels more cost-effective. So, for bookings that are not pre-paid, the travel agent is encouraged to book a PayCom hotel at the customer's destination.

With services such as these, smaller travel agents are able to compete on equal terms with the larger multiple agencies, many of which have developed in-house commission tracking systems of their own.

Functions of the agency management system

So, we have now explored BSP, which is one of the main reasons why a travel agent needs a back-office system, and we have considered some basic financial support systems. It is now time, therefore, to consider each of the basic characteristics of an agency management system in more detail. The following is an overview of the key functions that are embodied in accounting, MIS and marketing systems, which themselves should be included in any decent agency management system for a travel agent.

ACCOUNTING

This is the core of any agency management system. Accounting is the art of controlling a business by maintaining accurate book-keeping records, measuring and interpreting financial results, and communicating financial information to support managers and staff in running a business successfully and profitably. It is the set of applications that measures a travel agent's performance and enables management to control cash flows. To perform these accounting functions manually is a tiresome task that detracts from the amount of time the travel agent can spend selling. Far better to get a computer to do it all for you.

A travel agency management system is designed to do just that for you. The functions are usually sub-divided into sales ledger, purchase ledger and general ledger systems. All these ledgers are computer programs, which share common data bases and provide a set of standard reports, that represent a travel agent's set of books. But the books will not be created out of thin air; they need to be 'told' what is going on. It must therefore be remembered that the agency management system needs feeding with transactions and must be periodically reviewed. A good agency management system will usually generate many of these entries as by-products of the agency's day-to-day processes, which involve such tasks as issuing tickets and invoices, posting cash, cheques and card payments received from retail customers and making out cheques to suppliers.

An agency management system should therefore automate a great deal of these routine accountancy tasks. It should also be able to eliminate a lot of paper, which is costly, bulky and cumbersome as an information retrieval media. However, there is a rule that states that hard copy, i.e. paper, records of important transactions must be kept for five years for legal purposes. So it does not look like paper will be totally eliminated by agency management systems until the law is changed in this respect.

Sales ledger

The sales ledger is the part of an accounting system that is concerned with collecting moneys in return for the products and services that have been provided to customers of the business. This is handled in different ways depending upon the type of customer. In the travel business there are two main types of customers:

- **Retail customers** It is not often that retail customers are given credit. This is because their financial situation is often not known by the travel agent and even if a credit check could be done, it is not considered profitable to extend credit to retail customers. In any case there is no real demand for credit from travel agents from these customers who can, if they need credit facilities, always obtain them from a bank or card company. In general, retail customers represent a source of funding for travel agents because they invariably pay any moneys due in advance of it actually being handed over to the travel principal supplying them with the product or service booked.

  The sales ledger for retail customers therefore must be capable of generating a request for payment in advance of the date by which the principal is to be paid. This is not usually referred to as an invoice, although it is kind of difficult to explain the difference between a request for payment and an invoice. The sales ledger or sometimes the front-office part of the system, therefore needs to have the capability to generate a request for payment that can be sent to the customer. It is important to realize, however, that this request for payment does not set up any kind of receivable entry in the travel agent's books of account, i.e. the back-office system. So it is not possible to look at the sales ledger and see how many or how much is owed by retail customers.

  When a retail customer does pay the amount requested it must, however, go into some kind of suspense account until it is paid over to the travel principal involved. It is the value of this suspense account that contributes to a travel agent's positive cash flow. It must of course be remembered that this money represents cash in the bank that is held by the agent on behalf of the client.

- **Business customers** It is customary for business travel customers to be given credit accounts by the travel agent. This is mainly due to the fact that companies generally pay their bills on a regular cycle and would not expect to pay for travel services on an as needed basis. The travel agent does of course obtain credit references before extending credit to a business travel customer and will usually obtain a written agreement from the company. This should include a statement of the payment terms and the maximum period over which credit will be extended.

  When travel services are provided to a business travel customer an invoice is issued. If the travel agent is using an agency management system it will be this system that will print the invoice. An invoice is a document that states the amount due and describes the service or product provided by the travel agent to the company. As a by-product of issuing an invoice the agency management system automatically generates an accounting entry in the sales ledger. This entry will include the date the invoice was issued, the invoice number, the amount and a description of the service or product provided. The entry will be posted, i.e. entered into the sales ledger, under the main account code representing 'accounts receivable' and will be further segregated by a sub-account that is unique to the company to which the invoice was issued. The total of all these accounting entries represents the total amount owed to the travel agency by all its business travel customers. As you can see, business travel generates a negative cash flow.

These invoice entries remain in the 'accounts receivable' part of the sales ledger until they are paid by the customers. Periodically, usually monthly, the agency management system will produce a statement of account for each customer. This will list all the invoices that have been issued but have not yet been paid. Often this statement will show an aged analysis of the invoices outstanding. This is an exhibit of the outstanding invoices summarized by the length of time they have been outstanding. It is supposed to: (a) shame the company into paying up for the services provided and, (b) worry the travel agent if the aged analysis shows invoices over 60 days old.

Purchase ledger

The purchase ledger is the part of the accounting system that deals with payments to suppliers and other parties to whom the travel agency owes moneys. In the case of travel principals, the payment for services is usually required some time after the products are actually issued. Therefore a kind of accrual account is required into which is placed a record of a payment that is due to a supplier at a particular point in time. This is known as a purchase ledger. The total of all the entries in the purchase ledger is the amount that the travel agent owes its business partners; in other words its indebtedness.

Now there is little point in being too efficient here. If a particular travel principal's terms of business state that payment is due by a certain date in the future, then there is little point in the travel agent issuing a cheque in advance of this date. So a good purchase ledger allows payments due to be scheduled (or timetabled), so that they are made on time but not before time. This can make an important contribution to the travel agent's cash flow objectives. However, the travel agent does not always have a great deal of leeway for certain payments. Take, for instance, the issue of payment for airline ticket sales. There are very strict rules governing the format and timing of these sales. These rules are set by IATA (see Chapter 1) and are administered on its behalf by a wholly owned sub-group known in the UK as the BSP (see the lead-in section of this Chapter for more information on BSP):

- **BSP** One of the largest volumes of payments to travel suppliers is usually generated from the sale of airline tickets. The way in which air tickets are paid by travel agents to airlines is somewhat different from most suppliers, but the principle is the same as for any other travel product. A record of sales must be kept and used as the basis for reconciling the payment to the supplier. As previously described, the BSP is a central clearing house for all airline payments due from travel agents. BSP UK is operated by a third-party bureau service, which acts on behalf of IATA. Instead of all travel agents having to issue individual payments direct to the various airlines of the world for the airline tickets they issue, a single payment is instead made to BSP; which in turn pays the airlines individually on behalf of all UK travel agents. The introductory section of this chapter provides some more detailed information on BSP.

- **Other supplier payments** Payments to other suppliers are made from the purchase ledger in a more classical manner, which is similar to any other business. Items due for payment are logged in the purchase ledger and are assigned a 'pay-by' date. When this date is reached the agency management system should initiate payment. This means that the system will print a cheque, produce a remittance advice and update the purchase ledger to denote the occurrence of payment. In the case of payments made by travel agents to principals, the payments will usually be net of any commissions due to the agent. Sometimes a payment is made on receipt of a supplier's invoice. The processing for such cases is very similar to that described above.

In all cases where payment is made by a travel agent, there must be tight safeguards on the 'approval to pay' process. Usually, two signatures will be required on a supplier's invoice before payment is initiated. In the case of agency management systems, the senior manager should ideally be the one who controls the printing of the cheques. Once the cheques have been printed they should be countersigned by a senior member of the travel agent's management. The reason for this is of course the scope for an unscrupulous employee to issue cheques to themselves or an accomplice.

General ledger

The general ledger is sometimes known as the nominal ledger. It forms the core of the travel agent's accounts for financial reporting, tax and cash flow purposes. The sales and purchase ledgers are an integral part of the general ledger as are several other types of sub-account including, for example, the VAT account. However, the main products of the general ledger are the profit and loss account and the balance sheet. These two accounting exhibits are the ones needed to satisfy the fiscal authorities and the Inland Revenue:

- **Profit and loss accounts** The profit and loss account contains a summary of the expenses and the revenues generated by the business. It is therefore the means by which a company's operating performance is measured both for tax and reporting purposes. In most computerized back-office systems many of the entries in this account are automatically generated by the sales and purchase ledgers, but not all. This is where the travel agent's accountant comes into play. In order to be able to decide the category of certain items of expense or revenue, one needs some degree of training in accounting. No matter how clever and high tech the agency management system is, it cannot tell you what is the correct account caption for a certain book entry.
- **Balance sheet** The balance sheet is obtained by summarizing certain key accounts from the general ledger. It reflects the financial standing of the travel agency in terms of what it owns, how much it owes and how much is owed to it. The two main categories in a balance sheet are the assets and the liabilities. Assets are things that value the business, because they represent cash or items that are owned by the company that can be turned into cash. Examples are cash itself, buildings, moneys owed to the agency (e.g. accounts receivable), computer equipment and so on. Liabilities are the opposite. They are things that are owed by the business to others. Examples include loans, moneys held by the agency but which are owed to others (e.g. accounts payable to travel suppliers), and share capital. The total of all assets must equal the total of all liabilities as reported on a balance sheet.
- **Tax** An important role of an agency management system is to automate as many tasks as possible in preparing the annual tax return for a business. This became even more crucial with the introduction of the new 'pay and file' regime in late 1993. For accounting periods ending on or after 1 October 1993, every company is responsible for calculating its tax liability, paying the tax nine months after its year end, and submitting a detailed return within 12 months of the year end. So in place of the Inland Revenue assessment the responsibility for calculating and paying tax will lie with the company. Besides interest being charged on late payments of tax, the revenue will charge penalties if a company does not file its return on time. Although nine months may seem like a long time, a significant amount of effort is required to prepare the return and without a good agency management accounting system, it will be a challenge and a tiresome manual task.

MANAGEMENT INFORMATION

Management information or the MIS as it is often known, is critical to the successful running of a travel agency and is a key factor in the retention of its corporate business travel account customers. It should be obvious that you need information about how a business is performing if you are to control it and not let it control you. One place to get this information is the agency management accounting system. However, although financial information is one of the most important aspects of MIS, it is by no means the only one. There is a great deal of non-financial information that is critical to the effective running of the business. For example: (a) the number of nights your customers stay in certain hotels – this can be used to negotiate special commission deals, (b) the routes most frequently flown by each of your corporate customers – this can be used to negotiate upgrade arrangements for your business customers on certain frequently used routes.

As you will notice, these items of information are not strictly financial in nature. But they are often best recorded at the time a booking or a supplier payment is made. If management information is recorded separately then there will always be the problem of trying to tie it up with the original booking data that contain the core items of information concerning a transaction. So, we can begin to see the critical inter-relationship between the front-office system and the back-office system. The front-office system, although primarily concerned with reservations and client files must capture information in a format that is acceptable to the back-office system. Further, it must do this with the minimum amount of manual overhead. This is no easy task when one considers that the information that is needed by one business travel account is different from that required by another customer. Ideally, the front-office system should prompt the sales person for the required information needed for the customer being serviced. It should also generate as much information for the back-office as possible, as a by-product of the sales and ticketing process. All such information needs to be fed from the front-office system into the back-office system, ideally with no re-keying of the data, where it is stored on a data base.

Therefore, a back-office system is often the best vehicle to record MIS, even though this type of information does not always form an actual item of cost or revenue. The back-office data base is therefore usually used to record both financial and non-financial management information. To be effective this information needs to be stored so that it can be easily accessed and reported in a flexible way as needed by either the customer or the travel agency's manager. Let's take a closer look at the two main uses for MIS:

- **Corporate customers** MIS is really as much an integral part of a business travel service these days as an air ticket is. There are two reasons for this. The first is that it is a competitive benchmark by which customers compare different travel agents and their relative abilities to provide a first-class service. The second is that good MIS really is needed to negotiate favourable deals with travel suppliers. A few years ago when agency management systems started being used to process business travel, corporate customers were given mountains of computer printed reports containing MIS. No one read these reports because no one had the time to go through all the pages and extract the pertinent information that they needed. The next step was to provide customers with tailor-made reports many of which could be programmed to the customers' own special needs. Finally today, we are in the environment where the customer can be given a terminal connected directly into the agent's own agency management system. This can be used by the customer to access the travel agent's data base and obtain the information as and when it is needed in a format specified by the customer.

- **Supplier tracking** Supplier tracking, as the name implies, is a certain type of MIS that analyses the business done by the travel agent with supplier companies. The reason why this is important is because travel agents can often negotiate favourable deals if certain volumes of business are delivered to certain suppliers. For example, if your agency can book more than say 200 travellers into a certain hotel chain in any one year then that hotel chain may be willing to pay your agency a commission rate that is 1 per cent higher than the average. Of course, you could always just rely on the hotel chain to tell you what levels of business you had done with them. However, in the first place the hotel may not in fact know the answer to this question and in the second place you will be in a far stronger bargaining position if you have your facts at your fingertips as reported on a very professional looking agency management computer print-out. There are many other examples of MIS, which will pay for itself many times over by way of increased revenues from higher commission rates.

MARKETING

This is sometimes a misunderstood term, so it is worth explaining it in a little more detail. Marketing is the process by which a company distributes its products or services to its potential customer base. It does this by first of all identifying the types of customers that it will aim its products at. These may be high net worth individuals, i.e. rich people, people with average incomes who live in a certain part of the country, retired people or many other types. These customer types are given names by marketers, such as 'The Greys' who are people over about 50-years old, 'The Empty Nesters' who are people who either have no children or whose children have grown up and left home, and 'The Yuppies' who are young, upwardly mobile people. There are of course several other labels for different parts of the potential customer base.

Having decided the types of customers you are going after, the next decision is what kind of a product will appeal to them. So product design may be regarded as part of marketing. Then there is the kind of image that your business will want to give; this may, for instance, be a low cost image, a quality image or a young image. Then there are questions such as how best to reach your customer base; by advertising on television, radio, the local press, door drops or direct mail. It tends to be in this area, i.e. how to reach your customer, that agency management marketing systems have an important part to play. Because this is all very closely bound up with what types of customers have bought your products before, a good customer file is a cornerstone of most marketing campaigns:

- **Customer file** The customer file tends to be created and maintained by front-office sales staff. It represents a mine of extremely valuable information on people who have bought your products before. The agency management system should therefore have a customer file that records all relevant static information about the customer, e.g. name, address, post-code, sex, etc., as well as all the important dynamic information on the bookings actually made, e.g. the type of holiday last booked, when the booking was made, the resort used, etc. A good marketing system will have the ability to analyse the customer file and search for trends and common characteristics in previous purchasers.
- **Direct mail** One good way to reach customers is to send them a letter outlining the products and services you think will particularly appeal to them. A trawl through the customer file can, for example, select previous bookers who booked early last year and go for a particular destination, such as Majorca. The customer file for these people will provide their home addresses, and a good word processing or data base package can merge these into a pre-set standard letter. This can be personalized and mailed together with the appropriate holiday brochure, directly to the potential customer's home address, say two months from the month that the customer booked last season.
- **Telemarketing** This is a technique that has historically been used by large companies because the technology has been expensive and only justifiable for very large volumes. However, with the falling cost of technology and the widespread availability of computerized telephone systems, telemarketing is now within the reach of many smaller companies and travel agencies. Telemarketing is just like Direct Mail, as outlined above, except that instead of using the postal services to contact the customer, the telephone is used instead. The customer file or prospect file is trawled and a list of likely buyers is formed. This list is organized according to parameters like the telephone area code and presented to a telephone operator. Then sometime after 6.00 p.m. (in my case this is usually just when I am about to tuck into a nice dinner!), the potential customer is called and a predetermined telephone conversation is initiated by the travel agent caller. Telemarketing has been shown to be less costly than direct mail and more productive in terms of the number of sales made.

Well I hope that the preceding paragraphs have given you a brief introduction to the kinds of functions that are supported by agency management systems. I suppose that a lot of modern agency management systems cover a great deal more than just accounting. But for the average travel agent the accounting functions are surely the core ones upon which other more sophisticated applications can be built. Probably the best way to understand the relative mix of accounting and other functions in agency management systems is to evaluate some of the leading ones; that is specifically what I aim to do in the next section of this chapter.

Agency management systems

So, now it is time to be a little more practical and to consider the real world. The following sections contain an overview of some of the leading agency management systems on the market in 1997 (or at least in the advanced planning stages). This is by no means a comprehensive list and it certainly is not a list of recommended systems. It is simply a selected set of examples of fairly popular systems that are in widespread use or emerging new systems that could have an impact on the market in the not too distant future.

I have not stuck to any rigid way of describing these systems. Each one is evaluated from the perspective of its architecture, its functional strengths, the IT it uses and any unusual or innovative features that it portrays. I have not attempted to identify any weaknesses in these systems because my overriding concern is to be impartial and simply to present the products as examples of what are meant by agency management systems in general. Finally, there is no bias here: the sequence in which the systems is presented is entirely random.

I hope that by the time you have finished reading about all of these systems you will: (a) have gained a good understanding of the kind of systems that are available, and (b) identified the critical components that are important to automating a travel agency.

SABRE'S TRAVEL INFORMATION NETWORK (STIN)

STIN is an operating division of the Sabre Technology Group, which is owned by American Airlines. As a world leader in travel automation, STIN has a history of innovative deployment of IT for travel agents. Besides the Sabre CRS, which is described in Chapter 4, Sabre has for many years provided travel agents with the Agency Data System (ADS). This system is legendary in the USA, and even in the UK it is a well known and respected back-office system. Sabre's new technology agency management system called TravelBase was launched in 1992 and will over time, replace the ADS product. To start with then, let's consider the ADS system and then I will go on to cover the new high-tech TravelBase system in more detail.

ADS

ADS was originally developed by a relatively small travel agency in Tampa, Florida, in 1976. It was a highly successful system even at that time and had outgrown and overtaken the parent travel agency owned by John Annis, in which it had started. One or two major USA travel multiples were becoming interested in acquiring the ADS system for use throughout their national networks. In fact a great deal of attention was being paid to ADS at that time. United Airlines started the ball rolling in 1980 when it bought the software from ADS to use as the basis for its travel agency management system product. This was a substantial coup for ADS and most industry pundits thought that was the end of a happy travel industry fairy-tale.

However, the story did not end there; it was in actual fact only just starting. In 1981 American Airlines bought the ADS company, lock, stock and barrel! The original employees were retained to help with the product's support and continued development. American Airlines' strategy was similar in some ways to United Airlines – this was to use the ADS system as the basis for a travel agency management product. However, in the case of STIN, the system was to be linked into the Sabre reservations network. American Airlines spent a considerable amount of time, trouble and money on enhancing ADS.

One of the ways the system was enhanced was as the result of a joint development with American Express. This collaborative venture was aimed primarily at developing a multi-access reservations switch; but an important and integral sub-project was to work on a travel agency front-office system that would be of use to both companies. The end-result was a multi-access switch for American Airlines and an enhanced ADS system that became an industry-leading agency management system in the USA. For American Express the end products were: (a) an in-house multi-access airline switch, which was installed in Europe and is still in use today; and (b) a front-office systems blueprint, known as TRIPS, which was never totally implemented. American Airlines clearly benefited from the collaboration, and the fruits of the venture are still being enjoyed by it today.

The ADS system has therefore provided the travel industry with an excellent agency management systems product for over 15 years. ADS is used in around 5,500 agencies world-wide to provide accounting and management information systems. The main functions provided by the system are:

- **Interface** A key success factor in any travel agency management system is the ability to generate accounting entries automatically as a by-product of the airline ticketing process. The alternative is for the travel agent to key all the relevant ticket data into the agency management system, manually. This is not only laborious but gives rise to many errors caused by mis-keying. ADS has an in-built link to the Sabre reservations system. This means that whenever the agency uses Sabre to print an airline ticket, Sabre sends a specially formatted record into the ADS system where it is automatically posted to the appropriate account.
- **Sales ledger** ADS supports a sophisticated invoicing system and an excellent accounts receivable facility that has always been regarded as one of the product's main strengths. The UK version of the system generates the information needed to produce the BSP settlement report as a by-product of the invoice production process. The sub-system handles receipts and disbursements, invoice ageing, statement production and management information.
- **Purchase ledger** This contains a good accounts payable system, which includes disbursements, supplier statements, payable reports by supplier, payables queuing so that invoices are not paid too early, automatic posting of payments and cheque printing.
- **Nominal ledger** This is the general accounting facility, which is designed to support both large and small agencies. It allows the agent to set up a chart of accounts, record journal entries and maintain budgets along with comparisons to actual results. It produces a trial balance, a balance sheet and an income statement.
- **Hotel and car system** ADS tracks outstanding commissions due from suppliers and can issue invoices for amounts due to the travel agency. A flexible report generating program is also available to allow the agent to custom-design in-house reports.
- **The group system** This is a feature that is extremely helpful to those travel agents who design and operate their own group tours. It controls the booking of group resources and manages the allocation of travel services. It produces all necessary control reports including rooming lists, passenger manifests, vouchers and departure lists.
- **Management information** Besides a wide range of fairly standard management information reports, ADS also has a powerful report generator that allows travel agency staff to design their own custom-made reports. Each report may contain up to 90 different data elements from invoice, hotel, profit, payable and cash item data files. Finally, the ProfiTrak system allows an agency to track the amount of time spent on different tasks and different customers as well as commissions and revenues earned, in order to assess profitability.

The main ADS hardware platform is the Data General 32-bit series mini-computer, which is available in a range of models, each with a level of processing power depending upon the size and budget requirements of the travel agency. The system can, however, also be run on certain PC models and the Sabre LAN product that supports 'hot keying' between Sabre and ADS. The software, which is an integral part of the system, is kept up-to-date and is guaranteed at all times to support the current settlement requirements of IATA's BSP procedures. There is also a facility for ADS to download data from the system data files to an agent's own in-house PC for analysis by spreadsheet and word processing software. Besides full travel agency staff training, ADS has its own in-built on-line help facility and computer-assisted instruction, which includes several teaching modules.

By the late 1980s, however, Sabre started thinking about a future replacement for ADS. This was brought about mainly as the result of feedback from existing ADS customers. Although the system had been highly successful for many years in the USA, large customers were beginning to experience some constraints. This arose from the dual pressures of: (a) rapid business growth, and (b) the increasingly sophisticated PC technology that was becoming widespread in their businesses. It was against this backdrop that ADS began to look a little dated. In particular, ADS was perceived by travel agents to suffer from the following shortcomings:

- **Flat file structure** ADS uses a form of data storage that involves recording information in discreet chunks of data, which are stored sequentially on a disk or tape device. This was the standard way of storing data in the pre-1980s. Although it worked perfectly well, it was limited in terms of flexibility of access. It can be somewhat cumbersome to create a new report with a flat file and it is sometimes tricky to interrogate a flat file in a new and different way. A more effective way of managing data, known as relational data base technology, has recently become widespread. It solves many of the problems of flat files and is extremely flexible for both report generation and query handling.
- **Proprietary language** ADS uses a proprietary programming language, which is now infrequently used. So, it is more difficult to find programmers with a knowledge of this language and this has inhibited the development of the product. It also means that ADS customers find it extremely difficult to develop tailor-made applications of their own that feed off the ADS data base.
- **Proprietary operating system** ADS uses a proprietary operating system, which is bundled up with the hardware. It is not an 'open' operating system and the user base is not therefore considered to be an attractive market for which software suppliers can develop products. Consequently, the software products available for use in the ADS environment are somewhat limited.
- **Vendor dependent hardware** The ADS hardware is supplied by Data General and the ADS software will only run on Data General hardware. So it is not possible to obtain hardware from another manufacturer and still run the ADS software. Nowadays, it is possible for many items of software to run on a wide range of hardware from different manufacturers. Because many travel agents use different hardware platforms to Data General, this aspect of the system was seen as an unnecessary complication.

In summary then, ADS was perceived by some USA travel agency users to be a closed or proprietary system. A system that a travel agency could get locked into and from which it could not take advantage of the rapidly developing PC technologies and the enormous range of software packages available on the market. It was perhaps the move by most corporations and travel agencies into PCs that started to make the ADS system appear to look outdated. Coupled with this, the USA travel agency business was experiencing some fundamental shifts in its structure and culture. Fewer, yet larger, travel agency chains were beginning to emerge and these often spanned operations in different countries outside the USA. These larger agencies were experiencing restrictions imposed by the limited growth capabilities of ADS. Management information was becoming even more critical to travel agencies especially in the business travel field. In fact information was regarded by many as a strategic resource not just for travel agents but for the corporations they serviced. The older back-office systems that tended to split accounting and MIS functions looked increasingly outdated and did not meet the needs of travel agents and their customers for integrated management information.

Clearly Sabre needed seriously to consider developing the next generation of agency management systems. So in the late 1980s, Sabre began a market research exercise aimed at identifying the needs of travel agents for a new back-office systems product. Out of this preliminary research was born the TravelBase system.

TravelBase

The requirements for TravelBase were built on feedback from the travel industry and in particular those multi-vendor agencies that had global operations. Sabre commissioned its own independent market research into the emerging needs of travel agents for agency management systems. The findings were carefully analysed and a set of key business requirements began to be assembled. Also, the experience gained by Sabre of developing and supporting ADS over a period of several years was an important source of feedback from the trade. At the end of this requirements definition stage, the blueprint for Sabre's next generation of agency management systems was formulated:

- **Industry standard**   The system had to use hardware and software that were not only widely available in the computer market but that used established standards. One example of such standards were the 80 × 86 Intel processor chip and the IBM OS/2 operating system. Both were used widely within the PC world and were expected to evolve into common platforms in the future. The reason for the importance of standards was the need to provide travel agency customers with the widest possible choice of future options and to safeguard against leading customers down blind alleys of technologies that faded away in the future.
- **Relational data base**   As I mentioned above, a relational data base offers a high degree of flexibility in the management and usage of information. What customers needed was the capability to build a data base using known data elements and relationships and then to populate it with current information. They needed to do this safe in the knowledge that in the future they would be able to use the data base in a way that was not presently known. Furthermore, they wanted to be able to interrogate the data base with *ad hoc* queries in a quick and easy fashion without the need to write a complex computer program each time.
- **Flexible hardware configuration**   A modular and scalable approach to hardware choice was needed in order to cater for a wide variety of travel agency types and sizes. The system needed to operate equally well for a large independent agency as for a large global multiple. Scalable hardware also supports upgrade flexibility, which is vital for fast growing customers.
- **Networked system**   The system had to be able to take advantage of computing resources spread across: (a) a large single factory type of operation in one location, but with many workstations and computer peripheral devices; and (b) a geographically distributed travel agency with a heavy reliance upon telecommunications.
- **Customization features**   The system needed to be able to use a wide variety of other suppliers' software products. This 'openness' allows each travel agency user to add a specific set of software products to their system that they consider to be of particular relevance to their business. This can help to create a unique system that enables the agency to differentiate itself from its competitors.
- **Connectivity**   With a growing number of information sources, the ability for any new system to connect into other computers and networks was an important requirement. A travel agency might, for instance, have its own main-frame that needed to be integrated with the agency management system. In terms of external connectivity, if the new agency management system was to be truly flexible, then it needed to support interfaces to other CRSs besides Sabre.

From these industry specified business requirements sprang the new TravelBase product. It was launched at the STIN 'Networking '92' three-day event in the USA. This event was sponsored by Sabre, Marriott, Forte, Data General, Texas Instruments, Avis and American Airlines. Beta testing of the new product started early in 1993, i.e. preliminary testing in specific live user locations in a controlled environment. TravelBase is a true agency management system and spans the front-, middle- and back-office operations of a travel agency. It uses the latest PC-based technology, is modular in its construction and is based upon open or non-proprietary systems. TravelBase is a strategic Sabre product that will be around for many years to come. Although it will over time replace ADS, ongoing support will be provided for ADS users until the end of the decade.

The architecture of TravelBase uses client/server and distributed computing technologies that are accessed via a GUI, over a LAN. I have already covered these technical terms elsewhere in the book, but let me expand upon them in the context of TravelBase. The architecture may be summarised as follows:

- **Client workstation** Travel agency users access the TravelBase system via workstations that use a GUI. This GUI is provided by the IBM OS/2 operating system, which has a product known as Presentation Manager. Presentation Manager is like the IBM version of Windows and uses similar features, such as a mouse with simple point and click features, a desk-top with various icons and cut and paste facilities. The workstation is based on an 8 Mb 386 or 486 (SX or DX) PC running OS/2. (The RAM requirement is for a minimum of 8 Mb: if using third-party systems the RAM requirement could be anything from 12 to 20 Mb or more.)
- **Server** TravelBase is accessed and used via PC workstations that are connected into a LAN that itself has several types of servers: a file server, a data base server, an optional print server and an optional communications server. Each of these servers is explained in more detail as follows:
    - *File server* The TravelBase file server controls the flow of data and messages around the network and stores all the programs that make up the TravelBase system. It controls the dialogue between the workstations and the various programs that run on either the file server itself, the workstation or one of the other servers.

        The file server is based on a 16–20 Mb 386 or 486DX PC running OS/2. The software used to control the LAN is Novell Netware Version 3.11 (Version 4.0 in the future) and the operating system used to control the execution of the various programs is IBM's OS/2. The network cabling and message transmission protocols are based on Token Ring and TCP/IP (see under LAN below).
    - *Data base server* TravelBase uses a large computer with enormous storage and processing capabilities to hold all information on the travel agency business. The data base server is expected to be a computer that uses an open systems architecture. In other words it allows the user to choose from a variety of hardware suppliers for the central data base processing engine used to power the travel agency.

        The hardware is based on a classical or truly open system that uses reduced instruction set computer (RISC) processor chips and a UNIX operating system. One of the reasons for choosing RISC hardware is that it is declining rapidly in price. The data base software is based on a widely used product called SQL Server marketed by a company called Sybase. This is a relational data base that uses open systems standards and that supports SQL queries and flexible report generation.
    - *Print server* This is an optional server, but one which will probably be of significant benefit to larger users. As the name implies, it controls the printing of all reports generated by the TravelBase applications. Many types of printer may be connected to this server, including matrix printers and laser printers. These printers can be used to produce paper output for word processing, operational reports, management information reports, electronic mail messages and many other types of reports.

        The print server is based on an 8 Mb (minimum) 386 or 486SX PC running OS/2, which usually will have an optional serial port expansion card to allow several printers to be connected.
    - *Communications server* Again this is an optional server that controls all the external communication interfaces. Examples of such external interfaces are CRS connections including Sabre, of course, electronic mail networks and in cases where the head office of the travel agency is at a remote location, a connection to the remote file server using a local Token Ring extender.

        The communications server is based on an 8 Mbyte (minimum) 386 or 486 (SX or DX) PC running OS/2 with optional special purpose communication cards.
- **LAN** The TravelBase LAN uses Token Ring wiring and PC printed circuit cards to interconnect all workstations and servers. As mentioned above, the file server controls the passing of messages and data around the network and it uses Novell Netware to do this. Although Novell Netware is a proprietary product, it is the most widely used LAN operating system

in the world and has virtually become a *de facto* standard. The communications protocol used on the network is TCP/IP for data base access. Novell's IPX/SPX is used for file services and printing. (TCP/IP is a kind of cut down OSI protocol – and again is an open or non-proprietary standard).

All in all, this architecture supports a sophisticated level of travel agency automation. It allows a larger multiple travel agent to use a powerful central processor to hold all the data for the group and yet still enjoy the benefits of GUI workstations for access to data and applications. In other words TravelBase merges the dual strengths of PC and main-frame technologies into an integrated MIS. This allows the user to enjoy the benefits of: (a) the ease of use and widespread availability of PC technology, and (b) the power of a large central main-frame capable of storing large amounts of management information. It can therefore be seen that TravelBase is initially aimed very much at the larger travel agency user and, in particular, a travel agent that has a global operation. Although the smaller independent could still use the system, the sophisticated hardware and software technologies used by TravelBase are currently too costly for lower business volumes. As hardware costs drop and more end-user 'ease of use' features are added, the cost equation will change and TravelBase will begin to appeal to a wider user market.

Users access TravelBase via the OS/2 operating system. OS/2 is IBM's operating system for PCs. It is a multi-tasking operating system, which means that more than one program can execute, i.e. can run, simultaneously. So, for example, one workstation user can update a master file while running a data base query at exactly the same time. The operating system takes care of the complexities of multiple programs sharing processes in this kind of environment (the data file linking is handled by the data base server, not by the workstation). The functions of TravelBase are available via the OS/2 desk-top. This is a graphical screen controlled by OS/2's Presentation Manager GUI. The main icons on the desk-top will change as new releases of OS/2 are implemented, e.g. Version 2.1. They may also change under the control of the user who can configure them to their own liking. However, in order to give you some idea of what the main desk-top groups are, the following is the current state of play under OS/2 Version 1.3:

- **TravelBase applications** These are the processing functions that make up the TravelBase product and are explained in more detail below.
- **Client services** This is a set of functions provided by Sabre, including applications like automated credit card reconciliations and the provision of clients' statements on magnetic tape.
- **Administration** A set of TravelBase functions that allows the user to control things like systems security features, e.g. password maintenance, network administration, usage statistics as well as many other functions.
- **OS/2 applications** Provides access to other OS/2 control applications. This group of functions contains any applications the user wishes to install.
- **Main** This is the 'main menu' of OS/2, which allows the user to select the set of applications loaded onto the system. Applications such as electronic mail, word processing and spreadsheets. The choice of applications depends upon those that the agency has decided to load onto the system.
- **Utilities** A set of general purpose functions provided by OS/2 that are often needed to perform regular housekeeping tasks. The user can add their own set of utilities to this group.

Any one of these functions can be selected by a simple point and click operation using the mouse. Because of OS/2's multi-tasking capability, an application can be started and then minimized while it continues to run. So, for example, you could open the reporting window, select a particularly complex report to be run, start the program running and then select the minimize function. This has the effect of reducing the full screen containing the report application to a single icon no bigger than a quarter of an inch square, which is displayed at the bottom of the screen. You are then free to open another window and start another process off. Once again this process, when running, can be minimized and yet another application started. The power of multi-tasking opens up whole new dimensions of user productivity that conventional systems cannot provide.

The open features of the TravelBase architecture also provide some key benefits to users, especially when one considers the opportunities to use industry standard software to process TravelBase data. It is quite possible, for example, to use the Microsoft Excel spreadsheet software package to run against the Sybase data base without any special technical conversion programs being necessary. This enables the power of the system and the size and complexity of the TravelBase data base to be readily and widely available to a wide cross-section of the travel agency work force This means that usage of the system and the production of special reports and analyses are no longer solely within the domain of the IT department. This is especially important as new software tools come onto the market and new ways of using the information resource thereby become available.

The data maintenance functions of TravelBase are very flexible and easy to use because they make full use of OS/2's GUI and processing support features. The use of 'wild card' characters, for example, speeds up the search and replace tasks so often needed during regular data maintenance. Instead of having to specify individual fields, e.g. file names, the user can enter the common parts of a name field and use a wild card character to mean anything at all in that position. So, specifying the extraction of XY* using the wild card '*', i.e. an asterix, would in fact extract XY1, XY2, XYA, XYz, etc. This can be especially useful in selecting items from: (a) a pull-down menu, (b) a list of files, or (c) a list of items displayed on a screen. Also, when needing to update a range of panels of data from a pull-down list, the user simply has to highlight all of the entries to be updated. The system will then present each item in turn, one at a time, for updating and only go back to the main menu screen when all the highlighted items have been updated, e.g. the first screen will show Page 1 of 3 and this will be incremented as each is updated.

Another feature of data maintenance is the way in which TravelBase treats codes. There is a certain set of codes that is pre-set within TravelBase. These are the industry standard codes, which form the common language of travel agents and travel suppliers throughout the world. This set of standard codes is maintained by TravelBase and is regularly reviewed and updated as necessary. Other codes can be tailored by the end user to denote special items of data that are relevant to the user's own business environment.

A great deal of this functionality is provided by the OS/2 operating system. However, I do not propose to go into the details of IBM's OS/2 operating system here. This is a massive subject in itself, so I shall concentrate solely upon the 'TravelBase Applications' icon that appears on the OS/2 desk-top, as mentioned above. The set of functions behind this icon are:

- Sales Entry.
- Commission tracking.
- Cash management.
- Accounts payable.
- Accounts receivable.
- Statements.
- General ledger.
- The USA Airline Reporting Corporation (ARC)/BSP air settlements.
- Pre-programmed management information reports.
- TravelBase query and report writer.

The functions of TravelBase

I have covered most of these application areas before in other parts of the book and some of them in the earlier part of this chapter. So, what I propose to do here is to dive into a little more detail in some specific areas where TravelBase offers some unique features:

- **Sales entry (airlines)** TravelBase supports CRS interface records as part of the sales entry process. At present only Sabre and Worldspan interface records are supported but others, possibly Galileo, will be added in the future. In the case of Sabre, interface records may be consolidated at the headquarters level or fed into TravelBase at the local level.
  - *Headquarters level* Each time a ticket is printed, Sabre generates an interface record within the Sabre host main-frame computer. This may be either accumulated and downloaded to the travel agency head office once a day as a batch file, or it may be fed into

the head office computer as the ticket is printed. It is up to the travel agent to decide which option is to be used for the agency chain when the system is first set up.
  - *Local level* The interface record is received by the TravelBase data base in the location designated to receive accounting records. This may be within the local branch or at headquarters.
- **Accounts payable and accounts receivable** The choice of where these functions are performed lies with the TravelBase user. When the system is first set up by the travel agent, the distribution of processing functions around the network is decided upon. It may be, for example, that accounts payable will be a centralized head office function whereas accounts receivable is to be controlled within each travel agency office with overall control of MIS centralized in head office. This kind of flexibility, which enables a travel agent to decide where processing is to take place, is supported by TravelBase and enacted by specifying certain key parameters that control the overall use of the system.
- **ARC/BSP Air Settlements** Because TravelBase is designed for a global travel agency chain, it supports the BSP method used by IATA for the clearing and settlement of airline ticket sales (the American version of BSP is known as ARC and is very similar in operation to BSP, which is explained earlier in this chapter). Because the new UK BSP will take settlement information directly from CRSs, the functionality provided by TravelBase includes a reconciliation feature. This allows the agency to receive a reconciliation file from BSP (or alternatively data from the BSP billing analysis may be keyed manually into TravelBase). A TravelBase program then matches the BSP information to the data stored internally by TravelBase in order to flag any differences.
- **Global Customization** The master file structure of TravelBase allows the system to be tailored to meet the business environment of the country in which it is being used. Take VAT, for example. Although the basic concept of VAT is pretty much the same in many countries, the items to which the various rates apply, are often quite different. TravelBase allows users to specify to which products and financial transactions VAT is applied and whether it is included in the base price or is to be added as a separate amount.

  Another example of global customization is the way TravelBase stores foreign currency exchange rates. The system allows several different types of rate to be stored for each currency pair and for each rate there is a wide variety of descriptive fields to define how the rate is used. It is also possible for TravelBase to be configured so that it produces reports and analyses in different base currencies. This enables the system to be used by a travel agent that operates in more than one country. One can see evidence here of involvement in the requirements specification stage from at least one major global multiple travel agency chain.
- **Pre-programmed management information reports** TravelBase includes several standard reports which should satisfy the majority of most customers' and travel agents' needs for management information. These reports come with the basic system and can be called up for execution and printing at any time. They form the basis of a powerful library of business analysis tools, which are an integral part of TravelBase and which can be processed simultaneously thanks to the combination of OS/2's multi-tasking capabilities and the power of the data base server computer. The actual program that runs these reports is stored on the file server and executes on OS/2 workstations by querying the data base server.

  The reports produced by TravelBase can be distributed to customers using a variety of methods. One such method, for example, is to embed the reports in an electronic mail message. The resultant message can then be transmitted to the customer either directly or via a third-party network to which the TravelBase communications server is connected. The report could also be embedded in a document that was produced by a word processing package. All of these features help increase the image of the TravelBase travel agent as being a sophisticated and professional provider of travel services.

- **TravelBase query and report writer** This is probably one of the most powerful features of the system. The query and report writer capability is where TravelBase's open relational data base server architecture comes into play. TravelBase uses a powerful data base query and reporting tool, which is provided by a company called Knowledgeware (the product was formerly developed and marketed by a French company called Matesys). This company has produced a software package that allows users of Sybase relational data bases to make maximum possible use of the data they have stored. It does this in a way that hides much of the complexity of the technology from the end users who are, after all, primarily travel agents and not technologists. So, the software employs graphical methods to help the users define the kinds of special reports they require. Let's look at this in a little more detail.

The process starts with the system displaying all of the tables of data that are relevant to the query or special report required. Each table can be viewed and the fields involved in the query are selected as appropriate. The tool will actually draw lines between the boxes, which represent tables, itself because it understands the relationships between tables. The user then simply draws lines between the tables to signify the relationship between the data that will satisfy the query. Having done this the system will automatically construct the SQL command that is needed to select the data elements from the Sybase data base. To write the SQL command in the way usually carried out by a programmer would require substantial technical skill, whereas to do it via the Knowledgeware software is far easier. After finalizing the request, the query program thus constructed can be set to run while the user minimizes the application window and continues with another job.

All the time the query job is running the icon appears as an image of a little man at the foot of the screen, running on the spot. As soon as he 'breaks the winning tape', the query is finished. If the icon is then maximized the results of the query become visible on the full-size screen display. The results of the query will be displayed in a formatted way with basic column and row headings. The format can then be enhanced by the use of some clever graphical tools that enable a bit mapped graphical image to be embedded in the report. It is possible, for example, to add a logo to the report. This might be the logo of the travel agent alongside the logo of the corporate customer for whom the report is destined. Such logos could either be built up on the screen as needed or pulled in from a library of 'art-work' images. The resulting report is a highly polished product produced using only a minimum of human effort. It is important to mention, however, that this kind of sophisticated graphical report will need a good laser printer to reproduce the quality of the image accurately as it appears on the screen.

An important feature of the TravelBase report generator is a facility known as Direct Data Exchange (DDE). This is a technical feature, but one which it is important to understand in conceptual terms. Put simply, DDE provides the capability for the report that you generate to be automatically updated whenever other parts of the query change. In the olden days of the late 1980s (!!) whenever a query was selected from a data base and displayed on the screen, it was fine for a short period of time. But then as the query was updated, the original query fast became outdated and had to be run again using the updated file. With DDE this is no longer the case. The report as it appears on the system always reflects the latest state of the information within the query currently under construction. I trust you will appreciate the power of this feature. It means that you are always looking at and dealing with live data as it actually happens.

TravelBase support

I have talked a lot about the functions of the system and I trust you will agree that TravelBase is a sophisticated product. As such, an important consideration of running a system as complex as this is the training and support that is provided by the system's supplier. Sabre has made the financing and support options almost as flexible and as modular as the system itself. There is no hard and fast rule that controls the way in which TravelBase is maintained and supported. The level of support is customized to meet the requirements of the user. After all, some users are very sophisticated indeed and run their own internal help desk and IT department. Others have no IT infrastructure at all and rely solely on their suppliers to provide them with total support. So, the levels of support offered by Sabre are:

- **Hardware purchase or lease** The system may be purchased outright or it may be leased from Sabre. This will involve the travel agent making regular periodic payments to Sabre for a specified period of years.
- **Hardware maintenance** The travel agency may elect to make its own arrangements regarding the maintenance of the hardware that comprises the TravelBase system. Some travel agencies, for example, may have a third party maintenance deal that covers all the equipment installed in the agencies office networks and to which the TravelBase hardware could easily be added with very little incremental cost. Most have no such deal and need a full maintenance service to be provided by Sabre and its maintenance provider.
- **Software licence** The TravelBase software is owned by Sabre, and users must therefore pay a licence fee for usage of the system. This helps Sabre protect itself from unauthorized copying of the product and provides a revenue stream that helps recover some of the development and support costs.
- **Unbundled software** TravelBase is a product that comprises hardware and software. However, the system is provided in unbundled form, which means that you can in fact have one without the other. In other words you can have the software without the hardware (I suppose you could also have the hardware without the software, but this is unlikely to be a realistic scenario). Because the software is modular, it can be provided to travel agents on an 'as needed' basis. Some agents will require the whole system, while others will want selective parts of it. Naturally, there is a core element of TravelBase that all users must have, but some of the other modules are optional.

TravelBase can be tailored to suit a travel agent's own specific needs. It can also be used to produce tailor-made reports for customers as described above. Its flexibility is primarily achieved through some fairly complex parameter-driven options. So, the larger multiples who decide to use TravelBase will no doubt draw heavily upon the technical skills of their IT staff. However, the average travel agent does not have an in-house IT department and is going to be stretched to become sufficiently expert in the system to derive the maximum benefits possible. Sabre can of course provide some help here but STIN has recognized that it is not going to be able to respond to all travel agency requests for tailoring support and special projects for TravelBase, within a time frame acceptable for all users. So, Sabre foresees an opportunity for the use of specialist consultants in the field of TravelBase. These consultants would be experts in the use of TravelBase and could provide travel agents with the specialist technical support they need. Support that could include, for example, the development of special macros for Excel and Lotus, which use TravelBase data. It may even be possible that Sabre will over time build up a list of preferred consultants who have established a track record of competency in the TravelBase field.

It should be obvious by now that TravelBase is an extremely sophisticated product. If a travel agency is to get the most out of it, it will need to provide its staff with adequate training in how to use the system. Sabre is planning to offer three basic levels of training in TravelBase: Basic, Intermediate and Advanced. Each travel agency will need to determine its own training plan for its staff. I would offer the following guidelines for your consideration. Probably all staff will require the Basic training module, whereas only the business travel sales staff will attend the Intermediate course. The Advanced course should, however, be attended by at least two staff in the agency. These two staff would each be the kind of 'super user' who could represent a source of expertise in the office and could assist others in using the system. Those staff who have used a GUI such as Windows or OS/2 Presentation Manager before, will of course find the transition to TravelBase that much simpler. Sabre will make the TravelBase training courses available in a local training and support centre closest to the travel agent. In the UK this centre is located in Hounslow.

The TravelBase product has not even been fully launched (at time of writing in mid-1993) and yet Sabre is already considering some future developments. In the future, the Sabre reservations CRS access will probably be built into TravelBase. This will allow users to switch from one application to the other even more easily than with the Windows capabilities of Sabre. It will also mean that there will be a closer integration between other travel products and services booked via Sabre and recorded in TravelBase.

But, not all of the enhancements to TravelBase will be made by Sabre. Because TravelBase is an open system, it will be able to take advantage of the developments in the field of data base technology and related industry software packages. There is already a great deal of debate in this area, which is focusing mainly on the technical standards that control the connection of front-end tools such as word-processors and spreadsheets to back-end data bases such as Sybase and Oracle. The umbrella name for these are their open data base connectivity (ODBC) standards. The importance of these standards and the impact on end-user systems is evidenced by Microsoft's interest in the subject. Microsoft and other leading software suppliers are at present trying to agree the precise details of what these standards should be. So, assuming that agreement is reached on ODBC standards, a whole new set of software packages will become available for TravelBase users. These packages will be able to use the Sybase data base without any special conversion or pre-processing programs. Sabre is therefore monitoring these developments closely and is trying to balance today's needs of its customers against the future directions of the computer industry.

ICANOS

Icanos is the name for the IT solutions business that is a part of Travel Automation Services, a wholly owned British Airways subsidiary. Icanos' sister organization, Galileo UK, has successfully marketed the PAMS back-office system to travel agents for over 14 years. Although this product has served the travel agency population faithfully and well all these years, technology has moved on since PAMS first appeared and a new generation of agency management systems is now needed. A new product called TravelEdge with a broader-based appeal has therefore been added to the Icanos product portfolio and is due to be launched early in 1998. In this section I will first cover the existing PAMS product, then briefly describe what kind of product TravelEdge will be.

PAMS

PAMS is an Icanos PC product aimed principally at the business travel, back-office sector of the agency automation market. It has been a very successful product that has served its customers well for around 14 years. PAMS had its origins as the replacement for the original DPAS upon which it was modelled. It is therefore principally a business travel document printing and accounting system, which is currently used in around 350 UK travel agency offices, concerned principally with business travel. The PAMS systems are all PC-based and use the Olivetti range of PCs along with several different types of printer depending upon the users' precise requirements.

The PAMS product comprises two PCs: one is used for ticketing and the other is used for accounts processing. The PAMS PC ticketing system is one of the basic building blocks in the range. It comprises a PC and special software that is interconnected to other Galileo reservations PCs in the travel agency. With PAMS PC ticketing, any PC in the agency that is connected to the ticketing PC can initiate the printing of an airline ticket, an itinerary for a client and an invoice. When not being used to drive ticket printing, the PAMS PC ticketing system may also be used as an ordinary Galileo reservations terminal. It is the OPTAT airline ticket format that is used by PAMS. The second PC is dedicated to the processing of accounts and is connected to the PAMS PC ticketing system. The PAMS system comes in three models, each with a different level of functionality:

- **PR2000** This system has all the functions of the PAMS PC ticketing system, but in addition has some important sales reporting features. This system automatically prints BSP sales returns and sales reports. Additionally, the PR2000 PC produces basic client statistics, which are aimed very much at the business house sector of the travel agency market.

- **PR3000** This PC model builds upon the PR2000 and besides having all the features of that system it also has the basic accounting functions required by most agencies. It includes sales ledger processing, aged debtor analysis, credit control and customer statements. An agent using the PR2000 can easily upgrade to the PR3000 without changing hardware.
- **PR4000** This is Icanos UK's full travel agency back-office accounting system, which affords a high degree of financial control based on a powerful PC. Besides having all the functions of the PR3000, it also includes a full general or nominal ledger, purchase ledger, commission accounting, VAT accounting, banking, trial balance, budgeting, profit and loss accounts, balance sheets, customized nominal reports and automated multi-branching.

Finally, the PAMS systems have an optional report writer facility available, which represents a good management information system. This uses data that are automatically fed from PAMS systems and provides travel agency users with an easy to use report generation facility. The package includes a spreadsheet, a data manager, a tutorial, a time manager and a word processing capability. The package also has the ability to consolidate information from several branches into a central site for integrated reporting. Statistical analyses may be presented pictorially in graphs, pie charts and three-dimensional pictures.

PAMS and the future

PAMS is beginning to look increasingly outdated in the light of: (a) the rapidly changing travel agency market for back-office systems, and (b) the new GDS technologies. The original system was created to provide a back-office product aimed primarily at business travel agents. It therefore was not designed to meet the needs of the evolving mixed business/leisure travel agencies that are now becoming more numerous. Besides air ticket sales, business travel agents need to automate other related products, such as hotel accommodation and car rental. Most GDSs now provide a good level of access to these products. PAMS was not designed for the leisure/retail business, with its tailor-made holiday packages and the fundamentally different way in which the booking process is handled. Nevertheless, the leisure business is beginning to generate more scheduled airline ticket sales and is therefore seen as an important source of business by most carriers. Hence, from Icanos' viewpoint, it is important that an agency automation product is available for those business travel agents who also handle leisure/retail bookings.

There are other problems too. PAMS' screen formats, for example, are now looking a little archaic and the user/system dialogue is not considered particularly user-friendly in today's windows environments. This is especially true in comparison with its high tech competitors on today's travel agency automation market, which in many cases use windows-based GUIs to great effect. From a technical viewpoint the PAMS product is therefore showing its age. The PAMS operating system is a business operating system (BOS), which although ahead of its time during the 1980s, is now looking increasingly old-fashioned, non-standard and limited in terms of functionality in comparison to the latest windows-based PC operating systems. The original system was written using the Micro-COBOL programming language. Technical staff with skills in this area are now difficult to come by. Hence the cost of maintenance is high and the rate at which system enhancements can be brought to market is rather slow. For all of these reasons, Icanos decided that a new agency management system should be added to its portfolio of products.

TravelEdge

Icanos has developed the TravelEdge product for a wide cross-section of travel agents. It is a tried and proven travel agency software package that originates in the USA under the name of TRAVCOM. Over the period 1996 to 1997, Icanos contracted TRAVCOM to tailor the base product for use in the UK. It therefore now supports Sterling as a base currency, the full range of UK VAT codes, BSP and the vagaries of UK rail travel. TravelEdge is scheduled for its maiden launch in the UK during the early part of 1998 and over time, I would expect existing PAMS' users to migrate to this new agency management system. At the time of writing therefore, i.e. mid-1997, there is little detailed information that is publicly available from Icanos on the TravelEdge product, and so the following is very much a general overview.

The TravelEdge system is flexible and can be configured in a variety of ways depending upon the type of travel agency user. Its primary use is as a back-office accounts processor. As such, it is usually connected into the head office LAN of a multi-branch travel agency where it is fed automatically with all booking records generated by branches using the agent's chosen GDS (e.g. Galileo UK's GMIR). A multi-branch agency may therefore continue to use its Galileo point-of-sale PCs to make bookings and record transactions. The GCS takes these transactions and channels them to the agent's head office site where TravelEdge is installed. The agent's head office therefore receives transactions from all its outlying branches, which may then be processed by the TravelEdge agency management system. This enables both GMIRs, NVP MIRs and RESCON messages to be received by TravelEdge in a consistent format. All applications that comprise TravelEdge run within Microsoft Windows Version 3.11 and Windows 95.

At the centre, the underlying drive for TravelEdge's agency management system is the input of a Galileo booking record from a branch office, which can be used by the various accounting sub-systems to process ticket payment, settlement and billing functions as well as supporting special customer documentation. In TravelEdge, ticketing and itinerary printing functions are driven by the travel agent's chosen GDS. This is normally the GCS as delivered by Focalpoint or ET3000. Each of these systems can produce an MIR. Besides Galileo's MIR, TravelEdge can also support Eurostar's MIR, which is delivered via Galileo UK's NVP. Other suppliers' MIRs may well be planned for the future.

It is the entry of an automatically transmitted MIR into TravelEdge that acts as the primary trigger for several accounting functions. These functions include the sales, purchase and nominal ledgers, all of which are therefore updated in real-time as MIRs are received by TravelEdge. The receipt of MIRs is, however, carefully controlled within TravelEdge and it is only following their review and release by the agency's financial controller that the transactions enter the accounting system. TravelEdge also features an integrated MIS sub-system that enables a full suite of pre-formatted management reports to be produced as well as enabling *ad hoc* reports to be developed. The Crystal Reports software package can be integrated with TravelEdge to provide users with a flexible interrogation, query and reporting tool.

VOYAGER

Voyager is a popular agency management system that is now in its fourteenth successful year of operation. It is a system that was originally developed by the Voyager travel agency situated in the north of England, originally for itself and its own business. This is rather an interesting story and so I am going to spend a little time on explaining the background to the Voyager product and how the company was formed to sell and support it. Voyager itself started out as a small independent multiple travel agency based just outside Sheffield with four travel shops. Back in the early 1980s they were considering automation and started looking at various systems options available on the market.

By the time that the ABTA conference in Phoenix was held (back in 1982), Voyager was coming to the conclusion that it was unique because there was nothing on the market which met its automation requirements. There were two leading industry systems on display at the ABTA Phoenix conference and the Voyager travel agency staff took a careful look at them both. The choice at that time was severely limited, although there was a great deal of interest in ABTA's proposed Modulas system. Modulas was going to be an ABTA-backed travel agency system and it was being developed for ABTA by a Canadian travel company called Caltrav who had a successful track record of systems development in its own country. In the end, Modulas became too expensive and controversial. Part of the controversy was generated by the impossible task of agreeing upon a set of requirements that met the needs of every travel agent, large or small.

So, when Voyager surveyed the market in 1982, it concluded that unfortunately neither of the systems on display at Phoenix, nor for that matter any of the other systems on the market at the time, met its business requirements. This reinforced Voyager's view that its requirements for automation must somehow be unique and that a packaged solution was not going to be a viable option. The conclusion it therefore reluctantly came to was that it would develop its own agency management system using the relatively new concept of the PC.

Voyager started the development process by contracting a software house based in Sheffield to tailor-make a system for it. Unfortunately the software house went 'belly up' soon after commencing the project and Voyager was really no further forward in its quest for an agency management system. So, it took the parts of the system that had been written and that seemed to work satisfactorily, and went to work on the remaining parts of the system itself. Voyager concluded that it was impractical to take an accounting software package and turn it into a travel agency system. It was of the opinion that instead of trying to fit the needs of a travel agency into an accounting package, a better approach was to concentrate on the automated control of clients' moneys generated from point-of-sale processing. Consequently a completely new system was developed that was based on two key concepts. The first was that all customer funds were to be balanced on a weekly basis and the second was that all posting was to be done as a by-product of the point-of-sale booking process.

Finally, after a lot of hard work and the solving of many technical problems, a working product was produced. This was a single user system based on the Apricot Sirius PC. Soon after the system was installed, the local Sheffield Co-op expressed an interest and after a demonstration of the system, decided to install it throughout its branches. Voyager reasoned that if the system met the Co-op's requirements so successfully, then there was a reasonable chance that it would also meet other agents' needs for a fully automated system. So a direct mail shot of 35 agencies located across the north of England area was undertaken. This proved to be highly successful. Voyager received five responses of which four agencies actually purchased the product. Encouraged by this early success, Voyager decided to test the waters outside of the north of England area and held an exhibition of its systems products at the Royal Gloucester hotel in London. Within 15 months, Voyager had over 100 travel agency users of its system.

It was around this time that viewdata was becoming the accepted way to book holidays for leisure travel. Competitive products appeared on the scene that incorporated a viewdata link so that a single terminal could be used both as an agency management system and also as a viewdata terminal. So, Voyager management started some market research and product development. The result was an enhanced product and a fresh approach to the market. With the benefit of hindsight, there were three main factors that contributed to the subsequent success of the re-born Voyager product which resulted from this exercise:

1. **The Orbital rental scheme** A rental scheme was introduced, which was branded Orbital. In fact the scheme was so popular that all Voyager's customers are now fully on a rental basis. This scheme enables Voyager to derive an even cash flow yet does not involve any external finance company or interest charges to its customers. Among other things, this has enabled Voyager to enhance the system continually and to provide new versions of software free of charge to all Orbital customers. Additionally, new replacement hardware is provided to all Orbital customers every five years at no extra charge.

2. **Bundled support** Voyager took over all the support functions of the product including, for example, hardware maintenance, software maintenance, ongoing customer support and system enhancement. This enabled Voyager to provide an all encompassing service to its customers without the need to be dependent upon third parties. What's more, this bundled approach eliminated the old 'finger pointing' that I have talked about before. Voyager was clearly responsible for fault fixing, no matter what the nature of the fault was; customers liked this approach because they felt more secure with just one company to call whenever things went wrong.

3. **Apricot network support** The introduction of the Apricot LanStation product enabled Voyager to develop into a cost-effective multi-user system. At the same time, a facility was developed to enable the workstation PCs to support viewdata. So, from a travel agent's viewpoint, it was only marginally more expensive to use PCs in place of viewdata terminals. This was especially true when compared with the cost of a separate PC for agency management purposes and several separate terminals that could only be used for viewdata and nothing else.

Voyager grew in popularity and success. So much so in fact that Voyager Systems (Technical Division) Limited now supports the core business with Voyager Systems (Travel Division) Limited reduced to just a single travel agency outlet (Voyager Systems Limited is the overall holding group company). The single travel agency business owned by Voyager is as much a proving ground for the Voyager system as it is a business in its own right. This is an extremely effective way for Voyager Systems to keep in touch with the business and try out new innovations for the product. On the basis that small is beautiful, Voyager is a company of four directors and 20 staff. The company is at the forefront of travel agency automation and sponsors ABTECH (see Chapter 1). Voyager Systems has a number of travel-related products including: (i) a retail agent's system, called Voyager 2000; (ii) a full business travel system, called The Corporate Traveller, (iii) a system that links travel agency branches to head office, called Median; and (iv) a foreign exchange system, called Travelbank.

During 1995 Voyager formed a joint venture company based in Singapore that, over the period 1996 to 1997, installed systems in Singapore, Malaysia, Borneo and Thailand. Customers include travel agencies owned by Japan Airlines and Ken-Air Travel, the biggest foreign independent tour (FIT) operator in Singapore. The Voyager sites range in size from 20 to 70 or more workstations. Most sites are integrated with Abacus, which is the dominant GDS in many Far Eastern countries. The product that Voyager uses for this purpose is called Abacus Whizz.

All 1997 Voyager products have been developed as 32 bit applications for running under Microsoft Windows operating systems, such as NT4 Server, NT4 Workstation and Windows 95. The applications are written in Visual Basic 5 and Microsoft C++ Version 5. Voyager supports links to the major GDSs, including Galileo, Sabre, Worldspan, Amadeus and Abacus. In the UK, Voyager also supports Viewdata technology.

Voyager's agency management system

Voyager is a complete agency management system for a travel agent with a mix of business and leisure travel. For agencies who specialize in the business travel market, the Corporate Traveller product is especially suitable. Voyager has also been successful in the tele-sales area where each location uses between 30 and 100 workstations. This is somewhat of a specialist niche within the travel agency sector. A great deal of the success of Voyager is probably due to the fact that it is very much a point-of-sale system. The application is designed so that it compliments the natural tasks that a travel agent undertakes in dealing with many different types of clients.

Most of Voyager's business oriented products use a core reservations function. This has interfaces into all the major GDSs used throughout the world's travel industry. In the UK, the reservations function also supports a videotext link to the major third-party networks including AT&T, Leisurelink and Imminus. This means that a Voyager PC can be used just like a viewdata terminal and can access the holiday and charter seat reservation systems. Internet access is provided and all Voyager applications have been fully integrated with both Microsoft Exchange and Microsoft Office. The Windows environment in which Voyager operates enables several different windows to be active at any one time including, for example, a Voyager application module, a GDS screen, a viewdata screen and an Office document.

Unlike some systems, Voyager does not have its roots in an accounting package that has been adapted and modified for travel agency use. Although many of Voyager's competitors took this approach, Voyager considers that an accounting system software package in itself does not form a good base upon which to build an all encompassing travel agency management system. It is for this basic reason that Voyager does not use any bought-in application software. The entire system has been written and developed in-house. Below is an overview of the major functions of Voyager's travel agent system.

Voyager 2000's travel agents retail system

This product is aimed primarily at travel agents that have a high proportion of leisure travel business. Travel agents that handle this kind of business require a system that supports the processing of individual customers and allows a wide variety of products and payment methods to be handled. It consists of four main applications:

- **New Booking** The New Booking module has customized screens for each type of booking. Reservations may be initiated at any time by clicking on the appropriate icon. This provides access to the GDS being used by the travel agent or, alternatively, viewdata. There are many client support functions including client files and a full client history. Travel consultant functions include an inter-office messaging system, a calculator, a note pad and a calendar.

  Voyager provides full support for the sale of travel insurance. The system automatically calculates the customer's insurance premium based on factors such as the area of the world to be visited, the number of adults and children, the duration of the trip and the applicable insurance premium tax. Discounted and free insurance can be supported, as required by the agency. When the insurance sale is complete, the system automatically prints the customer's receipt, which becomes the actual insurance policy. Disks are prepared each month for use as part of the monthly settlement process undertaken with each participating insurance company, thus eliminating further manual calculations and administration.
- **Existing Bookings** This is a combination of an automated client file and a diary system. It allows the agency to produce confirmations, print changed itineraries, accommodate customer changes, collect outstanding balances, issue reminders to customers and prompt travel consultants about important actions that they must complete by pre-determined dates. Other diary triggered functions ensure that tickets are printed on the correct day and that the management of the agency can review the status of all bookings very easily.
- **Management Module** This module is heavily protected by rigorous password control because it enables the travel agent's payment functions to be initiated. Support is provided for BSP, BACS, cheque payments and customer refunds. Another important role that is fulfilled by this module is the maintenance of the travel agency's key parameters. A good example is the commission rate charged for certain products.
- **Reporting Module** This produces the daily, weekly and monthly reports that monitor the business of the agency. The reports produced by this module enable the cash to be balanced and the business controlled by means of certain checks and balances. A reporting facility is provided that enables users to customize reports to meet their own particular requirements. All reports may be viewed on the screen in textual or graphical form, which may then be optionally printed. A mailing function supports direct mail and customer communication programmes.

Voyager 2000's travel agents retail system with business house option

This product is similar to the Voyager 2000 retail system but, as the name implies, it also incorporates some specialized business travel functions. It is designed for those travel agents whose leisure business represents between 60 and 70 per cent of their turnover, with the balance being corporate house business. All accounts are integrated into the same data base as used for the retail product. Invoices may be printed for business travel bookings and regular statements are printed for companies. Bookings may be retrieved for examination by invoice number as well as several other key fields. An invoice may have multiple bookings and support is available for those companies that pay for their bookings using an IATA charge card. All the usual business travel support functions are included, such as settlement discounts and trade discounts. Back-office accounting functions that control the sales ledger, i.e. accounts receivable, and special business house reports, such as aged debtor analyses, are automatically produced by this product. As with any other software that runs within the Microsoft Windows environment, cutting and pasting between Voyager applications and standard office productivity tools is simple and straightforward.

Median – the millennium central product

This product supports multi-branch travel agents and allows them to communicate operational business transactions using out-of-hours dial-up communications. This can be used to transmit data from branches to an agency's head office for results consolidation purposes. It can also be used to send summarized information prepared at head office, to the outlying branches. Median's use of dial-up communications allows this to happen at a fraction of the cost of alternative dedicated leased line networks.

Median can be set to establish a dial-up connection automatically out of normal office hours when branches are closed and telecommunications costs are low. When connection is made, a two-way dialogue takes place: (a) transactions are transmitted from the branch and stored on the head office computer for consolidated processing, and (b) previously prepared information is distributed from the head office computer to the branch. Once an agency's data has been collected and stored on the head-office computer, Median can be used to make payments for all branches and complete the associated accounting entries. The system can also be used as a kind of off-line messaging service allowing branches to send notes to each other via head office.

Finally, Median provides an agency with a flexible reporting facility. Historical comparisons are supported, including week-on-week, month-on-month, branch-on-branch and several others. The reporting functions allow users to customize reports in a variety of ways and support a flexible information query facility. Support for Microsoft Access queries is also available for more complex queries.

The Corporate Traveller

This is a product that has been designed specifically for travel agents that focus primarily on business travel. It bears a resemblance to the retail product, but has five main functions instead of just four; and even this predominantly GDS-based business travel product also supports viewdata. The five functions are:

- **New Booking** A new booking may consist of several travel products and services for the same traveller. Once made firm, a high quality invoice is automatically printed showing full details of the booking. The system produces a fine level of detail as and when needed, by product; for example, a rail booking will show routings, class and fare calculations. Besides storing both company and customer profiles in textual format, The Corporate Traveller also supports the storage of photographs where appropriate, e.g. images of senior management may be stored along with their associated textual profile information. Voyager also supports an add-on feature that controls the distribution of incoming telephone calls and logs usage patterns.
- **Existing Booking** This function supports everything to do with an existing booking, including amendment histories, costing alterations, ticketing prompts and reminder calls. Itineraries may be integrated with documents created using Microsoft Word, thus providing a professional customer travel pack. There are standard customer communications for many situations, including cancellations, full and part ticket refunds, commission rebates and *ex gratia* payments. As already mentioned, Voyager supports the automatic control of commonly executed events, using a diary function. This ensures that all balances, final payments, ticket confirmations, etc., are actioned on-time.
- **Management Section** This contains a set of frequently used functions, including BSP reporting, BACS control, cheque payment and the production of sales return reports for rail tickets and coach companies such as National Express. These functions may be tailored to a travel agent's own requirements by means of various financial and executive settings,

parameters, trading terms, sales incentives and user access levels. Monthly business house statements are printed automatically with the option to produce individual ledger cards. Finally, the allocation of payments received from business house customers can be easily applied to open, i.e. unpaid, invoices using either the auto-allocation function or the individual apportionment routine.

- **Corporate Management** This is where Voyager stores the control parameters that govern how business house accounts are structured and reported. It allows a business house customer's internal departmental organization to be reflected in the accounts set-up within Voyager. This ensures, for example, that invoices are clearly and correctly headed and that invoices are charged to the correct location within a customer's account.
- **Reporting** The core reports that comprise this module are the daily and weekly audit reports, which enable the overall financial status of the travel agency to be balanced. Once this has been successfully accomplished, many other reports are available. Examples include an aged debtor analysis, business house analysis, passenger analysis by booking type and operator. Executive reporting provides historical analysis showing year-on-year comparisons of turnover, commissions, receipts, payments, bookings and head-counts. Future projections may be constructed from transactions already recorded within the system thus allowing forward sales, commissions and cash flow to be estimated. Many reports are pre-set and stored within a gallery from where they may be selected and customized for a particular customer. All reports may be either simply viewed on the screen or physically printed on paper. One-off enquiries may be fulfilled by using Microsoft Access queries on the Voyager data base.

Voyager technology

The Voyager software runs on Apricot PC products. A close business relationship with Apricot has been developed over the formative years of Voyager, as explained in the introduction to this section. Voyager is in fact a registered Apricot re-seller. Apricot itself were bought out by Mitsubishi around 1990. By restricting itself to the single source supplier for its hardware, Voyager is able to provide a comprehensive ongoing support service to its customers, which includes hardware as well as software maintenance services.

Agents are therefore actively discouraged from running the software on their own hardware because this prevents Voyager taking total responsibility for the whole system and increases the likelihood of 'finger pointing'. However, if a travel agent insists on having the Voyager software running on their own hardware, then Voyager will provide the software but only on the condition that a server unit is also supplied by Voyager. This allows for a clean demarcation between the agent's workstation hardware and Voyager's server system.

In terms of developing its software environment, Voyager is a member of the Microsoft Developer Group. This is a kind of club whose members are those software companies that base their products around Microsoft operating systems, such as Windows and Windows NT, as well as programming languages such as Visual Basic. Being a member of this group allows Voyager to receive advance copies of new software products and releases from Microsoft and also to receive additional technical support. Voyager runs in a LAN environment under Windows 95 with larger installations using Windows NT4.

ICC'S TRAVEL SYSTEMS

ICC markets the Concord system to UK travel agents. But before we dive into the details of the product it is worthwhile spending a little time on the background to the company and the life history of the agency management product. The travel agency systems part of the company had its origins back in the early 1980s when a small group of IT specialists started a computer consultancy business. The consultancy, which was actually formed in 1982, concentrated on several industry sectors, the main ones being local government, building, printing and travel agents. The company was successful in most of these areas and quickly grew to employ 25 full-time staff. They supplied ten local government locations with systems and over 100 travel agents with an early version of an automated back-office package called 'Travelpack'. The travel agency automation side of the business was particularly successful, due mainly to the Travelpack software package. It was around this time that ICC became IBM agents and began to supply a range of IBM computers, which nowadays includes, for instance, the AS/400 and System/36 minicomputers as well as many other products.

The next major event in the development of the company was its acquisition by Misys Plc. Misys is a diversified company with 14 separate businesses employing over 1,000 staff and which has a turnover of £76 million (1992). Among other things, Misys had developed a successful insurance broking system and was looking to mirror this success further in other fields. Following some careful analysis and market research, the travel agency automation business was identified as a prime target for expansion. So in 1988, Misys bought the fledgling general IT consultancy company known as ICC. Following the acquisition, ICC concentrated solely on its travel agency automation business and divested itself of the other industry sectors. This enabled all 25 of the original ICC staff to concentrate on the travel sector.

A. T. Mays became one of ICC's first major customers and installed a modified version of the single user Travelpack system in all their branches. ICC undertook the first set of modifications for A. T. Mays who then went on to tailor the system itself for operation throughout its network of 300 or so leisure travel agencies. ICC then began working with Hogg Robinson on the development of a front-office system for use in Hogg Robinson's travel agency branches. This multi-user system eventually was to become the basis for the Concord product. ICC continued to grow and staff numbers rose to 33. ICC agency management systems are now used in over 930 travel agency locations, of which 630 use the Concord product. The travel agency customers of ICC range from the small independent, through the miniple to the large multiple. (A miniple is simply a small multiple and comprises tens of agency outlets as opposed to hundreds, which is usually the case with a multiple.)

Concord

The growth and development of the Concord product is an interesting story. The current product was launched at the World Travel Market in 1989, but the idea for the software that forms the core of the system has its origins in the very first attempts at agency automation that took place in the early 1980s. At this time the state-of-the-art in travel agency automation was the Videcom airline reservations terminal supplied by Travicom, a wholly owned subsidiary of British Airways. The Videcom terminal was a dumb device; in other words it did not have any processing capabilities of its own as today's PCs do. The Videcom terminal was connected to a communications controller installed in the travel agency. The controller supported several Videcom terminals and was itself connected via a simple network of leased telephone lines into British Airways' computer reservations system. Travicom started out as a direct access system connected only to British Airways, but later grew into the Travicom multi-access switch. Besides providing reservation functions the Travicom system supported the automated printing of airline tickets and related documentation. Travicom eventually became the basis for Galileo UK, as described in Chapter 4.

ICC first developed an agency management system for travel agents based on the PC which in the early 1980s represented the leading edge of IT. This early solution to the agency automation problem used a combination of the single user Travelpack back-office system together with the TABS PC accounting package. This product used batch key entry for the capture of transactions. The challenge at this time was therefore to obtain an automatic feed from Travicom into the PC and thus eliminate the manual keying effort. The thrust of these developments was aimed at automating business travel operations that were characterized as being high volume, comparatively low margin and that had a relatively simple transaction profile. An automated interface to Travicom was the most effective way in which to capture sales information on the airline tickets generated by business travel operations. If the Travicom information could be fed automatically into the PC then this would save the manual key entry of the transactions with all the associated problems of staff costs and transcription errors.

The interface was accomplished by developing some special software that ran in the PC and emulated a Videcom terminal. The term emulation means that the PC software made the Travicom controller think that the PC was in fact a Videcom terminal. The successful interconnection of the Travelpack/TABS PC back-office accounting system and the Travicom reservations system was a crucial first step in the development of Concord. In fact, ICC was the first company to develop such an interface after CCL's DPAS back-office system.

The next focus of attention was to try and automate the non-air transactions. These transactions were recorded on paper forms and passed to the back-office system for keying, usually at the end of the day or next day. So, the next step was to support the entry of other non-air transactions at the point-of-sale. To do this effectively, a PC needed to be provided to each sales person working at the point-of-sale, and to support several PCs in this way a LAN was needed. Because in those days the concept of a LAN was fairly new and the technology was in its infancy, the interconnection was accomplished by using a minicomputer with direct connections to the PCs. This also provided the processing power required for high volume back-office operations. The IBM System/36 was used for this purpose and provided a powerful platform for larger users.

Although the development of this interface was a critical first step in the evolution of an agency management system, it left the front-office leisure travel aspects untouched. These poor people were still using the old Kalamazoo/Safeguard paper-based systems to record leisure travel transactions. The extension of Travelpack to the leisure travel front-office was therefore a major step forward in the growth of the product. A. T. Mays, who is particularly strong in the leisure travel market, was one of the first to use the system in the front-office. In these early days, the front-office of A. T. Mays was segregated in terms of how the customer was handled. The scene looked something like this:

- **At the point-of-sale** Customers spent time with sales people at the front counter who gave advice and closed the sale by placing an option with a tour operator using a videotex terminal. Viewdata was at this time fast becoming the primary leisure travel front-office reservations and information tool. Having placed an option with the tour operator using viewdata, the details of the sale were manually recorded. The customer was then asked to move across to a different part of the office to the cashier point where all cash for the agency was handled.

- **At the cashier point** The cashier also had a Travelpack PC and this was connected to A. T. Mays' central office. It therefore had access to the Concord data base and the sales transactions entered at the front counter. When the customer approached the cashier, the tour operator's system was again accessed and the option turned into a confirmed booking. The booking was automatically downloaded into the Travelpack system, which displayed a screen showing the details of the amount due from the customer. The cashier received the money and duly entered the details into the Travelpack PC. The original transaction was updated to reflect the amount paid by the customer.

This may seem long winded (and it was too!), but it met the requirements of A. T. Mays at the time. These requirements were intended to provide a system that: (a) offered the most secure way of handling cash in the agency because it involved dual control, i.e. computer and receipting control; and (b) provided a good way to keep sales persons free of administrative cash handling tasks, thus leaving them more time for sales-related activities. All in all it was a success because it met the requirements of A. T. Mays. As a result of this success, the system was further enhanced and rolled out to over 300 A. T. Mays travel agency branches. However, it was not the preferred solution for the majority of other UK travel agents. It was around this time that Hogg Robinson expressed more than simply a passing interest in the system.

Hogg Robinson wanted a system that was similar in many ways to the product used by A. T. Mays, but their requirements were quite different. So different in fact that a new system, completely re-written from scratch, was commissioned by them. The new system was undertaken as a joint venture between Hogg Robinson and ICC. It was this rewritten system that formed the skeleton of the actual Concord agency management product that is actively marketed today. About a year after the ICC/Hogg Robinson joint venture had started, ICC was bought out by Misys as described above in the opening paragraphs of this section. The acquisition was fortunate in that it provided a solid base and the capital infrastructure within which ICC could expand in the future and sustain a high rate of growth.

Today's Concord agency management system comprises many functions, which may be mixed and matched to meet the specific needs of a travel agency. An overview of the major components of the Concord system is presented under the classical groupings of travel agency operations in the front-, middle- and back-office as follows:

- **Front office** The front-office system supports the bookings of both business travel and leisure travel. Concord therefore provides interfaces to the reservation systems of the major CRSs including Galileo, Sabre and Worldspan. It also accesses those leisure travel reservation and information systems that are based on viewdata technology. In fact one of the first new functions that was added to the base product was an enhanced and more efficient viewdata front-end. This enables the Concord PCs at the point-of-sale to be used as viewdata terminals as well as providing agency management functionality. All of these travel industry systems may be accessed by a variety of telecommunication methods including Istel, MNS and Prestel. The Concord system has an integrated customer file into which booking information from these reservation systems can be automatically fed at the touch of a button.

  The booking process is made simple and easy to use by means of a split-screen facility. This enables the screen to be split vertically thus allowing CRS or viewdata reservation system screens to be displayed on the left with Concord screens containing customer files, for example, on the right. With just a single key depression the user may copy information from the reservation screen into the Concord booking files. Similarly, information may be retrieved from a booking file and formatted for inclusion in a supplier reservation message for transmission to a host system. Both these features are excellent productivity and quality enhancement aids that would benefit most UK travel agents. The front-office functions are also supported by automated screen prompts that anticipate a customer's requirements and lead a travel agency sales person through the various stages of a booking. All travel products and services transacted by a travel agent are supported by Concord.

  Finally, Concord automatically prints all the key documents driven by the booking process. This includes invoices, credit notes and receipts for customers. The system also has a correspondence facility that prints customer letters for both booking administration purposes and direct mail marketing.

- **Mid-office** The mid-office is not often discussed separately within travel agency automation circles. It really covers the management and administrative functions of the agency. However the term mid-office is a useful heading under which to group certain functions of Concord that are as important to the front-office as they are to the back-office. Functions such as: (a) a diary to control the life history of a booking and ensure that all the necessary tasks are carried out at the appropriate time; (b) customer files that contain background information on all previous customers for marketing and promotion purposes, as well as booking information on all current customers; and (c) transaction files that are generated mostly from customer files and that feed the various other sub-systems, such as accounting and vendor payments. Management information is an integral part of Concord and a comprehensive set of sales reports is available.

  The mid-office control system generates a transaction transmittal file (TTF), which may be input to a back-office system. Each agency usually has a different requirement for back-office processing and so the TTF file can be fed into a wide variety of back-office accounting systems. This is why Concord is flexible because it can support a wide variety of travel agents.

- **Back-office**   One of the most important parts of Concord's back-office functions is end-of-day processing. This is executed at the end of the business day and updates critical master files with the day's transactions. A variety of management and control reports are printed as a by-product of the end-of-day process. These include supplier payments, e.g. tour operator returns and BSP settlement reports, with automated cheque printing, business house accounts, aged debt analyses, profit/loss reports and a complete audit trail.

    Another basic Concord back-office function is the ability to support multiple branches from a larger regional agency or a headquarters location. This requirement arises from the advantages that a multiple agency can gain from centralizing all back-office operations in a single location. This approach removes non-productive administrative tasks from branches and allows them more time for selling. Links to the centre are therefore also a part of the back-office functionality. This supports the collection of transaction data from outlying branches and the distribution of reference information to the same branches. These sub-branches may be contacted by the main branch using either dial-up or leased data lines. So, in this environment, when you go home at night your computer carries on working for you. The head office system wakes up and dials up all the branches. Then while you are asleep it collects the day's transactions and processes the core functions. When you come in again the next morning, all data bases have been updated and the critical reports have been produced.

    Finally there are the underlying control functions, such as ticket stock control and cash control. Each system maintains information that mirrors the holding of tickets and cash by the agency and reports any discrepancies. The Concord back-office concept is based on the use of any one of a wide variety of back-office products. Besides the accounting packages provided by ICC, the system provides a standard interface file, i.e. a TTF, that may be used to feed a proprietary accounting package of the travel agent's choice. The choice of the most appropriate ICC back-office accounting approach depends to a large extent upon the size of the travel agency and the level of sophistication required. The main options are:

    - *Small agency*   The Books software package is not in itself an accounting system. Instead, it provides an intelligent feed from the Concord back-office, which may be remote or local, to a human accountant. The information thus provided should enable an accountant to produce a set of books for a small independent travel agency with a low volume of transactions and no IT or accounting expertise. In this scenario, no accounting software package is used and it is therefore very simple for the agent to implement the system.
    - *Large/medium agency or miniple*   Again, a PC-based software package product called Sage Sovereign is a well respected general purpose accounting system that has been specially adapted by ICC and Sage to meet the needs of a travel agency. It is a multi-currency and multi-branch system that incorporates the following accounting functions: purchase ledger (including BSP), nominal ledger, sales ledger (including statements), report generator, and banking.
    - *Very large agency or multiple*   A minicomputer product called the Compass General Accounting System (CGAS) is produced by the IMREX company. This is a sophisticated general purpose accounting system capable of processing high volumes. It is ideally suited to the large miniples or independent multiples and requires a reasonable degree of both IT and accountancy expertise to use the package effectively.

The Concord product therefore provides a package to suit all types and sizes of travel agents in both the leisure and business travel sectors. In addition to this, and very importantly, ICC provides a full spectrum of support to its customers. This ranges from provision of the system hardware, the Concord software, installation planning, installation itself, ongoing maintenance of both hardware and software, a programme of continual product enhancement, training and the support of an active Concord user group. ICC feels that it is only by providing its customers with a 'one stop shop' service such as this, that a high quality of service and control can be maintained.

The Concord technology

Concord is based on a modular approach. This means that the system is available in chunks of application programs and different models of hardware, each of which may be obtained and used separately. Naturally, there is a core set of such applications but the system may to a large extent be customized to individual travel agencies. There are two basic configurations, one for a leisure travel agency and one for a business travel agency. However, these two configurations may be mixed where the agency handles both types of business.

In a retail or leisure travel agency the Concord technology is configured with a file server PC and one or more workstation PCs connected together by a LAN. Each of the workstation PCs has its own set of Concord applications including, for example, client files and transaction recording. Additionally, each workstation may be connected to a videotex network for booking holidays and other products that are accessed via viewdata. There are two options: (i) each workstation may have a modem connected to it that enables it to dial-out across the telephone network to a viewdata host system; or (ii) each workstation may be connected to a communications device called a multiplexor, which in turn is connected into one of the videotex VANS, such as Istel or MNS (see Chapter 6). In a leisure travel environment such as this the file server PC would have two printers attached that produce letters and other reports. Finally, the server would have a modem connected to it that supports remote diagnostics. This means that if something goes wrong with the system, the ICC support centre can dial into the server to diagnose and in many cases actually fix the problem.

In a business travel agency the set-up is similar with the exception of the external interfaces. Instead of accessing videotex systems via individual dial-up modems or an office multiplexor, a CRS gateway device is used. The gateway may be connected from each workstation individually or via the office file server. This gateway is provided by the CRS, which may be Galileo, Sabre or Worldspan. Each CRS provides its own type of gateway device, which is called a Travipad by Galileo and is used to access its X25 communication network, and a Gateway PC by Sabre and Worldspan. This gateway device can also support one or more ticket printers.

For agencies that have a mix of leisure and business travel, a combination of both of the above configurations can be set up. The system is flexible enough to be able to support a wide variety of business mixes. In cases where the agency is part of a miniple or multiple, Concord provides a powerful head office processing capability. In this environment, one of the branches is designated the head office and has some special software running on the server. Some time after the close of business, the head office file server dials each branch in turn via the branch's own modem and collects the day's transactions. When all branches have been polled in this way, the headquarter's computer runs the processing routines for the entire network of agencies. This allows, for example, a single payment to be made to each supplier for all products purchased by all agencies in the group, a single cash flow picture to be assembled and a single set of books to be produced: and more importantly perhaps, it allows a single consolidated marketing data base to be maintained.

So, as you can see from the preceeding section, Concord is pretty flexible. In fact it can be configured to meet many different business environments, and the hardware on which the system can run may also be tailored to meet the needs of a wide range of sizes of agency. The software itself runs on several hardware platforms as described below:

- **Workstation** The Concord workstation is an Olivetti PC, which needs to be a 286 or higher with at least 1 Mb of RAM, no hard disk and an optional floppy disk drive. It communicates with a server PC located within the agency via a LAN. The workstation's software runs under the control of MS-DOS Version 5 or higher.
- **Server** The server is also an Olivetti PC but this model provides more processing power than the workstation. It is a 386 or 486 with at least 1 Mb of RAM and 100 Mb of hard

disk storage. The operating environment is MS-DOS Version 5 or higher running under Novel (see LAN below).

- **Back-office processor** The back-office processor may be either a PC or a minicomputer. It all depends upon the amount of processing and the sophistication of the travel agency's automation requirements as outlined above.
  - *Simple PC* A simple back-office product for a small agency would need only a normal PC running under the DOS operating system. This can be a 1 Mb 286, which uses the hard disk on the server.
  - *Sophisticated PC* If the agency is using the Sage Sovereign accounting software package then a larger and more powerful PC will be required. This will need to be a 386 or higher with 1 Mb running MS-DOS and using the hard disk on the file server. The precise details of the hardware configuration will depend upon whether the package is installed in a single-user site or as a multi-user system. Sovereign has been developed using Sage's own fourth generation programming language called Retrieve 4GL. This works with a full PC relational data base. This is a major factor in the flexibility of the system, which can be quickly and easily adapted to a travel agent's own specific requirements. It is Sovereign's System Manager that includes the user report generation facility and that enables the system to be tailored to a user's precise needs.
  - *Minicomputer* The IBM AS/400 or IBM System/36 are the two minicomputers upon which a sophisticated multi-branch back-office system is available for the headquarters of a large agency. The AS/400 minicomputer comes with its own integral relational data base and high-tech operating system. Although it is considered user-friendly in computer circles, it will probably require more support and expertise than a PC solution. The System/36 has now been replaced by the AS/400 although ICC still has several travel agent customers using this minicomputer.
- **LAN** Concord's systems talk to each other via a LAN based on Novel software running on the office server (see above). The cabling system of the LAN may be based on either the Ethernet or Token Ring topologies (A topology is simply a set of standards that describes how electronic signals are used to enable two or more computers to communicate with each other by means of cables that connect them together.)

TARSC

The Travel Agent Systems Resource Company (TARSC) is a UK agency management system for independent high street travel agents. TARSC has its origins back in 1982. It was around this time that Stewart Hall, a travel agent in Billericay and one of the founders of TARSC, developed a basic computer system that exploited the PC that were emerging onto the market at that time. With his background in the travel agency business, Stuart Hall had developed a system that automated many of the routine functions performed by the average high street travel agent. The system was at that time, however, fairly basic. It was, for example, a single-user system, i.e. one that could only be used by one person to do a single function at any time. Stuart Hall teamed up with Peter Healey and Roger Gibson of Vertical Systems who both had in-depth knowledge of the up and coming PC technology of the time and between them all they refined and developed the TARSC system.

The TARSC system established a reputation within the travel agency community that became fairly widespread. Around this time, i.e. circa 1985, Sony was looking for some marketing support to help increase sales of its KTX 1000 viewdata terminal. Sony decided that an alliance with a popular travel agency system would help it in these endeavours and so Sony approached TARSC. The result of the ensuing discussions was that Sony bought the marketing rights to the TARSC system. Under this umbrella, the number of TARSC users grew to around 123 within a matter of a few years. The system was sold via distributors including Vistek, Thorn EMI, Preview Data Systems and Livelink Data Systems. However, in 1988 Vertical Systems re-acquired the full rights to TARSC and a process of development and further promotion of the product began. In 1991 Kerry Costello joined as Marketing Director and sales started to grow strongly.

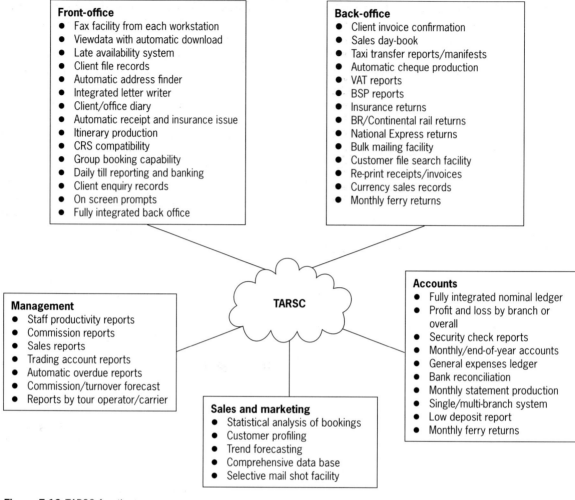

Figure 7.10 TARSC functions

The TARSC system

The TARSC system is suitable for both single and multi-branch travel agencies. It provides both a point-of-sale workstation system with reservations systems support, a sophisticated communications facility and a back-office processing system. TARSC comprises a set of computer programs that provides a wide range of automated functions to a travel agent. Figure 7.10 shows the general capabilities of the software:

The list of TARSC system functions gives an indication of the range and scope of the system. Furthermore, it is a flexible system that can be configured in a number of different ways to suit the needs of the travel agent. The TARSC product is appropriate for many sizes and types of travel businesses including: retail travel, business travel, group tours or tour operations. There are many features of the TARSC system that cannot be explained or covered fully in this book. Some of the more interesting features are those that are outlined below:

- **Accounting system** The accounting system is one that has been developed by TARSC. It does not use the functionality of a third-party system and has been designed for use specifically by travel agents. It also has the unique ability to continue to interact with historical

data stored on the system, even past the date of any event.
- **Printers** One or more printers are required to produce the customer documentation and the reports generated by the system. The choice of which type of printer to use is the user's, i.e. laser, inkjet and dot matix printers, all of which can be supplied by TARSC.
- **Report generation** There is a central data base facility that together with a report generator software package called IQ make the production of special reports a straightforward task.
- **Passwords** Access to the TARSC system is controlled by the use of passwords, as are most systems these days. Each password is associated with a set of functions that represents the set that are allowable to that user.
- **Electronic mail** The TARSC systems provide an e-mail capability, which uses the branch LANs and the multiple synchronized data base (MUSDAT) functionality (see below), to pass messages between the workstation users within the agency and provided a connection is made to a VANS, then other external parties may be contacted via e-mail.
- **Back-up** In the event of a branch file server failing for some reason, the workstations can be re-booted to operate in stand-alone mode. This means that if the file server becomes unavailable then the workstations can be set up to interface with the major reservation systems and sales transactions can continue to be recorded.

These product offerings all use various selections of modular components that make up the TARSC system. Generally speaking, the TARSC product comprises both the hardware and the software, bundled up as a package. This turn-key solution is the one recommended by the company. TARSC also supports an un-bundled version, which means that the travel agent is responsible for obtaining and maintaining the hardware and TARSC provides and supports the software. The technical details of the hardware and operating system environment for the TARSC products is as follows:

- **Hardware** File Server-Pentium 66 MHz PC with 16 Mb of RAM and 1.2 Gb hard disk running under the Novell 3.12 LAN operating system. Workstation – Pentium PC with 8 Mb of RAM and 1.2 Gb hard disk running under the Novell 3.12 LAN operating system.
- **Software** This is a set of over 1,000 TARSC application programs written in a fourth generation programming language entitled ODBS. This set of programs provides the agency management systems application functions that are shown in Fig. 7.10.
- **Communications** MUSDAT is the name of TARSC's front-end communications processor and is explained in more detail below. In summary it is a communications concentrator for multi-branch agencies and enables all communications carried out by branches to be routed through a single central site.

Now, because MUSDAT is one of the most powerful components of the TARSC system, I think it is worthwhile taking a little more time to explain it in more detail. This is because MUSDAT, like any telecommunications system, can have a significant impact on a travel agent's method of operation and operating cost profile.

MUSDAT

This product is the communications element of the TARSC system. MUSDAT comprises both hardware and software components. It is a product that is unique to TARSC. MUSDAT is a front-end processor, which is connected to an agency's outlying branches by either private leased communication lines or dial-up services. The front-end processor is in fact a set of multiple processors that are housed in a single cabinet. The precise number of processors will depend upon the number of branches and the number of workstations connected within each branch. There are two ways in which this may be effected:

- **Private leased lines** The front-end processor allows an agent to share the connections to travel principal systems (and a head office TARSC system), from a central point. This means that the workstations used in branches are able to communicate with viewdata booking systems via head office connections. Or, to take another example, all branch workstations can connect into the CRS used by the agency via the head office front-end processor. This enables CRS hosts like Galileo to be shared and this in turn allows the number of ports to be minimized for the entire branch network (the same remarks apply to the viewdata ports provided by VANS).

This approach means that the branch data is duplicated at the head office location. So, if the branch data were to be lost for any reason, e.g. a fire or flood, the data could be reconstituted quite quickly using the information stored at head office on the front-end MUSDAT processor.

Finally, it allows some of the processing to be provided by the head office computer. This means that the branch workstation can concentrate on its local functions, such as point-of-sale work, and the central processor can operate in the background supporting functions like the CRS interface, the updating of client files and other routine tasks.

- **Directional download** One of the features of most multiple branch agencies is the collection and distribution of data with head office. The MUSDAT system uses normal dial-up lines and some innovative technology to accomplish this very efficiently. The basic approach used in MUSDAT is based on detecting and transmitting only data that has changed since the last transmission. This, in conjunction with data compression techniques, allows for some very speedy and cost-effective communication sessions between head office and the file servers in branches: any number of branches can be supported by a single MUSDAT system.

Other TARSC products

I have tried to give you a summary of the main products offered by TARSC. However, there are others that are also available, some of which are designed for certain niche sectors of the travel business. After all, as I have mentioned before, no two travel agents are the same. Some may have a basic requirement for retail travel agency business functions, others may operate their own little tour business, others may have a flourishing group travel business. It is for these types of situations that TARSC has developed some speciality products:

- **Lates** This is a product that continually captures late availability information on holidays and charter flights from suppliers' systems. The captured information is stored centrally on the agent's head office server and can be downloaded to branches at any time during the day for local storage on a sales consultant's PC workstation. This saves each consultant from having to undertake their own trawl of suppliers' systems and therefore helps increase the agency's overall sales productivity while improving the service offered to customers.
- **Vocsoft** This is a viewdata emulation package for a PC. It is a piece of software that runs in a PC and that makes the PC act as though it were a viewdata terminal. A key feature is the ability to download data automatically into the client file. Vocsoft also works in concert with Lates and when one selects a holiday from Lates, the system will automatically dial up the appropriate tour operator's data base and find the particular holiday chosen by the customer.
- **Traveller** A small tour operator's package. This system provides automated support for travel agents who run their own tours. It provides functions such as passenger manifests, tour booking records, tour component usage and many other features.
- **Teleview** This is a teletext sales support package. It is designed for use by those agents that undertake advertising via one of the teletext services. The package supports the ability to download pages from teletext systems, present scripts to travel agency staff who are answering calls from would-be customers and records marketing information on all callers.
- **Titan** A business house software package. This system supports the functions required by travel agents who handle a large amount of business travel work. It includes some special management information reports frequently needed by customers in this field and a comprehensive range of analysis reports.
- **Tourstock** This is a package that supports the purchase and subsequent re-sale of tours. It provides sales capabilities from multiple terminals and allows for the taking of options and status reporting.

- **Tourmaster** A system designed for tour operators. It supports the integration of flights, accommodation and other products into a package for selling to customers.
- **Flightmaster** This is a system that TARSC has designed for the seat-only market. Many travel agents and some dedicated operators are in the business of selling airline seats available via consolidators, charter operators and other sources, to customers. Flightmaster automates many of the functions involved, e.g., manifests, printed tickets.
- **Groupmaster** A system specifically designed for those agencies that operate a group travel business. Very often this is a business that is closely allied to incentive travel but more generally it is one that addresses the need for groups of people to travel to a common set of destinations. Groupmaster provides the specialized functions required by this kind of business.
- **Seatmaster** A program designed for the specialist air charter seat wholesaler who sells flights either direct to the public or through retail travel agents or both. It includes automated ATOL return production, charter ticket printing and automated accommodation voucher printing.
- **Roomaster** This is a package that has been designed for accommodation-only operators who either sell directly to the general public or via the trade. Although it is primarily an accommodation product it also allows other ancillary products to be managed.
- **Clubmaster** This optional package is for travel agents who operate travel clubs. It includes full mailing facilities, membership card production, annual subscription collection and a client database.

OTHER AGENCY MANAGEMENT SYSTEMS

There are of course many other agency management systems on the travel agency systems market, and the list is continually growing. It has therefore not been possible to cover them all here at this point in time. For more information on other agency management systems in the UK, a good source of information is the ABTECH group (see Chapter 1).

Conclusions

I hope that with the review of these products, I have given you a good idea of the span of functions available in agency management systems. As you will no doubt have concluded, the variety of functionality provided by these systems is significant and evidently no two products are the same. That is why it is so important to understand your requirements before deciding which one to buy.

Epilogue

I hope you didn't cheat and start here! If you have read this book from the beginning and reached this point then you have probably gained a very good insight into how IT is used in the travel and tourism industries. But this is only a start. If I have succeeded at all, then all I will have done is to whet your appetite for the subject. You should be wanting to understand a lot more about IT in travel and tourism in even more detail. After all, I have only just scratched the surface. The only problem you will probably face is that there is not a lot of reading material on this specific subject. There are plenty of books on IT itself of course, but very few (if any) on IT in travel and tourism.

Nevertheless, you should by now have an inkling of the kind of technologies that are key to travel and tourism, based on what you have read and understood in this book. So, a good next step is to delve into those areas that will influence the travel and tourism technologies of the future – particularly the Internet and Intranet technologies. Keep abreast of travel and tourism automation products by asking suppliers for information on new and enhanced versions of their products and services: and of course, it is essential that industry practitioners keep up to speed by attending at least one technical course or seminar each year. After all knowledge doesn't get magically installed into our brains; it must be acquired by learning. It's nice to end up on the one thing that I think needs emphasizing without limit, and that is training. This book, coupled with an effective training programme is a powerful way to gain knowledge and to receive the training necessary to become proficient in the field of travel and tourism.

Appendix

AAT	agency accounting table	ATPCO	Airline Tariff Publishing Company
ABA	Australian Business Access	ATS	agent ticketing system
ABTA	Association of British Travel Agents	AVS	address verification system
ABTECH	ABTA's technology sub-group	AVS	availability status
ACM	agency credit memo	AXI	American Express Interactive
ACRISS	Association of Car Rental Industry Systems Standards	BA	British Airways
		BABA	book a bed ahead
ADM	agency debit memo	BABS	British Airways booking system
ADS	Agency Data Systems	BACS	Bank Automated Clearing Scheme
AEA	Association of European Airlines		
AGM	annual general meeting	BBR	basic booking request
AIM	Amadeus' instant marketing	BIOS	basic input/output system
AIS	Amadeus' information system	BK	booking
ALESA	Anasazi lodging enterprise system architecture	BL	British Leyland
		BOS	business operating system
AOL	America On-Line	BRBS	British Rail Business Systems
API	application program interface	BRL	British Rail
APN	airline passenger notice	BSP	bank settlement plan
APTIS	all purpose ticket issuing system	BT	British Telecom
ARC	Airline Report Corporation, USA	BTA	British Tourist Authority
ASM	ad hoc schedule messages	BTC	business travel centre
ASP	active server page	BTS	Business Travel Solutions
ATAC	airline tariff automated collection	CAA	civil aviation authority also Civil Aviation Authority, UK
AT&T	American Telephone and Telegraph (Company)		
		CATE	computer-aided timetable enquiry
ATB	automated ticket and boarding pass	CBI	computer-based instruction
		CBT	Certificate of Business Travel
ATC	air traffic control	CCL	Computer Communications Ltd
ATF	automated tariff filing	CD-ROM	compact disc read-only memory
ATLAS	AT&T's travel late availability search	CGAS	Compass General Accounting System
ATM	asynchronous transfer mode	CIP	carrier identification plate
ATM	automated teller machine	CIS	customer information system
ATOC	Association of Train Operating Companies	CoRRe	Centrally oriented ResReview Edition

CPF	central prices file	GMIR	Galileo's machine interface record
CRS	central reservation system		
CRS	computerized reservation system	GRS	Global Reference System
CRU	central reservations unit	GSA	general sales agents
CUG	closed user group	GTI	Global Trade Initiative
DAD	distribution access data base	GUI	graphical user interface
DCP	Data Call Plus	GWFS	Galileo workstation file server
DDE	Direct Data Exchange	HANK	Hotels Automated Network Know-how (System)
DEC	Digital Equipment Corporation		
DIMA	Department of Immigration & Multicultural Affairs, Australia	HCC	Hotel Clearing Corporation
		HDS	hotel distribution system
DPAS	Document Printing Agency System	HEDNA	Hotels' Electronic Distribution Network Association
EBES	European Board for EDI Standardization	HITIS	Hospitality Industry Technology Integration Standards
EBFS	enhanced booking file servicing	HOA	hotel availability
EC	European Commission	HOC	hotel complete (availability)
EC	European Community	HOD	hotel description
ECAC	European Civil Aviation Conference	HOI	hotel index
		HOM	hotel modification
EDI	electronic data interchange	HOR	hotel reference
EDIFACT	EDI for administration, commerce and transportation	HOU	hotel update
		HRS	hotel reservation system
EDI.RESCON	EDI for reservations and confirmation	HSSS	Hotel Systems Support Service (Ltd)
EDS	Electronic Data Systems	HTML	hyper-text mark-up language
EEG8	Expert Group No. 8	IATA	International Airline Transportation Association
EFT	electronic funds transfer		
ELVA	European local vendor access	IBE	Internet booking engine
EMEA	Europe, Middle East and Africa	ICC	Independent Computer Company
EPS	European passenger service	I-EDI	interactive EDI
ETA	electronic travel authority	IFITT	International Federation of Information Technology & Tourism
ETAS	electronic travel authority system		
ETB	English Tourist Board		
EU	European Union	IIS	Internet Information Server
EUK	Eurostar UK	ISAPI	Internet server applications program interface
FACETS	fully automated customer enquiry terminal system		
		ISDN	Integrated Services Digital Network
FIT	foreign independent tour		
FM	facilities management	ISO	International Standards Organization
FOS	flight operations system		
GBTA	Guild of British Travel Agents	ISP	Internet service provider
GCS	Galileo Central System	IT	independent tour
GDC	Global Data Connect	IT	information technology (ies)
GDP	gross domestic product	ITN	Internet Travel Network
GDS	global distribution system	JDS	joint distribution system
GEBTA	Guild of European Business Travel Agents	LAN	local area network
		MARSHA	Marriott's automated reservation system for hotel availability
GIS	General Information Systems		

MCO	miscellaneous change order	RAM	random access memory
MD8	Message Development Group 8	RAMP	Regional Applications and Messaging Platform
MIPS	millions of instructions per second	RCS	request capture system
MIR	machine interface record	RCT	rail combined ticket
MIS	Management Information System	RDMS	relational data base management system
MNS	Midland Network Services		
MPD	multiple purpose document	RESCON	reservation and confirmation
MS	Microsoft	RETOG	Retail Tour Operators' Group
MTBF	mean time between failures	RISC	reduced instruction set computer
MTT	Microsoft Travel Technologies	RPS	request processing system
MUSDAT	multiple synchronized data bases	RSP	Rail Settlement Plan
MVS	multiple virtual storage	RTB	regional tourist board
NAITA	National Association of Independent Travel Agents	Sabre	semi-automated business research environment
NDC	National Distribution Company	SAS	Scandinavian Airlines System
NDS	national distribution system	SCS	Sabre Computer Services
NITB	Northern Ireland Tourist Board	SITA	Societe Internationale de Telecommunication Aeronautiques
NMC	national marketing company		
NRES	National Rail Enquiry Service		
NVMIR	national vendor machine interface record	SME	small- to medium-sized establishment
NVP	National Vendor Platform	SMI	standard messaging interface
OCR	optical character recognition	SNCB	Societe Nationale des Chemins de Fer Belges
ODBC	open data base connectivity		
OPTAT	off-premises transitional automated ticket	SNCF	Societe Nationale des Chemins de Fer Francais
OSI	other service information	SO-ATB	surface operator automated ticket and boarding pass
PADIS	Passenger and Airport Data Interchange Standards		
		SOC	systems operation centre
PAMS	President Agency Management System	SPORTIS	super portable ticket issuing system
PAS	public access system	SQL	structured query language
PBA	productivity based agreement	SSIM	standard schedules information manual
PC	personal computer		
PCT	Private Communications Technology	SSL	secure socket layer
		SSM	standard schedules message
PDQ	past date quick	SSR	special service request
PF	programmable function	STARS	special traveller account record system
PIMMS	public information mailing management system		
		STF	sales transmittal form
PMS	property management system	STIN	Sabre's Travel Information Network
PNR	passenger name record		
POS	point-of-sale	STP	satellite ticket printing
POTS	plain old telephone service	TAAB	Travel Agency Advisory Board
PSTN	public switched telephone network	T&E	travel and entertainment
		TARSC	Travel Agent Systems Resource Company
QTVR	Quicktime virtual reality		
RAID	random array of inexpensive disks	TAS	Travel Automation Services

TAT	transitional automated ticket	UDID	user defined interface data
TESS	travellers' emergency service system	UFTAA	Universal Federation of Travel Agents' Association
Thisco	The Hotel Industry Switch Company	UK	United Kingdom
		UN	United Nations
TIC	tourist information centre	UPS	United Parcel Service
TIN	ticket invoice numbering	URL	uniform resource locator
TIS	tourist information system	USA	United States of America
TOC	train operating company	VAN	value-added network
TOD	ticket on departure	VANS	value-added network supplier
TOP	Thomson On-line Program	VAT	value-added tax
TPF	transaction processing facility	VDU	visual display unit
TRIPS	tourism resource information processing system	VFR	visits to friends and relatives
		VHS	video home system
TSG	Tribute Sales Guide	VIP	very important person
TSR	Terminate and Stay Resident	VM	virtual machine
TTG	Travel Technologies Group	VP	vice-president
TTI	Travel Technology Initiative	WAN	wide area network
TTS	transaction transmittal file	WSP	Worldspan
UCCCF	universal credit card charge form	WTP	World Travel Partners
		WTS	Worldspan Travel Suppliers

Index

A. T. Mays 19, 197, 366
ABA 55
Abacus 83, 151, 362
ABC Corporate Sales 99
ABC Electronic 259
ABTA 10, 23–4
   Travel Training Company 313
ABTECH 10
Access 337
Access One 218
accommodation booking services 254
accounting 325, 342
accounts receivable 343
ACM 329
acoustic coupler 277
ACRISS 157
ActiveX 267
ADM 329
ADS 348
ADVANTAGE 24
advertising 282
AEA 75
Aeroprint 328
Africa Flight Guide 101
aged analysis 344
agency management systems 348
Air France 109
AIRINC 59
airline billing analysis 330
Airtours 12, 19, 309
Alamo 127
ALESA 172
All Nippon Airways 162
allocation 46
Amadeus 109
Amadeus Pro 115
AMANET 110, 115
Amdahl 297
American Airlines 75, 132

American Express 23, 77, 222, 240, 312, 337
American Hotel & Motel Association 80
Ameritech 206
AMR Corporation 134
Amtrak 155
Anasazi Inc. 172
Ansett 116
answer back 63, 136
AOL 178
APEC card 55
APEX 88
API 49, 159
APN 72
Apollo 83, 116
   Travel 85
Apple 230
applets 289
Apricot 361
APTIS 89
ARC 327
archive 38
artificial intelligence 133
Ascom Autelea 91
ASM 66
assets 345
Associated Newspapers 279
assured booking 81
AT&T 218, 267, 285, 301, 338
ATAC 67
ATB 69
ATB2 15–16, 69, 332
ATF 67
ATLAS 287, 290
ATM
   asynchronous transfer mode 309
   automated teller machine 43
ATOC 88, 336

ATOL customer receipt 18, 19
ATPCO 67
ATS 91–3
auctioning 210
Aufhauser 218
Australia House 53
Australian Airlines 116
Australian Government 53
Autofile 10, 292
automated BSP reconciliation 332
automated reporting 330
automatic translation 260
Autoroute 39
availability and phone 61
availability display 7
availability reservisor 132
Avis 127
Avro 85, 289
AVS 201
AVS 60, 294
AXI 240

BAA plc 23
BABA 41
BABS 34, 64
back-office system (hotel) 79
back-office systems 324
backbone network 117
backwards compatible 17
BACS 328, 339
baggage tags 19
balance sheet 345
BALink 71, 294
bank card issuers 77
bank reconciliation 327
Barclays Bank plc 328
Batch EDI 278
bed & breakfast 36
Bell Corporation 301
Beta testing 351

INDEX 381

bias 108
biased display 7
Bildschirmtext 283
BIOS 29
Birmingham Cable 303
Black Diamond 259
Boeing 707 133
Book Hotel 169
book office 318
Booking by fax 36
booking
 conditions 200
 engine 180
 fee 183, 193, 223, 294
 file 123
 locator 59, 123
 screen 253
Bord Failte Eireann 44, 264
border management technology 55
Borland Paradox 37, 40
BOS 359
BRBS 88
British Airways 71, 116, 294, 358
British Car Rental 127
British Leyland 301
British Midland 71, 223
British Rail 39, 87
British Telecom 10
British Travel Centre 35
Brittany Ferries 12, 155, 292
broadcast 174
brochure distribution 260
browsers 177
BSP 11, 68, 191, 326
 history 327
 Plan Manager 327
BT 299
BTA 31, 255
 Events 42
BTC 316
bubble memory 90
bucket shops 317
Budget 127
bulk purchase 24
bulletin boards 174
bundled 361
bureau 317
bureau-de-change 6, 317
business travel 5, 239
buying power 24
by-pass 64

C++ 321
CAA 7–9, 23
cabotage 9
call centre 51
Caltrav 360
camera technology 271
Canterbury Chamber of Commerce 40
Canterbury TIC 40
CAPRI 97
car rental 157, 202, 207
 reservations 113
card
 authorization 76, 125
 fraud 76
 issuers 77
 standards 77
card index file 181
Carlson Wagonlit 23
Carte Blanche 222
cartel 22
cash advance 5
cash flow 343
CATE 88, 95
Cathay Pacific 152
CBT 23
CCL 324
central hosting 137
Central Reservations Unit 172
Ceres Securities 218
CGAS 369
Chameleon 117
Channel 4 279
charge card 6
chat session 197
chats and forums 210
CIP 328
CIS 95
Cisco 238
Citibank 340
City & Guilds 23
client file 125
clipboard 144
Club Med 148
Co-op 361
 Travel Care 19
co-operative processing 95, 107, 119, 135
coding systems 334
collateral 231
Columbus Group 267
Columbus Insurance 129

commission
 capping 186
 tracking systems 342
common booking record 18
common language 62–4
common ticket stock 335
communications server 352
Compaq 259
Compression Labs 274
Compuserve 148, 178
Computer Aided Instruction 128
Concert 299
Concord 366
confirmation advice 19
Congress Innsbruck 26
consolidated fares 200
consolidators 292
consumer activated kiosk 38
consumers 4
Continental Airlines 109
continuous presence 273
contracted rates 206
Corel Professional Photos 210
Cornell University 22
corporate
 card 6
 rate 81
corporate travel 149
CoRRe 321
Cosmos plc 12, 19, 85, 286
Covia 116
CPF 88
credit card 6
 billing analysis 330
CRS 107
 participation 60
 partition 63
 Screen Highlighter 319
Crystal Reports 360
CTV 42
CUG 25, 172
customer file 347
customization 210
cut and paste 130
Cyberseat 223

DAD 217
data base
 engine 45
 marketing 258
Data General 349
DDE 356
DEC Microvax 47

DEC VAX   34, 45
decision support technologies   134
dedicated terminals   104
defined viewership   172
Dell Computer Corporation   151
Delta Airlines   151, 337
demand side   48
Demon   255
denials   165
deregulation   56
destination information   209
Deutsche Telecom   313
Deutsche Universitats Verlag   193
DG14   35
dialogue   58
digital broadcasting   282
DIMA   53
Diners Card   222
direct access   62, 136
direct connect   63, 136
   availability   64, 84, 136
direct e-mail   180
direct mail   347
Discover   222
disintermediation   185
distressed inventory   290
distressed stock   218, 287
distribution costs   186
DPAS   324
drag and book   150
Dragon Air   152
drill down   35
DRS   34
dual control   367
dumb terminals   134

e-ticket Access Card   72
e-ticketing   71, 157
Easy Jet   224
Easy Res   99, 293
EBA   224
EBES   17
EC-BSP   337
ECAC   6, 9
EDI   15, 87, 148, 277, 306
   Message Sets   16
EDIFACT UN   15, 17
EDS   110
EEG8   15, 17
EFT   341
electronic
   booking form   129
   mail   276

publishing   260
ticketing   71, 125
travel agent   196
visa   54
e-mail   276
embedded chip   75
EMU   28
emulation   367
Encarta World Atlas   199
encryption   222
end transaction   123
ENTER   25, 264
EPS   89
Equinus   14
Esterel   237
ETA   53
ETAS   52
ETB   31, 255
ETB Internet experiment   256
Ethernet LAN   133
ETNA   39, 41
Euro   29
Eurocard   337
Eurodollar   127
Europcar   127
European Community   6
Eurostar UK Limited   12, 19,
   89–90, 119, 335–7
Excel   354
exchange line   285
Executive club card   72
Expedia   196
extranets   312

FACETS   88, 95
facilitator   272
Facilities Management   238
fare distribution   65
fare driven availability   112
fare driven display   294
fare quote   124
fares filing   67
fast batch mode   17
Fastrak   12, 304
Fax back   39
fax board   148
Fax Link   48, 51
Federal Express   197, 247
feedback   264
ferry
   bookings   155
   companies   291
   ticketing   306

Ferry#   292
firewall   48, 185
First Choice plc   12, 19
flat files   86
Flexicom   267
flight wizard   199
fly drive   202
fly rail   155
FM   304
follow me   273
food and beverage system   79
Forrester Research   178, 190
Forte   164
forum manager   197
frame relay   308
frames   249, 257
France Telecom   313
free sale   62
front desk support   78
front-office systems   319
full availability   61
funding   5
funds transfer system   54

Galileo   19, 115
   International   116
   UK   10, 18, 117
gateway   103
   city   9
Gateway Tours   85
GBTA   22
GCS   117
GDS   2, 107
   marketing   173
GEBTA   23
General Cable Corporation   303
general information   128
general ledger   345
General Telecom   303
geo-locating   157
ghost PNR   295
global mapping   242
Globus   85
Going Places   19
governments   2
GPT   274
Gradient Solutions   238
Grand Metropolitan Hotels   166
graphic design   179
graphical ticketing   146
Greek Government   19
Gsi   328
GTI initiative   278, 313

GUI 35, 47, 84
Gullink 48, 49
Gulliver 44, 264

Hakanson & Company 20
HANK 167
hard wired 302
HCC 340
HDS 4, 107, 162
    connections (type A & B) 163
headlines 174
HEDNA 19
HEDNA Net 20
Henry Utell 166
Hertz 127
high street travel agent 316
Hilton 75–7, 207
HITIS 80
hits 323
Hogg Robinson 23, 324, 366
hold and confirm 61
Holiday Systems Group 12
home page 179
hot links 261
Hotel and Booking Research Organization 163
Hotel and Travel Index 99
hotel
    booking 21
    commissions 340
    intranets 185
    wizard 201
Hotelbook 234
Hotelogic 210
HotelSpace 169, 296
Hoverspeed 12, 155, 292
HSSS 176
HTML 177
hub slashing 133
Hyatt 164
hyperlinks 180
hypertext links 210

I-EDI 17, 278
IATA 11, 23–4
    722 encoding standard 92
    accreditation 12
    licence plate 68
Iberia 109
IBM 75–7, 132
Icanos 117, 358
ICC Concord 365
ICC Travel Systems 251

ICL VME 86, 286
IER 124
IFARES 7
IFITT 25
Immigration 53
Imminus 12, 285, 303
Imminus Talking Windows 306
implant 316
imprinter 328
IMREX 369
in-room technology 78
inbound 40
Independent Tour 193, 293
independents 317
INFINI 162
informative pricing 112
Inland Revenue 345
inplant 316
inside access 86
inside availability 84, 122
inside links 126
Institute of Travel & Tourism 10
Integra 42
intelligent agent 192
intelligent ticketing 75
Interactive @ Brann 256
interactive
    dialogue 17
    EDI 278
    mapping 234
    Teletext 282
    TV 189
Intercity 90
interlining 73–4
International Fund for Ireland 44
Internet 177
Internet Business Ireland 267
Intranet 177, 185, 288
intranet island 312
invoice 343
invoice/itinerary 141
Irish Ferries 12
Irish Tourist Board 44, 264
ISDN 35, 48, 271
ISDN-2 274
ISDN-30 274
ISO9735 standard 278
ISP 230, 300
Istel 142, 301
ITN 216
ITV 279
IVN Communications 210

JAL AXESS 169
Japan Airlines 362
Java applets 257
JCB card 222
JDS (Joint Distribution System) 89
Journey Planner 97
Jupiter Communications 190

Kalamazoo 367
kanji 145
Karsten Karcher 193
Ken Air Travel 362
KLM 116
knowledgeware 356

last room availability 83
last seat availability 59
late availability scanning 287
late booking 287
late sale matrix 287
Lazy Susan 132
Le Shuttle 89
legacy systems 58
leisure Travel 4
level of participation 136
liabilities 345
Lilliput 48–9
line noise 19
Line One 300
Link Initiative 10
Livelink Data Systems 371
load and balance 133
lodge card 6, 247
London Tourist Board 36
long record 334
look to book ratio 204
lost bookings (hotel) 81
Lufthansa 71, 74, 109
Lunn Poly 18–19

Ma Bell 301
Magellan Geographix 210
magnetic strip 69
magnetronic reservisor 132
mail box 276
management information 248, 325, 345
mapping data base 202
Mariott International Hotels 82
market gateways 259
marketing 178, 325, 347
    on the GDS 173

Marriott 164, 207, 229
MARSHA 82, 229
MARSHA III 84
Mastercard 222, 337
Matesys 356
MCI 299
MD8 17, 278
meeting planning 272
meeting types 271
mega-carriers 57
merchandising 210
Mercury 274, 305
meshed network 117
message
  flow 60
  handler 18
  processing engine 48
  router 48
metal oxide strip 70
method-1 330
Microsoft 196
  Access 34, 86
  Development Group 365
  Merchant Server 223
  NT 4.0 49
middle-office 368
Midland Bank 303
Midnet 304
millennium 16, 29
miniples 315
Minitel 46, 52, 283
MIR 92, 130
Misys plc 366
Mitsubishi 365
MNS 303
Moby Lines 155
Modulas 360
Monarch 85
Mondex 76
Montreal 11
MPD 329
MTBF 305
MTT 240
multi-access 62, 136
multi-media data base 35
multi-modal ticket 75
multi-rate access 172
multi-tasking 144
multiples 315
multiplexor 302
Murdoch Electronic
    Publishing 164
Murdoch magazines 167

MUSDAT 373
MVI Limited 279
MVS 110

NAITA 23
NatWest 337
  Streamline 10
negotiated fares 244
negotiated rate 81
net/net rate 81
Netscape
  Commerce Server 222, 225
  Live Wire 217
  Navigator 213, 226
networks 270
new intermediaries 196
NewPage Systems Ltd 238
News & Views 14
News America Inc. 167
newsgroup 180
NITB 44
NMC 109
nominal ledger 345
non-homogeneous PNR 114
non-serviced accommodation 32
North Sea Ferries 292
Northwest 151
Novell 86, 145
Novus 225
NPC 116
NRES 90, 95
NVMIR 119
NVP 118

OAG 66, 99
  Agents Gazeteer 101
  Air Travel Atlas 101
  Cruise & Ferry Guide 101
  Direct 66
  Flight Disk 103
  Genesis 66
  Guide to International
    Travel 101
  Holiday Guides 101
  Hotel Disk 103
  Pocket Flight Guide 101
  Rail Guide 101
  Travel Planner 101
  World Airways Guide 100
ODBC 358
on request 61
on-the-fly Web pages 218, 267
open skies 6, 56, 67

Open systems 19
Operation Unison 110
OPTAT 68
option 201
OS/2 353
outbound 40
outsource 86
outsourcing 179
Owners Abroad 10

P&O
  European Ferries 155
  Ferries 12, 137, 292
  North Sea Ferries 155
PADIS 17
pages on the fly 218, 267
palm top 104
PAMS 94, 358
PARS 83, 151
PAS 43
Passenger Agency Services 12
Passenger Services 12
passive segments 151
Passport number 54
payment 5
PBA 151
PCT 249
Pegasus Systems Inc. 216
Pentium 51
PF keys 140
Philips Electronics 279
Picturetel 274
PIMMS 37, 260
pixel 281
plating 328
PMS 46, 78, 83
PNR 59, 88
  code 59
  indexing 63
  locator code 136
  referencing 64
point-of-sale
  assistants 319
  systems 79
polling 58
port 285
positive acknowledgement 153
preferred rate 81
Prestel 46, 283–4
Preview Data Systems 371
Preview Travel 216
principal display 112
print server 352

Private label   170
profiles   140
profiling   182
profit and loss account   345
Project Rome   240
promotional rate   81
proprietary system   350
pseudo city code   65
PSTN   48
Public service kiosk   51
pull marketing   181
purchase ledger   344
push marketing   181

QTVR   230
quality control   252
queues   322
queuing   126
Quickfare   91

rack rate   81
Radio Rentals   284
RAID   51
rail   87
Rail Planner   95
Rail regulator   88
Railtrack   87, 297
RAMP   215
Random House   206
RCS   54
RCT   93
reciprocity   7
record return   111
Redwing   12
Reed Elsevier plc   98
Reed Travel Group   66, 98
Reed Travel Training   99
regulatory
    issues   7
    structure   2
relationship marketing   181
remote authoring   218
request   46
request and reply   132
ResAssist   251
RESCON   16–19, 286
Resolution 100   7
retail outlet   4
RETOG   14, 19
RFP   21
RISC   352
Ritz Carlton   82, 207
rooming list   86

route planning engine   267
router   309
Rover Group   301
RPS   54
RSP   88, 94, 336
RTB   32

Sabre   10, 64, 83, 131, 348
    Interactive   206
SAGE   131, 369
sales conversion ratio   288
sales ledger   343
Sally Ferries   292
SAS   71, 74
Scandinavian Seaways   292
scanning bureaux   287
scripts   130, 140
Sea France   292
Sealink   284
seamless   84
    availability   64, 136
    connectivity   82, 136, 163
search engine   183, 257, 264, 268
seat
    assignment   323
    inventories   56
    map   125, 245, 252
seat-only air   292
secondary carrier specific
        display   111
self-service check-in   72
self-service kiosks   78, 189
sell and report   61
SEMA Group   86–8, 95
server farm   310
serviced accommodation   32
session   120
shared access   65
SHERE   91
short record   333
sign-on   174
Singapore Airlines   152
Single Payment Scheme   338
SITA   53, 59
    AIRFARE   68
    FAIRSHARE   67
    PRICECHECK   68
SITA Sahara   169
Sky   310
smart card   75
SME   34–6, 45, 261
SME bookings   254
SNCB   155

SNCF   119, 155
SO-ATB   335
SoftKlones Talking Windows   306
software robots   192, 321
Sony   283, 371
special services   124
Speedwing   118
Split screen   368
SPORTIS   90
Springer-Verlag Wien New York
    28
Sprint   50, 313
SQL Server   49
SSIM   66
SSL   222, 225
SSM   66
standardization (hotels)   80
START   16, 237
Stena Line   12, 19, 155, 292
STF   329
STIN   348
store and forward   59
STP   74, 158, 331
Streamline Transtrac   337
structured messages   18
studios   271
Summit International Hotels   170
Sun   222, 238
Sun Sparc Station   118
Sun World   19
super long record   334
super servers   107
suppliers   2
supply side   48
surfing the net   183
Swiss Air   116
Switch   337
Sybase   352, 354
System Aid Technology Ltd   12
System One   83, 109–10

T&E market   5
T&E voucher   248
TAAB   110
TABS   366
Take Off   282
talk back   282
targeted marketing   181
TARSC   371
TAS   117, 358
TCP/IP   226, 352
tele-sales   189
Telecommunications   299

telemarketing 347
Teletext Limited 279
teletourism 35
Texas Instruments 124
The Cable Corporation 303
The Sharper Image 218
Thisco 84, 216
Thistle Hotels 170
Thomas Cook 19, 23, 303
Thomson Holidays 18–19, 279, 304
Thorn EMI 371
Thrifty 127
TIC 39, 255
ticket on departure 188
ticket stock 326
ticket stock control 333
ticketed carrier 191
tiffany cards 132
TIMATIC 54, 128
Tirol Werbung 25
TIS 25
TOC 336
TOD 63, 73
token ring 352
toll-free telephone 37
TOP 279, 286
topology 371
touch vision 43
tour operator settlement 338
tour operators 84
Tourama 85
tourism 195, 254
Tourism Forecasting Council 53
tourism life cycle 30
tourist authorities 2
tourist offices 4
TPF 82, 110, 230
traffic audit 259
Train Operating Companies 87
Transnet 169
transparent link 293
Transtrac 337
TRAVCOM 359
travel
   agent billing analysis 330
   agents 4, 315
   counsellor 142
   industry intranet 312
   manager 241
   policy 125, 150, 241, 323
   requisition 141
Travel & Tourism Research Limited 308
Travel Edge 358
Travel Technology Group 319
Travel Training Company 10
Travel weekly 99
TravelBase 350
Travellog Systems 12
TravelNet 99, 216, 253
Travelocity 206
Travelpack 366
TravelWeb 216
Travicom 117, 366
Travipad 370
Tribute 89, 95
trip planning 242
trip wires 222
TRIPS 33, 41, 349
TSG 89–91, 119
TSR 320
TTG 251
TTI 12
TWA 151
type A & B HDS connections 163
Tyrolean Tourist Board 25

UCCCF 329
UFTAA 99
UIC 93
Ultraswitch 164, 216
UN EDIFACT EEG8 14
unbiased displays 7
unbundled 357
UNICORN 16–19, 50, 155, 292
Unison 167
United Airlines 116, 133, 348
UNIX 86
UPS 218
URL 179
US Department of Transportation 9, 234, 298, 342
Utell International 99, 164, 296
UUNet/Pipex 230

validator 328
VAN 279
vertical integration 85
Vertical Systems 371
VFR 5
Via Rail 155
Vicinity Corporation 207
Videcom 366
video-conferencing 35, 270
   benefits 275
   standards 273
videotex 46, 86, 283
   screen maps 18
viewdata 283
virtual brochure 257
virtual white board 36
visa 52
Visa 222
Visit Britain 256
Vistek 371
voice response system 38
Voyager Systems 10, 360
VS 110
Vtel 274

wait list 322
Warsaw convention 72–4
Weather Services Corporation 210
Web award 26
Web site 179
weekend rate 81
Weissmann Travel Reports 99, 269
what if scenarios 253
whiteboard 272
Win PIMMS 37
Working Together 14
World Travel & Tourism Council 10
World Travel Guide On-Line 267
Worldspan 197, 211
   View 297
   corporation 206
WTP (World Travel Partners) 197

X25 48, 303
X400 310

Y2K 28
Yahoo 183, 264
yield management 79, 81
Yorkshire Cable 303